MODERN ASPECTS OF ELECTROCHEMISTRY

No. 44

Modern Aspects of Electrochemistry

Topics in Number 43 include:

- Mathematical modeling in electrochemistry using finite element and finite difference methods
- Modeling atomic systems of more than two bodies
- Modeling impedance of porous electrodes, in order to permit optimal utilization of the active electrode material
- Multi-scale mass transport in porous silicon gas
- Physical theory, molecular simulation, and computational electrochemistry for PEM fuel cells, with emphasis on fundamental understanding, diagnostics, and design
- Modeling of catalyst structure degradation in PEM fuel cells
- Modeling water management in PEM fuel cells
- Modeling electrochemical storage devices for automotive applications

Topics in Number 42 include:

- The electrochemistry and electrocatalysis of Ruthenium in regards to the development of electrodes for Polymer Electrolyte Membrane (PEM) fuel cells
- Breakthroughs in Solid Oxide Fuel Cell (SOFC) anodes and cathodes leading to improved electrocatalysis
- Electrocatalysis of the electrochemical reduction of CO_2 on numerous metals
- The interfacial phenomena of electrodeposition and codeposition, and the need for new theoretical analyses of the electrode-electrolyte interface
- Advantages of scanning tunneling microscopy (STM) in understanding the basics of catalysis, electrocatalysis and electrodeposition
- The role of electrochemistry in emerging technologies including electrodeposition and electroforming at the micro and nano levels, semiconductor and information storage, including magnetic storage devices, and modern medicine

MODERN ASPECTS OF ELECTROCHEMISTRY
No. 44

Modelling and Numerical Simulations II

Edited by

MORDECHAY SCHLESINGER
University of Windsor
ON, Canada

Series Edited by

RALPH E. WHITE
Series Editor
University of South Carolina
Columbia, South Carolina

CONSTANTINOS G. VAYENAS
Series Editor
University of Patras
Patras, Greece

Editor
Mordechay Schlesinger
Department of Physics
University of Windsor
Windsor ON N9B 3P4
Canada
msch@uwindsor.ca

Series Editors
Ralph E. White
University of South Carolina
Columbia, South Carolina

Constantinos G. Vayenas
University of Patras
Patras, Greece

ISBN 978-0-387-49584-2   e-ISBN 978-0-387-49586-6
DOI 10.1007/978-0-387-49586-6
Springer Dordrecht Heidelberg London New York

Library of Congress Control Number: 2009926033

© Springer Science+Business Media, LLC 2009
All rights reserved. This work may not be translated or copied in whole or in part without the written permission of the publisher (Springer Science+Business Media, LLC, 233 Spring Street, New York, NY 10013, USA), except for brief excerpts in connection with reviews or scholarly analysis. Use in connection with any form of information storage and retrieval, electronic adaptation, computer software, or by similar or dissimilar methodology now known or hereafter developed is forbidden.
The use in this publication of trade names, trademarks, service marks, and similar terms, even if they are not identified as such, is not to be taken as an expression of opinion as to whether or not they are subject to proprietary rights.

Printed on acid-free paper

Springer is part of Springer Science+Business Media (www.springer.com)

Preface

The present volume is the second in a two-volume set dealing with modelling and numerical simulations in electrochemistry. Emphasis is placed on the aspect of nanoelectrochemical issues.

It seems appropriate at this juncture to mention the now-growing body of opinion in some circles that George Box was right when he stated, three decades ago, that "All models are wrong, but some are useful". Actually, when the statement itself was made it would have been more appropriate to say that "All models are inaccurate but most are useful nonetheless". At present, however, the statement, as it was made, is far more appropriate and closer to the facts than ever before. Currently, we are in the midst of the age of massively abundant data. Today's philosophy seems to be that we do not need to know why one piece of information is better than another except through the statistics of incoming and outgoing links between information and this is good enough. It is why, both in principle and in practice, one can translate between two languages, without knowledge of either. While none of this can be ignored, and it may even be true that "All models are wrong and increasingly you can succeed without them" the traditional approach of scientific modelling is still the order of the day. That approach may be stated as hypothesize – measure – model – test. It is in this light that the present volume should be viewed.

Again, as in the case of the previous volume, it is worth noting that the demarcation lines between disciplines are no longer as clear as they used to be in the past. This positive state of affairs may be looked upon as one of the hallmarks of twenty-first-century science, enabling desired cross-fertilization between related and even not so related fields. This volume and the previous one are examples of this trend.

The reader is presented with ten chapters written by 21 experts in the fields of modelling in electrochemistry and its many subfields. The first chapter deals with the subject of modelling in electrochemistry in general. The second and third chapters take up issues dealing with optics as related to applications in nanoelectrochemistry. The fourth, fifth and sixth chapters refer to surface electrochemistry. The last of these introduces the subject of Monte Carlo simulations and thus establishes the connection to the mathematically related topics of the next two chapters, which deal with the mathematics of corrosion and density functional theory, respectively. The final two chapters discuss acoustic microscopy and current distribution in electrochemical cells.

As in the previous volume, the chapters are independent in that they may be read in any order that suits the reader. Omitting a given chapter simply because the reader is familiar with the subject matter and reading others should be of benefit no less.

Thanks are due to the 21 authors who made the volume possible.

Mordechay Schlesinger
University of Windsor
Windsor, ON, Canada

Contents

Preface v

Contributors xix

Chapter 1

NUMERICAL MODELING OF CERTAIN ELECTROCHEMICAL PROCESSES

Nader G. Zamani

I.	Elementary Aspects of Electrochemical Reaction	1
II.	A Simple Mathematical Model	3
	1. Boundary Conditions	6
III.	Application in Cathodic Protection	8
	1. Iteration Process	10
IV.	Analytical Solution to Two Benchmark Problems	11
	1. Corrosion Cell 1	12
	2. Corrosion Cell 2	13
V.	Application in Electrodeposition	14
VI.	Analytical Solution to a One-Dimensional Electrodeposition Problem	16
VII.	General Framework of Numerical Approximation	19
	1. Finite-Difference Method	19
	2. Finite-Element Method	20
	3. Boundary-Element Method	22
VIII.	Implementation of the Finite-Difference Method in Cathodic Protection...........................	23
	1. Choosing a Lattice	23
	2. Discretization	24
	3. Mesh Equations	25
	4. Solving the Mesh Equations	26
IX.	Implementation of the Boundary-Element Method in Cathodic Protection...........................	31

X.	Numerical Implementation of the Boundary-Element Method	33
	1. Iteration Procedures	34
XI.	Concluding Remarks	37

Chapter 2

NEAR-FIELD OPTICS FOR HEAT-ASSISTED MAGNETIC RECORDING (EXPERIMENT, THEORY, AND MODELING)

William A. Challener and Amit V. Itagi

I.	Near-Field Transducers for Heat-Assisted Magnetic Recording	53
II.	Modeling Techniques	58
III.	Near Field Compared With Far Field	59
IV.	Figures of Merit	60
	1. Far-Field Transmittance	61
	2. Peak Field or Field Intensity	62
	3. Percent Dissipated Power in the Recording Medium	63
	4. Temperature Rise in the Recording Medium	64
V.	Mechanisms for Enhancement of the Figure of Merit	64
	1. Localized Surface Plasmon Resonance	64
	2. Lightning Rod Effect	70
	3. Dual-Dipole Resonance	72
VI.	Comparison Of Near-Field Transducers	73
	1. Circular Aperture	75
	2. Tapered Rectangular Aperture	76
	3. Bow-Tie Aperture	79
	4. C Aperture or Ridge Waveguide	82
	5. Triangle Antenna	83
	6. Bow-Tie Antenna	88
VII.	Antenna and Aperture Relationship	91
VIII.	Near-Field and Far-Field Relationship	92
	1. Radiation from Antennas	96
	2. Radiation from Apertures	97
	3. Numerical Modeling	99

IX.	Photonic Nanojets 104
X.	Conclusion 109

Chapter 3

SYMMETRY CONSIDERATIONS IN THE MODELLING OF LIGHT–MATTER INTERACTIONS IN NANOELECTROCHEMISTRY

Chitra Rangan

I.	Introduction 114
II.	Optical Properties of Metal Nanoparticles 115
	1. Macroscopic Theories 115
	2. Discrete-Dipole Approximation 116
	3. Optical Properties of Nanoparticles on a Surface ... 117
	4. Towards an Optical Method of Surface Electrochemistry............................... 118
III.	Optical Manipulation of Semiconductor Quantum Dots . 120
	1. Atomic Model of Semiconductor Quantum Dots ... 120
	2. Pseudospectral Method 124
	3. Finite-Difference Method 125
IV.	Summary 128

Chapter 4

APPLICATIONS OF COMPUTER SIMULATIONS AND STATISTICAL MECHANICS IN SURFACE ELECTROCHEMISTRY

P. A. Rikvold, I. Abou Hamad, T. Juwono, D. T. Robb, and M. A. Novotny

I.	Introduction 132
II.	Molecular Dynamics Simulations of Ion Intercalation in Lithium Batteries............................... 133
	1. Molecular Dynamics and Model System 133
	2. Simulations and Results....................... 134

III.	Lattice-Gas Models of Chemisorbed Systems 136
IV.	Calculation of Lattice-Gas Parameters by Density Functional Theory 137
V.	Monte Carlo Simulations 142
	1. Equilibrium Monte Carlo 142
	2. Kinetic Monte Carlo 143
VI.	Electrochemical First-Order Reversal Curve Simulations 144
VII.	Conclusion 146

Chapter 5

AC-ELECTROGRAVIMETRY INVESTIGATION IN ELECTROACTIVE THIN FILMS

Claude Gabrielli and Hubert Perrot

I.	Introduction 151
II.	General Considerations 153
	1. Thermodynamics 153
	2. Swelling 155
	3. Conductivity 156
III.	Electrochemical Approach of Electroactive Materials . . 157
	1. Models of the Charge Transport Through the Electroactive Film 158
	(i) Compact Model (Diffusion–Migration Model) 158
	(ii) Porous Model (Transmission Line Model) 161
	2. Calculation of the Impedance 163
	(i) Two-Species Problems 163
	(ii) Three-Species Problems 167
	(iii) Applications of Impedance Analysis 172
IV.	Coupled Electrochemical and Gravimetric Approach for Electroactive Materials 174
	1. Cyclic Voltammetry and Quartz Crystal Microbalance 174
	2. AC Electrogravimetry 183
	(i) Steady State 190
	(ii) Dynamic Regime 190

	(iii) Electrochemical Impedance 191
	(iv) Mass/Potential (Electrogravimetric) Transfer Function 195
	(v) Diagnostic Criterion 196
	(vi) Simulations 198
V.	Experimental 205
	1. Basic Microbalance Concepts 205
	2. AC-Electrogravimetry Aspects 207
	3. Dynamic Characterization of the Frequency/Voltage Converter 209
VI.	Examples of Applications of AC Electrogravimetry 210
	1. Prussian Blue 210
	(i) Film Preparation 212
	(ii) Voltammetric and Mass/Potential Curves 212
	(iii) Electrogravimetric Transfer Function and Electrochemical Impedance 213
	2. Polypyrrole 218
	(i) Film Preparation 218
	(ii) Voltammetric and Mass/Potential Curves 219
	(iii) Electrogravimetric Transfer Function and Electrochemical Impedance 219
	3. Complex Polymeric Structures 226
	(i) Electrode Preparation 226
	(ii) Electrogravimetric Transfer Function and Electrochemical Impedance 227
VII.	Conclusion 231

Chapter 6

MONTE CARLO SIMULATIONS OF THE UNDERPOTENTIAL DEPOSITION OF METAL LAYERS ON METALLIC SUBSTRATES: PHASE TRANSITIONS AND CRITICAL PHENOMENA

M. Cecilia Giménez, Ezequiel P. M. Leiva, and Ezequiel Albano

I.	Introduction: Some Basic Aspects of the Underpotential Deposition Phenomenon 240
II.	Some Thermodynamics on the UPD Phenomenon 246

III.	Description of the Monte Carlo Simulation Method and the Model for Metal Deposition 251
	1. The Lattice Model 251
	2. The Grand Canonical Monte Carlo Method........ 252
	(i) Change of Occupation 253
	(ii) Diffusion 254
	(iii) Calculation of the Coverage 255
	3. Interatomic Potential: The Embedded-Atom Method...................................... 255
	4. Surface Defects.............................. 257
	5. Energy Tables 258
IV.	Adsorption Isotherms 259
	1. Systems Studied and Adsorption Energies......... 259
	2. Evaluation of Adsorption Isotherms for Defect-Free Surfaces 262
	(i) UPD Compared with OPD: First-Order Phase Transitions 262
	(ii) The Influence of Temperature on the Isotherms 265
	3. Study of the Influence of Surface Defects 266
	(i) Isotherms Corresponding to UPD Systems: The Effect of Kinks and Steps as Compared with the Complete Monolayer 266
	(ii) Isotherms Corresponding to OPD Systems: The Formation of Surface Alloys 269
	4. Comparison with Experiments 272
V.	Dynamic Response of AG Monolayers Adsorbed on AU(100) Upon an Oscillatory Variation of the Chemical Potential 273
	1. Dynamic Phase Transitions: Basic Concepts 273
	2. Simulation Method........................... 274
	3. Dynamic Response of the Coverage Degree 275
	4. Dynamic Phase Transitions 277
VI.	Conclusions 283

Chapter 7

TOPICS IN THE MATHEMATICAL MODELING OF LOCALIZED CORROSION

Kurt R. Hebert and Bernard Tribollet

I. General Introduction 289
II. Pitting Corrosion 290
   1. Introduction 290
   2. General Structure of Pitting Models 292
   3. Review of Recent Models 293
      (i) Models of Pit Growth 293
      (ii) Models of Pit Growth and Repassivation 298
   4. Transport in Concentrated Electrolyte
      Solutions 305
   5. Concluding Remarks 311
III. Galvanic Coupling at the Interface Between
    Two Metals 312
    1. Theoretical Description of the Currents
       and Potentials at the Interface of Two Metals 312
    2. Application to the Al–Cu Coupling 314
       (i) Mathematical Model 314
       (ii) Experimental 318
       (iii) Experimental Results and Discussion 318
       (iv) Comparison Between Theoretical
              Calculations and Experimental Observations... 320
    3. Conclusions 323
IV. Impedance in a Confined Medium 323
    1. Introduction 323
    2. Experimental 325
    3. Theory 327
    4. Results and Discussion........................ 333
    5. Conclusions 337

Chapter 8

DENSITY-FUNCTIONAL THEORY IN EXTERNAL ELECTRIC AND MAGNETIC FIELDS

Ednilsom Orestes, Henrique J. P. Freire, and Klaus Capelle

I.	Scope of this Chapter 341	
II.	Elements of the Quantum Mechanics of Many-Electron Systems 343	
	1. Hamiltonians and Wave Functions 343	
	2. Density Matrices and Density Functionals......... 347	
	3. Functionals and Their Derivatives 352	
III.	The Hohenberg–Kohn Theorem..................... 353	
	1. Enunciation and Discussion of the Hohenberg–Kohn Theorem................ 353	
	2. A Simple Example: Thomas–Fermi Theory 359	
IV.	The Exchange–Correlation Energy 361	
	1. Definition of the Exchange–Correlation Energy 361	
	2. Interpretation of the Exchange–Correlation Energy . 364	
	3. Selected Exact Properties of the Exchange–Correlation Energy 365	
V.	The Kohn–Sham Equations 367	
	1. Self-Consistent Single-Particle Equations and Ground-State Energies 368	
	2. Single-Particle Eigenvalues and Excited-State Energies...................................... 371	
VI.	An Overview of Approximate Exchange–Correlation Functionals....................................... 375	
	1. Local Functionals: LDA........................ 376	
	2. Semilocal Functionals: GEA, GGA and Beyond.... 379	
	3. Orbital Functionals and Other Nonlocal Approximations: Hybrids, Meta-GGA, SIC, OEP, etc. 381	
	4. Performance of Approximate Functionals: A Few Examples.............................. 386	
VII.	External Electric And Magnetic Fields 390	
	1. Magnetic Fields Coupling to the Spins: SDFT 390	
	2. Brief Remarks on Relativistic DFT 393	

	3.	Magnetic Fields Coupling to Spins and Currents: CDFT 393
	4.	Electric Fields 398
	5.	Polarization and Magnetization 399
VIII.	Outlook 401

Chapter 9

ACOUSTIC MICROSCOPY APPLIED TO NANOSTRUCTURED THIN FILM SYSTEMS

Chiaki Miyasaka

I.	Introduction 409	
II.	Principle of the Scanning Acoustic Microscope 412	
	1. Imaging Mechanism 412	
	2. Description of Acoustic Lens 415	
	(i) Piezoelectric Transducer 415	
	(ii) Buffer Rod 416	
	(iii) Lens 416	
	(iv) Acoustic Antireflection Coating 418	
III.	Resolution 419	
IV.	Principle Of Quantitative Data Acquisition 422	
	1. $V(z)$ Curve 422	
	2. Phase Change 424	
	3. Theory of the Surface Acoustic Wave Velocity Measurement 427	
	4. Optimizing Measurement Precision 428	
V.	Contrast 429	
	1. Reflectance Function 429	
	2. Reflectance Function for Layered Media 430	
	3. Contrast Enhancement Caused by Discontinuities .. 436	
	4. Computer Simulation 440	
	5. Experimental Result 444	
VI.	Conclusion 448	

Chapter 10

CURRENT DISTRIBUTION IN ELECTROCHEMICAL CELLS: ANALYTICAL AND NUMERICAL MODELING

Uziel Landau

I.	Introduction and Overview	451
II.	Significance of Modeling the Current Distribution	452
III.	Experimental Determination of the Current Distribution	453
IV.	Analytical Derivation of the Current Distribution	454
	1. The Current Density	454
	2. Material Balance	455
	3. Boundary Conditions	457
	4. General Solution Procedure	459
	5. Thin Boundary Layer Approximation	460
V.	Common Approximations for the Current Distribution	462
	1. Primary Distribution: $\eta_\Omega \gg \eta_s + \eta_c$	463
	2. Secondary Distribution: $\eta_\Omega + \eta_s \gg \eta_c$	466
	3. Mass Transport Controlled Distribution: $\eta_c \gg \eta_\Omega + \eta_a$	470
	4. Tertiary Distribution: $\eta_c \sim \eta_\Omega \sim \eta_s$ ("Mixed Control")	472
VI.	Scaling Analysis of Electrochemical Cells	473
VII.	Transport Effects On Kinetically Controlled Systems	477
VIII.	Comparison of Analytical and Numerical Solutions	478
IX.	A Simplified Solution Algorithm	479
X.	Numerical Procedures for Solving the Laplace Equation	480
	1. The Finite-Difference Method	481
	(i) Methods of Solving the Finite-Difference Equation	483
	2. The Finite-Element Method	485
	3. Boundary-Element Methods	485
	4. Orthogonal Collocation	486
XI.	Numerically Implemented Solutions for the Current Distribution	487
XII.	Determination of the Current Distribution in Special Applications	489

1. Multiple Simultaneous Electrode Reactions,
   Including Alloy Codeposition and Gas Coevolution . 489
2. Moving Boundaries in Deposition and Dissolution
   Applications 491
3. Electropolishing, Leveling, and Anodizing 492
4. Current Distribution on Resistive Electrodes....... 492
5. Current Distribution in the Metallization
   of Through-Holes, Blind Vias, and Trenches....... 493

Index 503

List of Contributors, MAE 44

Ezequiel Albano
Instituto de Investigaciones Fisicoquímicas Teóricas y Aplicadas (INIFTA), UNLP, CONICET, Casilla de Correo 16, Sucursal 4, 1900 La Plata, Argentina, Ezequielalb@yahoo.com.ar

I. Abou Hamad
HPC2, Center for Computational Sciences, Mississippi State University, Mississippi State, MS 39762-5167, USA

Klaus Werner Capelle
Departmento de Fisica e Informatia, Instituto de Fisica de Sao Carlos, Universidado de Sao Paulo, Caixa Postal 780, 13560-970 Sao Carlos, Sao Paulo, Brazil, capelle@if.sc.usp.br

William Challener
Seagate Technology, Pittsburgh, PA 15222-4215, USA, william.a.challener@seagate.com

Henrique J.P. Freire
Departmento de Fisica e Informatia, Instituto de Fisica de Sao Carlos, Universidado de Sao Paulo, Caixa Postal 780, 13560-970 Sao Carlos, Sao Paulo, Brazil, freire@if.sc.usp.br

Claude Gabrielli
UPMC Université Paris 06, UPR 15, LISE, 4 place Jussieu, 75252 Paris Cedex 05, France, claude.gabrielli@upmc.fr
CNRS, UPR 15, LISE, 4 place Jussieu, 75252 Paris Cedex 05, France

M. Cecilia Gimenez
Physics Department, San Luis University, Chacabuco 917, C.P. 5700, San Luis, Argentina, cecigime@unsl.edu.ar

Kurt R. Hebert
Department of Chemical and Biological Engineering, Iowa State University, Ames, IA 50014, USA, krhebert@iastate.edu

Amit Itagi
Seagate Technology, Pittsburgh, PA 15222-4215, USA, amit.itagi@seagate.com

T. Juwono
Centre for Materials Research and Technology, School of Computational Science, and Department of Physics, Florida State University, Tallahassee, FL 32306-4350, USA

Uziel Landau
Department of Chemical Engineering, Case Western Reserve University, Cleveland, OH 44106, USA, uziel.landau@case.edu

Ezequiel P.M. Leiva
Chemical Sciences Faculty, Córdoba National University, Haya de La Torre and Medina Allende, C.P. 5000, Córdoba, Argentina, eleiva@mail.fcq.unc.edu.ar

Chiaki Miyasaka
University of Windsor, Windsor, ON, Canada
The Pennsylvania State University, University Park, PA N9B 3P4, 1682-5901, USA, cmiyasaka1@yahoo.com

M.A. Novotny
HPC2, Center for Computational Sciences, Mississippi State University, Mississippi State, MS 39762-5167, USA
Department of Physics & Astronomy, Mississippi State University, Mississippi State, MS 39762-5167, USA

Ednilsom Orestes
Departmento de Química e Física Molecular, Instituto de Química de São Carlos, Universidade de São Paulo, Caixa Postal 780, 13560-970 São Carlos, São Paulo, Brazil

Departamento de Física e Informática, Instituto de Física de São Carlos, Universidade de São Paulo, Caixa Postal 369, 13560-970 São Carlos, São Paulo, Brazil, eorestes@if.sc.usp.br

Hubert Perrot
UPMC Université Paris 06, UPR 15, LISE, 4 place Jussieu, 75252 Paris Cedex 05, France CNRS, UPR 15, LISE, 4 place Jussieu, 75252 Paris Cedex 05, France

Chitra Rangan
Department of Physics, University of Windsor, Windsor, ON N9B 3P4, Canada, rangan@uwindsor.ca

P.A. Rikvold
Center for Materials Research and Technology, School of Computational Science, and Department of Physics, Florida State University, Tallahassee, FL 32306-4350, USA, rikvold@scs.fsu.ed National High Magnetic Field Laboratory, Tallahassee, FL 32310-3706, USA

D.T. Robb
Department of Physics, Clarkson University, Potsdam, NY 13699, USA

Bernard Tribollet
Laboratoire Interfaces et Systèmes Electrochimiques, UPR 15 du CNRS, Université Pierre et Marie Curie, 75252 Paris Cedex 05, France

Nader G. Zamani
Department of Mechanical Engineering, University of Windsor, Windsor, ON N9B 3P4, Canada, zamani@uwindsor.ca

1

Numerical Modeling of Certain Electrochemical Processes

Nader G. Zamani

Department of Mechanical Engineering, University of Windsor, Windsor, ON, Canada

Summary. This expository chapter deals with the basic mathematical models which arise in some electrochemical processes pertaining to the galvanic corrosion phenomena. The elementary model discussed is sufficient as a preliminary tool in designing cathodic protection systems and their reverse effect, namely, electroplating. After the model is introduced, different numerical approaches for obtaining an approximate solution are discussed. The mathematical content is deliberately kept at the elementary level for it to be accessible to general readers. The discussion is limited to galvanic aqueous corrosion and therefore atmospheric factors are ignored.

I. ELEMENTARY ASPECTS OF ELECTROCHEMICAL REACTION

The basic principle behind corrosion can be explained in terms of the reaction between a hypothetical metal M placed in an ionic solution as shown in Fig. 1. The chemical reaction due to the electron exchange can be represented by the following formula

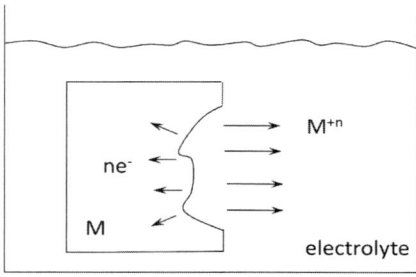

Non-conducting vessel

Figure 1. Degradation of base metal.

$$M \to M^{n+} + ne^-. \qquad (1)$$

Here n is the number of electrons lost and M^{n+} is the positive metal ion produced. Furthermore, e^- refers to a single electron. This process leads to an electric current in the ionic solution. The net result is the degradation of the metal M.

One can utilize this process for two different purposes. The first application is protecting a metal from corrosion, whereas the second is to employ the reverse effect for plating the degrading metal on another metal. The former application is referred to as cathodic protection and the latter is known as electroplating.

Figure 2 shows two bars made of zinc and copper immersed in an electrolyte (seawater). Needless to say, after some time, the zinc bar shows signs of corrosion. In this situation an electrical potential difference is established between the two metals which can be measured and is in agreement with the table of galvanic series in seawater.[1] Here, the zinc and copper bars acts as an anode and a cathode, respectively. The simple experiment demonstrates the basic idea behind the concept of cathodic protection. The zinc bar (anode) is being sacrificed to protect the copper bar (cathode). This method of cathodic protection is known as the sacrificial anodic method.

The sacrificial anodic protection method has been known since ancient times, dating back to 100 BC when Pilny the Roman employed it to prevent the corrosion of bronze and iron. In 1823, Sir Humphry Davy was commissioned by the Royal Navy to investigate the corrosion of copper used in the hulls of wooden battleships.[2]

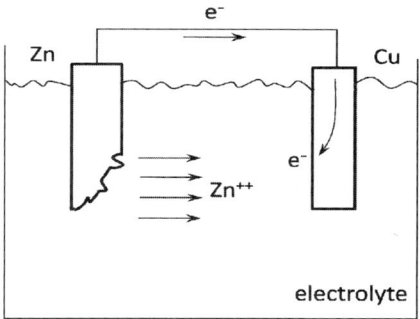

Figure 2. Zinc and copper in galvanic series.

II. A SIMPLE MATHEMATICAL MODEL

The mathematical model developed in this section is based on the conservation of charge. Consider a cube element of the electrolyte as shown in Fig. 3. The sides of the cube have dimensions Δx, Δy, and Δz, respectively. In the absence of charge generation (within the cube), the inflow and outflow of charges must be equal. Therefore, given the charge density vector $\vec{i} = (i_x, i_y, i_z)$, one can write

$$i_x \Delta A_x - \left(i_x + \frac{\partial i_x}{\partial x}\Delta x\right)\Delta A_x + i_y \Delta A_y - \left(i_y + \frac{\partial i_y}{\partial y}\Delta y\right)\Delta A_y$$
$$+ i_z \Delta A_z - \left(i_z + \frac{\partial i_z}{\partial z}\Delta z\right)\Delta A_z = 0. \qquad (2)$$

Simplifying the expression leads to

$$-\left(\frac{\partial i_x}{\partial x}\Delta x \Delta y \Delta z\right) - \left(-\frac{\partial i_y}{\partial y}\Delta y \Delta x \Delta z\right) - \left(-\frac{\partial i_z}{\partial z}\Delta z \Delta x \Delta y\right) = 0. \qquad (3)$$

Finally, dividing through by $\Delta x \Delta y \Delta z$ leaves us with the charge continuity:

$$\frac{\partial i_x}{\partial x} + \frac{\partial i_y}{\partial y} + \frac{\partial i_z}{\partial z} = \text{div}(i) = 0. \qquad (4)$$

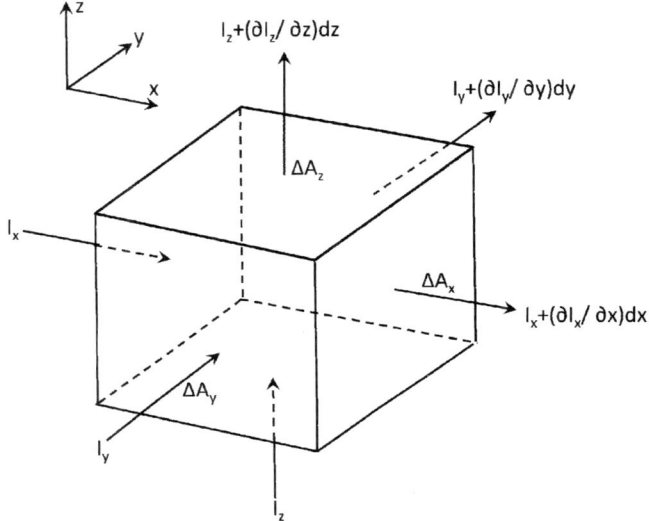

Figure 3. Control volume used for charge conservation.

However, the charge density vector is proportional to the gradient of the electrical potential ψ. Therefore,

$$(i_x, i_y, i_z) = -\sigma \left(\frac{\partial \psi}{\partial x}, \frac{\partial \psi}{\partial y}, \frac{\partial \psi}{\partial z} \right). \tag{5}$$

In (5), the constant σ is the conductivity of the electrolyte and has units of per ohm meter or amperes per volt meter.

Substituting the components of (i_x, i_y, i_z) in (5) and dividing through by σ, one obtains the celebrated Laplace equation:

$$\frac{\partial^2 \psi}{\partial x^2} + \frac{\partial^2 \psi}{\partial y^2} + \frac{\partial^2 \psi}{\partial z^2} = 0. \tag{6}$$

Equation (6) may be written in terms of the electrochemical potential ϕ, which is related to the electrical potential Ψ according to $\psi = c - \phi$. For more details, see the Appendix.

This substitution leads to the same type of partial differential equation described by (7):

$$\frac{\partial^2 \phi}{\partial x^2} + \frac{\partial^2 \phi}{\partial y^2} + \frac{\partial^2 \phi}{\partial z^2} = 0. \tag{7}$$

From this point on we will deal only with the electrochemical potential ϕ and refer to it simply as the "potential."

In the event that sources responsible for charge generation are present, (7) is modified accordingly and takes the following form. This is commonly referred to as the Poisson equation:

$$\frac{\partial^2 \phi}{\partial x^2} + \frac{\partial^2 \phi}{\partial y^2} + \frac{\partial^2 \phi}{\partial z^2} + S = 0. \tag{8}$$

In this equation, S represents the strength of the source (or sink). Implicit in the above derivation is the assumption that the steady-state condition prevails and therefore no time variation is considered.

Equations (7) and (8) arise in many areas of science and engineering, such as fluid dynamics, heat transfer, elasticity, and electrostatics. The significance and simplicity of the Laplace equation has led to a great deal of mathematical research into its solution.

The domain in which the solution to (7) is being sought is either finite or infinite. For example, if the electrolyte is in a bounded container, the solution domain is finite. On the other hand, if one is investigating the cathodic protection of a ship in the open sea, the solution domain is infinite. The most common method for obtaining the solution to the potential function ϕ is the concept of separation of variables.[3] In this method, the function ϕ is assumed to be decoupled as shown below:

$$\phi(x, y, z) = X(x)Y(y)Z(z). \tag{9}$$

The functions $X(x)$, $Y(y)$, and $Z(z)$ are to be found once additional information is provided. There are other analytical methods such as the conformal mapping technique that can be employed. The main difficulty with such analytical approaches is associated with irregular domains. Solutions can be found in simple domains such as circular, rectangular, and elliptical regions in two dimensions. With some additional effort, problems in simple three-dimensional domains can also be arrived at. In view of these facts, a numerical solution becomes necessary. These issues will be discussed in later sections.

1. Boundary Conditions

In electrochemical modeling, there are a variety of boundary conditions that can be specified.[4] If the value of the potential function is known at a point, the condition takes the form below, where ϕ_{constant} is a constant:

$$\phi(x, y, z) = \phi_{\text{constant}}. \tag{10}$$

The above condition is commonly referred to as the Dirichlet boundary condition. The known potential value is usually selected from the electromotive force series table.

The situation where the current is specified at a given point is known as the Neumann boundary condition. Mathematically speaking it is represented by

$$-\sigma \frac{\partial \phi}{\partial n} = i_e. \tag{11}$$

At well-painted surfaces (also called "insulated surfaces") the current normal to the surface is zero. Therefore, $i_e = 0$ and (11) reduces to

$$-\sigma \frac{\partial \phi}{\partial n} = 0. \tag{12}$$

At the exposed surfaces, the current in the normal direction is no longer a fixed value but depends on the local potential value. This type of boundary condition is the mixed or Robin type. Symbolically, the mixed boundary condition is displayed below:

$$-\sigma \frac{\partial \phi}{\partial n} = i_0 g(\phi - \phi_0) = i_0 g(\eta). \tag{13}$$

The term $\eta = \phi - \phi_0$ is known as the overpotential, where ϕ_0 is the electrode equilibrium potential. The function $g(\eta)$ has the expression described in (14) and with this choice of the function, (13) is called the "Butler–Volmer equation."[3]

$$g(\eta) = e^{\left(\frac{\gamma F}{RT}\right)\eta} - e^{-\left(\frac{(1-\gamma)F}{RT}\right)\eta}. \tag{14}$$

In this equation, R is the universal gas constant, T is the absolute temperature of the electrode, F is the Faraday constant, and γ is a symmetry parameter ((4) is a consequence of the Tafel equation stating that $\eta = a + b \log(i)$). In the event that the Butler–Volmer equation does not adequately model the physics of the problem, the experimental curve for the function g has to be used. This curve is known as the polarization curve. Figures 4 and 5 display a typical polarization curve for plain carbon steel in seawater. Figure 4 represents the scenario where the metal acts as the anode, whereas Fig. 5 corresponds to case where it acts as a cathode. Note that the potential is measured with respect to the silver–silver chloride electrode.

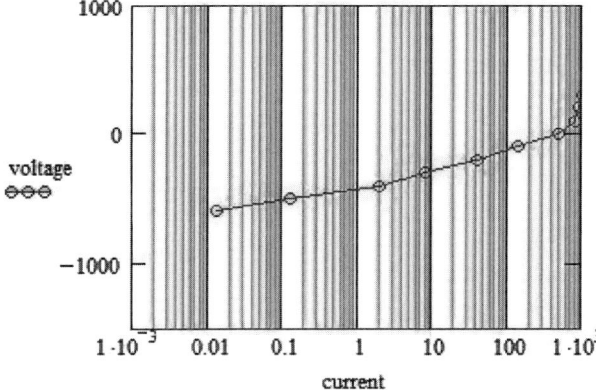

Figure 4. Anodic branch (millivolts vs. milliamperes).

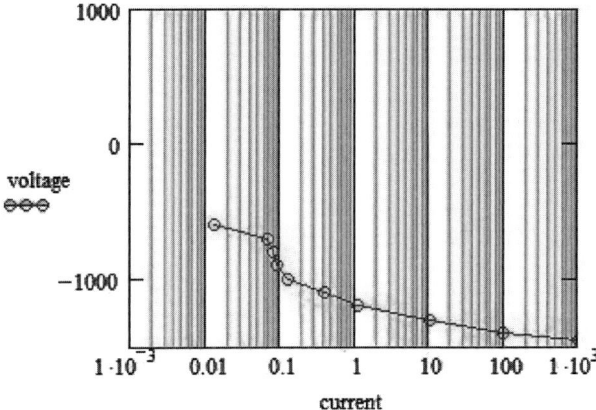

Figure 5. Cathodic branch (millivolts vs. milliamperes).

Finally, in an impressed current system (used in cathodic protection systems), the magnitude of the current density is specified.

$$-\sigma \frac{\partial \phi}{\partial n} = i_{\text{constant}}. \qquad (15)$$

The boundary conditions associated with electrochemical processes are collectively associated with the subject of electrode kinetics.[5–7]

III. APPLICATION IN CATHODIC PROTECTION

In the discussion to follow, we have ignored any physical variables (such as the conductivity σ) and replaced the expression for the polarization curve with the mathematical function $f(\phi)$.

Cathodic protection is a method for protecting metals against corrosion. There are two techniques to achieve this objective. The first approach is to use a sacrificial anode (a less noble metal) and consume it to protect another metal. This technique has been utilized for centuries in marine structures. As pointed out earlier, the galvanic coupling between the two metals results in a current density flowing in the electrolyte as shown in Fig. 6a. In the second approach, known as an impressed current system, the current density is artificially created using an inert electrode. This is depicted in Fig. 6b. Mathematically speaking, the inert electrode can be viewed as a current source where the value of i is assumed to be a known constant at a point. In either case, the intent is to ensure that the

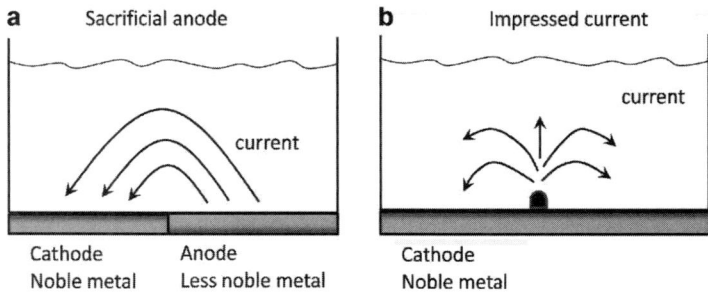

Figure 6. Cathodic protection methods.

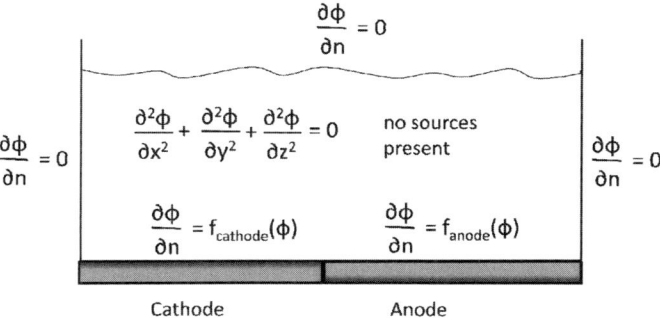

Figure 7. A bounded sacrificial anode cathodic protection system.

potential on the noble metal is lowered below a threshold potential. Sometimes this threshold value is referred to as the protection potential.

The schematic in Fig. 7 depicts a bounded sacrificial cathodic protection system (bounded by the container). In this figure, it is assumed that the container is nonconducting and therefore the current density is zero on the container boundary. The governing partial differential equation and the associated boundary conditions are displayed in the figure.

The domain under consideration (electrolytic domain) may also be infinite. This is clearly the case in marine applications. A fictitious two-dimensional version of this situation is displayed in Fig. 8. Here, the structure to be protected is the well-painted ship hull but with bare areas (cathode) being present. The hull is to be protected with anodic sections. The governing partial differential equation and the associated boundary conditions are depicted in the schematic. In the case of an infinite electrolyte, two auxiliary constraints are included. The behavior of the far-field potential ϕ_∞ is given by (16):

$$\phi(r) = O(1/r^2) + \phi_\infty. \tag{16}$$

Here, ϕ_∞ is the unknown potential at infinity.

The second constraint is based on the statement that the inflow of current equals the outflow of current; therefore,

$$\oiint i \, dS = 0. \tag{17}$$

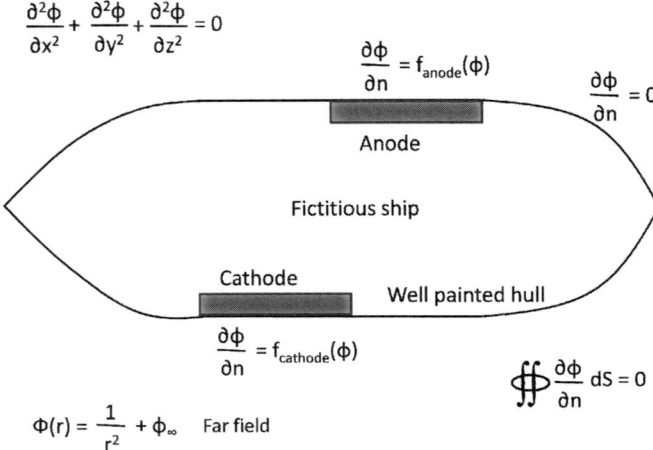

Figure 8. An infinite sacrificial anode cathodic protection system.

In terms of the potential function, (17) can be rewritten as

$$\oiint \frac{\partial \phi}{\partial n} dS = 0. \tag{18}$$

The variable n in this expression represents the outward unit normal to the ship hull.

1. Iteration Process

Owing to the nonlinear boundary conditions, the governing boundary value problem has to be solved iteratively. Although the equation is solved numerically (the details will be discussed in a later section), we disregard this issue and assume that the solution to the linearized problem is somehow obtained. The most general format for the boundary value problem associated with a cathodic protection system is described below.

$$\nabla^2 \phi = 0 \quad \text{in } \Omega \text{ (within electrolyte)}, \tag{19}$$

$$\phi = c_1 \quad \text{on } \Gamma_1 \text{ (known potential)}, \tag{20}$$

$$\frac{\partial \phi}{\partial n} = c_2 \quad \text{on } \Gamma_2 \text{ (an inert electrode)}, \tag{21}$$

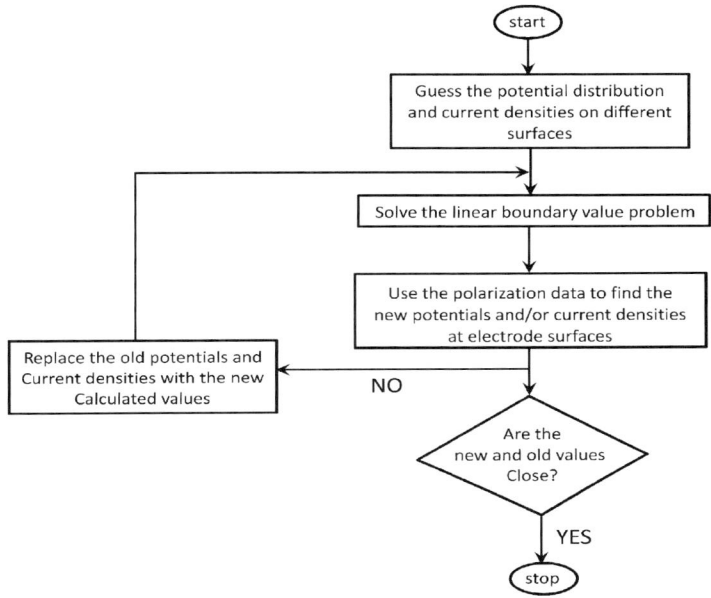

Figure 9. Successive iteration flowchart.

$$\frac{\partial \phi}{\partial n} = f(\phi) \quad \text{on } \Gamma_3 \text{ (polarization behavior)}. \qquad (22)$$

For well-painted surfaces, (21) applies where c_2 is zero. The boundary of Ω consists of the union of the boundaries in (20)–(22), $\partial \Omega = \Gamma_1 + \Gamma_2 + \Gamma_3$.

The flowchart in Fig. 9 adopted from[4] describes the iteration process through the calculations.

IV. ANALYTICAL SOLUTION TO TWO BENCHMARK PROBLEMS

It was pointed out earlier that the analytical solutions of corrosion cell problems are rather complicated (if not impossible). To demonstrate this claim, two relatively simple geometries are presented below.

1. Corrosion Cell 1

The cell is assumed to have two-dimensional characteristics as displayed in Fig. 10. The three sides of the boundary are assumed to be well painted and therefore the current densities in the normal direction are zero. On the bottom segment of the cell, the anode and cathode are placed side by side. The dimensions of the cell are described by the parameters a, b, and c as shown in the figure. The polarization curves are linear to further simplify the solution. These polarization curves are depicted in Fig. 11, where

$$L_a = L_c = L = 1.$$

The variables L_a and L_c define the slopes of the linearized anodic and cathodic polarization curves as defined by (23) and (24) below.

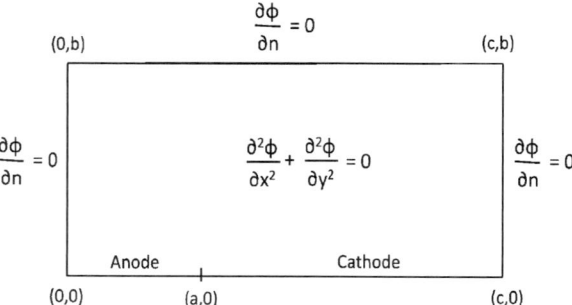

Figure 10. Corrosion cell 1.

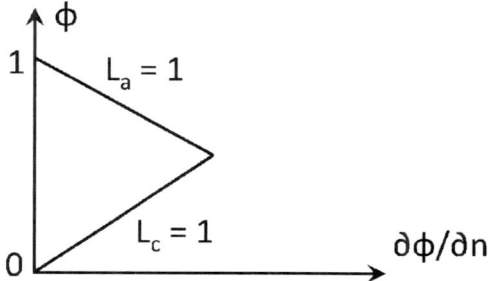

Figure 11. Linear polarization curves.

The analytical expressions for the polarization boundary conditions are given by (23) and (24):

$$\frac{\partial \phi}{\partial n} = \frac{\phi - 1}{L_a} 0 \le x \le a, y = 0, \quad (23)$$

$$\frac{\partial \phi}{\partial n} = \frac{\phi}{L_c} a \le x \le c, y = 0. \quad (24)$$

The analytical solution of the problem[8,9] can be obtained using the method of separation of variables and is represented by (25):

$$\phi(x, y) = a + \frac{2}{\pi} \sum_{n=1}^{\infty} \frac{\sin \sin (n\pi a) \cos \cos (n\pi x) \cosh [n\pi (b - y)]}{n [\cosh \cosh (n\pi b) + n\pi L \sinh (n\pi b)]}. \quad (25)$$

Keep in mind that the solution (25) is based on the assumption $L_a = L_c = L = 1$. Once again, the variables L_a and L_c define the slopes of the linearized anodic and cathodic polarization curves as defined by (23) and (24).

2. Corrosion Cell 2

The geometry in this problem is slightly more complicated than that for the previous cell. The bottom segment considered to be the cathode is a cosine curve, whereas the top edge is the anode (at a constant potential). The cell is depicted in Fig. 12. The shape of the cathode is described by (26):

$$y(x) = 0.15 [1 - \cos(\pi x)]. \quad (26)$$

The polarization boundary condition is once again assumed to be linear:

$$\frac{\partial \phi}{\partial n} = h_1(x)\phi + h_2(x). \quad (27)$$

Assuming that the components of the outward unit normal are described by (n_x, n_y),

$$h_1(x) = -\pi n_x \tan(\pi x), \quad (28)$$

$$h_2(x) = -\pi n_y \cos\left(e^{-\pi y} - e^{(y-2H)}\right). \quad (29)$$

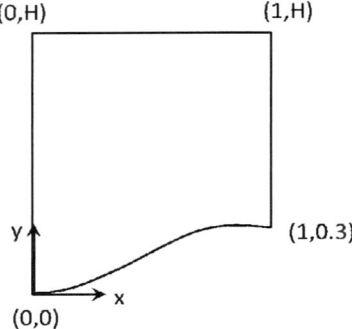

Figure 12. Corrosion cell 2.

The other boundary condition are as follows:

$$\phi(x, H) = 0 \quad 0 \le x \le 1 \quad y = H, \tag{30}$$

$$\frac{\partial \phi}{\partial n} = 0 \quad 0 \le y \le H \quad x = 0, \tag{31}$$

$$\frac{\partial \phi}{\partial n} = 0 \quad 0 \le y \le H \quad x = 1. \tag{32}$$

This problem was solved in[10] analytically with a closed-form solution given by (33):

$$\phi(x, y) = \cos(\pi x) \left(e^{-\pi y} - e^{\pi(y-2H)} \right). \tag{33}$$

V. APPLICATION IN ELECTRODEPOSITION

The process of electrodeposition is the reverse of the corrosion phenomenon. The intent in electrodeposition is to deliberately consume that material which is supplied by the anode to plate it on the cathode.[10–13] A major difference between the two applications is that the domain occupied by the electrolyte is constantly changing as the plating process proceeds. Such problems are classified as the moving-boundary problems. Therefore, the nonlinearities are twofold. Both the nonlinear polarization boundary condition and the changing domain contribute to this effect.

The governing partial differential equation is still the Laplace equation:

$$\frac{\partial^2 \phi}{\partial x^2} + \frac{\partial^2 \phi}{\partial y^2} + \frac{\partial^2 \phi}{\partial z^2} = 0. \tag{34}$$

In electrodeposition, it is reasonable to assume that the anode maintains a constant potential; therefore, the boundary condition on the anodic surface is quite simple:

$$\phi(x, y, z) = \phi_{\text{anode}} = \text{constant}. \tag{35}$$

On the well-painted surfaces (insulated surfaces), the current density vanishes:

$$\frac{\partial \phi}{\partial n} = 0. \tag{36}$$

On the cathodic surfaces, the Butler–Volmer equation prevails. The associated boundary condition is

$$-\sigma \frac{\partial \phi}{\partial n} = i_0 \left[\exp\left(\frac{-\alpha_C \tilde{n} F}{RT}\phi\right) - \exp\left(\frac{\alpha_A \tilde{n} F}{RT}\phi\right) \right]. \tag{37}$$

The parameters in (37) are as follows: i_0 is the exchange current density, $\tilde{n}$ is the number of electrons involved in cathodic reaction, F is the Faraday constant, T is the absolute temperature, R is the universal gas constant, α_A is the anodic kinetic parameter, and α_C is the cathodic kinetic parameter

The instantaneous growth of the cathode (in terms of the outward normal growth function $h(t)$) is given next:

$$\frac{dh}{dt} = \left(\frac{M}{\tilde{n} \rho F}\right) \left[\exp\left(-\frac{\alpha_C \tilde{n} F}{RT}\phi\right) - \exp\left(\frac{\alpha_A \tilde{n} F}{RT}\phi\right) \right]. \tag{38}$$

The additional parameters introduced in (38) are as follows: t is time, $h(t)$ is the outward normal growth of the cathode, M is the molecular weight, and ρ is the density of the electrolyte.

Nondimensionalizing the variables according to the scheme suggested in[11] leads to considerable simplification of the initial-boundary-values problem. The final nondimensional form is provided by (39)–(43):

$$\frac{\partial^2 \phi}{\partial x^2} + \frac{\partial^2 \phi}{\partial y^2} + \frac{\partial^2 \phi}{\partial z^2} = 0 \text{ (in the electrode)}, \tag{39}$$

$$\phi = \phi_A \quad \text{(on the anode)}, \tag{40}$$

$$\frac{\partial \phi}{\partial n} = \zeta \left(e^{-\alpha_C \phi} - e^{\alpha_A \phi} \right) \quad \text{(on the cathode)}, \tag{41}$$

$$\frac{\partial \phi}{\partial n} = 0 \quad \text{(on insulated surfaces)}, \tag{42}$$

$$\frac{dH}{d\tau} = \zeta \left(e^{-\alpha_C \phi} - e^{\alpha_A \phi} \right) \quad \text{(cathode growth)}. \tag{43}$$

The mathematical description of the initial-boundary-value problem above is complete, once the initial condition has been specified. It is assumed that the initial height is zero; therefore,

$$H = 0 \text{ at } \tau = 0. \tag{44}$$

To predict the cathode shape, the initial-boundary-value problem described by (39)–(44) has to be integrated in time.

Clearly, obtaining an analytical solution is extremely difficult in three dimensional and even two-dimensional geometries. In the next section, a simple one-dimensional problem is treated in detail.

VI. ANALYTICAL SOLUTION TO A ONE-DIMENSIONAL ELECTRODEPOSITION PROBLEM

To demonstrate the difficulties associated with solving initial-boundary-value problems arising in electrodeposition, a fictitious one-dimensional problem is treated next. The fictitious electrolyte is initially present between $x = 0$ and $x = L$ as shown in Fig. 13. The point $x = 0$ corresponds to the location of the anode and the point $x = L$ corresponds to the initial position of the cathode. Note that

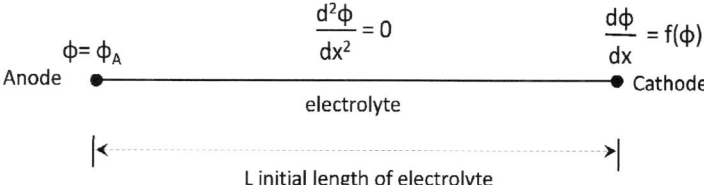

Figure 13. One-dimensional geometry of the electrodeposition problem.

Numerical Modeling of Certain Electrochemical Processes

because of the plating process, the cathode grows and the distance between the electrodes (i.e., the length of the electrode) is reduced. If this distance is denoted by $s(t)$, and $h(t)$ represents the outwards normal growth of the cathode, the following relationship is true:

$$s(t) = L - h(t). \tag{45}$$

The partial differential equation reduces to an ordinary differential equation describing the variation of the potential function $\varphi(x)$ within the electrolyte:

$$\frac{d^2\phi}{dx^2} = 0. \tag{46}$$

The boundary conditions are

$$\phi(0, t) = \phi_A \quad \text{(on the anodes)}, \tag{47}$$

$$\frac{d\phi}{dx} = \zeta \left(e^{-\alpha_c \phi(h,t)} n - e^{\alpha_A \phi(h,t)} \right) \quad \text{(on the cathode)}, \tag{48}$$

$$\frac{dh}{dt} = \zeta \left(e^{-\alpha_c \phi(h,t)} - e^{\alpha_A \phi(h,t)} \right) = \frac{d\phi}{dx} \quad \text{(cathode growth rate)}, \tag{49}$$

$$\frac{ds}{dt} = L - \int_0^t \frac{ds}{d\tau} d\tau \quad \text{(rate of electrolyte length change)}, \tag{50}$$

$$s(0) = L \quad \text{(initial electrolyte length)} \tag{51}$$

Note that (51) is equivalent to the following condition:

$$\frac{ds}{dt} = -\frac{dh}{dt} = -\frac{d\phi}{dx}. \tag{52}$$

The general solution to (47) is of the form

$$\phi(x, t) = C_1(t)x + C_2(t). \tag{53}$$

Imposing the boundary conditions at the cathode and anode allows one to calculate the constants of integration:

$$C_1(t) = \zeta \left(e^{-\alpha_C [C_1(t)s(t) + \phi_A]} - e^{-\alpha_A [C_1(t)s(t) + \phi_A]} \right), \tag{54}$$

$$C_2(t) = \phi_A. \tag{55}$$

In view of the fact that $C_1(t) = -(ds/dt)$, (54) can be rewritten as

$$\frac{ds}{dt} = -\zeta \left\{ \exp\exp\left[\alpha_C \left(s\frac{ds}{dt} - \phi_A\right)\right] - \exp\left[-\alpha_A \left(s\frac{ds}{dt} - \phi_A\right)\right] \right\}. \tag{56}$$

The expression on the right-hand side of (56) can be linearized by taking the first term of the Taylor series expansion in terms of $s(ds/dt) - \phi_A$. Performing the linearization and some algebra, one arrives at an explicit expression involving ds/dt:

$$\frac{ds}{dt} = \frac{r\phi_A}{1+rs}. \tag{57}$$

Here, the parameter r is given by $r = \zeta(\alpha_A + \alpha_c)$.
The exact solution for (57) can easily be obtained:

$$s(t) = \frac{1}{r}\left[-1 + (1 + 2rA + 2r^2\phi_A t)^{1/2}\right]. \tag{58}$$

Since $s(0) = L$ the constant A is calculated from

$$A = L + \frac{r}{2}L^2. \tag{59}$$

Combining the results, we have

$$C_1(t) = -\frac{ds}{dt} = \frac{-\phi_A}{(1 + 2rA + 2r^2\phi_A t)^{1/2}}, \tag{60}$$

$$\phi(x, t) = \phi_A \left(1 - \frac{rx}{(1 + 2rA + 2r^2\phi_A t)^{1/2}}\right), \tag{61}$$

$$s\frac{ds}{dt} = \phi_A \left(1 - \frac{1}{(1 + 2rA + 2r^2\phi_A t)^{1/2}}\right). \tag{62}$$

Once again, the above approximations are based on the critical assumption that the entity $s(ds/dt) - \phi_A$ is small. For example, if we expect $(ds/dt) - \phi_A \le \epsilon$, it is straightforward to show that it translates to the following constraints:

$$t\varepsilon^2 r^2 - \epsilon \left(1 + 2rA + t^2\varepsilon^2 r^4\right)^{1/2} \le \phi_A \le 0. \quad (63)$$

Naturally, analytical solutions to two-dimensional and three-dimensional problems become intractable.

VII. GENERAL FRAMEWORK OF NUMERICAL APPROXIMATION

Essentially, there are three numerical techniques for solving elliptic partial differential equations.[14–16] In the case of cathodic protection and electrodeposition, the governing partial differential equation is very simple, namely, the Laplace equation. These techniques are classified as follows:

(a) Finite-difference method (FDM)
(b) Finite-element method (FEM)
(c) Boundary-element method (BEM)

The general description of these methods is provided in the present section. More detailed information on the finite differences and boundary elements will be provided in later sections of the article. It will become clear that in electrochemistry applications (of interest to us), the BEM could be the most efficient numerical technique.

1. Finite-Difference Method

The domain under consideration (the electrolyte) is covered with a grid. For the sake of simplicity, we use a two-dimensional Cartesian grid. The discrete version of the Laplace equation at each grid point is represented by (64):

$$\phi_{i+1,j} + \phi_{i-1,j} + \phi_{i,j+1} + \phi_{i,j-1} - 4\phi_{i,j} = 0. \quad (64)$$

The subscripts (i, j) refer to the location (x_i, y_j) where discretization takes place. Furthermore $\phi_{i,j}$ is the approximation to

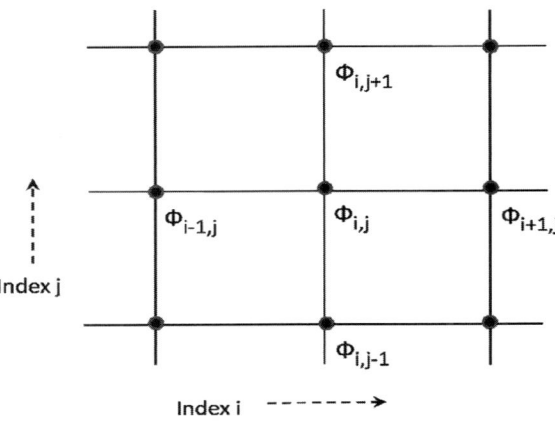

Figure 14. The finite-difference grid.

the unknown potential $\phi(x_i, y_j)$. These are symbolically depicted in Fig. 14. Sufficient numbers of equations are written at the grid points and once the boundary conditions have been taken into account, a system of algebraic equations is obtained. This system can be solved using direct or iterative methods available in numerical linear algebra. In the case of nonrectangular geometries, the procedure is more complicated. However, in principle, it is possible to map such regions to a rectangular one using techniques such as grid generation. In such situations, although the domain is simplified considerably, the governing partial differential equation takes a different form owing to the coordinate change.

2. Finite-Element Method

In the FEM, the variational formulation (or weak formulation) of the boundary value problem is employed. As a concrete example, suppose the strong form of the boundary value problem is described by (65)–(67):

$$\Delta^2 \phi = 0 \quad \text{in } \Omega, \tag{65}$$

$$\phi = \text{prescribed value on } \Gamma_1, \tag{66}$$

$$\frac{\partial \phi}{\partial n} = \text{prescribed value on } \Gamma_2. \tag{67}$$

It can be shown (although it is neither trivial, nor intuitive) that the solution to the above boundary value problem, among all functions satisfying (66), also minimizes the functional $J(\phi)$ described by (68):

$$J(\phi) = \frac{1}{2} \iint_\Omega (\nabla \phi)^2 \mathrm{d}\Omega - \int_{\Gamma_2} \frac{\partial \phi}{\partial n} \mathrm{d}S. \tag{68}$$

An alternative description is to state that among all functions, satisfying (66), the exact solution of the boundary value problem satisfies the following condition:

$$\iint_\Omega \nabla \phi \nabla \psi \mathrm{d}\Omega = \int_{\Gamma_2} \frac{\partial \phi}{\partial n} \psi \mathrm{d}S. \tag{69}$$

The function ψ is arbitrary but vanishes on Γ_2.

The electrolyte is discretized with different types of elements available in the FEM. A typical element shown in Fig. 15 is a three-noded triangular type which is a linear approximation to $\phi(x, y)$ within its interior. The global system of equations is obtained by the assembly process of contributions from individual elements. Upon applying the boundary conditions, one can solve the resulting system of algebraic equations using the direct or iterative methods. Owing to the nature of the FEM, complicated geometries can easily be handled. The only drawback (as in the FDM) is the fact that the electrolyte needs to be discretized. Keep in mind that in cathodic protection and electrodeposition, the value of the potential on the bounding surfaces is of primary interest.

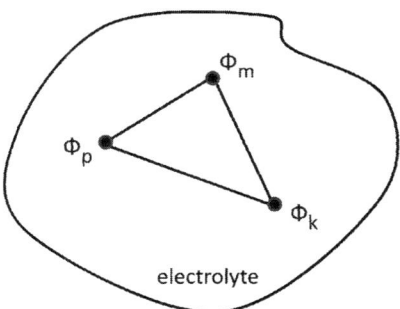

Figure 15. A three-noded triangular element in the finite-element method.

3. Boundary-Element Method

The basic principle behind the BEM is to convert the partial differential equation into an integral equation using the classical methods of applied mathematics. This technique requires the Green function associated with the differential operator (in this case, the Laplace operator). After the appropriate manipulations, (70) is obtained:

$$c(p)\phi(p) = \oint_\partial^\Omega \phi(p')\frac{\partial G(p, p')}{\partial n}\mathrm{d}S + \oint_\partial^\Omega \frac{\partial \phi(p')}{\partial n} G(p, p')\mathrm{d}S. \quad (70)$$

Since for the sake of simplicity we have limited our domain (electrolyte) to two-dimensional geometries, the integrals in (70) are in fact line integrals. In the three-dimensional case these become surface integrals. Furthermore, the variable of integration is p', while p is held fixed.

The coefficient $c(p)$ in (70) depends on the location of $p = (x, y)$ where the potential $\varphi(x, y)$ is being evaluated. This value is given below:

$$c(p) = \begin{cases} 1 & \text{if } p \text{ is in the interior of } \Omega \\ 0.5 & \text{if } p \text{ is on the smooth part of } \partial\Omega \\ 0 & \text{if } p \text{ is in the exterior of } \Omega \end{cases} \quad (71)$$

In the BEM, the boundary is divided into panels (patches or elements on the boundary) as depicted in Fig. 16. On these panels,

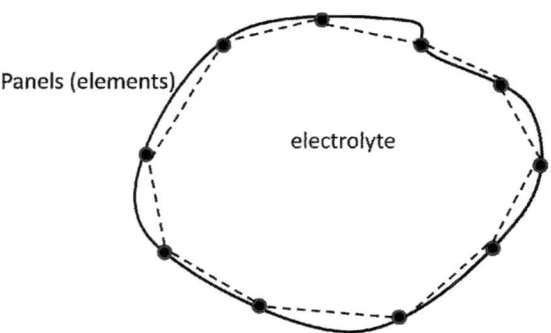

Figure 16. The boundary is discretized with panels (boundary elements).

the integral equation (70) is discretized. The result is a system of algebraic equations which can be solved once the boundary conditions have been taken into account. A major advantage of the BEM is that the body of electrolyte need not be discretized. On the other hand, the resulting system of equations is fully populated. This is in contrast to the FDM and the FEM, where the matrices are sparse. The BEM has been widely used in aerodynamics applications. In that field, the method is also known as the panel method (or the source distribution technique[17]).

VIII. IMPLEMENTATION OF THE FINITE-DIFFERENCE METHOD IN CATHODIC PROTECTION

The discussion in this section is limited to a two-dimensional rectangular domain. The system under consideration is a rectangular container filled with an electrolyte.[5] The container's vertical walls are insulators. The electrolyte's surface is open to air. The bottom of the container consists of two metals in contact with each other exactly as in the geometry shown in Fig. 25 in the Appendix. Metal 1 on the left (cathode) has a higher equilibrium electromotive force and is thus more noble than metal 2 on the right (anode). The system boundary is a rectangle; there are insulators on three sides (left, right, top), and metals on one side (bottom). The FDM has been used for the calculation of the potential φ in the interior and on the boundary.

The steps in the FDM are:

1. Choosing a lattice
2. Discretization
3. Writing the mesh equations
4. Solving the mesh of equations

1. Choosing a Lattice

A rectangular lattice is a family of vertical and horizontal lines in the xy plane. The point of intersection of lines is often called a "node." Each rectangle formed by two adjacent vertical lines and two horizontal lines in the lattice is called a "mesh." Each side of the mesh is called a "link." A rectangular lattice is said to be uniform if all of the horizontal links are equal to a constant b. If $a = b$, then we have a square lattice; however, the condition $a = b$ is not essential and

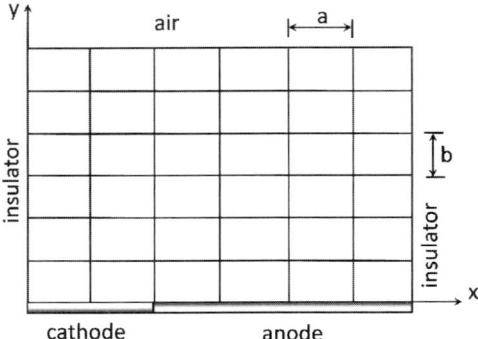

Figure 17. The corrosion cell and its finite difference method lattice.

it will not be required here. Finally, a rectangular lattice is said to be centered if the origin $(x, y) = (0, 0)$ is one of the lattice points. In a centered uniform rectangular lattice, the lattice points are the points for which $x = ma$ for $m = 0, \pm 1, \pm 2, \ldots$ and $y = nb$ for $n = 0, \pm 1, \pm 2, \ldots$. Henceforth the term "lattice" will be used to mean a centered uniform rectangular lattice as shown in Fig. 17.

2. Discretization

Once a lattice has been chosen, it must be decided how to approximate the various derivatives of ϕ by difference quotients. In the discussion that follows, reference will be made to the diagram shown in Fig. 18. The lattice points have been given the abbreviations E (east), S (south), W (west), N (north), and C (center). The values of the function at these points are denoted by ϕ_E, φ_S, ϕ_W, ϕ_N, and ϕ_C respectively. Thus, for example, if $C = (ma, nb)$, then $\phi_C = \phi(ma, nb)$, $\varphi_N = \phi(ma, (n+1)b)$, etc.

The FDM is affected by replacing the derivatives of φ at a lattice point such as C with their approximations as follows:

$$\frac{\partial^2 \phi}{\partial x^2} \approx \frac{\phi_E - 2\phi_C + \phi_W}{a^2}, \qquad (72)$$

$$\frac{\partial^2 \phi}{\partial y^2} \approx \frac{\phi_N - 2\phi_C + \phi_S}{b^2}. \qquad (73)$$

Numerical Modeling of Certain Electrochemical Processes

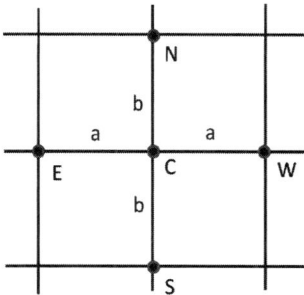

Figure 18. Nodes used in approximating partial derivatives.

The so-called five-point approximation of the Laplacian operator at C is

$$\Delta\phi \approx \frac{\phi_E - 2\phi_C + \phi_W}{a^2} + \frac{\phi_N - 2\phi_C + \phi_S}{b^2}. \qquad (74)$$

The first partial derivatives (at C) will be approximated by their centered difference quotients:

$$\frac{\partial\phi}{\partial x} \approx \frac{\phi_E - \phi_W}{2a}, \qquad (75)$$

$$\frac{\partial\phi}{\partial y} \approx \frac{\phi_N - \phi_S}{2b}. \qquad (76)$$

Centered difference quotients are more accurate than one-sided difference quotients.

3. Mesh Equations

The letter C will be used as before to denote a typical lattice point and the letters E, N, W, and S will be used to denote the four neighboring nodes as shown in Fig. 18. We now write the five-point approximation of the Laplace equation at each node C in the interior or on the boundary of the system:

$$\frac{\phi_E - 2\phi_C + \phi_E}{a^2} + \frac{\phi_N - 2\phi_C + \phi_S}{b^2} = 0 \qquad (77)$$

From the above equation, the potential at C is calculated in terms of ϕ_E, ϕ_N, ϕ_W, and ϕ_S:

$$\phi_C = \frac{\delta}{2(1+\delta)}(\phi_E + \phi_W) + \frac{1}{2(1+\delta)}(\phi_N + \phi_S). \quad (78)$$

In the (78), the variable δ is defined by $\delta = (b/a)^2$. Upon defining two other variables $\delta_1 = \delta/2(1+\delta)$ and $\delta_2 = 1/2(1+\delta)$, one can further reduce (78):

$$\phi_C = \delta_1(\phi_E + \phi_W) + \delta_2(\phi_N + \phi_S). \quad (79)$$

The constants δ, δ_1, and δ_2 have been defined for convenience; they satisfy $\delta_1 + \delta_2 = 1/2$. Furthermore, if $a = b$, then $\delta = 1/2$ and $\delta_1 = \delta_2 = 1/4$, so (79) simplifies to

$$\phi_C = \frac{\phi_E + \phi_N + \phi_W + \phi_S}{4}. \quad (80)$$

In general, a and b need not be equal; therefore, (79) has been used throughout.

4. Solving the Mesh Equations

After having written the Laplace equation in the discretized form at every node C in the system, we arrive at the set of mesh equations. The next step is to solve these equations and find the values of ϕ at every system node. The solutions to the mesh equations will only be approximately equal to the exact value of; this is because the mesh equations themselves are only approximations to field equation.

The mesh equations may be solved either directly or iteratively, and within each of these two categories there are several possibilities. The method discussed here is iterative. By an "iteration" we mean a "lattice iteration," to be described presently. Iterations have simple structure and low storage requirements; moreover, the iterations are all identical in structure (stationary), and they are in general nonlinear. The iterative method begins by initializing the values of ϕ at all nodes in the system to reasonable but otherwise arbitrary values. An arbitrary pair of nodes is then chosen in the system as the first node and the last node. The first lattice iteration begins as follows.

We start marching through the lattice in fixed and simple manner, beginning at the first node, visiting each and every node in the system once. With use of the Laplacian formula (79), at each step (i.e., at each node) a new value of ϕ at that node is calculated in terms of ϕ at the four neighboring nodes. Once this has been done, we move to the next node and update its potential by the same procedure. The first lattice iteration is completed as soon as the procedure has been carried out for the last node. Second, third, and higher lattice iterations are not only possible, but usually quite necessary.

In the process of marching through the lattice, nodes C for which some of the four neighboring nodes, N, E, S, and W are outside the system will be encounted; these nodes outside the system are called the "fictitious nodes." Without any boundary conditions, fictitious nodes would lead to an undetermined set of equations. It is, however, possible to eliminate the fictitious nodes with the help of the boundary conditions, thereby reducing the number of unknowns to the number of equations available.

For each node C in the system, the elimination of its neighboring fictitious nodes (if it has any) changes the basic mesh equation at C. The change depends on the location of C. For the system in Fig. 17, there are 11 cases; these cases can now be examined with reference to Fig. 19, which shows how the different cases have been enumerated. The polarization functions of the cathode (metal 1) and the anode (metal 2) are denoted by f and g, respectively.

Case 1 (the interior)

This is the simplest case. Node C is in the interior of the system, its four neighboring nodes all belong to the system, and there are no fictitious nodes in the neighborhood of C. The value of ϕ_C is calculated in terms of ϕ_E, ϕ_N, ϕ_W, and ϕ_S according to the equation below:

$$\phi_C = \delta_1 (\phi_E + \phi_W) + \delta_2 (\phi_N + \phi_S). \tag{81}$$

Case 2 (the top, excluding the corners)

For this case, node C can be anywhere at the interface between the electrolyte and the air, except the corners. The only fictitious node is N. To eliminate ϕ_N, the boundary condition for insulators is used with the unit inward normal vector $\vec{n} = (0, -1)$. From (82) it follows that $\phi_N = \phi_S$.

$$0 = \frac{\partial \phi}{\partial n} = -\frac{\partial \phi}{\partial y} = -\frac{\phi_N - \phi_S}{2b}. \tag{82}$$

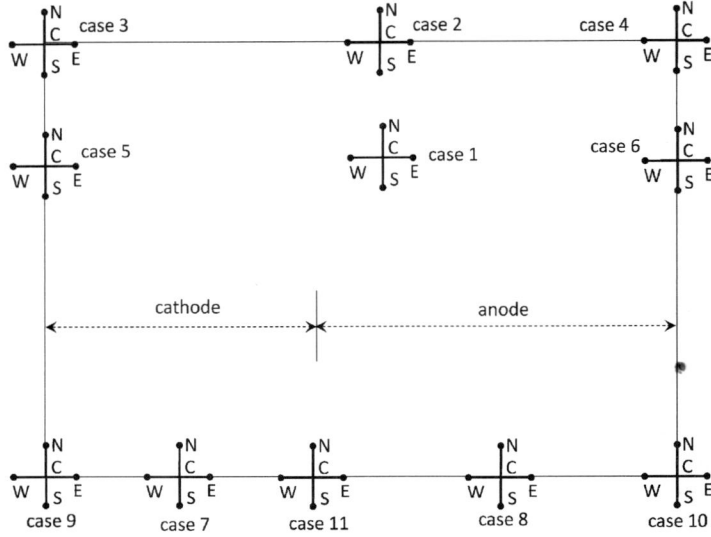

Figure 19. Fictitious nodes for different nodes on the boundary (lattice not shown).

In view of (82), the mesh equation takes the following form:

$$\phi_C = \delta_1 (\phi_E + \phi_W) + 2\delta_2 \phi_S. \tag{83}$$

Case 3 (the top-left corner)

Here, there are two fictitious nodes, N and W. To eliminate φ_N, the insulator boundary condition with $\vec{n} = (0, -1)$ gives $\phi_N = \phi_S$, just as in the previous case. To eliminate ϕ_W, again the insulator boundary condition with $\vec{n} = (1, 0)$ gives

$$0 = \frac{\partial \phi}{\partial n} = \frac{\partial \phi}{\partial x} = \frac{\phi_E - \phi_W}{2a}. \tag{84}$$

Therefore, $\phi_W = \phi_E$ and the overall mesh equation is described below:

$$\phi_C = 2\delta_1 \phi_E + 2\delta_2 \phi_S. \tag{85}$$

Case 4 (the top-right corner)

Here, the fictitious nodes are N and E. To eliminate ϕ_N and ϕ_E, the insulator boundary can be applied twice with $\vec{n} = (0, -1)$ and

$\vec{n} = (-1, 0)$, respectively. The results are $\phi_N = \phi_S$ and $\phi_E = \phi_W$. Therefore,

$$\phi_C = 2\delta_1 \phi_W + 2\delta_2 \phi_S. \tag{86}$$

Case 5 (the left wall, excluding the corners)

Here, node C can be anywhere on the left wall except at the very bottom or the very top. To eliminate the fictitious node W, once again the insulator boundary condition is used with $\vec{n} = (1, 0)$. It follows that $\phi_W = \phi_E$, from which we get

$$\phi_C = 2\delta_1 \phi_E + \delta_2 (\phi_N + \phi_S). \tag{87}$$

Case 6 (the right wall, excluding the corners)

Here, node C is anywhere on the right wall except at the endpoints. To eliminate the fictitious node E, the insulator boundary condition with $\vec{n} = (-1, 0)$ can be used to obtain $\phi_E = \phi_W$, and then

$$\phi_C = 2\delta_1 \phi_W + \delta_2 (\phi_N + \phi_S). \tag{88}$$

Case 7 (cathode, excluding its endpoints)

For this case, C may be anywhere on the cathode, except at its endpoints. The only fictitious node is S. To eliminate ϕ_S, the metal boundary condition $\sigma (\partial \phi / \partial n) = f(\phi)$ with $\vec{n} = (0, 1)$ should be used. The approximation is described by (89):

$$\sigma \frac{\partial \phi}{\partial n} = \sigma \frac{\partial \phi}{\partial y} = \sigma \frac{\phi_N - \phi_S}{2b} = f(\phi_C). \tag{89}$$

It follows that $\phi_S = \phi_N - (2b/\sigma) f(\phi_C)$, which when substituted into the basic mesh equation gives

$$\phi_C + \frac{2b\delta_2}{\sigma} f(\phi_C) = \delta_1 (\phi_E + \phi_W) + 2\delta_2 \phi_N. \tag{90}$$

Case 8 (anode, excluding its endpoints)

The conditions in this case are similar to those in case 7; the only difference is that here, the anode's polarization function g'' should be used instead of the cathode's. The result is

$$\phi_C + \frac{2b\delta_2}{\sigma} g(\phi_C) = \delta_1 (\phi_E + \phi_W) + 2\delta_2 \phi_N. \tag{91}$$

Case 9 (cathode, its left endpoint)

The fictitious points are S and W. To eliminate ϕ_W, the insulator boundary condition with $\vec{n} = (1, 0)$ is used to get $\phi_W = \phi_E$. To eliminate ϕ_S, the metal boundary condition with $\vec{n} = (0, 1)$ is used to get $\phi_S = \phi_N - (2b/\sigma) f(\phi_C)$. Substituting ϕ_W and ϕ_S in the basic equation gives

$$\phi_C + \frac{2b\delta_2}{\sigma} f(\phi_C) = 2\delta_1 \phi_E + 2\delta_2 \phi_N. \tag{92}$$

Case 10 (anode, its right endpoint)

The fictitious points are S and E. The elimination of ϕ_s and ϕ_E leads to $\phi_S = \phi_W$ and $\phi_S = \phi_N - (2b/\sigma)g(\phi_C)$. Therefore, the resulting mesh equation is

$$\phi_C + \frac{2b\delta_2}{\sigma} g(\phi_C) = 2\delta_1 \phi_W + 2\delta_2 \phi_N \tag{93}$$

Case 11 (the cathode–anode junction)

This case is similar to cases 7 and 8. The difference is that instead of the polarization functions f or g alone, the polarization function $f + g$ should be used. This is because at the junction, each electrode is sending its own current vertically upward into the solution (the actual direction is negative for the cathode), so the total current density from the junction into the solution is the algebraic sum of the contributions from both. The result is

$$\phi_C + \frac{2b\delta_2}{\sigma} [f(\phi_C) + g(\phi_C)] = \delta_1 (\phi_E + \phi_W) + 2\delta_2 \phi_N. \tag{94}$$

Equations (81)–(89) give the potential ϕ_C explicitly in terms of the potentials at the nonfictitious neighboring nodes of C. Equations (90)–(94) also give the potential ϕ_C; however, to find ϕ_C from these latter equations it is necessary to use a numerical technique since the polarization functions f and g can be nonlinear.

Cases 1–11 have been incorporated without any change into a Fortran code named COR_CELL.[5] The code is simple but system-dependent; shape and boundary conditions both affect each line of the central part of the program. Modifications will be necessary at various places in the code before it can be used for other geometries or boundary conditions. The user who introduces the changes must have some familiarities with the FDM. The user must also know how the FDM is adapted to the features of the system in question. The purpose of investigating cases 1–11 for the corrosion cell problem was to demonstrate how geometric and boundary conditions can be dealt with, and to make it easier for the reader to understand the organization of the code.

Keep in mind that several iteration cycles may be required to arrive at a converged solution.

IX. IMPLEMENTATION OF THE BOUNDARY-ELEMENT METHOD IN CATHODIC PROTECTION

Once again, as in Sect. VIII, the discussion is limited to a bounded two-dimensional domain. However, the domain is no longer restricted to a rectangular one and is assumed to be arbitrary in shape.[18] The boundary of the domain Ω denoted by $\partial\Omega$ consists of Γ_A, Γ_C, and Γ_I; therefore, $\partial\Omega = \Gamma_A \cup \Gamma_C \cup \Gamma_I$. These segments represent the anodic, cathodic, and insulated boundaries. The field equation is the Laplace equation:

$$\frac{\partial^2 \phi}{\partial x^2} + \frac{\partial^2 \phi}{\partial y^2} = 0. \qquad (95)$$

The boundary conditions are given by (96)–(100) below.

$$\frac{\partial \phi(Q)}{\partial n_Q} = f_A[\phi(Q)] \quad Q \in \Gamma_A \text{ (anode)}, \qquad (96)$$

$$\frac{\partial \phi(Q)}{\partial n_Q} = f_C[\phi(Q)] \quad Q \in \Gamma_C \text{ (cathode)}, \qquad (97)$$

$$\frac{\partial \phi(Q)}{\partial n_Q} = 0 \quad Q \in \Gamma_I \text{ (insulation)}. \qquad (98)$$

Here $f_A[\phi(Q)]$ and $f_C[\phi(Q)]$ represent the anodic and cathodic polarization curves. Furthermore, n_Q is the unit outward normal to the boundary at the point Q under consideration, as displayed in Fig. 20.

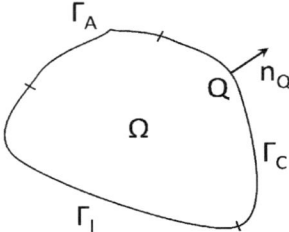

Figure 20. The domain under consideration and different segments.

Additional possible boundary conditions are shown below. These conditions are applicable depending on the problem at hand:

$$\frac{\partial \phi(Q)}{\partial n_Q} = q_A \quad Q \in \Gamma_A \text{ (impressed current condition)}, \quad (99)$$

$$\phi(Q) = \phi_A \quad Q \in \Gamma_A \text{ (nonpolarizable anode)}. \quad (100)$$

The classical methods of applied mathematics (Green's theorem specifically), allow us to calculate the potential at an arbitrary point P in the interior of $\partial\Omega$ according to the following equation:

$$2\pi\phi(P) = -\oint_{\partial\Omega} \left[\phi(Q)\frac{\partial}{\partial n_Q}\left(\frac{1}{r_{PQ}}\right) - \left(\ln\frac{1}{r_{PQ}}\right)q(Q)\right]d\Gamma_Q. \quad (101)$$

If the point P belongs to the boundary of Ω, the counterpart of the equation is revised according to (101):

$$\pi\phi(P) = -(\text{PV})\oint_{\partial\Omega} \left[\phi(Q)\frac{\partial}{\partial n_Q}\left(\frac{1}{r_{PQ}}\right) - \left(\ln\frac{1}{r_{PQ}}\right)q(Q)\right]d\Gamma_Q. \quad (102)$$

In these equations, r_{PQ} stands for the distance between the points P and Q, and (PV) represents the integration in terms of the Cauchy principal value. Furthermore, $q(Q)$ is the flux at point Q on the boundary.

In cathodic protection modeling, (102) is the relevant form. Once the unknowns $\phi(Q)$ and $q(Q)$ on the boundary have been evaluated, the potential at an arbitrary interior point can be calculated with the aide of (101). This, however, is rarely needed as the surface potential and fluxes are of primary interest.

Several properties of the integral equation (102) will be used and are listed below:

(a) The condition of solvability of the interior potential problem is

$$\oint_{\partial\Omega} q(Q) d\Gamma_Q = 0.$$

(b) If the potential $\phi(Q)$ is specified (i.e., the Dirichlet boundary condition), the flux $q(Q)$ on Γ can be uniquely determined from (102) and it automatically satisfies condition (a).
(c) If the flux $q(Q)$ on Γ and satisfies condition (a), the potential $\phi(Q)$ is not unique, but any two solutions differ by a constant.
(d) Finally, if condition (a) is violated, no solution for the potential $\phi(Q)$ can be obtained.

It is worth mentioning that the term $\ln(1/r_{PQ})$ in the above integral equations is the Green function for the two-dimensional Laplacian operator. For the case of the Laplacian operator in three dimensions, the Green function is different and is expressed by $1/r_{PQ}$. In either case, the Green function is singular when the points P and Q coincide. Therefore, the BEM involves calculating singular integrals.

X. NUMERICAL IMPLEMENTATION OF THE BOUNDARY-ELEMENT METHOD

The BEM is based on covering the boundary of the domain under consideration with elements (or panels) where the degrees of freedom are the potential and flux. The degrees of freedom associated with the jth panel (element) are ϕ_j and q_j.

The discrete version of the integral equation (102) at the mid-side point of each element is described by (103):

$$\sum_{j=1}^{N} H_{ij}\phi_j = \sum_{j=1}^{N} G_{ij}q_j \qquad i = 1, 2, \ldots, N. \qquad (103)$$

In (103), N is the total number of elements and i refers to the field point at which the integral equation has been discretized.[16] The expressions for H_{ij} and G_{ij} for a two-dimensional domain are given below:

$$H_{ij} = \begin{cases} \pi + \int_{s_{1j}}^{s_{2j}} \frac{\partial}{\partial n_Q}\left(\ln \frac{1}{r_{PQ}}\right) d\Gamma_Q \text{ if } i = j \\ \int_{s_{1j}}^{s_{2j}} \frac{\partial}{\partial n_Q}\left(\ln \frac{1}{r_{PQ}}\right) d\Gamma_Q \text{ if } i \neq j, \end{cases} \quad (104)$$

$$G_{ij} = \int_{S_{1j}}^{S_{2j}} \ln \frac{1}{r_{PQ}} d\Gamma_Q. \quad (105)$$

The integrals H_{ij} and G_{ij} can be evaluated using a local coordinate system. The expanded form of (103) is displayed by (106):

$$\begin{bmatrix} H_{11} & \cdots & H_{1N} \\ \vdots & \ddots & \vdots \\ H_{N1} & \cdots & H_{NN} \end{bmatrix} \begin{Bmatrix} \phi_1 \\ \vdots \\ \phi_N \end{Bmatrix} = \begin{bmatrix} G_{11} & \cdots & G_{1N} \\ \vdots & \ddots & \vdots \\ G_{N1} & \cdots & G_{NN} \end{bmatrix} \begin{Bmatrix} q_1 \\ \vdots \\ q_N \end{Bmatrix}. \quad (106)$$

The matrix H happens to be singular whereas matrix G is invertible. Although it is not the case in corrosion problems, if the boundary conditions are linear, upon applying the known boundary conditions (i.e., known potentials and fluxes), one can solve the linear system of equations for the unknown degrees of freedom.

1. Iteration Procedures

Recall that the boundary of the domain under consideration was the union of Γ_A, Γ_C, and Γ_I, which represented the anodic, cathodic, and insulated surfaces. Assuming that f_A and f_C denote the anodic and cathodic polarization curves, the symbolic expressions for the boundary conditions are described by (107)–(109):

$$q_A = f_A(\phi_A), \quad (107)$$
$$q_C = f_C(\phi_C), \quad (108)$$
$$q_I = 0. \quad (109)$$

Substituting these equations in the expanded form (106), where the degrees of freedom have been grouped together, leads to following system of nonlinear transcendental equations:

$$\begin{bmatrix} H_{11} & H_{12} & H_{13} \\ H_{21} & H_{22} & H_{23} \\ H_{31} & H_{31} & H_{33} \end{bmatrix} \begin{Bmatrix} \phi_C \\ \phi_A \\ \phi_I \end{Bmatrix} = \begin{bmatrix} G_{11} & G_{12} & G_{13} \\ G_{21} & G_{22} & G_{23} \\ G_{31} & G_{32} & G_{33} \end{bmatrix} \begin{Bmatrix} f_C(\phi_C) \\ f_A(\phi_A) \\ 0 \end{Bmatrix}. \tag{110}$$

In this system, the unknowns are Q_C, Q_A, and ϕ_I, representing the groups of degrees of freedom on the three surfaces. A useful rearrangement of the system (110) is displayed below:

$$\begin{bmatrix} H_{11} & H_{12} & -G_{13} \\ H_{21} & H_{22} & -G_{23} \\ H_{31} & H_{31} & -G_{33} \end{bmatrix} \begin{Bmatrix} \phi_C \\ \phi_A \\ 0 \end{Bmatrix} = \begin{bmatrix} G_{11} & G_{12} & -H_{13} \\ G_{21} & G_{22} & -H_{23} \\ G_{31} & G_{32} & -H_{33} \end{bmatrix} \begin{Bmatrix} f_C(\phi_C) \\ f_A(\phi_A) \\ \phi_I \end{Bmatrix}. \tag{111}$$

A further rearrangement of (111) is obtained through multiplying by the inverse of the matrix of the right-hand side, which is symbolically written below:

$$\begin{bmatrix} A_{11} & A_{12} & A_{13} \\ A_{21} & A_{22} & A_{23} \\ A_{31} & A_{31} & A_{33} \end{bmatrix} \begin{Bmatrix} \phi_C \\ \phi_A \\ 0 \end{Bmatrix} = \begin{Bmatrix} f_C(\phi_C) \\ f_A(\phi_A) \\ \phi_I \end{Bmatrix}. \tag{112}$$

Since the unknown vector ϕ_I is decoupled from the first two equations, we can concentrate on the reduced system (113):

$$\begin{bmatrix} A_{11} & A_{12} \\ A_{12} & A_{22} \end{bmatrix} \begin{Bmatrix} \phi_C \\ \phi_A \end{Bmatrix} = \begin{Bmatrix} f_C(\phi_C) \\ f_A(\phi_A) \end{Bmatrix}. \tag{113}$$

In the remaining portion of this section, three different methods are discussed for iteration purposes.[19]

The first method to be discussed is the Jacobi iteration. This is achieved according to the following scheme:

$$\phi_C^{n+1} = A_{11}^{-1} \left[-A_{12} \phi_A^n + f_C \left(\phi_C^n \right) \right], \tag{114}$$

$$\phi_A^{n+1} = A_{22}^{-1} \left[-A_{21} \phi_C^n + f_A \left(\phi_A^n \right) \right]. \tag{115}$$

Here n and $n+1$ represent the nth and $(n+1)$th iterations, respectively. The iteration procedure proceeds as follows:

(a) Make a guess of ϕ_C^n and ϕ_A^n.
(b) Calculate $f_C \left(\phi_C^k \right)$ and $f_A \left(\phi_A^k \right)$ from the cathodic and anodic polarization curves.

(c) Find ϕ_C^{n+1} and ϕ_A^{n+1} from (114) and (115).
(d) Repeat steps (a)–(c) until convergence is achieved.

The more sophisticated version of the above procedure is the Gauss–Siedel method. This method is based on the following iteration procedure:

$$\phi_C^{n+1} = A_{11}^{-1}\left[-A_{12}\phi_A^n + f_C\left(\phi_C^n\right)\right], \tag{116}$$

$$\phi_A^{n+1} = A_{22}^{-1}\left[-A_{21}\phi_C^{n+1} + f_A\left(\phi_A^n\right)\right]. \tag{117}$$

The main difference from the Jacobi iteration is that as soon as a new value is found, it is used immediately. The steps for this algorithm are described below:

(a) Make a guess of ϕ_C^n and ϕ_A^n.
(b) Calculate $f_C\left(\phi_C^k\right)$ and $f_A\left(\phi_A^k\right)$ from the cathodic and anodic polarization curves.
(c) Find ϕ_C^{n+1} and ϕ_A^{n+1} from (116) and (117).
(d) Repeat steps (a)–(c) until convergence is achieved.

The third technique discussed is the Newton–Raphson method. To describe the procedure, we define the two auxiliary functions $F_1(\phi_C, \phi_A)$ and $F_2(\phi_C, \phi_A)$:

$$\left\{\begin{array}{c} F_1(\phi_C, \phi_A) \\ F_2(\phi_C, \phi_A) \end{array}\right\} = \left\{\begin{array}{c} f_C(\phi_C) \\ f_A(\phi_A) \end{array}\right\} - \left[\begin{array}{cc} A_{11} & A_{12} \\ A_{21} & A_{22} \end{array}\right]\left\{\begin{array}{c} \phi_C \\ \phi_A \end{array}\right\}. \tag{118}$$

We now write the system (118) explicitly as

$$F_1(\phi_C, \phi_A) = f_C(\phi) - A_{11}\phi_C - A_{12}\phi_A = 0, \tag{119}$$

$$F_2(\phi_C, \phi_A) = f_A(\phi_A) - A_{21}\phi_C - A_{22}\phi_A = 0. \tag{120}$$

The Jacobian matrix J associated with the above system is given by

$$J = \frac{\partial(F_1, F_2)}{\partial(\phi_C, \phi_A)} = \left[\begin{array}{cc} f_C'(\phi_C) - A_{11} & -A_{12} \\ -A_{21} & f_A'(\phi_A) - A_{22} \end{array}\right], \tag{121}$$

where $f_C'(\phi_C) = df_C/d\phi_C$ and $f_A'(\phi_A) = df_A/d\phi_A$.

Therefore, the Newton–Raphson method takes the following form:

$$\left\{ \begin{array}{c} \phi_C \\ \phi_A \end{array} \right\}_{n+1} = \left\{ \begin{array}{c} \phi_C \\ \phi_A \end{array} \right\}_n - J^{-1} \left\{ \begin{array}{c} F_1(\phi_C, \phi_A) \\ F_2(\phi_C, \phi_A) \end{array} \right\}_n. \quad (122)$$

The iteration proceeds as follows:

(a) Make a guess of ϕ_C^n and ϕ_A^n.
(b) Calculate $F_1(\phi_C^n, \phi_A^n)$ and $F_2(\phi_C^n, \phi_A^n)$.
(c) Calculate $f'_C(\phi_C^n)$ and $f'_A(\phi_A^n)$ from the cathodic and anodic polarization curves.
(d) Calculate the Jacobian matrix.
(e) Repeat steps (a)–(e) until convergence is obtained.
(f) Calculate ϕ_I from the last row of (112), i.e., $\phi_I = A_{31}\phi_C + A_{32}\phi_A$.

XI. CONCLUDING REMARKS

In this expository article, the basic mathematical model of some simple electrochemical processes was discussed. The model is based on the concept of conservation of charge within the electrolyte. The boundary conditions, on the other hand, are problem-specific. The subject of electrode kinetics is central to the proper specification of the boundary conditions. In their most general form, the conditions are nonlinear, leading to a nonlinear boundary value problem. This is closely tied to the nonlinear polarization curves. The analytical solution of the mathematical model is formidable and for moderately simple two-dimensional regions is impossible to obtain. The only feasible approach is numerical simulation. The use of high-speed digital computers is an essential tool in solving such problems.

The most common techniques for numerical simulation are the FDM, the BEM, and the FEM. Each of these techniques has its advantages and limitations. In this article, finite differences and boundary elements were briefly discussed. The application of the FEM in electrochemistry has been reviewed by Schlesinger recently.[20] Since the objective of the present article was not to present a comparative study of the available techniques, no specific examples were considered. However, on the basis of the available literature, at least in the context of cathodic protection simulation, the BEM may be the most efficient numerical approach.

APPENDIX. ELECTRODE KINETICS IN CATHODIC PROTECTION

This section reviews parts of electrochemistry relevant to corrosion and cathodic protection. For a more complete account, one must consult books on physical chemistry or corrosion.[5–7]

Currents

Under favorable conditions, at every point on the surface of a metal immersed in an electrolyte, oxidation (anodic) and reduction (cathodic) reactions occur simultaneously. The anodic reaction results in a current density flowing out of the metal and into the electrolyte. The cathodic reaction at the same point is the density current flowing from the electrolyte into the metal. The former is called the "anodic current density" (denoted by i_{anodic}) and the latter is called the "cathodic current density" (denoted by i_{cathodic}). The net current density out of the metal and into the electrolyte is then

$$i_{\text{net}} = i_{\text{anodic}} - i_{\text{cathodic}}. \tag{123}$$

At any point on the wet metal surface i_{anodic} and i_{cathodic} are directly proportional to, or can be defined as, the oxidation and reduction rates. Equilibrium at a point is defined as a condition in which the oxidation and reduction rates at that point are equal, or when i_{anodic} equals i_{cathodic}, or when $i_{\text{net}} = 0$.

As a vector quantity, the electrical current density has a direction as well as magnitude. At each point on the wet metal surface, the components of the current density normal to the surface and pointing to the electrolyte will be taken to be i_{net}.

This property will be used later to write the boundary conditions.

Potentials

The electrolyte and metal begin to interact chemically once they are in contact. As mentioned above, this interaction is in the form of oxidation (positive metal ions migrating from the metal into the electrolyte) and reduction (positive metal ions returning from the electrolyte to the metal and becoming metal atoms). At any given point on the wet surface of the metal surface, an electric field and hence an electrical potential difference is developed across an infinitesimally thin layer at the metal–electrolyte interface. We will refer to this electric field as the interface field and to the accompanying electric potential difference as the interface field or the electromotive force. The direction and magnitude of the interface field (the polarity and size of the interface potential) depend on several factors, including the electrolyte ionic concentration. In corrosion problems, it is often the case that the metal becomes slightly negative with respect to the neighboring points in the electrolyte.

Potential Measurements

By a system is meant here a structure comprising one or more types of metal and electrolyte which are in contact with one another in an arbitrary but fixed manner. Thus, a system may be a container with one or more electrodes and electrolytes, or a much larger structure, such as a ship. In addition to the currents and potentials along the metal surface, there is current and potential distribution throughout the entire system. The potential and current density at every point of the system are denoted by ϕ and i respectively. Their units are volts and amperes per square meter.

There is usually interest in knowing the electrical potential at different points in the system, particularly at points near or adjacent to the metals or the electrodes involved. However, it is impossible to measure the true electrical potential difference between a metal and the solution which is in contact with it. This is because a second metallic electrode is required, and a second electrode introduces its own interface potential into the readings, even when those readings are taken by an ideal voltmeter which draws no currents. So, there is no method of measuring either the actual electromotive force of an electrode or the true electrical potential at any point inside the solution. But, by arbitrarily assigning a potential to one electrode, one can measure or otherwise determine the potentials elsewhere in the system with respect to this standard. Applications vary from indoor laboratory work to underwater measurements at offshore petroleum establishments. A few of the reference electrodes in common usage are the hydrogen electrode, the saturated calomel electrode (SCE), the silver–silver chloride electrode, and the glass electrode.

The potential, current, and their signs (positive or negative) can be the source of much error and disagreement; therefore, we must adhere to a rigid set of conventions if hopeless confusion about the sign and interpretation, especially of ϕ, is to be avoided. To begin with, a distinction must be made between what we call here electrochemical potential and the electrical potential difference in its original sense as defined in physics. The former has proven useful and convenient in the field of electrochemistry and its applications, such as corrosion engineering. To measure ϕ, one connects the positive terminal of an (ideal) voltmeter to the trunk or body of the metals in the system; the negative terminal of the voltmeter goes to a probe (or sensor) such as the SCE; the probe is placed at the point at which ϕ is to be measured. To measure ϕ elsewhere in the system, we simply move the probe to the new spot. The positive terminal of the voltmeter remains fixed at some point on the metallic bulk of the electrodes. Using this procedure, the measurements taken at points sufficiently close to the surface of a metal are the electromotive force of the metal at those points. Equally important, such measurements are consistent with the conventions of the International Union of Pure and Applied Chemistry which imply that the baser a metal, the lower its electromotive force.

If we let ψ denote the actual electrical potential difference between a point in the system and the electrical ground, which is usually a metallic bulk, then ϕ and ψ are related though

$$\psi = c - \phi. \tag{124}$$

The constant c in (124) depends on the type of the sensor used for measuring φ (SCE, etc.). It may be added here that metals, owing to their high conductivity, allow little or no potential variation throughout their interiors (or bulk). It is for this reason that the metallic bulk is frequently a natural and convenient choice for a point of reference or a point of zero electrical potential (also called the "electrical ground").

Equation (124) does establish a relation between ϕ and ψ, at least for geometries where an electrical ground is easily identifiable. However, (124) may not be used to find ψ from the knowledge of ϕ obtained from measurements. This is because c is not known. Also note that while both ϕ and c depend on the type of measuring probe, their difference, ψ, does not. This is to be expected since the true electrical potential difference ψ should not depend on our choice of the probe.

Polarization Equations

At each point on a metal surface which is in contact with the electrolyte, i_{anodic} and i_{cathodic} are given in terms of the electrochemical potential Φ according to the Tafel–Butler–Volmer polarization equations below:

$$i_{\text{anodic}} = i_0 \exp\left(\frac{\phi - \phi_0}{\alpha}\right), \tag{125}$$

$$i_{\text{cathodic}} = i_0 \exp\left(\frac{\phi - \phi_0}{-\beta}\right). \tag{126}$$

The net current density is then

$$i_{\text{net}} = i_0 \left[\exp\left(\frac{\phi - \phi_0}{\alpha}\right) - \exp\left(\frac{\phi - \phi_0}{-\beta}\right)\right]. \tag{127}$$

The current densities i_{anodic} and i_{cathodic} are not independent of each other as they satisfy (128):

$$i_{\text{anodic}}^{\alpha} i_{\text{cathodic}}^{\beta} = i_0^{\alpha + \beta}. \tag{128}$$

The parameters i_0, ϕ_0, α, and β depend on factors such as the temperature, electrolyte ion concentration, and the type of reference electrode used for potential measurements. During a single experiment, however, these entities are assumed to be constant.

The constant ϕ_0 is the equilibrium potential or the reversible potential of the electrode; its sign and magnitude depend on the reference electrode. The difference $\eta = \phi - \phi_0$ is called the "overpotential" or the "overvoltage" and it represents the deviation of ϕ from the equilibrium potential ϕ_0. Thus, $\eta = \phi - \phi_0$ may take on both positive and negative values at different points on the surface of an electrode.

The positive constant i_0 is called the "exchange current density." Equations (125) and (126) imply that i_0 is the current flowing across a unit area of the electrode in each direction (metal to solution and solution to metal) at the reversible potential (where $\eta = \phi - \phi_0 = 0$).

The positive constants α and β are given by $\alpha = RT/\gamma F$ and $= RT/(1-\gamma)F$. Here R is the gas constant, T is the absolute temperature, and F is the Faraday constant. The dimensionless constant γ is in the interval $(0, 1)$; it is called the "symmetry factor," though that name is sometimes also used for α and β. The constants α and β are not independent of each other as they satisfy (129):

$$\frac{1}{\alpha} + \frac{1}{\beta} = \frac{F}{RT}. \tag{129}$$

In this paragraph we will consider some of the properties of the relation between ϕ and i_{net}. Equation (127) motivates the definition of a polarization function f as follows:

$$f(\phi) \stackrel{\text{def}}{=} i_0 \left[\exp\left(\frac{\phi - \phi_0}{\alpha}\right) - \exp\left(\frac{\phi - \phi_0}{-\beta}\right) \right]. \tag{130}$$

Then $i_{\text{net}} = f(\varphi)$ and

$$\dot{f}(\phi) = i_0 \left[\frac{1}{\alpha} \exp\left(\frac{\phi - \phi_0}{\alpha}\right) + \frac{1}{\beta} \exp\left(\frac{\phi - \phi_0}{-\beta}\right) \right], \tag{131}$$

$$\ddot{f}(\phi) = i_0 \left[\frac{1}{\alpha^2} \exp\left(\frac{\phi - \phi_0}{\alpha}\right) - \frac{1}{\beta^2} \exp\left(\frac{\phi - \phi_0}{-\beta}\right) \right]. \tag{132}$$

According to (131), i_{net} is a strictly increasing function of ϕ. The slope $\dot{f}(\phi)$ of the graph of i_{net} versus ϕ is a minimum at the inflection point ϕ_1 given by

$$\phi_1 = \phi_0 + \frac{2\alpha\beta}{\alpha + \beta} \ln\left(\frac{\alpha}{\beta}\right). \tag{133}$$

Note that $\phi_1 = \phi_0$ iff $\alpha = \beta$, i.e., iff $\gamma = 1/2$. Finally, the graph of f is concave up for $\phi > \phi_1$, and concave down for $\phi < \phi_1$. Moreover, $i_{\text{net}} > 0$ for $\phi > \phi_0$, $i_{\text{net}} = 0$ for $\phi = \phi_0$, and $i_{\text{net}} < 0$ for $\phi < \phi_0$.

A linearized polarization function is sometimes used as an approximation to the nonlinear function f. The particular linearization function here is

$$f_{\text{linear}}(\phi) = i_0 \left(\frac{1}{\alpha} + \frac{1}{\beta}\right)(\phi - \phi_0). \tag{134}$$

This linearization is expected to give good results for small values of $|\phi - \phi_0|$.

POLARIZATION CURVES

The graphs of f and f_{linear} are shown in Fig. 21 for the cases $\alpha < \beta$, $\alpha > \beta$, and $\alpha = \beta$. The graph of f resembles that of a hyperbolic sine function. It is exactly a hyperbolic sine function iff $\alpha = \beta$ for (127) and (130) simplify to

$$i_{\text{net}} = f(\phi) = 2i_0 \sinh\left(\frac{-\phi_0}{\alpha}\right). \tag{135}$$

Practitioners in the field seem to prefer other variations of these graphs. A first variation (not shown) is where we sketch i_{net} or i_{net}/i_0 versus the overpotential $\eta = \phi - \phi_0$. Sketching versus η is equivalent to a horizontal shift of the graph in Fig. 21 to the origin, and sketching i_{net}/i_0 instead of i_{net} simply amounts to a vertical scaling of the same graph.

A second variation is where ϕ is treated as a function of i_{net}. This variation is readily available since the polarization function is invertible, irrespective of the relation between α and β. The resultant graphs are mirror images of those in Fig. 21 with respect to the bisector of the first quadrant. This is displayed in Fig. 22.

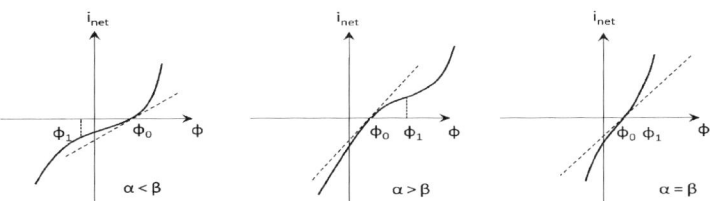

Figure 21. The graph of i_{net} versus ϕ.

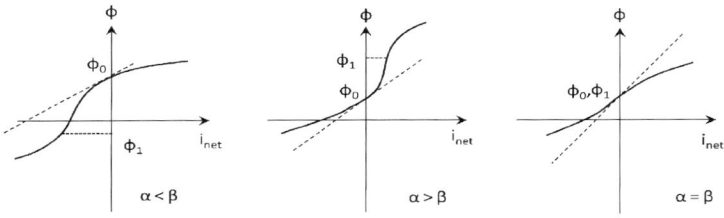

Figure 22. The graph of ϕ versus i_{net}.

It is clear from the graph in Fig. 22 and (125) and (126) that for $\phi > \phi_0$, we have $i_{\text{anodic}} > i_0 > i_{\text{cathodic}}$. In other words, for $\phi > \phi_0$, the anodic reaction dominates the cathodic reaction and i_{net} is positive, that is, a net current is flowing out of the given point on the electrode. The part of the graph corresponding to $\phi > \phi_0$ ($i_{\text{net}} > 0$) will be referred to as the oxidation branch. In Fig. 21, the oxidation branch is the part of the graph above the ϕ-axis. In Fig. 22, the oxidation branch is to the right of the ϕ-axis. Similarly, the inequalities $i_{\text{cathodic}} > i_0 > i_{\text{anodic}}$ hold for $\phi < \phi_0$. In this case the cathodic reaction dominates the anodic reaction and i_{net} is negative, that is, a net current is flowing into the given point on the electrode. The part of the graph for which $\phi < \phi_0$ ($i_{\text{net}} < 0$) is called the "reduction branch." In Fig. 21, the reduction branch is to the left of the ϕ−axis.

A third variation occurs when in Fig. 22 the reduction branch is sketched with ϕ versus $-i_{\text{net}}$. In Fig. 22, this is equivalent to a 180° rotation of the reduction branch about the ϕ-axis (or drawing the mirror image of the reduction branch with respect to the ϕ-axis) while keeping the oxidation branch fixed. This variation is shown in Fig. 23.

The graphs in Figs. 21–23 are all different representations of the same equation, namely, (127), which is a relationship between i_{net} and ϕ. There is another type of graph which differs from the previous ones. The idea is to sketch not the relation between i_{net} and ϕ, but the relation between i_{anodic}/i_0 and $\eta = \phi - \phi_0$, and also the relation between i_{cathodic}/i_0 and overvoltage $\eta = \phi - \phi_0$ both on the same coordinate system. Therefore, the new graphs are obtained from (125) and (126) instead of (127). The result is shown in Fig. 24. This figure also includes the graph where a semilogarithmic scale is used. The outcome is two straight lines with slopes α and $-\beta$.

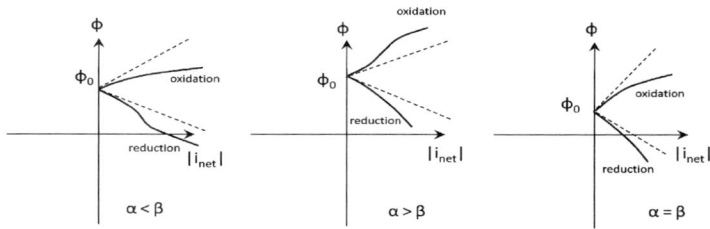

Figure 23. Drawing oxidation and reduction branches on the same side.

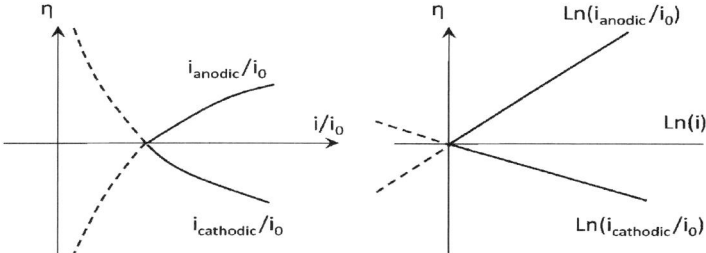

Figure 24. Overvoltage versus anodic and cathodic currents.

A Corrosion Cell

Polarization curves can sometimes be helpful in providing insight into a specific problem. As an example, consider what may be the most basic corrosion system, namely, a cell consisting of two small (microscopic) pieces or grains of metal in contact with each other and with an electrolyte. The small size of the cell allows us to assume that the potential variation on the surface of each metal is zero, i.e., the potential ϕ on the surface of each piece is constant.

There are two possibilities for the galvanic contact between the electrodes:

Case 1. The point of contact between electrodes is at the surface of the electrodes.
Case 2. The point of contact between the electrodes is in the interior of the electrodes.

The geometry displayed in Fig. 25a corresponds to case 1. In this configuration the point of contact between the two metals also polarizes to a potential ϕ. Since this point is common to both metals and each metal surface is at a constant potential, it follows that both metal surfaces are at the same potential, namely, ϕ. Implicit in this statement is that there is no jump in potential in the direction along the surface of the electrodes, an assumption that is correct and can be justified on physical grounds, but it has not been done here.

The current density on the surface of metal 1 is constant and it is given by $i_1 = f_c(\phi)$, where f_c is the polarization function of metal 1. Similarly, the current density on the surface of metal 2 is constant and it is given by $i_2 = f_a(\phi)$, where f_a is the polarization

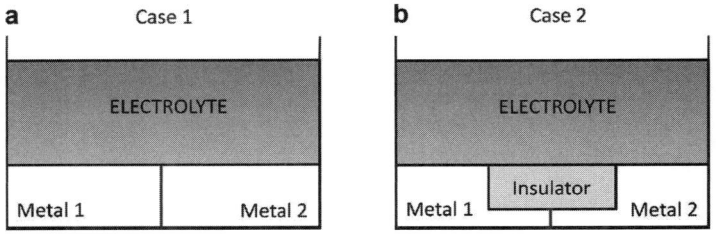

Figure 25. Two cases of a simple corrosion cell.

function of metal 2. On the other hand, i_1 and i_2 satisfy $i_1 + i_2 = 0$. Therefore,

$$i_2 = -f_c(\phi) = f_a(\phi). \tag{136}$$

Consequently, the following expression is correct:

$$\phi = (f_c)^{-1}(i_2) = f_a^{-1}(i_2). \tag{137}$$

To find (ϕ, i_2) graphically using (136), we intersect the graphs of f_a and $-f_c$. To find (i_2, ϕ) graphically using (137), we intersect the graphs of f_a^{-1} and $(-f_c)^{-1}$.

The governing equations are (see (134) for linear curves)

$$\phi - \phi_c = r_c i_1 = -r_c i_2, \tag{138}$$
$$\phi - \phi_a = r_a i_2. \tag{139}$$

Linear polarization has been assumed for both electrodes.

In (138) and (139), ϕ_c and ϕ_a are the equilibrium potentials of metals 1 and 2, where it has been assumed that $\phi_c > \phi_a$. Note that $i_2 > 0 > i_1$ and $\phi_c > \phi > \phi_a$. Therefore, metal 1 has been operating on the cathodic (reduction) branch of its polarization curve and metal 2 has been operating on the anodic (oxidation) branch.

In a corrosion cell of the type under consideration, the metal with higher reversible potential is called the "cathode," while the one with the lower reversible potential is called the "anode." The steady-state potential is somewhere between the two reversible (or equilibrium) potentials. A net current flows out of the anode and into the cathode. Thus, the anodic piece is attacked by the electrolyte and the cathodic piece is protected. The rate of attack is directly proportional to the current density i_2. The constants $-r_c$ and r_a in

Numerical Modeling of Certain Electrochemical Processes

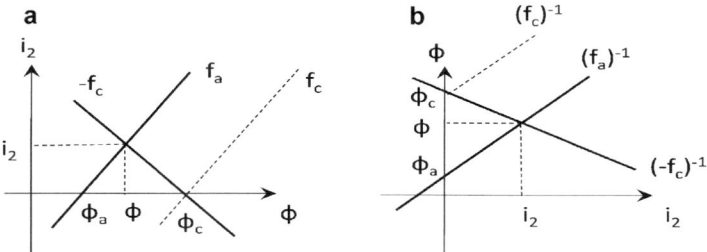

Figure 26. Steady-state potential and current density for case 1.

(138) and (139) are the slopes of the lines in Fig. 26b. For linear polarization, exact solutions are easily obtainable and are given by

$$\phi = \frac{r_a}{r_a + r_c}\phi_c + \frac{r_c}{r_a + r_c}\phi_a, \qquad (140)$$

$$i_2 = -i_1 = \frac{\phi_c - \phi_a}{r_c + r_a}. \qquad (141)$$

Equations (139) and (140) confirm our earlier observations. As a convex combination, ϕ is always between ϕ_c and ϕ_a, and $\phi_c > \phi_a$ implies $i_2 > 0 > i_1$.

From the inequalities $\phi_a < \phi < \phi_c$ and the statement following (134) about the accuracy of the linear model, it can be seen that if the difference $\phi_c - \phi_a$ is too large, then the original exponential curves must be used to obtain reasonably accurate results. The steady-state potential and current density for the small corrosion cell in Fig. 25a using nonlinear polarization functions are shown in Fig. 27.

As Fig. 27 indicates, the error due to linearization can be considerably greater for the current density than it can be for the potential. Linearization error for the current density can be consequential since the rate of corrosion depends on the current density. If the current density obtained from the linear model is substantially lower than its true value, then the actual corrosion rate will be severely overestimated and this will lead to overly optimistic guesses about, for example, the lifetime of a metallic machine part. Therefore, it seems advisable to work with the exponential model, especially when the difference $\phi_c - \phi_a$ is large. Since solving nonlinear equations may be impossible or difficult to carry out manually, one might consider the following approach based on Figs. 26 and 27. First, find the

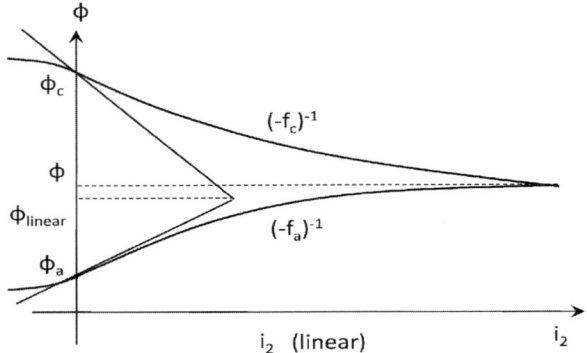

Figure 27. Nonlinear polarization functions and their linearizations (case 1).

steady-state potential from the linearized model, i.e., (140). Second, substitute the potential thus obtained in the original (nonlinear) polarization functions for each of the metals. And third, choose the greater answer as an estimate of the steady-state current density.

We end the above discussion with a comment about the effect of the conductivity of the electrolyte. It was assumed at the beginning of this section that the potential over the entire surface of each electrode is constant. Strictly speaking, that assumption is valid only when the electrolyte conductivity is infinite, or its resistivity is zero. In reality, of course, no electrolyte has zero resistivity. The results obtained in (140) and (141) and Figs. 26 and 27, however, are important and valuable because they provide worst-case (hence, safe) approximations to practical corrosion situations.

We now turn to case 2 in Fig. 25b. Here too we shall assume the dimensions of the cell are so small that each piece of metal has a constant potential over its entire surface. Call them ϕ_1 and ϕ_2. Since the two surfaces do not meet at a point on the electrolyte boundary, we may not assume $\phi_1 = \phi_2$. This is in contrast with case 1, and it is necessary to take into account the conductivity or the resistivity of the electrolyte as shown in Fig. 28.

If (i_1, ϕ_1) and (i_2, ϕ_2) are the steady-state current density and potential pairs for metals 1 and 2 ,respectively, they must satisfy

$$i_1 = f_c(\phi_1), i_2 = f_a(\phi_2), i_1 + i_2 = 0, \phi_1 - \phi_2 = r i_2, \quad (142)$$

Numerical Modeling of Certain Electrochemical Processes

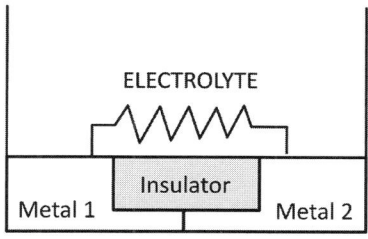

Figure 28. Electrolyte conductivity taken into consideration for case 2.

where r is directly proportional to the resistivity of the electrolyte. Eliminating i_1 from (142), we find

$$i_2 = -f_c(\phi_1) = f_a(\phi_2) = \frac{\phi_1 - \phi_2}{r}. \tag{143}$$

Equation (143) represents a system of three equations in the three unknowns ϕ_1, ϕ_2, i_2.

Unlike case 1, it is difficult to graphically solve for ϕ_1, ϕ_2, i_2 using a two-dimensional graph, even if f_c and f_a are linearized. We then proceed to solve (143) analytically for the linear case. The result is

$$\phi_1 = \frac{r_c}{r + r_a + r_c}\phi_a + \frac{r + r_a}{r + r_a + r_c}\phi_c, \tag{144}$$

$$\phi_2 = \frac{r + r_c}{r + r_a + r_c}\phi_a + \frac{r_a}{r + r_a + r_c}\phi_c, \tag{145}$$

$$i_2 = \frac{\phi_c - \phi_a}{r + r_a + r_c} = -i_1. \tag{146}$$

As convex combinations, both ϕ_1 and ϕ_2 are between ϕ_c and ϕ_a. A simple test shows that $\phi_2 < \phi_1$; therefore $\phi_a < \phi_2 < \phi_1 < \phi_c$ and $0 < i_2 < i_{max}$, where i_{max} is the current density corresponding to $r = 0$. For $r = 0$, it follows that $\phi_1 = \phi_2$. For $r = \infty$, we get $\phi_1 = \phi_c$, $\phi_2 = \phi_a$, and $i_2 = 0 = i_1$. These results are shown in Fig. 29.

Figure 29 is the counterpart of Fig. 26b for case 1. An equivalent of Fig. 26a is not drawn here. Equations (144)–(146) are the counterparts of (140) and (141). Even though cases 1 and 2 deal with

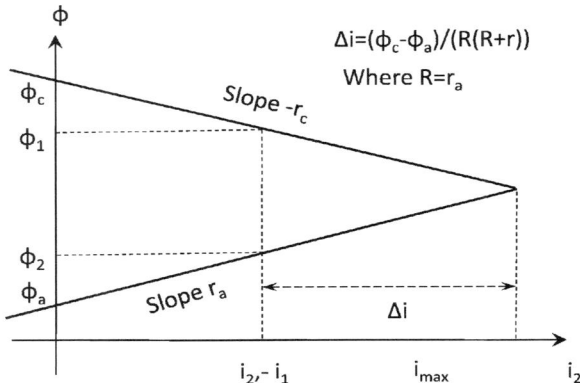

Figure 29. Steady-state potential and current for case 2.

different geometries, the effect of electrolyte resistivity is apparent from a comparison between (139) and (142) and a comparison between Figs. 26b and 29. Note that if the electrolyte is not a very good conductor, i.e., $r \gg 0$, then the corrosion current density will be less than r_{max}.

As a precautionary measure it is common to assume that the corrosion rate is maximum, that is, the cell current density equals r_{max} even when $r \gg 0$.

What was said earlier for case 1 about the effect of linearization holds true here as well. It should be noted though that attack on metal 2 is a maximum for the geometry in Fig. 25a. Therefore, conservative answers for the potential and current density for the present case may be obtained from the results for case 1. For more accurate answers, we can use (144) and (145) to find the potentials and then substitute them in nonlinear (exponential) polarization functions to find the current density.

REFERENCES

[1] D.R. Askeland, The Science and Engineering of Materials, Third Edition, PWS Publishing company, Boston, MA, 1994.
[2] J.H. Morgan, Cathodic Protection, The McMillan Company, New York, 1960.
[3] G.F. Carrier and C.E. Pearson, Partial Differential Equations, Theory and Techniques, Academic Press, New York, 1976.

[4] N.G. Zamani, J.F. Porter, and A.A. Mufti, A Survey of Computational Efforts in the Field of Corrosion Engineering, International Journal of Numerical Methods in Engineering, vol. 23, pp. 1295–1311, 1986.

[5] M. Mobed and N.G. Zamani, COR_CELL Version I, MWR Technical Report, FDRI-TR-94–02, 1994.

[6] M.G. Fontana and N.D. Greene, Corrosion Engineering, McGraw-Hill Science, New York, 1978.

[7] K.B. Oldham and J.C. Mayland, Fundamentals of Electrochemical Science, Academic Press, New York, 1994.

[8] E. Kennard and J. Waber, Mathematical Studies on Galvanic Corrosion, Equal Coplanar Anode and Cathode with Unequal Polarization Parameters, Journal of Electrochemical Society, vol. 117, pp. 880–885, 1970.

[9] J.T. Waber and B. Fogan, Mathematical Studies on Galvanic Corrosion IV, Influence of Electrolyte Thickness on The Potential and Current Distributions of Coplanar Electrodes using Polarization Parameters, Journal of Electrochemical Society, vol. 103, pp. 64–72, 1956.

[10] E. Hume, R. Brown, and W. Dean, Comparison of Boundary and Finite Element Methods for Moving Boundary Problems Governed by a Potential, International Journal for Numerical Methods in Engineering, vol. 21, pp. 1295–1314, 1985.

[11] R. Alkire, Predicting Electrode Shape Change with Use of Finite Element Method, Journal of Electrochemical Society, vol. 125, pp. 1981–1988, 1978.

[12] R.H. Nilson and Y.G. Tsuei, Free Boundary Problems of ECM by Alternating Field Technique by Inverted Plane, Computer Methods in Applied Mechanics and Engineering, vol. 6, pp. 265–282, 1975.

[13] N.G. Zamani, J.M. Chuang, and C.C. Hsuing, Numerical Simulation of Ecectrodeposition Problems, International Journal of Numerical Methods in Engineering, vol. 24, pp. 1479–1497, 1986.

[14] G.E. Forsythe and W.R. Wasow, Finite Difference Methods for Partial Differential Equations, Wiley, New York, 1960.

[15] K.J. Bathe, Finite Element Procedures, Prentice-Hall, Englewood Cliffs, NJ, 1996.

[16] C.A. Brebbia, J.C.F. Telles, and L.C. Wrobel, Boundary Element Techniques, Theory and Applications in Engineering, Springer, Berlin, 1984.

[17] J.J. Bertin and M.L. Smith, Aerodynamics for Engineers, Prentice-Hall, Englewood Cliffs, NJ, 1979.

[18] J.M. Chaung, N.G. Zamani, and C.C. Hsuing, "Some Computational Aspects of BEM Simulation of Cathodic Protection Systems", Journal of Applied Mathematical Modeling, vol. 11, pp. 371–379, 1987.

[19] J.M. Chaung, Ph.D. Dissertation, TUNS, Halifax, Nova Scotia, 1986.

[20] M. Schlesinger, Modern Aspects of Electrochemistry, volume 43, Springer, Berlin, 2008.

2

Near-Field Optics for Heat-Assisted Magnetic Recording (Experiment, Theory, and Modeling)

William A. Challener and Amit V. Itagi

*Seagate Technology, 1251 Waterfront Place, Pittsburgh, PA 15222, USA,
william.a.challener@seagate.com; amit.itagi@seagate.com*

I. NEAR-FIELD TRANSDUCERS FOR HEAT-ASSISTED MAGNETIC RECORDING

One application of near-field transducers (NFT) is in heat-assisted magnetic recording (HAMR). HAMR is similar to conventional magneto-optical (MO) recording in that the data are stored in magnetic bits on a disk by heating the area of the bit with a laser beam in the presence of an external field to set the magnetic orientation of the bit as it cools. The optical head in conventional MO recording is mounted on an actuator and optical feedback signals are used to maintain a constant spacing between the head and the recording medium, which is generally on the order of tens or hundreds of nanometers. Also, for conventional MO recording the applied magnetic field is very small (approximately 0.02 T), typically generated by a large fixed external magnet, and the laser energy rather than the magnetic field is modulated with the input data stream. On the other hand, for HAMR the integrated optical–magnetic head is mounted on a slider, which flies over the surface of the recording medium at 10 nm or less. The applied field for HAMR is highly localized, very large in magnitude (up to 1 T or more), and generated by a miniature recording pole positioned within tens of nanometers of the optical spot. For HAMR the magnetic field from the pole is modulated with the input data stream, while the laser energy on the medium can remain constant.

Conventional magnetic recording technology records magnetic bits with down-track and cross-track dimensions less than 100 nm. Areal recording densities of up to 400 Gb/in.^{2} have been demonstrated. Unfortunately, it is difficult with conventional recording technology to achieve substantially larger densities. As the storage density increases, the area of each bit decreases, but to maintain the same level of signal-to-noise ratio, the number of magnetic grains within each bit must not decrease. Therefore, greater areal densities require smaller magnetic grains. The magnetic grain diameter is presently on the order of 10 nm. As the volume of a magnetic grain is reduced, it reaches a point where the magnetic orientation of the grain becomes thermally unstable. Essentially, the average thermal energy within the grain, which is proportional to $k_B T$, becomes comparable to the magnetic anisotropy energy, $K_u V$, where k_B is Boltzmann's constant, T is the absolute temperature, K_u is the magnetic anisotropy constant of the grain, and V is the volume of the grain. This has been termed the "superparamagnetic limit" of magnetic storage density. Although it is possible to increase the stability of the recording medium by increasing the magnetic anisotropy of the recording material, eventually the applied magnetic recording field from the recording head is insufficient to switch the magnetic state of the medium. It requires new technologies to achieve areal densities beyond this point. In HAMR the magnetic anisotropy of the medium is momentarily reduced to enable recording by raising its temperature. The recorded bit is then quickly cooled back to its high-anisotropy state at ambient temperature to stabilize it. In this manner, extremely high magnetic anisotropy materials such as FePt can be recorded in a HAMR system, thereby potentially enabling areal storage densities in the range of 1–40 Tb/in.^{2}.[1,2]

At these storage densities, the recorded domains are only tens of nanometers in length and width. Hence, the optical spot used to heat the recording medium in HAMR must be an order of magnitude smaller than the optical wavelength of low-cost and high-power semiconductor lasers, which is in the range of 650–830 nm. A conventional lens can only focus light to a spot size defined by the diffraction of light from the clear aperture of the lens. The diffraction limit for the focused optical spot is given by

$$d = \frac{0.5\,\lambda}{n \sin \theta}, \qquad (1)$$

where d is the full-width spot diameter at the half maximum point (FWHM), λ is the wavelength, n is the refractive index of the medium in which the light is focused, and θ is the half angle of the cone of focused light. In other words, conventional optics are able to focus light to a spot size of approximately a half wavelength. For example, the new Blu-ray technology operates at a wavelength of 405 nm and a numerical aperture (equivalent to $n \sin \theta$) of 0.85, which corresponds to a focused optical spot size of approximately 240 nm. Although this is a very small optical spot, it is still much too large for use with HAMR. In a sense, HAMR replaces the difficulty of surpassing the superparamagnetic limit with the difficulty of focusing light below the diffraction limit.

A solid immersion lens[3] (SIL) is a somewhat unconventional focusing optic that is able to bring light to a focus inside a transparent high-index material, resulting in a spot size that is n times smaller in diameter than that for light brought to a focus in air by a lens with the same numerical aperture, where n is the refractive index of the SIL. Such an optic may be an essential part of a HAMR disc drive. However, this spot size is still at least twice as large as that required for a 1-Tb/in.^{2} HAMR areal storage density. Therefore, a HAMR disc drive requires a new approach for concentrating light energy into a spot smaller than the diffraction limit. Such devices are possible by making use of the "near field," that is, by concentrating energy that consists of both propagating and nonpropagating components. Because the nonpropagating components are evanescent – they decay exponentially with the distance from their source – the NFT can only generate a sub-diffraction-limited spot within a distance that is much smaller than a wavelength. Fortunately, even in a conventional disc drive, the recording head flies within 20 nm of the recording surface, well within the near field of a HAMR transducer.

NFTs are becoming popular in various areas of spectroscopy (e.g., surface-enhanced Raman spectroscopy with a lithographically defined surface of gold or silver nanoparticles of various shapes).

NFTs may also become useful for extremely high-density optical lithography[4] and optical imaging.[5] However, the requirements for the NFT in HAMR are substantially greater than those for NFTs in spectroscopy, lithography, or near-field imaging. In all cases, the NFT must concentrate optical energy into a spot much smaller than the diffraction limit, or in the time-reversed sense, scatter or transmit light from an optical region much smaller than the diffraction limit. For HAMR, however, the efficiency of the NFT is also of primary importance. A NFT which confines the light energy to a 20-nm spot but which only conducts one part in 10^5 of the incident laser power into this spot is not useful for HAMR even though it might work for spectroscopy. To make use of low-cost, commercial semiconductor lasers for HAMR disc drives, the NFT coupling efficiency must be approximately 5%. Although this may seem like a very small efficiency, it should be remembered that the efficiency of light transmission through a near-field tapered optical fiber with a 50-nm aperture is only approximately 0.001%.[6] Thus, the HAMR NFT must have a power coupling efficiency into the recording medium that is orders of magnitude greater than the transmission efficiency of tapered optical fibers.

This immediately raises the question whether it is correct to make a comparison between the transmission efficiency of a NFT and its power coupling efficiency into a recording medium. Transmission efficiency is a far-field property, while coupling efficiency is a near-field property. Is it possible that a very tiny aperture with a far-field transmittance of 10^{-5} could nevertheless in the near field couple optical power efficiently into a recording medium? Are the far-field and near-field properties of NFTs related and, if so, in what way? Of course the far-field transmittance of a NFT is only defined in the *absence* of a recording medium. When a recording medium is placed within the near field of a transducer, does that significantly affect the optical properties of the NFT itself? What is the best way to judge the merit of a NFT?

In the literature for NFTs a variety of approaches have been reported for judging the merit or efficiency of NFTs. One popular efficiency measure is the value of the enhancement of the electric field in the vicinity of the NFT relative to that of an incident plane wave. Another figure of merit (FOM) is the amount of power that is transmitted through a NFT aperture relative to the incident power on the NFT integrated over the surface of the NFT aperture. This FOM assumes that the transmitted power (a far-field quantity) is directly related to the near-field coupling efficiency of the transducer. It is not simple to apply this FOM to NFT *antennas*. A sensible FOM for HAMR is the power coupling efficiency, i.e., the ratio of the power dissipated within the optical hot spot of the recording medium to the total power in the incident beam. This FOM is generally somewhat more difficult to compute than the other FOMs because it requires an incident *focused* light beam with a well-defined power rather than a simple incident plane wave.

With an appropriate FOM, it is possible to study, optimize, and compare different NFT designs in detail theoretically.[7] The NFTs can generally be categorized as either antennas or apertures, although there are some NFT designs that incorporate aspects of both. Several mechanisms can be identified in these different designs that enhance the coupling efficiency of light into the recording medium. For example, in most cases the NFT is chosen to support surface plasmons that resonate in the incident optical field and thereby greatly enhance the optical field amplitude in the near field of the transducer. Often these NFT designs incorporate sharp tips to further increase the field amplitude via the lightning rod effect. A small gap between two regions of the NFT can be used to enhance field amplitudes via the dual-dipole effect. Other mechanisms are designed for more efficiently funneling energy from the incident beam into the active region of the NFT.

The outline of this chapter is as follows. In Sect. II we discuss the modeling techniques employed in this study. In Sect. III the difference between the near field and the far field is considered and it is argued that any FOM based on far-field quantities is not appropriate for HAMR. Various FOMs are considered in Sect. IV as they relate to HAMR. Several mechanisms that may be employed by NFTs for enhancement of the coupling efficiency are discussed in Sect. V. In Sect. VI these mechanisms are studied for a variety of transducer designs and a FOM is used to compare them. Because both antennas and apertures may be useful for HAMR, we discuss the relationship between these different transducer approaches in Sect. VII. The relationship between the far field and the near field, especially in so far as far-field measurements may be used to characterize NFTs, is discussed further in Sect. VIII. Finally, the means for efficiently illuminating the NFT is an important topic which we address in Sect. IX in a discussion on photonic nanojets.

II. MODELING TECHNIQUES

Analytical solutions for the electromagnetic fields can be obtained for only a small set of physical objects which generally exhibit some form of symmetry. A solution to the problem of the scattering of plane waves by spherical or ellipsoidal objects was found by Mie.[8] Many useful insights can be obtained from this semianalytical theory and we make use of it in this chapter to discuss the local field enhancement due to the surface plasmon resonance of metallic spheres. However, in general it is not possible to study the wide variety of NFT designs analytically. We have found that the scattered field finite difference time domain (FDTD) technique[9] is well suited to our transducer studies. In this technique, the incident electric field is defined analytically throughout the computation space, but the scattered field is computed numerically in the time domain at specific points throughout the computational space on a Yee cell lattice as shown in Fig. 1. As can be seen from this figure, the individual electric and magnetic field components are specified at different points within each cell.

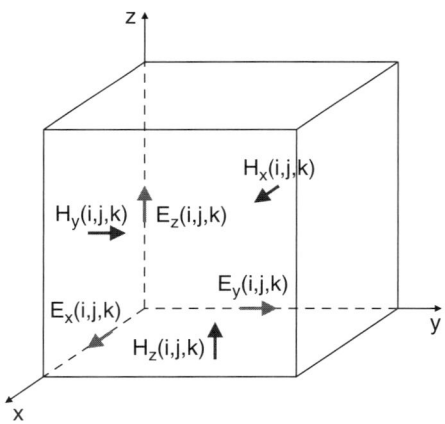

Figure 1. The finite difference time domain (FDTD) computation space is composed of Yee cells which define the locations of the electric and magnetic field components on the cell edges and cell faces, respectively.

The Yee cells must be chosen sufficiently small that the numerical approximation is accurate. In practice, for modeling surface plasmon phenomena at optical wavelengths for highly conducting metals, we have found that a cell size of $(2.5\,\text{nm})^3$ is generally reasonable. The cell size and the computational resources in turn limit the size of the computation space. At the boundaries of the space, appropriate boundary conditions must be implemented so that scattered fields do not get reflected back into the computation space. For the simulations in this chapter, we used either reradiating boundary conditions or perfectly matched layers.[10] The size of the Yee cell also determines the maximum size of the time step which can be used to avoid numerical instability. The Courant time[10] is an upper limit on the step size, but in practice it is found that somewhat smaller time steps are required for stability. Smaller Yee cells require shorter time steps. For plane wave scattering problems, it is generally necessary to run the simulation for five or more complete periods of the wave to reach nearly steady state conditions. At integral values of the time step the scattered electric field at each Yee cell is updated from the incident electric field, the scattered electric field, and the scattered magnetic field. At half-integral time steps the scattered magnetic field is updated from the scattered electric field. In the scattered field FDTD technique, as opposed to the total field FDTD technique, the update equations are significantly more complex for materials that include optical losses. We model metals as Debye materials in the FDTD calculation with a separate set of Debye parameters for each wavelength.

III. NEAR FIELD COMPARED WITH FAR FIELD

It is not unusual to find articles on NFTs that begin with the classic result of Bethe[11] for the far-field transmission efficiency of a circular aperture. Bethe was able to solve analytically for the light transmitted through a circular aperture in an infinitesimally thin perfectly conducting sheet. He discovered that when the aperture diameter is small compared with the wavelength of the incident light wave, the transmission efficiency is given by

$$T = \left(\frac{64\pi^2}{27}\right)\left(\frac{d}{\lambda}\right)^4, \qquad (2)$$

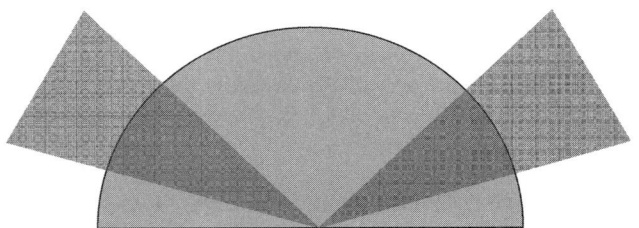

Figure 2. Example of solid immersion lens illuminated beyond its cutoff angle to illustrate the difference between far-field and near-field properties.

where T is the ratio of the power per unit area transmitted through the aperture to the power per unit area incident upon the aperture and d is the diameter of the aperture. Obviously, the fourth-power dependence on the ratio of the diameter to the wavelength causes the transmitted power to fall drastically with aperture diameter. This discouraging result convinced many people that it was impossible to efficiently conduct optical energy into volumes much smaller than λ^3.

However, it is not difficult to demonstrate that far-field measurements are not necessarily a measure of near-field efficiency. Perhaps the simplest example is to consider a SIL that is illuminated only for angles greater than $\sin^{-1}(1/n)$, where n is the index of refraction of the SIL, as shown in Fig. 2. In this case no light is transmitted into the far field – it is all internally reflected at the bottom interface of the SIL. This optical transducer would fare very poorly with any FOM that is based on a far-field property. However, if any object such as a recording medium is placed adjacent to this surface, the light energy in a highly concentrated spot at the focus of the SIL will be coupled into the medium. The near-field coupling efficiency is not zero and in fact may be quite respectable. This simple example, therefore, demonstrates that there is not necessarily a one-to-one correspondence between far-field transmittance and near-field coupling efficiency.

IV. FIGURES OF MERIT

How does one know if a particular NFT is promising for use in a specific application? To optimize a particular NFT design or to make comparisons between different NFTs, it is important to have a

FOM. A variety of FOMs have been used in the literature to judge the performance of NFTs. Examples include far-field transmittance, peak field intensity in the neighboring medium, percent dissipated power in the medium, and temperature rise in the medium. We shall consider each of these in turn and discuss their advantages and limitations.

1. Far-Field Transmittance

For a NFT that is an aperture, the simplest theoretical and experimental procedure for evaluating NFT efficiency is to calculate or measure the far-field transmittance. The total power transmitted through the aperture must be normalized in some manner. The incident beam in a theoretical calculation is frequently a plane wave; however, the incident power in a plane wave is infinite. Because only a finite amount of power is transmitted through an aperture, the transmittance of the aperture as a ratio of transmitted power to incident power is exactly zero for a plane wave; therefore, the transmittance of an aperture for a plane wave is not a useful FOM. However, there is a finite amount of power in a plane wave in the cross-sectional area of the aperture. A popular FOM is the ratio of the transmitted power from a plane wave incident upon the NFT (or the absorbed power of the medium next to the NFT) to the power/area of the plane wave multiplied by the cross-sectional area of the aperture. For periodic arrays of NFTs, the FOM is the ratio of the transmitted power to the power/area of the plane wave multiplied by the area of a unit cell. The power/area for a plane wave is

$$S = \frac{|E_0^2|}{2\eta} \left(\text{W/m}^2\right), \tag{3}$$

where E_0 is the amplitude of the incident plane wave in volts per meter and η is the impedance of the medium of propagation in ohms. For free space,

$$\eta = \sqrt{\frac{\mu_0}{\varepsilon_0}} \cong 377\,\Omega. \tag{4}$$

Unfortunately, as discussed in the previous section, it is easily demonstrated that the far-field transmittance is not necessarily related to the near-field coupling efficiency for a NFT meant to be used for HAMR. Although this FOM may be appropriate for NFTs designed for some applications, it is not appropriate for designing or comparing HAMR transducers and will not be used in this study.

2. Peak Field or Field Intensity

Another common FOM in studies of NFTs is the ratio of the peak electric field amplitude in the vicinity of the NFT to the electric field amplitude of an incident plane wave. This is a particularly appropriate FOM for designing NFTs for surface-enhanced Raman scattering. The Raman signal from various organic compounds is experimentally found to be enhanced by many orders of magnitude[12,13] when the organic molecules are attached to rough silver surfaces or to gold or silver nanoparticles of different shapes. Because the Raman effect is a two-photon process, the intensity of the scattered light is proportional to the fourth power of the electric field in the vicinity of the molecule. Indeed, the amplification of the Raman spectrum is so large that individual molecules can be detected.[14,15] By optimization of the NFT design for the greatest field enhancement, arrays of NFTs on a substrate can be optimized for surface-enhanced Raman scattering. On the other hand, the local field enhancement of the NFT when it is suspended in free space or some other dielectric medium is not a particularly appropriate FOM for HAMR. This is because the field enhancement from a NFT in free space can be significantly different from the field enhancement in the presence of any metallic or lossy medium. This will be demonstrated in the studies of triangle and bow-tie antennas discussed in Sect. V.

The $|E|^2$ field intensity in a lossy medium is directly proportional to the dissipated power. In particular,

$$P_{\text{diss}} = \frac{1}{2}\text{Re}(\sigma)\,|E|^2, \tag{5}$$

where σ is the complex optical conductivity of the lossy material. The optical conductivity is directly related to the complex optical dielectric constant of the lossy material,

$$\sigma = -\frac{\text{i}2\pi c\varepsilon_0}{\lambda}\left(\frac{\varepsilon}{\varepsilon_0} - 1\right)(\Omega\text{m})^{-1}, \tag{6}$$

where c is the speed of light and ε_0 is the permittivity of free space, 8.854×10^{-12}. Therefore, the peak field intensity within the medium normalized by the incident field intensity is closely related to the total power dissipation within the medium and can be a useful FOM for HAMR studies of transducers. It will be used for the studies discussed in Sect. V. There are two disadvantages of this FOM, however. First of all, the spatial distribution of the field intensity in the recording medium can differ greatly for different NFTs. The total power dissipated in the medium is proportional to the $|E|^2$ field intensity integrated over the volume of the medium, not just the peak $|E|^2$ at some point within the medium. A second issue with this FOM is more subtle. The peak field intensity in the incident beam is a function of the wavelength and polarization of the incident focused beam. If two NFTs couple light into a medium with the same peak $|E|^2$ FOM, but at two different wavelengths, then the NFT which operates at the shorter wavelength will be more efficient at coupling power into the medium. At the shorter wavelength, the focusing optics will generate a smaller spot with dimensions proportional to λ^2. Therefore, for the same peak field amplitude of the incident beam, there is more optical power in the vicinity of the transducer at the shorter wavelength to be coupled into the medium. An alternative way of looking at this is that for a given optical power in the incident beam, the field intensity at the focus of the beam is proportional to $1/\lambda^2$. Therefore, shorter-wavelength operation of a NFT is an advantage. This factor is not explicitly taken into account in a FOM based solely on $|E|^2$.

3. Percent Dissipated Power in the Recording Medium

Although every FOM has certain advantages and disadvantages, one of the best optical FOMs for HAMR is based on the total optical power dissipated within a certain region of the recording medium. Once a calculation has been performed for the electric field around the NFT and within the medium, it is straightforward to apply (5) to determine the dissipated power within any region of the computational space. This FOM does account for the wavelength dependence of the focused spot. For high-density HAMR storage, the bit cell will be smaller than 50 nm. Therefore, the NFTs considered in Sect. V will be evaluated on the basis of the percentage of the power in the incident beam that is dissipated within a circular area of 50 nm in diameter in the recording medium.

4. Temperature Rise in the Recording Medium

Finally, if the thermal properties of the recording medium are known, then thermal models may be applied to convert the dissipated power within the medium into a corresponding temperature profile. The FWHM size of the thermal spot and its peak temperature for a given input optical power to the transducer are directly related to the capability of the NFT for HAMR. Because this FOM depends on the specific thermal properties of the multilayer film stack in the recording medium, and these properties are often not known with precision, this FOM is of limited usefulness.

V. MECHANISMS FOR ENHANCEMENT OF THE FIGURE OF MERIT

As mentioned in Sect. I, there are several different mechanisms that operate in a well-designed NFT for enhancing the FOM. For the purposes of this article either the peak $|E|^2$ intensity in the medium or the dissipated power in the medium will be chosen as the FOM for studying these enhancement mechanisms for HAMR. Depending on the specific NFT design, the order of importance of these mechanisms may vary, but in general the best NFTs will combine most or all of these mechanisms. The ones we will consider in this section are localized surface plasmon resonance (LSPR), the lightning rod effect, and the dual-dipole effect.

1. Localized Surface Plasmon Resonance

Small metallic particles are well known to exhibit LSPRs.[16] Surface plasmons are collective excitations of surface charge which under suitable conditions can be excited by an external optical field. *Localized* surface plasmons (LSPs) are oscillations of surface charge on a finite structure with fields that decay exponentially from the surface of the structure in both directions normal to the surface. The structure may be composed of a metal surrounded by a dielectric, or it may be composed of a dielectric surrounded by a metal. Examples include metallic nanoparticles and nanobubbles embedded in metals. Nanoholes in metal films also support LSPRs even though a hole is not entirely surrounded by the metal film. The surface plasmon resonance wavelength is determined by the size, shape, and material of the structure and the surrounding medium.

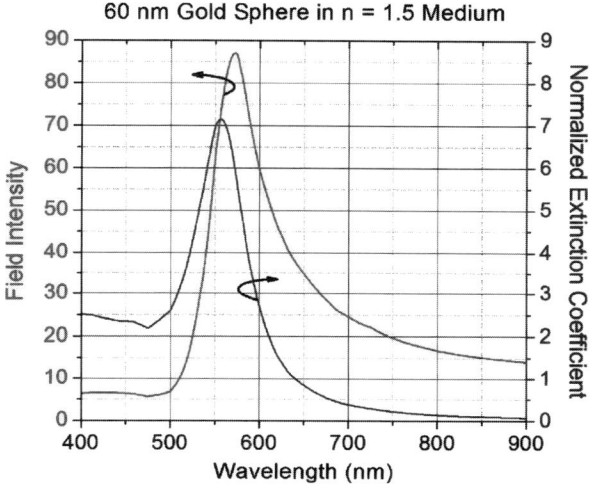

Figure 3. The extinction coefficient and the peak $|E|^2$ field intensity at the surface of a 60-nm gold sphere for an incident plane wave of unit amplitude are computed from Mie theory using the refractive indices from references.[19,20]

At resonance, the nanoparticles absorb the incident optical energy much more efficiently and generate enhanced electric fields at their surfaces from the oscillating surface charge. The enhanced absorption from LSPR of silver and gold nanoparticles embedded in glass has been used since medieval times to make stained glass windows with yellow and red colors.[17] In Fig. 3 the extinction coefficient for a 60-nm gold sphere is shown calculated from Mie theory[8,18] and graphed along with the electric field intensity at the surface of the particle. The LSPR is observed at approximately 550 nm by the peak in both the extinction coefficient and the field intensity at the surface. It should be noted that the peak $|E|^2$ field intensity at the surface of the sphere is more than 70 times larger than the field intensity of the incident plane wave. A plot of the field intensity in the neighborhood of the sphere is shown in Fig. 4.

The resonance wavelength of LSPs is determined in part by the refractive index of the surrounding medium. This is illustrated in Fig. 5 by plotting the peak field intensity for the 60-nm gold sphere versus wavelength for several different surrounding dielectrics. As the index of the dielectric increases, the resonance shifts towards longer wavelengths.

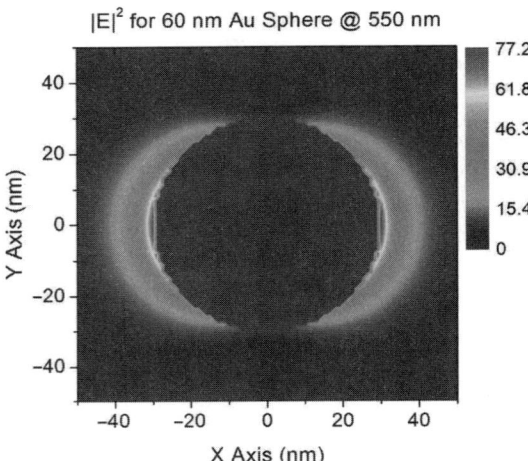

Figure 4. Contour plot of the field intensity in the xy plane of a 60-nm gold sphere embedded in a dielectric of index 1.5 when excited by a plane wave of unit amplitude which is polarized along the x axis at a wavelength of 550 nm. The points in the plane are computed with an increment of 1 nm.

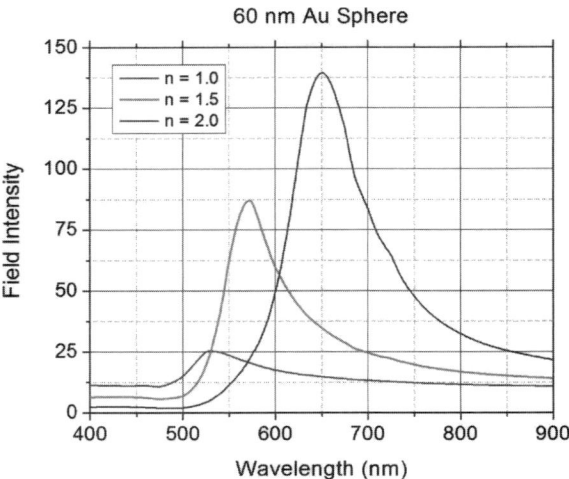

Figure 5. Effect of the surrounding dielectric index on the surface plasmon resonance wavelength of a 60-nm gold sphere. Increasing the dielectric index shifts the resonance to longer wavelengths and enhances the peak field intensity.

Near-Field Optics for Heat-Assisted Magnetic Recording

Although a nanosphere is an excellent structure for illustrating the surface plasmon resonance enhancement of electric fields, it is not a particularly well designed NFT for HAMR. A better NFT for HAMR is the triangle antenna. This structure, which can also be considered to be a nanoparticle, exhibits a LSPR. A plane wave incident upon the triangle antenna and polarized along its length can drive surface currents back and forth along the antenna. For appropriate antenna dimensions, the antenna becomes a resonant structure of oscillating surface currents which is a LSPR. As previously stated, it is not possible to compute the resonance fields analytically or semianalytically for most NFT structures, which have much lower symmetry than spherical nanoparticles. Therefore, in this article such calculations were carried out with the scattered field FDTD numerical approach.[9,10,21] In the FDTD calculation a plane wave of unit amplitude is incident onto the triangle antenna propagating in the $-z$ direction. A plot of the peak $|E|^2$ at the tip of a triangle antenna versus wavelength is shown in Fig. 6. The LSPR occurs at a wavelength of 775 nm. At this wavelength the peak field intensity, as computed

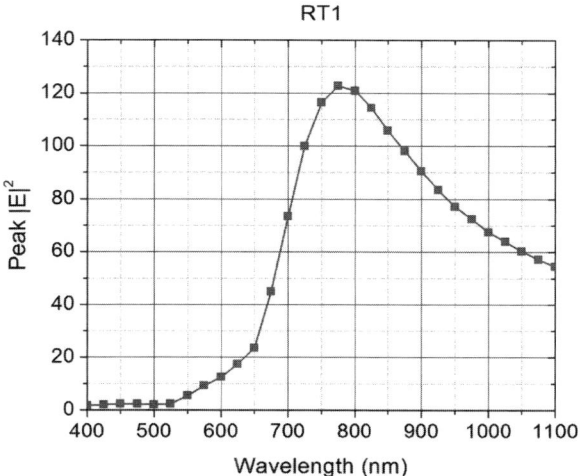

Figure 6. Peak field intensity at the tip of a gold triangle antenna embedded in free space versus wavelength for excitation by an incident plane wave of unit amplitude polarized along the length of the antenna. The antenna has an apex angle of 45°, a length of 200 nm, and a thickness of 80 nm.

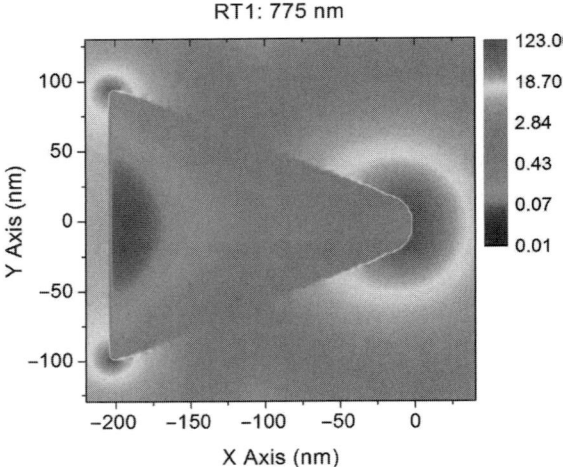

Figure 7. Local field intensity for a gold triangle antenna with a length of 200 nm, radius of curvature at the apex of 20 nm, apex angle of 45°, and thickness of 80 nm. The incident plane wave of unit amplitude is polarized along the x axis. The FDTD cell size is $(2.5\,\text{nm})$.[3]

by FDTD with a cell size of $(2.5\,\text{nm})$,[3] is enhanced by over a factor of 120! The field intensity at this wavelength is plotted in the xy plane through the center of the antenna in Fig. 7, showing that the peak field intensity occurs at the edge of the apex of the antenna as would be expected from the lightning rod effect (to be discussed in the next section). A plot of the field intensity along the x axis through the center of the apex in Fig. 8 demonstrates the characteristic exponential decay of the field strength on either side of the edge of the antenna.

The LSPR is also affected by the dimensions of the nanoparticles. In Fig. 9, the peak intensity is plotted for the triangle antenna versus the length of the antenna. The moral of the story is that if NFTs are designed properly, their dimensions, their optical properties and those of the surrounding materials will all be chosen so as to maximize the field enhancement in the recording medium by operating at the resonance of the LSP. Although the isolated nanoparticles considered in this section give theoretical field intensity enhancements of over 2 orders of magnitude, it should

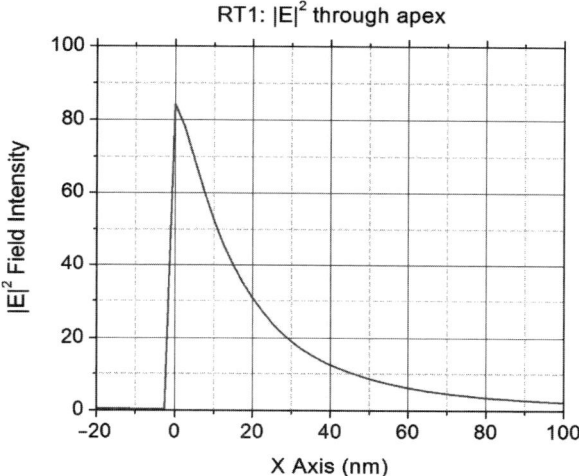

Figure 8. Field intensity computed along the x axis of the triangle antenna in Fig. 6 showing the exponential decay characteristic of the field from surface plasmons. The decay for negative x into the gold antenna is of course much faster than the decay into the surrounding dielectric.

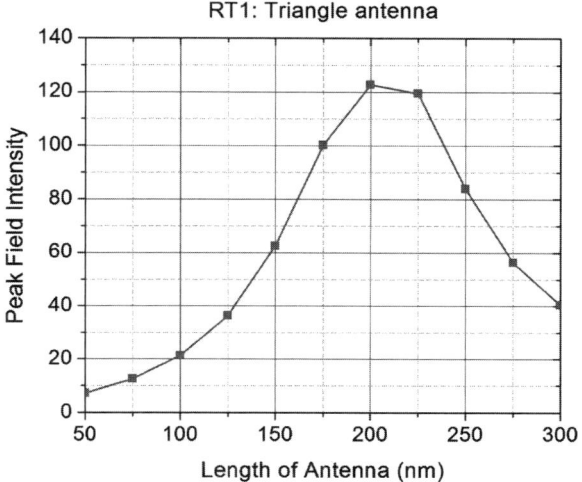

Figure 9. Field intensity at the apex of the triangle antenna as a function of the antenna length computed for plane wave excitation at a wavelength of 775 nm.

be remembered that these values are not very relevant to HAMR. When the NFTs are in the presence of lossy metallic materials like the recording medium, the field within the medium is shielded and greatly reduced. Moreover, the power absorption of the medium greatly reduces the Q of the resonance, leading to much smaller field enhancements.

2. Lightning Rod Effect

The lightning rod effect refers to the well-known fact that sharp metallic objects tend to generate very large localized fields.[22,23] Electric field lines must terminate normally to the surface of a perfect conductor. This effect tends to concentrate the field lines at any sharp points of highly conducting materials.[24] This is a shape effect, not a resonance effect, and therefore does not have any particular wavelength dependence. It may or may not be associated with a LSPR. For example, as a spherical gold nanoparticle is pulled into an ellipsoidal shape, the LSPR splits into resonances at two different wavelengths. One of the resonances shifts towards shorter wavelengths with increasing obliquity and one shifts towards longer wavelengths. The longer-wavelength resonance corresponds to surface charge oscillating along the long axis of the ellipsoid and it is found that the fields at the tips of the ellipsoid at the resonance get stronger as the end of the ellipsoid gets narrower and sharper.[25] This effect is shown in Fig. 10. The lightning rod effect can generate extremely large field enhancements.

The triangle antenna also provides an excellent illustration of the lightning rod effect. In this case the FDTD technique is used to compute the fields at the apex of the antenna as the radius of curvature at the apex is varied. All calculations are carried out with a cell size of $(2.5\,\text{nm})^3$. The results are graphed in Fig. 11. The peak field at the apex for this particular antenna design and within the accuracy of the FDTD calculation is somewhat smaller than the absolute peak field as can be seen from Fig. 7. Clearly it is beneficial to design the NFT with a sharp point(s) to both enhance the field intensity and localize it within the recording medium.

A contour plot of the field intensity for the antenna with a 5-nm radius of curvature is shown in Fig. 12 for comparison with the plot for a 20-nm radius of curvature in Fig. 7.

The strong effect on field enhancement of the lightning rod effect leads directly to a remark which, although obvious, nevertheless

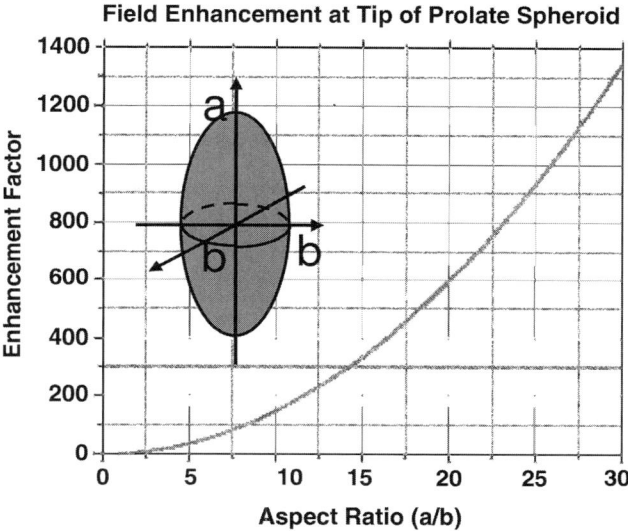

Figure 10. Field enhancement at the tip of a prolate spheroid as a function of its aspect ratio.

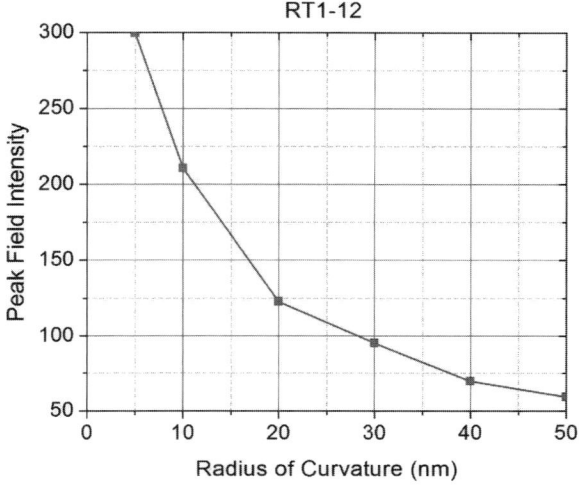

Figure 11. Peak field intensity at the apex of a triangle antenna versus radius of curvature. The antenna is 200 nm long, 80 nm thick, with a 45° apex angle. The incident plane wave has a wavelength of 775 nm.

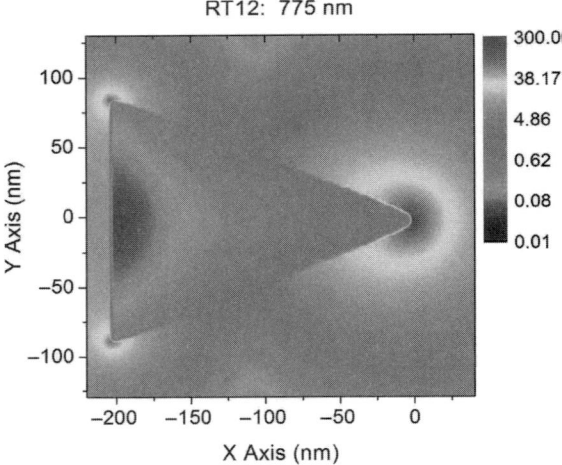

Figure 12. Local field intensity for a gold triangle antenna with a length of 200 nm, radius of curvature at the apex of 5 nm, apex angle of 45°, and thickness of 80 nm. The incident plane wave of unit amplitude is polarized along the x axis. The FDTD cell size is $(2.5\,\text{nm})^3$.

seems to be often neglected in the literature. In particular, the peak field intensity is also necessarily a function of the cell size used in the numerical simulation. It is well known that the electric field amplitude at the edge of a semi-infinite perfectly conducting straight edge has a logarithmic divergence.[26] If this were modeled numerically with a finer and finer mesh, the peak field amplitude would be found to continuously increase. Therefore, when comparisons are made between different NFTs using numerical calculations of peak field amplitude, care should be taken to ensure that the same numerical algorithm and same cell size are being employed in the comparison. Otherwise the results are meaningless.

3. Dual-Dipole Resonance

A third technique for field enhancement is the dual-dipole effect. In this case, two resonant particles are brought close enough together to interact with each other. In the gap region between the two particles, the field can become much more intense than that from either

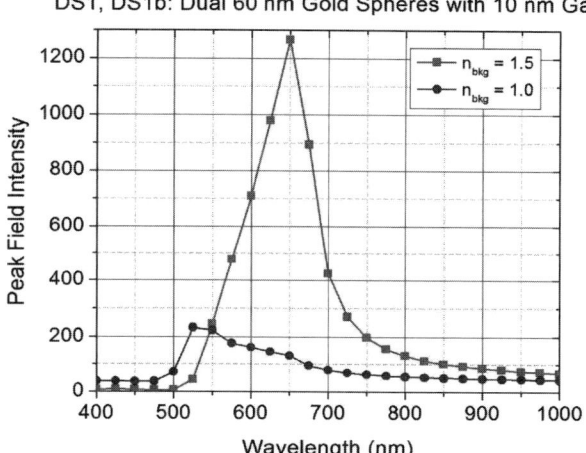

Figure 13. Peak field intensity in the 10-nm gap between two 60-nm–diameter gold spheres for two different background dielectric indices.

particle separately. As a simple example, we first consider the case of two 60-nm gold particles with an incident plane wave polarized along the axis connecting them. The peak field intensity is plotted versus wavelength for a 10-nm gap between the spheres in Fig. 13. There is clearly a resonance wavelength at 650 nm for excitation of the LSPs on the spheres. The field intensity distribution at this wavelength is plotted in Fig. 14. The peak field intensity of approximately 1,200 is in the region between the two spheres. The peak field intensity in the gap between the spheres is plotted versus gap distance in Fig. 15. The intensity falls very rapidly with increasing gap distance.

VI. COMPARISON OF NEAR-FIELD TRANSDUCERS

In this section the results of the previous two sections are combined to compare several NFT designs that have been suggested for use in data storage. In particular, the triangle antenna and the bow-tie antenna are compared with the circular aperture, the tapered rectangular aperture, the bow-tie aperture, and the C aperture. All NFTs are illuminated by a highly focused beam using a SIL with a refractive index of 1.5 to obtain an optical spot size with dimensions

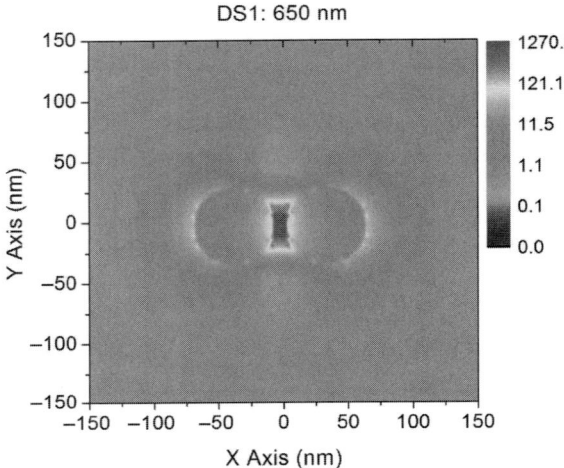

Figure 14. Field intensity distribution at resonance for the dual gold spheres showing the large field enhancement in the gap between the spheres.

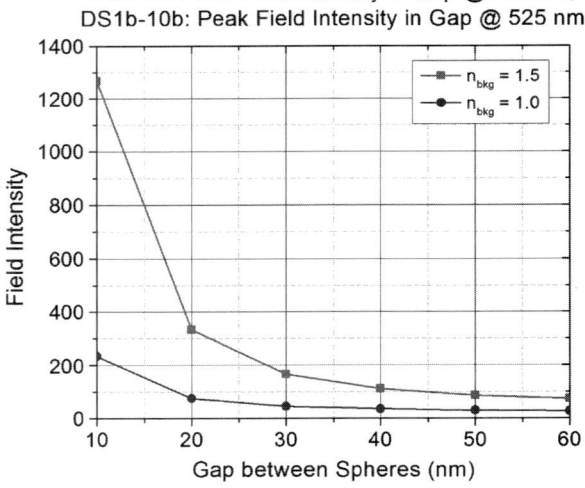

Figure 15. Field intensity in the gap between dual gold spheres as a function of gap distance for a background dielectric with index 1.5 at a wavelength of 650 nm, and a background dielectric with index 1.0 at a wavelength of 525 nm.

of 0.49λ × 0.38λ as calculated using the stationary-phase approximation in the Richards–Wolf theory.[27] A simple recording medium consisting of 10 nm of cobalt laminated to a 100-nm gold heat sink is placed 7.5 nm below the NFT. The separation distance of the NFT from the medium is determined by several considerations. At terabit per square inch storage densities, the down-track distance between magnetic transitions is only approximately 10–15 nm. With such small spacing between transitions, it is necessary for the magnetic reader to fly extremely close to the surface of the medium. Moreover, the fields generated in the medium by the NFT are primarily evanescent fields. If the medium is spaced too far from the NFT, the amplitude of these fields is too small to couple power efficiently.

As previously demonstrated, an efficient NFT should make use of a LSPR effect. This effect requires a metallic surface that is highly conductive at optical frequencies. There are relatively few metals that satisfy this criterion. Silver and aluminum can support LSPs throughout the visible region. Gold and copper can support LSPs in the near infrared region and slightly into the red region of the visible spectrum. These elements and their alloys are the only reasonable choices for NFTs in device applications. However, pure silver, copper, and aluminum all have problems with corrosion. This leaves gold as the material of choice for the NFT and, therefore, gold is used for all NFT comparisons in this section.

The FOM for making the NFT comparisons is the peak field intensity within the recording medium. The FWHM of the spot size within the top layer of the recording medium is required to be 50 nm or less for a realistic HAMR storage device at terabits per square inch densities. The minimum dimension allowed within the NFT structure is 20 nm for all NFTs. This ensures that no NFT design is given an "unfair" advantage in the comparison by making use of the lightning rod effect to a greater extent than the other designs. The cell size in the FDTD calculations is (2.5 nm).[3] With these restrictions, it is possible to make reasonable comparisons of the NFTs. However, it should be remembered that if the desired FWHM optical spot size within the transducer is specified, then a better FOM is the dissipated power within this area.

1. Circular Aperture

The circular aperture in an opaque film is the simplest NFT. It has traditionally been given a poor rating as a NFT based in large part on

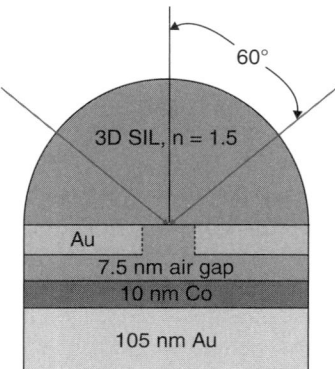

Figure 16. Circular-aperture near-field transducer and recording medium.

the discouraging far-field transmittance of such apertures as found both theoretically[8] and experimentally.[6] However, as has been previously discussed, the far-field transmittance of an aperture does not necessarily correlate to its near-field power coupling efficiency. The geometry of this NFT is shown in Fig. 16.

The peak field intensity in the medium exhibits a LSPR and a maximum value at a wavelength of approximately 650 nm regardless of hole diameter as shown in Fig. 17. Unfortunately, there are two problems with this NFT. The peak field intensity within the medium is extremely small and the dissipated power within the medium spreads over an area that is much larger than the hole, as shown in Fig. 18. On the other hand, the dissipated power within a 50-nm-diameter cylinder in the medium is 0.14%, which is much larger than the value of 10^{-5} that might be expected for the far-field transmittance based on the theory of Bethe.[8] By filling the hole with a high-index dielectric, one can reduce the optical spot size and increase the field intensity in the medium.

2. Tapered Rectangular Aperture

The efficiency of the circular aperture can be improved significantly by tapering the side walls. Moreover, because the circular aperture produces an oblong dissipated power spot along the direction of the incident polarization as shown in Fig. 18, it makes sense to widen

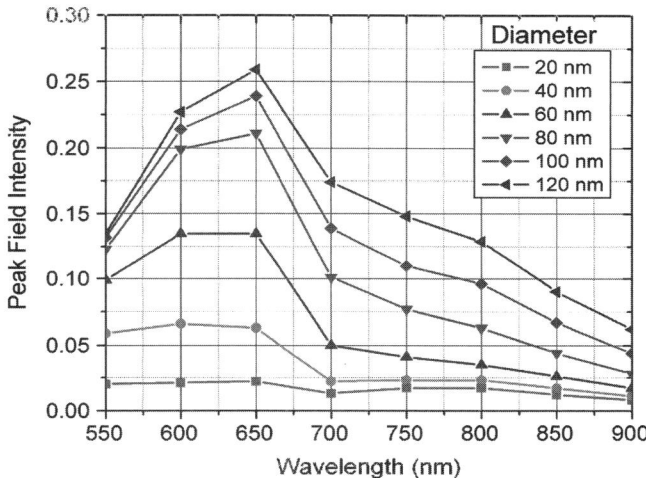

Figure 17. Peak field intensity in the recording medium versus wavelength for circular apertures of various diameters. The gold film is 40 nm thick.

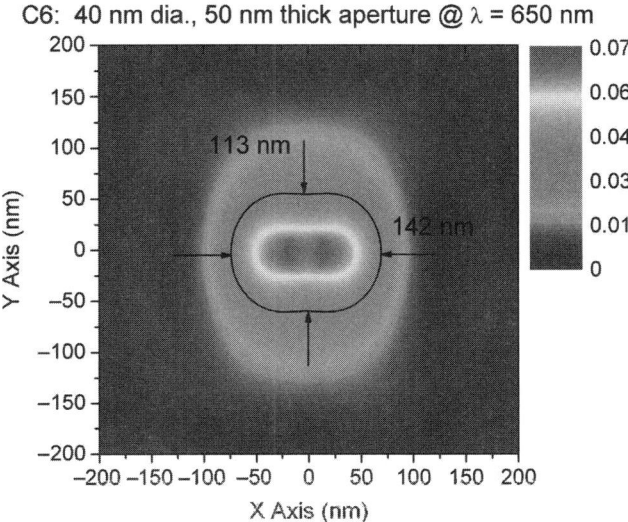

Figure 18. Field intensity within the recording medium for a circular aperture in a gold film with a 40-nm diameter and a thickness of 50 nm. (Reprinted from Ref. [7]. Copyright 2006 with permission from the Institute of Pure and Applied Physics.)

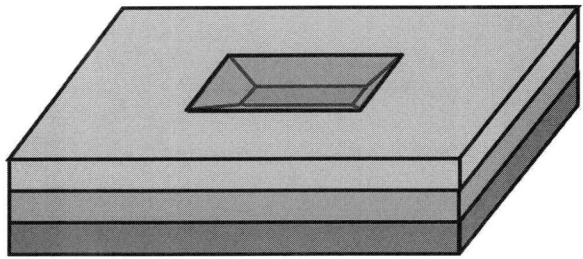

Figure 19. Tapered gold rectangular aperture. The aperture is filled with the glass of the solid immersion lens with refractive index 1.5.

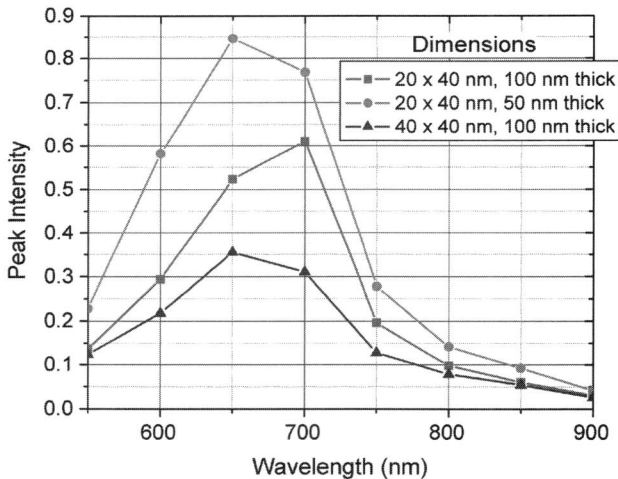

Figure 20. Field intensity in the medium versus wavelength for the tapered gold rectangular aperture.

the aperture in the orthogonal direction to obtain a more circular dissipated power spot in the medium. This can be easily accomplished with a tapered rectangular aperture as shown in Fig. 19. Furthermore, if the aperture is filled with a high-index material, like the glass of the SIL, additional optical power can be concentrated within the aperture.

The peak field intensity within the recording medium as a function of wavelength is shown in Fig. 20 for several different aperture

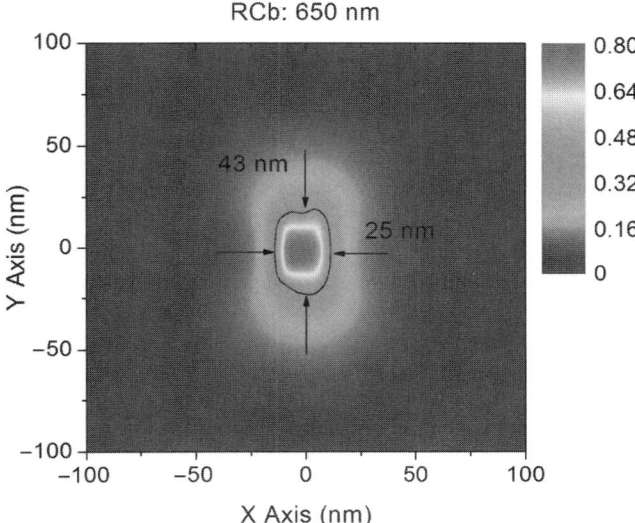

Figure 21. Cross section of the field intensity in the medium for the 20 nm × 40 nm rectangular aperture in a 50-nm-thick gold film.

dimensions. The 20 nm × 40 nm rectangular aperture in the 50-nm-thick gold film with a 45° slope to the side walls generates the largest field intensity in the medium at a wavelength of approximately 650 nm. The field intensity within the medium is shown in Fig. 21. The total power dissipated in the central 50 nm of the medium is 0.92%. Moreover, the optical spot within the medium is smaller than the desired 50-nm FWHM. This is a substantial improvement over the air-filled circular aperture with straight side walls.

3. Bow-Tie Aperture

An aperture in the shape of a bow tie, also called a "bow-tie slot antenna" is shown in Fig. 22. This aperture is essentially a rectangular aperture with a constriction in the center. When it is illuminated with light polarized across the gap as shown in the figure, a LSPR is excited which oscillates surface charge into the two tips in the center. The sharpened tips enhance the field strength in this region via the

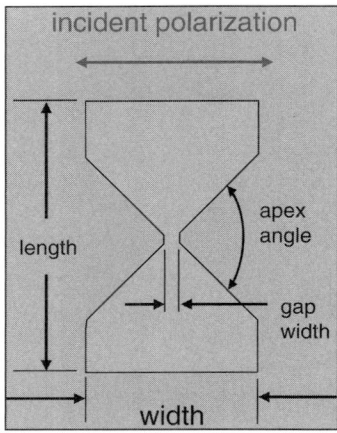

Figure 22. Bow-tie aperture.

lightning rod effect. Moreover, the two tips separated by a small gap provide field enhancement via the dual-dipole effect. Therefore, all three field-enhancement mechanisms are present in this NFT.

With so many dimensions to specify for this aperture, the optimization process is lengthy. Variation of the length of the aperture with wavelength indicates an optimum length of 300 nm or greater although the LSPR wavelength is approximately 800 nm and only weakly dependent on aperture length. Variation of the aperture width gives a similar result for the optimum value and the optimum thickness is approximately 80 nm. As the gap is made narrower, the field intensity increases in the gap via the dual-dipole effect. There is some variation in efficiency with apex angle, but values in the range of 60°–90° are good.

The wavelength dependence of the peak field intensity in the medium is plotted in Fig. 23 for an aperture with a length of 300 nm, a width of 290 nm, a thickness of 80 nm, a gap of 20 nm, and an apex angle of 90°. There is a narrow LSPR at 725 nm. The field intensity in the medium at this wavelength is plotted in Fig. 24. The FWHM optical spot size in the medium is somewhat larger than the desired 50 nm. The percentage of power delivered to a 50-nm cylinder in the medium is 1.7%. This NFT is not as successful at confining the optical energy as some of the other designs.

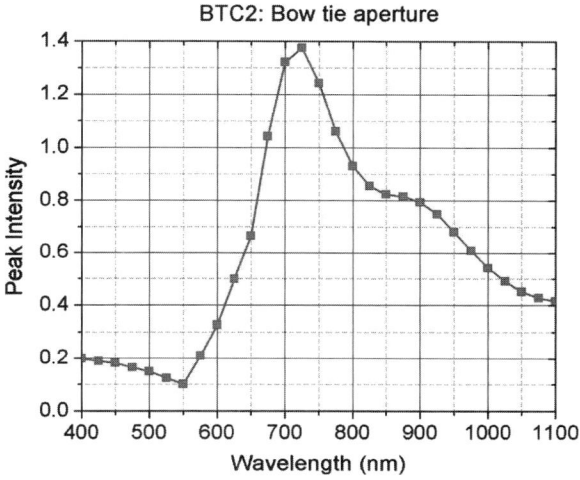

Figure 23. Peak field intensity in the medium versus wavelength for a bow-tie aperture with dimensions given in the text.

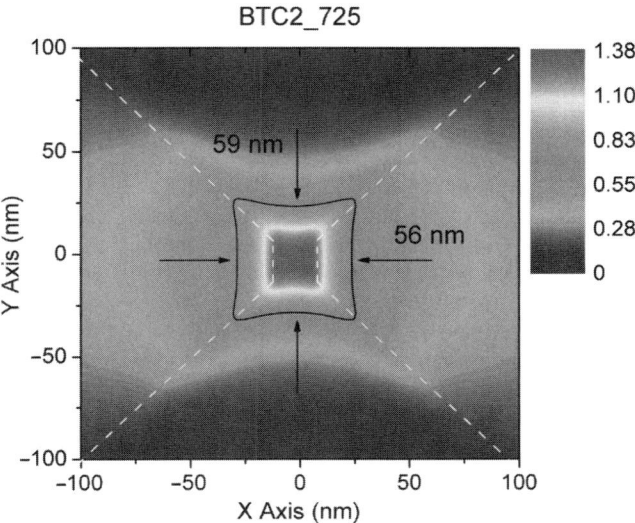

Figure 24. Field intensity from the bow-tie aperture in the medium at a wavelength of 725 nm. (Reprinted from Ref. [7]. Copyright 2006 with permission from the Institute of Pure and Applied Physics.)

4. C Aperture or Ridge Waveguide

Next, the C aperture or ridge waveguide is considered. The C aperture was originally proposed by Shi et al.[28] but they originally considered apertures in perfectly conducting metal films in the absence of a recording medium and for an incident plane wave. Although these calculations indicated 3 orders of magnitude greater field intensities from the C aperture than from a square aperture, these results are not directly relevant for HAMR. Many additional studies have been made which include the effects of real metals and focused incident beams.[29–34] The C aperture is shown in Fig. 25.

The ridge waveguide is a well-known geometry for transporting microwaves. Like the bow-tie aperture, the C-aperture length can be less than the cutoff dimension for a rectangular aperture. For a rectangular aperture in which the incident field is polarized parallel to the short dimension, the field amplitude tends to zero at the short edges of the aperture and is maximum in the central region. The ridge in the center of the C aperture squeezes the field and thereby further enhances the field strength between the ridge and the opposite side. This can also be considered a dual-dipole effect, where the opposite side serves as an image surface to the ridge. For a C aperture in a real metal there is also a LSPR. Finally, the ridge itself enhances the field via the lightning rod effect. Therefore, this NFT also makes use of all the field-enhancement mechanisms. Propagating surface plasmon polaritons can be excited between the bottom of the SIL and the aperture. In principle these surface plasmons may siphon energy away from the LSP within the aperture, thereby reducing the coupling efficiency. However, with clever engineering these surface plasmons can actually be made to contribute additional energy to the LSP.[34]

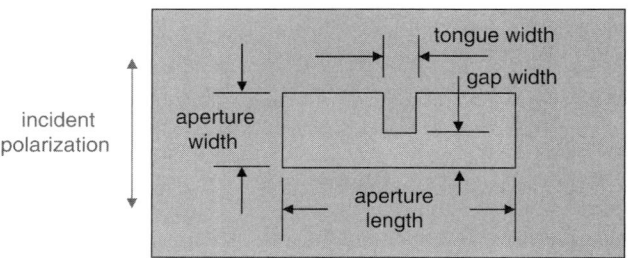

Figure 25. Dimensions of the C aperture.

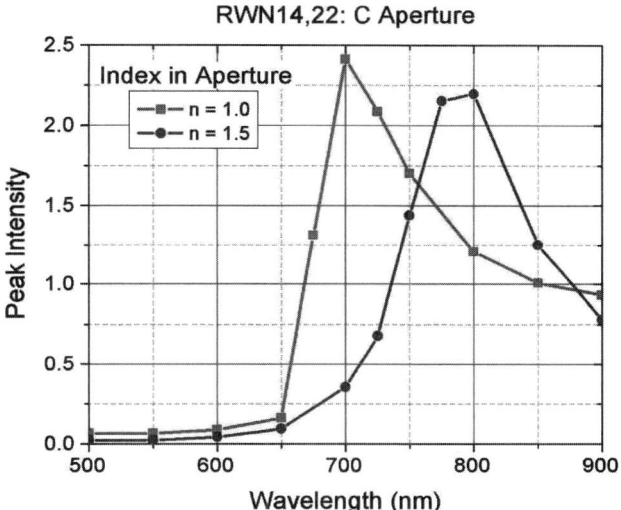

Figure 26. Peak field intensity in the medium versus wavelength for the C aperture. The results are plotted for an aperture filled with air ($n = 1$) and an aperture filled with glass ($n = 1.5$).

There are again many dimensions to be optimized for this NFT. A length of approximately 300 nm is found to be near optimum. The LSPR occurs at approximately 700 nm and the width is optimized at 55 nm for a ridge that is 20 nm wide and has a gap of 20 nm. The optimum thickness is approximately 100 nm. The wavelength dependence of the field intensity in the medium is shown in Fig. 26. As the index of the material inside the aperture increases, the resonance is found to shift towards longer wavelengths. A plot of field intensity within the medium in Fig. 27 shows that the light is very well confined. This NFT delivers 2.1% of the incident power into the central 50-nm region of the recording medium.

5. Triangle Antenna

Antennas have also been proposed as NFTs for HAMR. The simplest antenna design may be the triangle, as shown in Fig. 28. The lightning rod effect was demonstrated in Sect. IV.2 for a triangle antenna in free space. This antenna also exhibits a LSPR. It does not

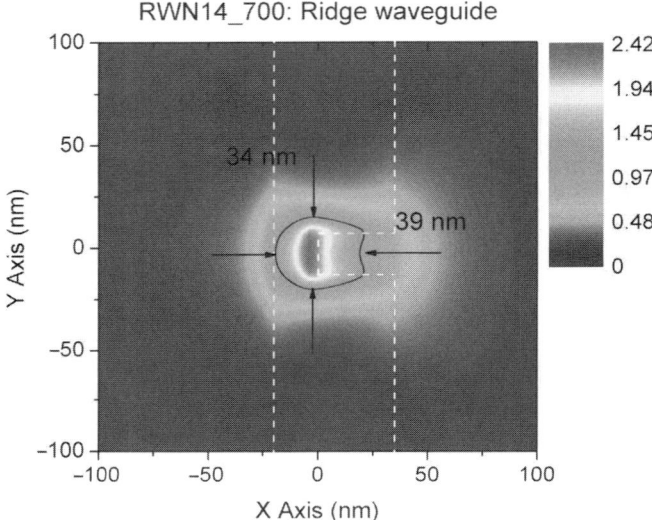

Figure 27. Field intensity within the medium from the C aperture. (Reprinted from Ref. [7]. Copyright 2006 with permission from the Institute of Pure and Applied Physics.)

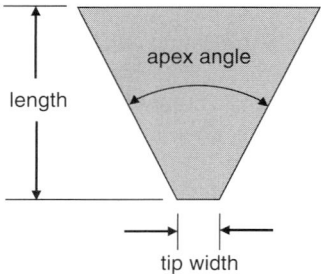

Figure 28. Dimensions for the triangle antenna.

make use of the dual-dipole effect for field enhancement. When the antenna is adjacent to a lossy metallic recording medium, however, it behaves very differently. The LSPR wavelength is a very sensitive function of antenna length. A length greater than 150 nm places the resonance at wavelengths greater than 900 nm. A 100-nm antenna

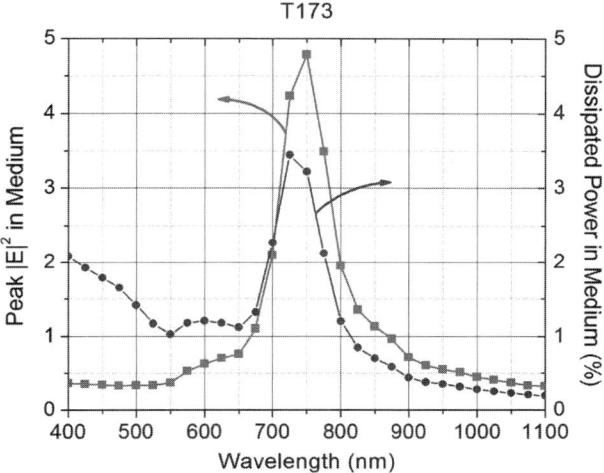

Figure 29. Wavelength dependence of field intensity and dissipated power in the medium from a 100-nm-long triangle antenna with a 30° apex angle.

has a resonance at a wavelength of approximately 800 nm. Confinement of the optical spot is difficult to achieve, however, with large apex angles, so an apex angle of 30° is chosen. The wavelength dependence of the peak field intensity in the medium is shown in Fig. 29 for a 100-nm-long antenna that is 50 nm thick. Although the LSPR occurs at 750 nm, the field intensity within the medium is not confined, as shown in Fig. 30. By operating the antenna at shorter wavelengths, one obtains better field confinement at the expense of field intensity, as shown in Fig. 31. The dissipated power within the medium at a wavelength of 650 nm is approximately 1.1%. However, the field intensity in the medium tends to spread out away from the tip of the antenna even at this wavelength.

It is interesting to compare these results with calculations of the triangle antenna in free space. Plots of the extinction, scattering, and absorption cross sections for the 100-nm triangle antenna in free space along with the peak field intensity at the apex are shown in Fig. 32. The resonance occurs at 675 nm, significantly shifted from the resonance wavelength in the presence of the medium. The peak field intensity occurs at the apex of the antenna, and is clearly

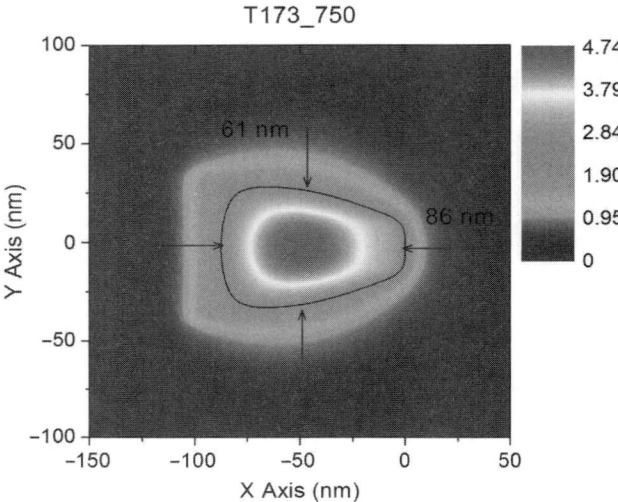

Figure 30. Field intensity in the medium from a 100-nm-long triangle antenna with a 30° apex angle at a wavelength of 750 nm. (Reprinted from Ref. [7]. Copyright 2006 with permission from the Institute of Pure and Applied Physics.)

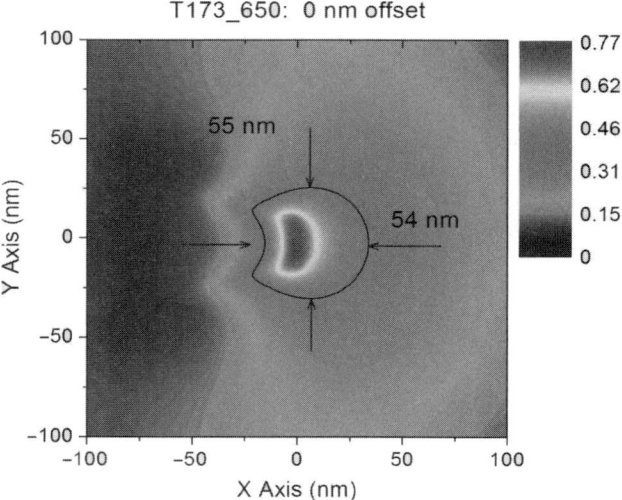

Figure 31. Field intensity in medium from a 100-nm-long triangle antenna with a 30° apex angle at a wavelength of 650 nm.

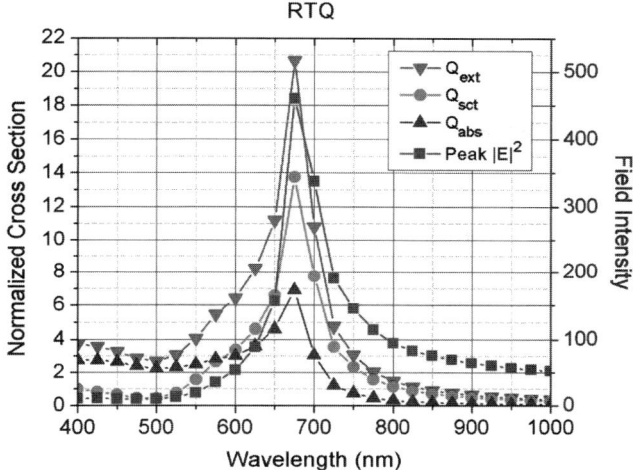

Figure 32. Extinction, scattering, and absorption cross sections normalized by the area of the antenna for the 100-nm triangle antenna on a glass substrate as a function of wavelength. The peak $|E|^2$ field intensity versus wavelength is also plotted.

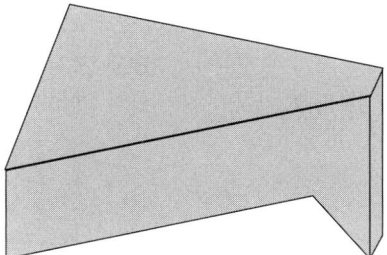

Figure 33. "Beaked" triangle antenna.

not useful for predicting the distribution of dissipated power in the medium. This clearly exhibits the unreliability of using peak field intensity for an antenna or aperture in free space as a FOM for HAMR.

One way in which the problem of lack of confinement of the coupled power to the medium can be solved is to cant the antenna so that only the tip is close to the medium. Another approach is to add a small "beak" at the end of the antenna as shown in Fig. 33.[35]

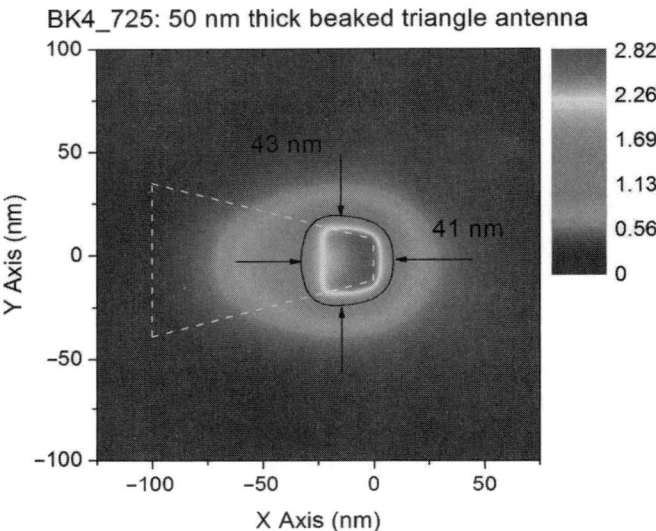

Figure 34. Field intensity in the medium from a beaked triangle antenna at 725 nm. (Reprinted from Ref. [7]. Copyright 2006 with permission from the Institute of Pure and Applied Physics.)

If a beak with a 20-nm width, length, and height is added to the triangle antenna, then the resonance wavelength is slightly shifted to 725 nm, but the field intensity within the medium at resonance is much better confined, as shown in Fig. 34. The dissipated power in a 50-nm cylinder in the medium is 2.9%.

6. Bow-Tie Antenna

The bow-tie antenna was first proposed as a NFT by Grober et al.[36] In the microwave frequency range, the bow tie is a well-known antenna design. As shown in Fig. 35, the antenna is composed of two triangular metallic plates with a narrow gap between them. This NFT is the complement of the bow-tie aperture. All three NFT enhancement mechanisms are clearly present in the design. Optimizing the antenna design proceeds along lines similar to those for the triangle antenna. An antenna that is 200 nm long, 50 nm thick, with a 20-nm gap, 20-nm apex width, and 30° apex angle exhibits two LSP resonances at approximately 625 nm and 750 nm as shown in Fig. 36.

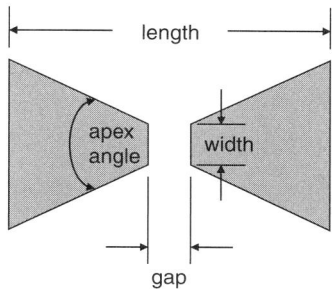

Figure 35. Bow-tie antenna.

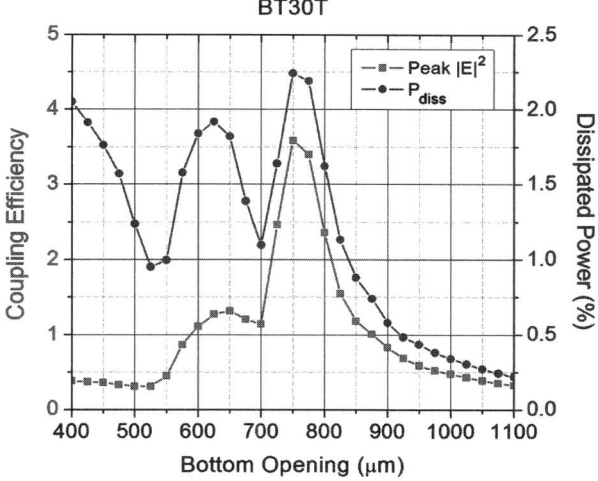

Figure 36. Field intensity in the medium and dissipated power versus wavelength for the bow-tie antenna.

As in the case of the triangle antenna, however, the long-wavelength resonance with the highest field intensity corresponds to an unconfined spot, as shown in Fig. 37. The shorter-wavelength resonance, on the other hand, does generate a small spot in the medium, as shown in Fig. 38, and delivers 1.9% of the incident optical power to the medium at a wavelength of 625 nm. Again, to obtain a confined spot at the peak of the resonance curve, the bow-tie antenna

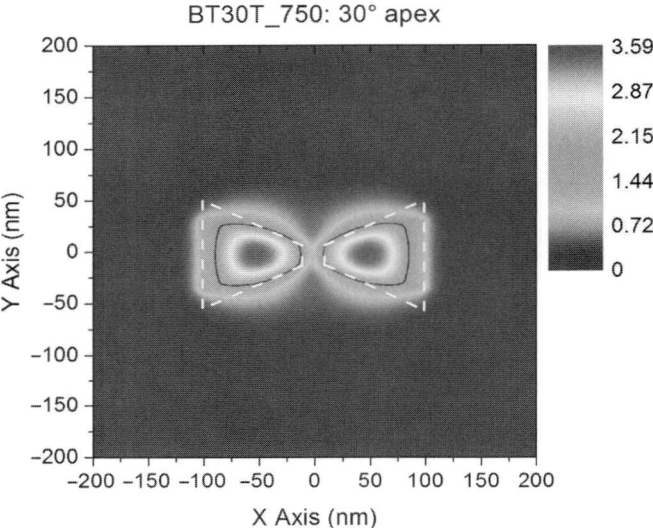

Figure 37. Field intensity in the medium for a bow-tie antenna at a wavelength of 750 nm.

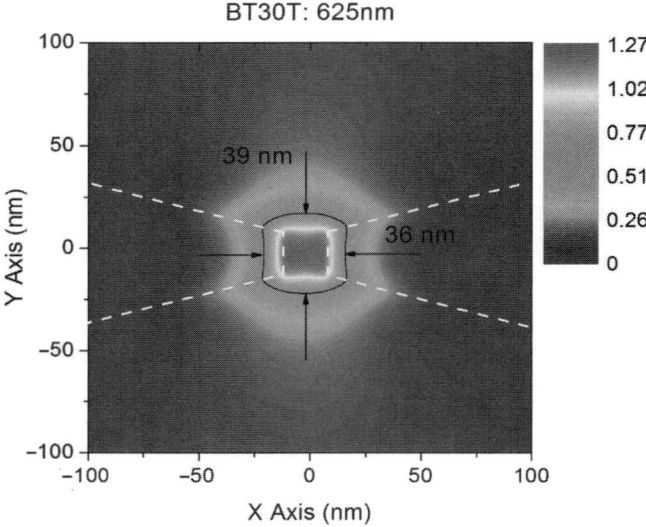

Figure 38. Field intensity in the medium for a bow-tie antenna at a wavelength of 625 nm.

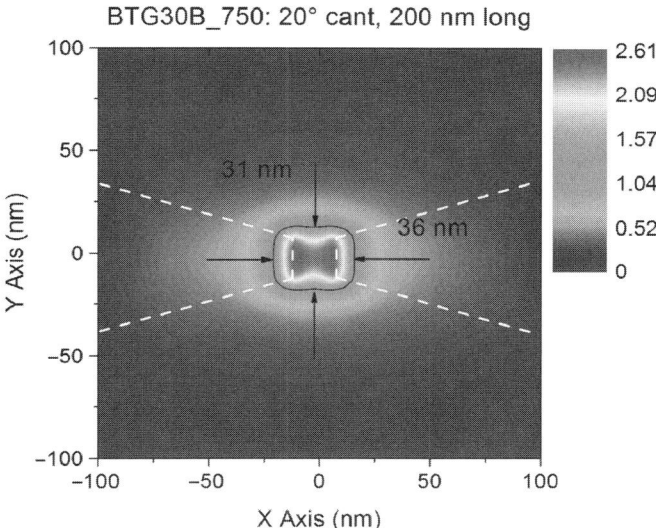

Figure 39. Field intensity in the medium for a bow-tie antenna canted at 20° at a wavelength of 750 nm.

can be canted so that only the high-field region between the tips is in close proximity to the medium. A 20° cant of the two antenna halves generates a much smaller spot at the resonance wavelength of 750 nm, as shown in Fig. 39. The canted bow tie delivers 2.1% of the incident power into the central 50 nm of the medium.

VII. ANTENNA AND APERTURE RELATIONSHIP

In the earlier sections we considered different near-field structures. These structures were one of three types: apertures, antennas, and hybrid structures. The apertures have a finite dielectric opening in a metal thin film. The resonant near field of interest is located within and in the vicinity of the opening. Antennas are finite metallic structures located in an infinite dielectric region. The resonant near field of antennas is located around the metallic structure. Then there are hybrid structures such as a metal-coated, tapered optical fiber, which on one hand do not have an aperture and on the other have metal going to infinity. In such hybrid structures, the fact that the metal goes to infinity is not important. In fact, the resonant field of interest is in a geometrically localized region around the metal – the region around the tip in the case of the tapered fiber. If the metal is terminated at a certain distance (a distance on the order of the decay length of the associated surface plasmons), the field in the geometrically localized region does not change. Thus, these hybrid structures can be converted into an aperture or an antenna structure without considerably altering the physics of the near field. Thus, we assume that all the near-field structures of our interest are either of the aperture or of the antenna type. The calculation of the cross sections goes along different lines for the two types. Hence, this classification is needed.

If we interchange the dielectric in the aperture opening and the thin film metal, we get a complementary structure, which is an antenna. Is there any relation between the resonance properties of the two structures? If we assume that the aperture metal film is infinitesimally thick, and that the metal is a perfect electrical conductor (PEC), the aperture and the complementary antenna structure are connected by a form of Babinet's principle. Suppose that the aperture is illuminated by an incident electric field $\vec{E}_i$. The interaction with the aperture will set up a total electric field $\vec{E}_1$. Now, suppose that the complementary antenna structure is illuminated by an incident magnetic field that is vectorially equal to the incident electric field in the aperture case. Thus, the incident magnetic field in the antenna case is $\vec{E}_i$. In this case, let $\vec{H}_2$ be the total magnetic field. The particular form of Babinet's principle states that

$$\vec{E}_1 + \vec{H}_2 = \vec{E}_i. \tag{7}$$

Of course, in the case of a real metal film that is not infinitesimally thick, the principle is not expected to be perfectly satisfied.

VIII. NEAR-FIELD AND FAR-FIELD RELATIONSHIP

It is difficult to design experiments to characterize the near field of the structures. Any probe such as the scanning near field optical microscope, which probes the near field directly, could end up altering the structure of the near field. This could have an effect of shifting the wavelength of the desired resonance. Experiments that account for the light radiated in the far field can also be designed.

But then, is the amount of far-field radiation a good measure of the near-field enhancement? We discuss this connection in the following section.

The source of radiation, in the classical electromagnetic theory, is an accelerated charge. For time-harmonic fields, electrical current serves as the source. There is a considerable amount of literature on the radiation properties of apertures and antennas at radio and microwave frequencies. At these frequencies, the penetration of the fields into a metal is small. Thus, it is frequently quite acceptable to model these structures by assuming the metals are PECs. At optical frequencies, a significant portion of the incident energy can be dissipated in the metal. In addition, typical metals exhibit surface plasmon resonances at optical frequencies. Associated with a surface plasmon is an oscillating charge distribution on the surface of the structure, localized within the skin depth of the metal.

In the absence of sources outside a closed surface, the tangential electric and magnetic fields on the surface uniquely define the field distribution outside the surface. In particular, the tangential fields can be interpreted as electric and magnetic currents on the surface. The equivalent currents replace the physical sources.[37] The fields generated by the physical and the hypothetical sources are the same outside the surface. Inside the surface the field due to the hypothetical sources is zero. This theorem is used to calculate the far-field radiation pattern from the near-field FDTD simulation. We apply the theorem in the special case of the region outside the surface being a homogeneous dielectric medium. In FDTD simulations the infinite domain is converted into a finite computational domain using matching boundary conditions. This is true even in the case of stratified media (e.g., thin film structures) that extend to infinity. Thus, in the rest of the discussion, we assume that the domain of interest is infinite.

Figure 40 shows a scatterer embedded in a hypothetical closed surface. The surface currents are defined on the closed surface. The equations that connect the surface currents to the tangential fields and govern the radiation from the currents can be found in the popular FDTD texts.[19]

The Poynting vector has units of power per unit area. When the normal (outward) component of the Poynting vector is integrated over a closed surface, it represents the electromagnetic power leaving the surface. For monochromatic fields the Poynting vector oscillates harmonically about a direct-current offset. The frequency of the

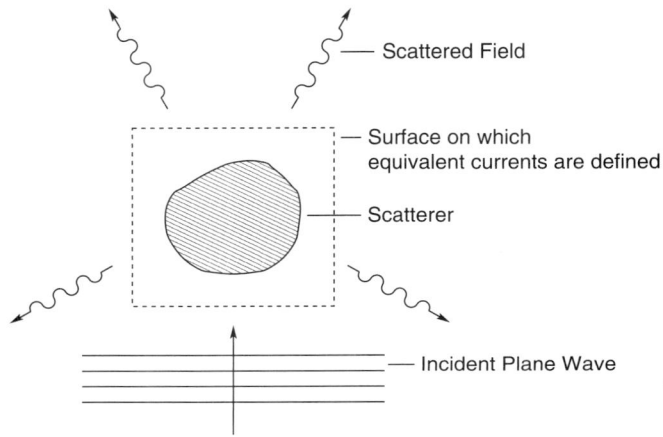

Figure 40. Scattering geometry.

oscillation is twice that of the oscillating fields. The average value of the Poynting vector over an oscillation period is the measure of the net power flow across the surface in one direction. In lossless regions, the divergence of the Poynting vector is zero. Thus, in accordance with the Gauss divergence theorem, if we choose a closed surface in a lossless region, the surface integral of the energy flux is zero. The arbitrariness in the choice of the surface in the equivalence theorem does not change the net energy flux through the surface. Let the time-harmonic electric and magnetic field (at frequency ω) at a point be given by

$$\vec{E} = \vec{E}_o \exp(-i\omega t) \qquad (8)$$

and

$$\vec{H} = \vec{H}_o \exp(-i\omega t), \qquad (9)$$

respectively. The vector quantities $\vec{E}_o$ and $\vec{H}_o$ contain the amplitude and phase information, and are hence complex. The time-averaged Poynting vector is given by

$$<\vec{S}> = \frac{1}{2}\text{Re}\left(\vec{E}_o \times \vec{H}_o^*\right). \qquad (10)$$

Here, Re and $*$ stand for the real part and complex conjugation, respectively. In the rest of the discussion we will only be interested in the time-averaged Poynting vector for monochromatic radiation. The only aspect of (10) that we carry forward is the linear dependence of the Poynting vector on the electric and magnetic fields. Hence, we simplify the notation by dropping the angular brackets and the subscript o on the fields. We represent the bilinear form by

$$\vec{S} = (\vec{E}, \vec{H}). \tag{11}$$

Henceforth, Poynting vector refers to the time-averaged Poynting vector.

The fields $\vec{E}$ and $\vec{H}$ can be decomposed into the incident field (indicated with subscript i) and the scattered field (indicated with subscript s). The incident field is the field that would have been present if the scatterer were absent. This assumes that the optical source excitation is the same. We have been vague in our definition of the scatterer. To be specific, we choose a geometrical arrangement as our starting point. This is the incident geometry, and the field is the incident field. We then alter the geometry. The change is small enough so that the optical source can still be assumed to be unperturbed. In particular, the change that we make would either be placing a microscopic particle (the antenna) in the geometry, or punching a hole in a metal film (the aperture). The difference between the field in the changed geometry and the field in the incident geometry is defined as the scattered field. In the context of the equivalence theorem, the change that we make is done inside the hypothetical surface. No matter how the incident geometry is defined, we assume that the region outside the hypothetical surface is lossless and homogeneous. With this decomposition of the fields, the Poynting vector is given by

$$\vec{S} = \vec{S}_i + \vec{S}_s + \vec{S}_c, \tag{12}$$

where

$$\vec{S}_i = (\vec{E}_i, \vec{H}_i), \tag{13}$$

$$\vec{S}_s = (\vec{E}_s, \vec{H}_s), \tag{14}$$

and

$$\vec{S}_c = (\vec{E}_i, \vec{H}_s) + (\vec{E}_s, \vec{H}_i). \tag{15}$$

$\vec{S}_i$ and $\vec{S}_s$ are the Poynting vectors of the incident and the scattered field, respectively. $\vec{S}_c$ is an interference term. In a homogeneous dielectric, $\vec{S}$, $\vec{S}_i$, and $\vec{S}_s$ are divergenceless; hence, $\vec{S}_c$ is also divergenceless. The total energy flux from the volume inside the hypothetical surface is the integral of the component of the Poynting vector along the outward normal, over the closed surface. Let us represent this integral by I and the integral of the three terms on the right-hand side by I_i, I_s, and I_c. Thus,

$$I = \oiint_A \vec{S} \cdot d\vec{A}, \tag{16}$$

and

$$I = I_i + I_s + I_c. \tag{17}$$

Here, A represents integral over the closed surface. If the closed surface is distorted such that the volume swept in distorting the surface is in a homogeneous dielectric medium, then owing to the divergenceless property, the integrals I, I_i, I_s, and I_c remain invariant.

1. Radiation from Antennas

Let the incident geometry be the infinite free space and the incident field be a plane wave. We use spherical polar coordinates such that the polar and the azimuthal angles are denoted by θ and ϕ, respectively. The polar angle is measured with respect to the $+Z$ axis. The azimuthal angle is measured with respect to the $+X$ axis in the XY plane. The incident plane wave is propagating along the $\theta = 0°$ direction ($+Z$ direction), and the polarization of the incident beam is along the ($\theta = 90°$, $\phi = 0°$) direction ($+X$ direction). Since the Poynting vector is divergenceless in this medium, $I_i = 0$.

In the far field, the radiation field in a certain direction appears locally like a plane wave propagating in that direction. The plane wave in a particular direction can further be decomposed into two mutually orthogonal polarizations. An analysis using Green's function indicates that only the plane wave propagating in the same direction as the incident wave and possessing the same polarization can contribute to the term I_c.[18] Thus, I_c is proportional to the appropriately polarized scattered field radiation in the direction of the incident beam. Energy conservation considerations indicate that I is precisely the negative of the power being absorbed inside the hypothetical surface, averaged over a field oscillation. We denote this quantity by I_a. In fact, I_s and I_a divided by $\left|\vec{S}_i\right|$ are called the scattering and the absorption cross sections, respectively. Their sum, the total or "extinction" cross section, is thus directly proportional to the field strength of the scattering in the direction of the incident wave. This is more commonly known as the "optical theorem."[16] A cross section has physical dimensions of area, and can be normalized by the physical cross section of the antenna to obtain a dimensionless normalized cross section. If we consider all possible slices of the antenna normal to the propagation direction, the physical cross section of the antenna is the area of the slice with the largest cross section.

2. Radiation from Apertures

Consider a plane polarized wave incident normally on a metal film of finite thickness. The incident energy is transmitted across the film, reflected back, or absorbed in the film (see Fig. 41). Considering that the plane wave is infinite in extent, each of the three energy contributions is infinite. Let the field distribution in the presence of the film be termed the "incident field." Let us now etch an aperture of finite cross section in this film. Let the difference of the field after and before etching the aperture be termed the "scattered field." We follow an analysis similar to the case of the antennas. The aperture is the source of the scattered field. Owing to the loss in the metal, this scattered field decays inside the metal with increasing lateral distance from the aperture. Since the metal film is infinite in its plane, the hypothetical surface used in defining the equivalent currents has to wrap around the metal at infinity. We define the surface (see Fig. 42) to be $S1 - S2 - S3$ on one side, and $S4 - S5 - S6$ on the other. The expression for the power flux, (17), is applicable here. However, the

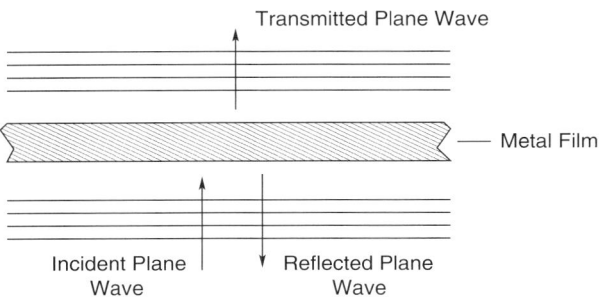

Figure 41. Incident geometry.

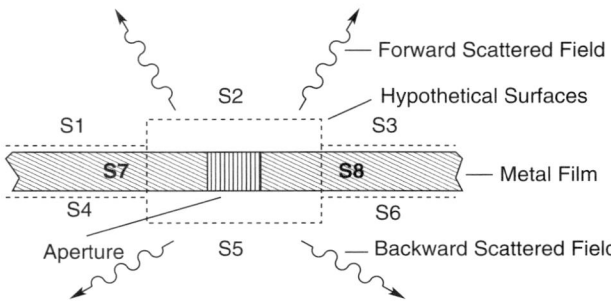

Figure 42. Aperture geometry.

key difference from the antenna case is that I_i is no longer zero. In fact, the principle of energy conservation implies that I_i and the net power absorbed inside the closed surface before etching the aperture, I_{af}, add up to zero. Note, both are infinite quantities. Also, I and the net power absorbed inside the closed surface after etching the aperture, I_{aa}, add up to zero. Hence,

$$I_s + (I_{aa} - I_{af}) = -I_c. \tag{18}$$

The scatterer (aperture) is finite in extent. Moreover, owing to the loss in the metal, fields decay in the film away from the aperture; hence, I_s, $I_{aa} - I_{af}$, and I_c are finite quantities. The definition of the scattering cross section is analogous to the antenna case. However, in the definition of the absorption cross section, we replace I_a with $I_{aa} - I_{af}$. In the case of the antenna, the term I_c was stated to be directly proportional to the radiation intensity in the forward direction. For the aperture, a Green's function analysis similar to the antenna case indicates that the contribution to I_c from the surface $S1 - S2 - S3$ is proportional to the radiation intensity in the forward direction. Similarly, the contribution from the surface $S4 - S5 - S6$ is proportional to the radiation intensity in the backward direction. Thus, I_c is a linear combination of the radiation intensity in the forward and backward directions.

To calculate the radiation pattern of the apertures using FDTD, we need to define the equivalent currents on the hypothetical surface. To overcome the difficulty of dealing with an infinite surface, we choose the closed surface to be $S2 - S7 - S5 - S8$. The assumption is that the surfaces S7 and S8 are chosen far away from the aperture, so that the scattered field is negligible on them. Our claim is as follows: it is possible to choose surfaces S1, S3, S4, and S6 infinitesimally close to the metal film surface and surfaces S7 and S8 sufficiently far from the aperture, such that the scattered field on the surfaces S1, S3, S4, S6, S7, and S8 is infinitesimally small. This is possible owing to the dissipation in the metal. Thus, the equivalent currents are essentially present only on S2 and S5. Thus, in the FDTD code, the radiation pattern can be calculated exactly as in the antenna case – by using the closed surface S2 − S7 − S5 − S8.

3. Numerical Modeling

We apply the concepts discussed in the last few sections to the case of a C aperture in aluminum. The thickness of the aluminum film is chosen to be 100 nm. The dimensions of the C aperture are as follows: aperture length 155 nm, aperture width 70 nm, tongue width 25 nm, and gap width 25 nm. The incident field is X-polarized. The XZ plane is a mirror symmetry plane for the C aperture. The surrounding dielectric is assumed to be free space. The normalized scattering and absorption cross sections as a function of wavelength are shown in Fig. 43.

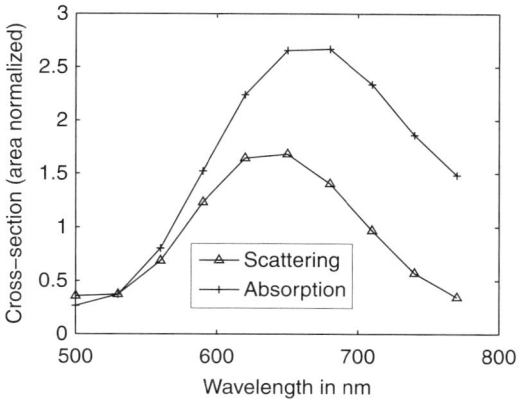

Figure 43. Cross sections of the C aperture (normalized by the area).

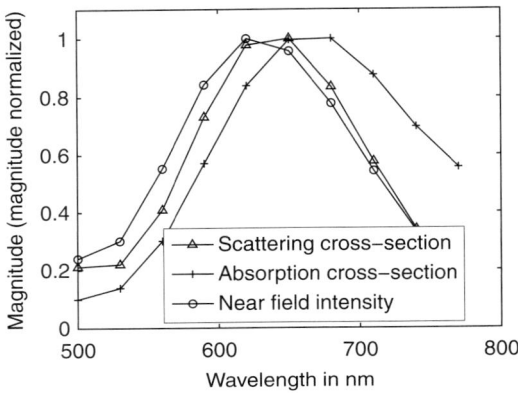

Figure 44. Cross sections and near-field intensity of a C aperture (magnitude normalized).

The near-field intensity is calculated at a point, in the gap, 5 nm beyond the transmission side of the aperture. The cross sections are normalized with respect to the physical area of the aperture. For comparison with the near-field intensity, we normalize the cross sections such that the peak cross section is unity. To distinguish this from the area normalization, we call this the "magnitude normalization." The cross sections and the near-field intensity are shown in Fig. 44.

The three quantities have been magnitude-normalized in this plot. Geometrically, the C aperture is a ridge waveguide of finite extent. For a PEC waveguide of the same cross section, the cutoff wavelength for the lowest-order transverse electric mode is around 500 nm. As one approaches the cutoff wavelength from shorter wavelengths, the longitudinal wave vector decreases in magnitude. Hence, for the same length of the waveguide, the field has a larger number of transverse traversals in the aperture. On the other hand, if one moves away from the cutoff wavelength towards longer wavelengths, the longitudinal wave vector becomes imaginary, indicating evanescent decay. Hence, the strongest resonance is expected to be at the cutoff wavelength. Three things about the aluminum C aperture are different from the PEC waveguide: the metal can support surface plasmons, the incident field (field in the metal film before the aperture is etched) has a Fabry–Perot resonance, and leaky modes that

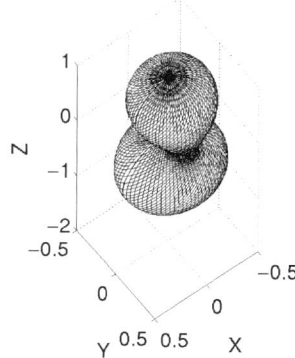

Figure 45. Radiation pattern of a C aperture.

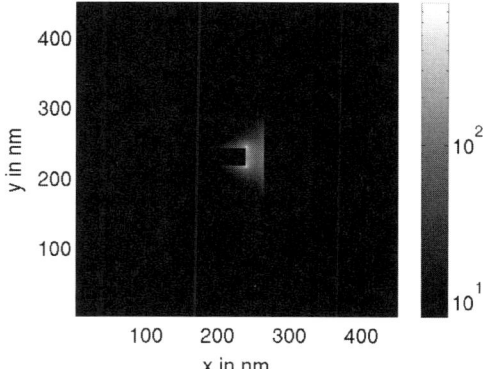

Figure 46. Near-field intensity of the C aperture.

are not seen in an infinite waveguide can be excited in the case of a waveguide whose length is a fraction of the wavelength. One or more of these effects could cause a shift in the resonance wavelength. In fact, we observe a resonance at approximately 650 nm. The radiation pattern of the aperture at a wavelength of 650 nm is plotted in Fig. 45. The corresponding near field intensity is shown in Fig. 46.

An electric dipole is induced in the gap of the C aperture. The far-field radiation pattern of the dipole is expected to have a

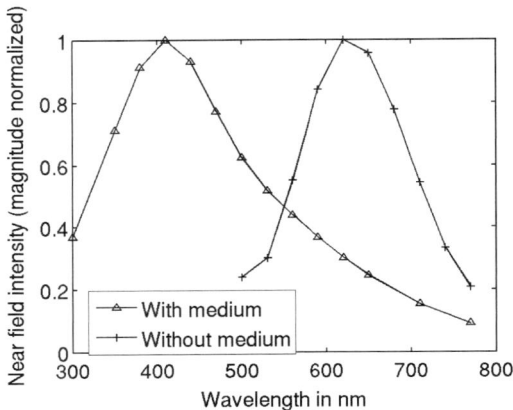

Figure 47. Effect of the medium on the near-field resonance of a C aperture.

doughnut shape with X as the cylindrical symmetry axis. However, the presence of the infinite metal in the xy plane is expected to quench the radiation pattern in that plane. This would cause a pinching of the radiation pattern in the xy plane. This is seen in Fig. 45.

To see the effect of the medium, we place a cobalt film 5 nm from the aperture. The magnitude-normalized near-field intensity with and without the medium is plotted in Fig. 47. A considerable shift in the resonance wavelength in seen. Thus, the medium loads the C aperture.

We then consider the antenna structure complementary to the C aperture – the C antenna. Even though we do not have a PEC antenna, we would like to test the agreement with Babinet's principle. Instead of rotating the polarization of the incident beam by 90°, we rotate the antenna structure by 90°. The normalized cross sections of the C antenna are shown in Fig. 48. The resonance wavelength is the same as that of the C aperture; however, the scattering cross section is much more enhanced in this case. If we assume that a resonance enhancement of the electric field has an associated enhancement in the magnetic field, and vice versa, then Babinet's principle suggests that the complementary structure should also have a resonance in the same spectral region. An enhanced magnetic field of opposite phase

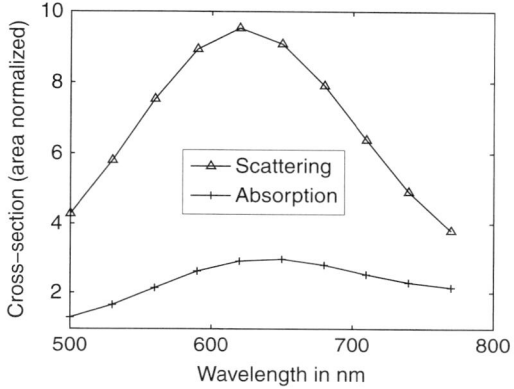

Figure 48. Cross sections of the C antenna (normalized by the area).

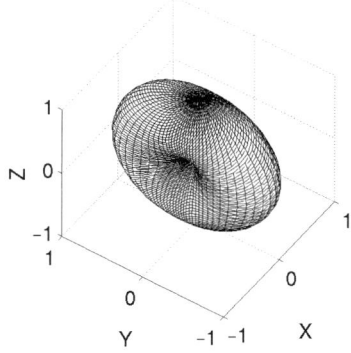

Figure 49. Radiation pattern of the C antenna.

would be needed to nullify the enhancement in the electric field of the complementary structure.

The radiation pattern of the antenna at a wavelength of 650 nm is plotted in Fig. 49. This radiation pattern shows the cylindrical symmetry of the doughnut-shaped dipolar radiation pattern.

IX. PHOTONIC NANOJETS

Up to this point we have been considering NFTs for applications such as HAMR. The transducer itself, however, is only one part of a complete device for recording. The transducer will not be effective unless it is situated at the position of a large field amplitude from the incident laser beam. This is generally accomplished by focusing the beam onto the transducer. A simple objective lens may be quite satisfactory for this purpose. In this section we discuss techniques for highly concentrating an incident beam into a "nanojet," i.e., a narrow beam of energy with an extended path length. In principle, nanojet optics could form one part of the complete system for near-field recording.

In the geometrical optics description of conventional lens focusing, the focus is the point where all the light rays converge. In Fig. 50, the focusing of a plane wave by a lens is shown.

The focal point is situated in the middle of a sphere. If the refractive index of the sphere is greater than unity and the sphere is truncated to a hemisphere (part on the right of the dashed line is removed), we end up with a SIL. The key feature of this geometry is that all the rays converge to a point – the focal point. If, instead, the lens is removed from the system such that all parallel incident rays fall directly on the sphere (Fig. 51), then all the rays will not converge to a single point.

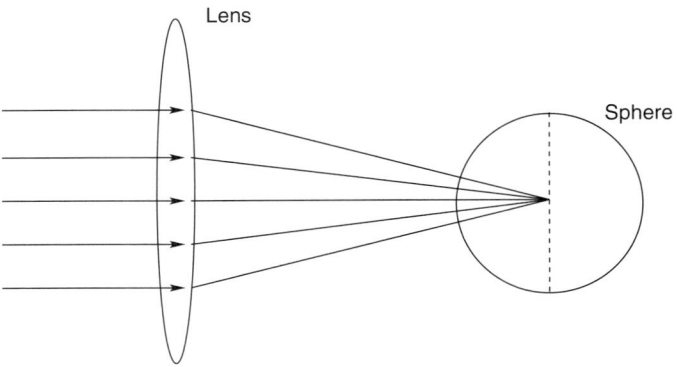

Figure 50. Focusing by a lens.

Near-Field Optics for Heat-Assisted Magnetic Recording

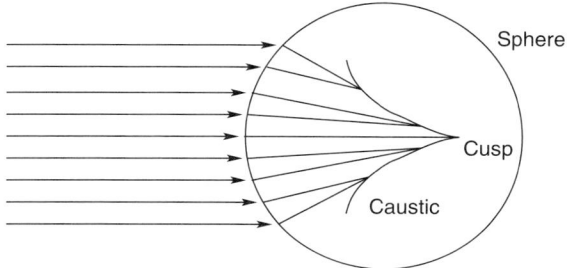

Figure 51. Focusing to a caustic.

Nonetheless, for an appropriate choice of the sphere refractive index, there exists a surface such that several rays converge at every point of the surface. In other words, the focal point degenerates into a surface. This surface is termed a "caustic." Owing to the symmetry of the problem, the caustic has a cylindrical symmetry. The caustic has a cusp at the point where the caustic intersects the symmetry axis. Our discussion so far has been based on geometrical optics. When one goes to a complete electromagnetic description, the focal point of a lens does not have a field singularity. Nonetheless, there is a focal region of high field concentration. Similarly, for the caustic one ends up with a region of high field concentration in the neighborhood of the geometrical caustic. In addition, in the electromagnetic description, the wavelength adds a length scale to the phenomenon. Thus, for a fixed radius and refractive index of the sphere, the caustic region will depend on the wavelength of light (in free space).

In Fig. 51, the cusp of the caustic is shown to lie inside the sphere. In such a situation, the cusp can be pushed to the surface of the sphere by reducing the refractive index of the sphere. In the geometrical optics description, this will happen when the refractive index of the sphere is twice that of the surrounding medium (assumed to be free space here). In the physical optics description, the choice of the refractive index ratio depends on the ratio of the radius of the sphere to the wavelength. Typically, it is found to be smaller than 2. When the cusp region is chosen close to the sphere surface, an interesting phenomenon of "photonic nanojet" emerges. On the free space side of the surface, an intense optical-jet-like region is generated. A two-dimensional FDTD model of this phenomenon is shown in Fig. 52.

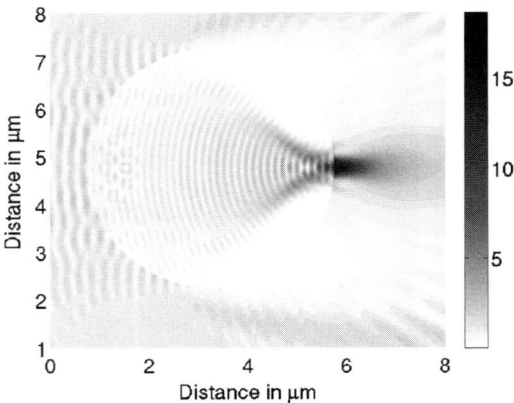

Figure 52. Two-dimensional photonic nanojet intensity modeled using the FDTD method.

The photonic nanojets display two remarkable features. Even though the spot size is comparable to a high-numerical-aperture diffraction-limited SIL spot, the depth of focus is much larger. For a comparable spot size from a lens, the depth of focus would have been much smaller. Secondly, from Fig. 52, the decay length of the two-dimensional photonic nanojet (distance between the peak field and the $1/e$ field in the longitudinal direction) is larger than the wavelength. The spot size of the nanojet at its waist is marginally larger than the size of a spot generated by a two-dimensional lens of unit numerical aperture[38] (see Fig. 53).

The angular spectrum (spatial frequency content) of the photonic nanojet is shown in Fig. 54. The amplitude distribution of the angular spectrum alone does not explain the long decay length of the photonic nanojet. The bathtub-shaped phase distribution plays a key role. Different spatial frequencies gain different phases on propagation. This dephasing causes spot divergence. The bathtub shape counteracts the typical dephasing factor that decreases with increasing magnitude of the spatial frequency.

When a nanoparticle is placed in the light path, light is scattered. Assuming that the incident light is a plane wave, the light that radiates back towards the source is termed the "backscattered light." When nanoparticles are illuminated by a plane wave, the intensity of backscattered light is small compared with the intensity

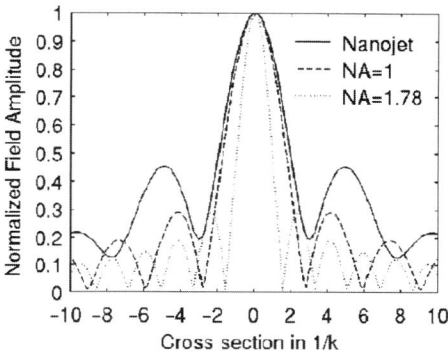

Figure 53. Comparison of the two-dimensional nanojet spot size with that of lens focusing. k is the wave vector in free space. (Reprinted from Ref. [38]. Copyright 2005 with permission from the Optical Society of America.)

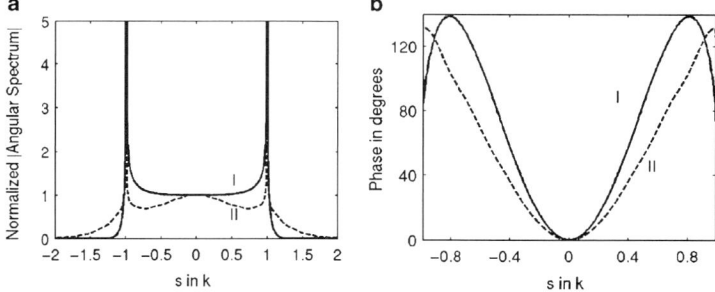

Figure 54. Angular spectrum of a two-dimensional nanojet. s and k are the spatial frequency and the free-space wave vector, respectively. (Reprinted from Ref. [38]. Copyright 2005 with permission from the Optical Society of America.)

of the incident light. If instead, a lens is used to focus light onto the nanoparticle, the backscattered light intensity increases by a few orders of magnitude. However, if the nanoparticle is placed in the photonic nanojet, the backscattering increases by several orders of magnitude.[39] In the two-dimensional case, the effect is still seen, but it is not as pronounced. The enhanced backscattering for the two-dimensional case is shown in Fig. 55. The effect of the particle size on the back-cattering enhancement is shown in Fig. 56.

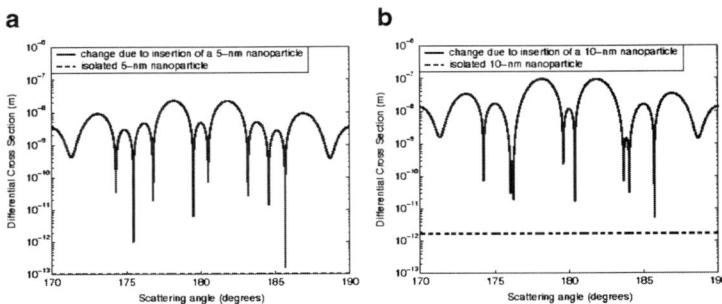

Figure 55. Differential cross section of a particle placed in a nanojet. (Reprinted from Ref. [39]. Copyright 2004 with permission from the Optical Society of America.)

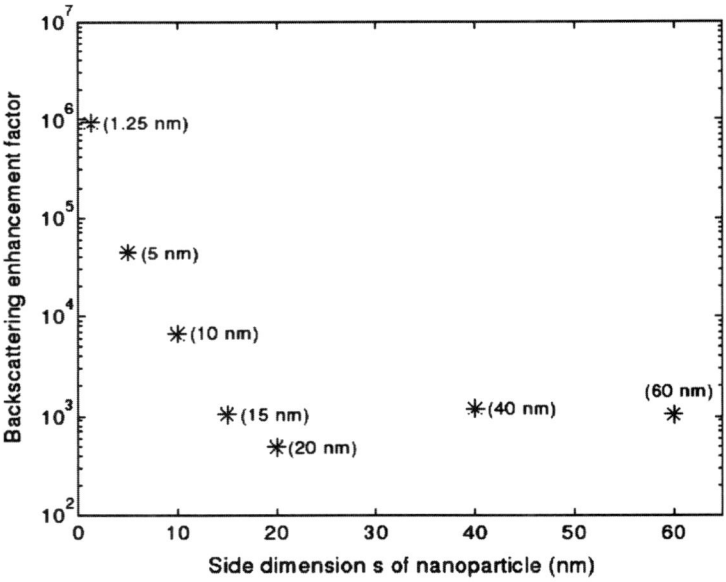

Figure 56. Backscattering enhancement in the nanojet as a function of particle size. (Reprinted from Ref. [39]. Copyright 2004 with permission from the Optical Society of America.)

Chen et al.[40] have argued that while the large intensity of the nanojet provides a lenslike enhancement of the backscattering, it is the coordination between the backscattering and the modes of the nanojet-creating sphere that generates the superenhancement of the nanojet.

X. CONCLUSION

A variety of mechanisms have been discussed for enhancing the efficiency of NFTs for use in HAMR. These include the LSPR effect, the lightning rod effect, and the dual-dipole effect. Several common FOMs for NFTs have been discussed and it has been shown that peak $|E|^2$ field intensity within the recording medium, or even better, the dissipated power within the recording medium are the best FOMs. On the other hand, far-field transmittance or even peak field amplitude in the absence of a recording medium are not useful for judging the merits of NFTs for HAMR. Several transducer designs have been analyzed theoretically and compared using a standard geometry that approximates the situation found in HAMR. The results are summarized in Table 1. Surprisingly large power coupling efficiencies can be obtained theoretically for the best transducer designs, lending credibility to the engineering challenge of building such a data storage device.

Our study of the C aperture indicates that the resonance wavelength for an aperture of finite length can be shifted from the cutoff

Table 1.
Summary of near-field transducer (*NFT*) performance. The peak $|E|^2$ intensity is normalized by that of the incident beam.

NFT design	λ_{res}	Peak $|E|^2$	P_{diss}	FWHM spot size (nm^2)
Circular aperture	650	0.07	0.14%	113×142
Rectangular aperture	650	0.80	0.92	43×25
Bow-tie aperture	725	1.38	1.7	59×56
C aperture	700	2.42	2.1	34×39
Triangle antenna	650	0.77	1.1	55×54
Beaked triangle	725	2.82	2.9	43×41
Bow-tie antenna	650	1.41	1.4	39×36
Canted bow tie	750	2.61	2.1	31×36

FWHM full width at half maximum

wavelength of the corresponding PEC waveguide, especially at optical frequencies. The far-field cross sections and the near field intensity have resonances at around the same wavelength. The strong currents associated with the near-field enhancement are also responsible for the absorption and far-field radiation. It might be possible to come up with a current distribution of certain orientation and phase relationship such that the far-field radiation is small. Whether there is a geometry in which this current distribution can be excited by a plane wave is an open question. Complementary aperture/antenna systems that seem to be resonating in completely different modes can still have similar resonance properties in accordance with Babinet's principle.

Finally, we have briefly considered one interesting optical technique for exciting the NFT via a nanojet.

ACKNOWLEDGMENTS

We would like to acknowledge many useful conversations with our Seagate colleagues Ed Gage, Eric Jin, Terry McDaniel, and Chubing Peng during the course of this work.

REFERENCES

[1] T. W. McDaniel, W. A. Challener, K. Sendur, *IEEE Trans. Magn.* **39** (2003) 1972.
[2] T. W. McDaniel, *J. Phys. Cond. Matter* **17** (2005) R315.
[3] S. M. Mansfield, G. S. Kino, *Appl. Phys. Lett.* **57** (1990) 2615.
[4] L. Wang, S. M. Uppuluri, E. X. Jin and X. Xu, *Nano Lett.* **6** (2006) 361.
[5] A. Bouhelier, M. R. Beversluis and L. Novotny, *Appl. Phys. Lett.* **83** (2003) 5041.
[6] M. Ohtsu and H. Hori, *Near-Field Nano-Optics,* Kluwer, New York 1999, p. 128.
[7] W. A. Challener, E. Gage, A. Itagi and C. Peng, *Jpn. J. Appl. Phys.* **45** (2006) 6632.
[8] G. Mie, *Ann. Phys.* **25** (1908) 377.
[9] K. S. Kunz and R. J. Luebbers, *The Finite Difference Time Domain Method for Electromagnetics*, CRC, Boca Raton, FL, 1993.
[10] A. Taflove and S. Hagness, *Computational Electrodynamics,* Artech House, Boston, MA, 2000.
[11] H. Bethe, *Phys. Rev.* **66** (1944) 163.
[12] M. Fleischmann, P. J. Hendra and A. J. McQuillan, *Chem. Phys. Lett.* **26** (1974) 163.
[13] M. Moskovits, *J. Chem Phys.* **69** (1978) 4159.
[14] K. Kneipp, Y. Wang, H. Kneipp, L. T. Perelman, I. Itzkan, R. R. Dasari and M. S. Feld, *Phys. Rev. Lett.* **78** (1997) 1667.
[15] S. Nie and S. R. Emory, *Science* **275** (1997) 1102.
[16] C. F. Boren and D. R. Huffman, *Absorption and Scattering of Light by Small Particles,* Wiley-Interscience, New York, 1983.

[17] http://www.physics.ohio-state.edu/~stroud/optics.ppt#2
[18] M. Born and E. Wolf, *Principles of Optics*, Pergamon, Oxford, 1975, chap. 13.
[19] D. W. Lynch and W. R. Hunter, "Gold" in *Handbook of Optical Constants of Solids*, E. D. Palik (ed.), Academic, San Diego, 1998, 286.
[20] R. A. Innes and J. R. Sambles, *J. Phys. F: Met. Phys.* **17** (1987) 277.
[21] W. A. Challener, I. K. Sendur and C. Peng, *Opt. Exp.* **11** (2003) 3160.
[22] J. Gersten and A. Nitzan, *J. Chem. Phys.* **73** (1980) 3023.
[23] P. F. Liao and A. Woakun, *J. Chem. Phys.* **76** (1982) 751.
[24] G. T. Boyd, Th. Rasing, J. R. R. Leite and Y. R. Shen, *Phys. Rev. B* **30** (1984) 519.
[25] P. F. Liao and A. Wokaun, *J. Chem Phys.* **76** (1982) 751.
[26] A. Sommerfeld, *Math. Ann.* **47**, (1896) 317.
[27] B. Richards and E. Wolf, *Proc. Roy Soc. London Ser. A* **253** (1959) 358.
[28] X. Shi, L. Hesselink, and R. L. Thornton, *Opt. Lett.* **28** (2003) 1320.
[29] E. Jin, et al., *Proc. IMECE'03* (2003) 1.
[30] K. Sendur, W. Challener and C. Peng, *J. Appl. Phys.* **96** (2004) 2743.
[31] X. Xu, et al., *Proc. SPIE* **5515** (2004) 230.
[32] E. Jin and X. Xu, *Jpn. J. Appl. Phys.* **43** (2004) 407.
[33] E. Jin and X. Xu, *Appl. Phys. Lett.* **86** (2005) 111106.
[34] K. Sendur, C. Peng and W. Challener, *Phys. Rev. Lett.* **94** (2005) 043901.
[35] T. Matsumoto, T. Shimano, H. Saga and H. Sukeda, *J. Appl. Phys.* **95** (2004) 3901.
[36] R. D. Grober, R. J. Schoelkopf and D. E. Prober, *Appl. Phys. Lett.* **70** (1997) 1354.
[37] S. A. Schelkunoff, *Bell System Tech. Jour.* **15** (1936) 92.
[38] A. V. Itagi and W. A. Challener, *J. Opt. Soc. Am. A* **22** (2005) 2847.
[39] Z. Chen, A. Taflove and V. Backman, *Opt. Exp.* **12** (2004) 1214.
[40] Z. Chen, A. Taflove, X. Li and V. Backman, *Opt. Lett.* **31** (2006) 196.

3

Symmetry Considerations in the Modelling of Light–Matter Interactions in Nanoelectrochemistry

Chitra Rangan

Department of Physics, University of Windsor, Windsor, ON N9B 3P4, Canada
rangan@uwindsor.ca

Summary. The modelling of light–matter interactions at nanometre length scales is becoming increasingly important in modern nanoelectrochemistry. The ability to fabricate extremely sophisticated nanostructures in the laboratory that cannot be described analytically has driven the need for modelling. Advances in scientific computation techniques, and the availability of computing resources have also led to cost savings in several industries, such as the automotive industry. One of the most important considerations for choosing the optimal numerical technique for a problem is *symmetry*. This rather old-fashioned consideration has tremendous effects on both the accuracy and the efficiency of numerical methods. We show how symmetry considerations play a major role in modern scientific computation. Examples of supported and unsupported quantum dots and quantum dot clusters are presented.

I. INTRODUCTION

The ability to fabricate nanostructures on surfaces and to exploit their optical, chemical and electrical properties has defined the frontier of surface science.[1] These properties are increasingly being used in the development of technologies in several industries. Examples are nanostructured catalysts in the automobile industry, nanoplasmonic sensors in the biomedical industry and nano-interconnects in the electronics industry. Often, computational modelling is used to advance research in these applications, and the big advantage is cost savings. Despite advances in techniques of scientific computation, and sheer computing power, the true gain in R&D time depends on the choice of the optimal numerical method. A critical criterion in the choice of numerical technique is the symmetry of the problem, which significantly affects both the accuracy and the efficiency of the numerical method used.

With advances in nanotechnology, quantum dots are ubiquitous in surfaces with myriad applications in electrochemistry. In this chapter, we discuss the numerical modelling of quantum dots – both the spherically symmetric (metal or semiconductor) dots embedded in a matrix (usually a dielectric) and the hemispherical metal dots (a.k.a. nanoparticles or thin-film islands) supported by a substrate (again, usually a dielectric). We are particularly concerned about the optical properties of quantum dots, since the interaction of light with quantum dots is used widely for several applications in surface electrochemistry – from characterization of the thickness and quality of thin films, to the development of surface sensors – as well as in nanoelectronics and quantum computing.

Quantum dots are probed with light in two ways – with white light to extract macroscopic information such as size and refractive index, and with coherent, laser light to manipulate the quantum wavefunction of the bound electrons. Therefore, we present the computational methods used in the modelling of two categories of experiments – white light interaction with metallic unsupported and supported quantum dots, and laser interaction with spherical semiconductor quantum dots. In the former category, we are interested in the macroscopic effects of the quantum dot (and the substrate) on the absorption, transmission and reflection of the incident white light. In the latter category, we are interested in the quantum wavefunction and energy of the quantum dot that can be probed and manipulated by a laser.

II. OPTICAL PROPERTIES OF METAL NANOPARTICLES

In this section, we describe theoretical methods that describe the macroscopic optical properties of metal nanoparticles (a.k.a quantum dots). Recently, silver and gold nanoparticles have found tremendous use in biological assays, detection, labelling and sensing because of their sensitive optical spectra. While some works in the literature refer to these as 'quantum dots', in optical absorption experiments their quantized energy structure is not probed. The spectrum is a probe of the localized surface plasmon phenomenon, a collective electronic excitation that is localized in spatial extent owing to the small size of the nanoparticle compared with the wavelength.

1. Macroscopic Theories

For a single spheroidal nanoparticle with dimensions much smaller than the wavelength of light, the absorption spectrum can be calculated to experimental accuracy using the well-known Mie theory.[2] The incident light sets up the localized surface plasmon oscillation, and the induced potential is to a good approximation a dipole. The spectrum of the re-radiated light is calculated, and this has a peak whose position depends on the size, shape and composition of the nanoparticle.

A collection of nanoparticles embedded in a dielectric medium is modelled by effective medium theories such as the Maxwell–Garnett[3,4] theory where each nanoparticle is treated as a dipole, and the medium is treated as homogeneous with effective dielectric properties. This model provides qualitative agreement with experimental absorption spectra, but applications such as sensing and catalysis demand greater agreement between theoretical predictions and experimental results.

In-between the two limits is the interesting regime where one must study electrochemistry produced by nanoclusters.[5,6] Nanoparticles linked by ligands show a spectrum completely different from that when they are apart. This phenomenon is the basis of several biosensing schemes. The analytical theory of the optical properties of dimers is challenging. The coupled-dipole approximation (where each nanoparticle is modelled as a dipole and their interaction is dipole–dipole) is limited to very small nanoparticles. In practice, nanoparticles of dimension 20–50 nm have a significant quadrupolar contribution in their induced potentials. Recently, Klimov et al.[7] completed a beautiful mathematical treatment of the localized surface plasmon eigenmodes of a nanoparticle dimer, but did not connect their results to optical experiments.

2. Discrete-Dipole Approximation

One numerical method that is suitable for the study of small clusters ($N = 2 - 10$) of nanoparticles (10–30 nm) is the well-known discrete-dipole approximation (DDA). Developed by Draine and Flatau[8] for modelling atmospheric phenomena, the DDA relies on the approximation of a continuous material by a discretized cubic grid of N point dipoles. One of the limitations of the method is the faithful representation of target surfaces. This problem could be circumvented by increasing the dipole density in high-curvature surface regions, but this means giving up the use the of the fast Fourier transform algorithm, which requires equally spaced grid points.

Each dipole is uniquely described by its grid location r_i and polarizability α_i. The polarizabilities are calculated from the complex dielectric function ε_i of the material, using the Clausius–Mossotti relation:[9]

$$\frac{\varepsilon_i - 1}{\varepsilon_i + 2} = \frac{n_d \alpha_i}{3}, \tag{1}$$

where n_d is the number density of the array. The polarizabilities give a relation between the polarizations of the dipoles and the local electric field (the incident field plus the fields of all other dipoles). From the above equation, one may construct a system of $3N$ complex, linear equations from which the polarization may be extracted. After solving for the polarization, one may use it to construct the near-field and far-field optical properties of the target. We are interested in the extinction cross-section of the particles, or the sum of the absorption and scattering cross-sections.

The validity criterion for the DDA is the long-wavelength approximation: $|m|kd < 1$, where m is the complex refractive index, k is the wavenumber and d is the grid spacing. We choose the grid spacing to be small enough so as to satisfy this criterion. The DDA calculations for a coated gold *sphere* (both in air and in an aqueous medium) compared with the calculation from Mie theory[2] agree extremely well.

A second advantage of this method is that we can plot contour maps of the evanescent field around a metal nanoparticle. This gives us the ability to extract optoelectrical information at an extremely small surface. A limitation of this method is that computational resources (memory and time) place a limit on the size of the nanoparticles and/or the number of particles in a cluster.

3. Optical Properties of Nanoparticles on a Surface

In modern sensing applications, nanoparticles are immobilized on a surface so they present the maximum detection surface to the analyte. The sensing signal is the optical absorption spectrum. This configuration is well known to researchers in the surface science community as surface quantum dots or supported thin-film islands, and their optical properties have been studied for a while. Specifically, the Marton–Schlesinger[10] method and the Bedeaux–Vlieger[11] methods have provided both quantitative calculations of the optical properties of nanoparticles on a surface. One big advantage of the latter method is the effect of the substrate is naturally built into the formalism (see Fig. 1). A limitation of these methods is that the

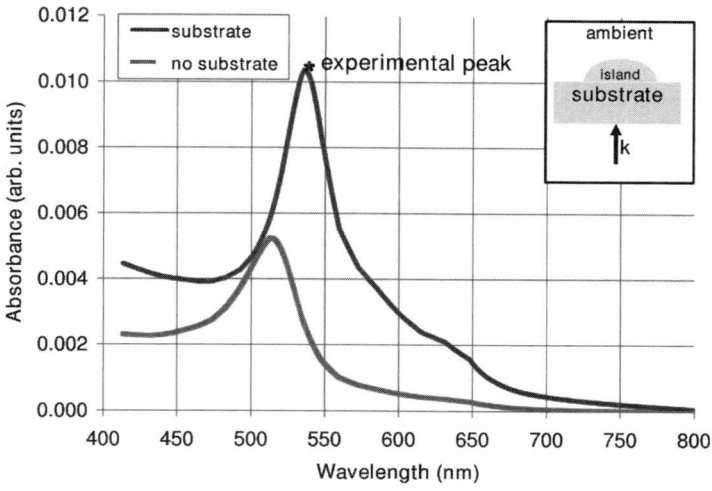

Figure 1. Calculated absorption spectrum of a hemispherical gold nanoparticle of radius 7 nm with and without a SiO_2 substrate. *Inset*: Geometry of calculation.

Optical spectrum of hemispherical gold nanoparticle dimer of
radius 7nm, and center to center separation 17.5nm

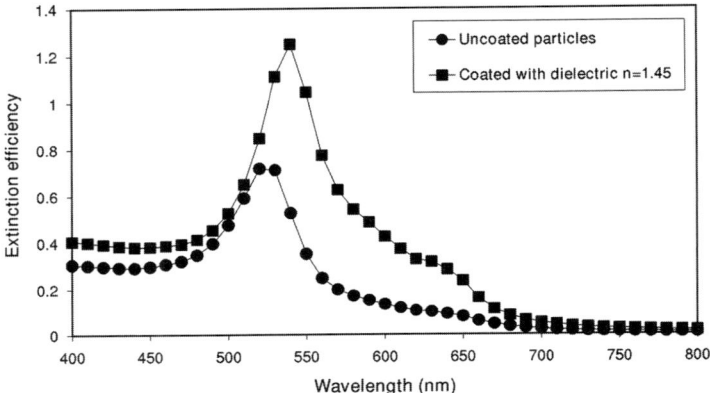

Figure 2. Calculated localized surface plasmon resonance spectrum of a dimer of hemispherical gold nanoparticles of radius 7 nm and interparticle centre-to-centre distance 17.5 nm. A dielectric coating of $n = 1.45$ and thickness 1.75 nm leads to a second absorption feature at 650 nm.

electric field itself cannot be mapped, and more complex structures (such as nanoparticles made of concentric shells of materials) cannot be modelled. Note that later modifications of this method[12] have made it possible to visualize the multipolar potential, yielding more physical insights into this problem.

Despite the limitations of the DDA method, it is best suited for applications where it is important to know the local electric field on the surface, or when the nanoparticle itself has a composite structure. For example, the second feature in the two-nanoparticle absorption spectrum shown[13] in Fig. 2 can be explained by plotting the electric field map. The map reveals that the presence of a dielectric can mediate the overlap of evanescent fields, an effect that was hitherto unknown.[14]

4. Towards an Optical Method of Surface Electrochemistry

In concluding this section, we would like to present a teaser of an idea. The holy grail of electrochemistry is to determine the

Symmetry Considerations

electrochemical potentials that determine and control chemical reactions. The more complex the molecule's geometry, the more difficult this process is. It would be exciting to develop an easy and inexpensive method for this determination. With this goal in mind, consider the following experiment. Molecule B is a long molecule that has ligands that bind to gold. Clusters of gold nanoparticles can be created in solution by linking individual nanoparticles via these linkers. Molecule A is a reducing agent that reduces a specific bond in the linker, creating monomers. During this process, the peak of the absorption spectrum changes by approximately 100 nm as seen in Fig. 3.[15] An exciting recent finding[15] is that if one uses various types of reducing agents, the rate of the spectral change depends on the size of the reducing agent. From an electrochemical viewpoint, apart from stearic hindrance, the rate of reduction would depend on the reduction potential. Thus, the rate of change of the spectrum, once it has been calibrated, can provide an optical (and inexpensive) means of determining the reduction potentials in a class of reactions!

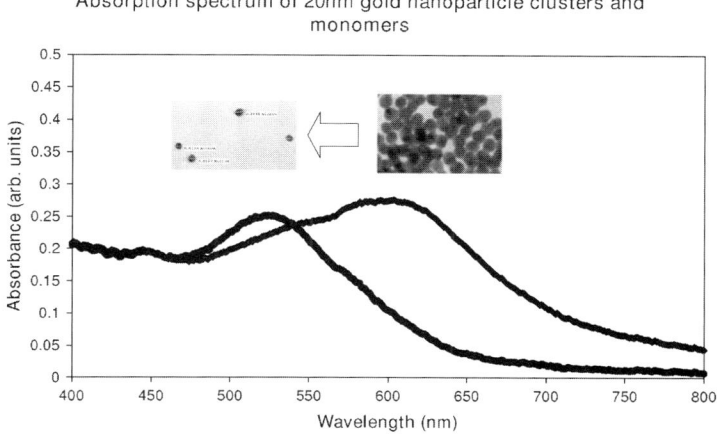

Figure 3. Measured absorption spectrum of a colloid of gold nanoparticles of radius 10 nm. When linked by molecule B, they are clustered with an absorption spectrum in the blue. When reducing agent molecule A is added, monomers are formed, and the absorption spectrum turns red.

III. OPTICAL MANIPULATION OF SEMICONDUCTOR QUANTUM DOTS

In this section, we describe the modelling of experiments in which laser light (both single-frequency and broadband) is used to probe and manipulate the quantum wavefunction of a semiconductor quantum dot. As in the case of naturally occurring atoms, quantum dots have discrete bound electronic states, and hence are referred to as 'artificial atoms'. As explained by Kastner,[16] "Modern techniques of lithography make it possible to confine electrons to sufficiently small dimensions that the quantization of both their charge and their energy are easily observable. In fact, there is a close analogy between the confined electrons inside an single electron transistor (SET) and an atom". Recent developments in optical technologies have enabled the probing of this electronic structure using single-frequency (continuous-wave laser) or broadband (pulsed laser) coherent light. The experiments involve the exact determination of the energy levels (spectroscopy) or the precise control of the electronic wavefunction (quantum control), much in the ways of atomic, molecular and optical physics or physical chemistry. The applications range from nanoelectronics to solid-state quantum optics.

1. Atomic Model of Semiconductor Quantum Dots

As described above, a semiconductor quantum dot can be modelled with good accuracy as a hydrogen-like atom. An excellent introduction to the quantum hydrogen atom is presented in chapter II of volume 43 of *Modern Aspects of Electrochemistry*, as well as in classic texts.[17]

Briefly, the electron in a hydrogen atom lies in a spherically symmetric 'Coulomb' potential due to the positively charged nucleus (in atomic units with $e = m_e = \hbar = 1$):

$$V(r) = -\frac{1}{r}. \qquad (2)$$

The Schrödinger equation that describes the stationary states of an electron in this potential is

$$-\frac{\nabla^2}{2}\psi + V(r)\psi = E\psi. \qquad (3)$$

Symmetry Considerations

Recall the well-known result that symmetries in a problem are associated with conserved quantities. Because of the spherical and reflection symmetries in the potential, it immediately follows that angular momentum and parity are conserved quantities in this system. In spherical coordinates, the above equation is separable into its radial and angular components and the radial equation corresponding to an angular momentum ℓ is

$$-\frac{1}{2r^2}\frac{d}{dr}\left(r^2\frac{d\psi_{n\ell}}{dr}\right) + V_{\text{eff}}^{\ell}(r)\psi_{n\ell}(r) = E_n\psi_{n\ell}(r), \qquad (4)$$

where the effective potential is

$$V_{\text{eff}}^{\ell}(r) = -\frac{1}{r} + \frac{\ell(\ell+1)}{2r^2}. \qquad (5)$$

The eigenvalues $E_n = -\frac{1}{2n^2}$ and eigenvectors $\Psi(\mathbf{r})$ which are the products of the radial functions $\psi_{n\ell}(r)$ and spherical harmonics $Y_{\ell m}$ have long been known. An excellent pictorial representation of the radial wavefunctions can be found in the text *Theoretical Atomic Physics*.[18]

Since the energy eigenvalues depend only on n, they are degenerate with respect to both ℓ and m. For each value of n, ℓ can vary from 0 to $n-1$, and for each value of ℓ, m can vary from $-\ell$ to $+\ell$. Degeneracies in energy are associated with conserved quantities and symmetries in a system. The degeneracy in m is characteristic of a central force field, for which the potential depends only on the radial distance. The ℓ degeneracy is characteristic of the Coulomb field, as distinguished from other central force fields.[19] This degeneracy, sometimes referred to as 'accidental degeneracy' in the literature, is associated not with a geometrical symmetry but with a dynamical symmetry – represented by the $O(4)$ group, of which the angular momentum vector and the Runge–Lenz vectors are generators.[20] In equivalent classical terms, the angular momentum and the Lenz vector are constants of motion for an electron in a Coulomb potential.

More relevant to our understanding of quantum dots are alkali-metal atoms. Alkali-metal atoms are similar to hydrogen since they have one valence electron. The behaviour of the outer electron may be understood as a single electron moving in the combined potential of the nucleus and the inner-shell electrons, i.e. the core. This combined potential is central but only approximately of the Coulomb form because of the size of the core and multielectron effects within it. This prevents the states with the same quantum number n from having the same energy (as in the Coulomb degeneracy in field-free hydrogen). In spite of this difference, the energy levels of field-free alkali metals can be calculated from a relation that is very similar to that for hydrogen,

$$E_{n\ell} = \frac{-1}{2(n - \mu_\ell)^2}, \tag{6}$$

where μ_ℓ is the quantum defect.[21] This arises owing to the core penetration of the wavefunction where the potential is not Coulombic. Recall from (5) that the higher the angular momentum number, the further out is the centrifugal potential barrier. The radial wavefunctions with smaller ℓ values penetrate the core (and feel the effects of the inner electrons), while those with $l > 2$ hardly penetrate the core at all. Thus, the quantum defect depends on the ℓ quantum number, being large for the $\ell = 0$ or s states and small for states with $\ell > 2$. Therefore for states with high n or ℓ, the approximation of the nuclear potential as Coulombic is a very good one.

To translate the physics of atoms to that of quantum dots, it is necessary to modify the mass of the electron using an effective electron mass.[22] Calculation of the quantized energy values of a quantum dot can be accomplished numerically by solving the Schrödinger equation of the hydrogen-like atom with an effective electron mass.[23]

Owing to rapid technological developments in the last two decades, quantum dots are increasingly being subjected to external fields, and often to rapidly changing external fields. It is now possible to dynamically manipulate the quantum wavefunction of a quantum dot. Indeed, such systems are being considered as candidates for quantum computing! It is useful therefore to have methods of modelling such processes.

Again, the quantum dot is modelled as a hydrogenic atom with an effective electron mass that simply scales the calculation. The potential experienced by an atomic electron in a static electric field ε_s in the z direction is

$$V(\mathbf{r}) = -\frac{1}{r} + \varepsilon_s z. \tag{7}$$

The system exhibits competing symmetries – spherical symmetry due to the Coulomb potential and axial symmetry due to the Stark potential (applied electric field). Parity and angular momentum, ℓ, are no longer conserved; only the magnetic quantum number, m, is a good quantum number in this system. However, the z-component of the Runge–Lenz vector (besides L_z) is a constant of the motion in this system.

An atom in a strong magnetic field is another example of a system which exhibits competing symmetries (spherical symmetry due to the Coulomb potential and cylindrical symmetry due to the magnetic field). The Hamiltonian for the hydrogen atom in a magnetic field in the z direction[24] (one atomic unit of magnetic field is 2.35×10^5 T) is

$$H = \frac{\mathbf{P}^2}{2} - \frac{1}{r}A(r)\mathbf{L} \cdot \mathbf{S} + \frac{\mathbf{L} \cdot \mathbf{B}}{2} + \mathbf{S} \cdot \mathbf{B} + \frac{\mathbf{B}^2}{8}(x^2 + y^2). \quad (8)$$

The $\mathbf{L} \cdot \mathbf{S}$ term is negligible except at very low values of the magnetic field; i.e. for the magnetic field strengths that we are interested in, the spin and angular momentum are decoupled and the contribution of spin may be ignored. The paramagnetic terms, which are linear in $\mathbf{B}$, add a constant energy to the Hamiltonian, yielding an overall phase factor in the time-dependent wavefunction, and may also be ignored. This Hamiltonian also conserves parity; thus, for the Coulomb–diamagnetic problem, the magnetic quantum number m as well as parity are conserved quantities. For each value of m, the unperturbed Hamiltonian is $(n - |m|)$-fold degenerate. The degeneracy of the ℓ states, $|m| \leq \ell < n$, is then lifted by the diamagnetic potential, which is quadratic in the magnetic field. The Schrödinger equation representing the diamagnetic atom is not separable in any coordinate system, in contrast to its counterpart, an atom in an electric field, which separates in parabolic coordinates.

We will look at numerical methods to compute both the energy levels of a quantum dot as well the dynamics of the quantum wavefunction in the presence of an external field. There are a variety of methods to choose from, and in this section we show that by using the symmetry properties one can greatly enhance the accuracy and efficiency of the calculation.

2. Pseudospectral Method

Here, we present a method that is highly efficient and accurate at solving the eigenproblem of the hydrogen-like atom when only a limited number of eigenstates need be known. The radial, time-independent Schrodinger equation (4) for a hydrogen atom is a second-order differential equation. The numerical method presented is a pseudospectral method (also known as 'collocation method'). First, the dependent variable of the differential equation is expanded in a basis of orthogonal functions that is truncated at some order N. In this method, the residual is set to zero at each of the N collocation points (defined for that basis). This produces a matrix eigenvalue problem whose eigenvalues are the same as that of the differential equation eigenproblem.[25]

While expanding in a basis of orthogonal functions is fairly easily understood, care must be taken in choosing an appropriate set of basis functions. In this case the symmetry of the basis functions chosen must match that of the problem, as seen below. The importance of symmetry in the problem is beautifully presented by the choice of basis for the radial coordinate. Consider two choices of the basis for the radial coordinate – a Fourier basis and a Laguerre basis. That is, the radial functions can be expanded in a basis of Fourier functions (sines and cosines) or Laguerre functions. The collocation points can be loosely thought of as the nodes of the basis functions.

In the Fourier basis, the collocation points (think of nodes of sines and cosines) are equally spaced from $-\infty$ to $+\infty$. The problem of representing a Coulomb potential on a uniform grid is well known; since the potential is steep near the origin, the wavefunctions are highly oscillatory there, and need a finely spaced grid to be represented there. This can be overcome by mapping the exact potential on a nonuniform grid to a transformed, uniform collocation grid. But another problem is the radial hydrogen wavefunctions have a domain from 0 to ∞, whereas the domains of the Fourier functions are from $-\infty$ to $+\infty$. The problem of the domain matching is overcome by expanding the radial collocation grid to negative values to $-\infty$, and choosing only those solutions that go to 0 at $r = 0$.

The Laguerre basis suits the symmetry of the problem, because its collocation points (think of nodes of the Laguerre functions) lie between 0 and ∞, and the collocation grid spacing is nonuniform. Of course, we have prior intuition that this basis is better suited because the analytic solutions of the hydrogen atom Schrödinger equation are Laguerre polynomials.

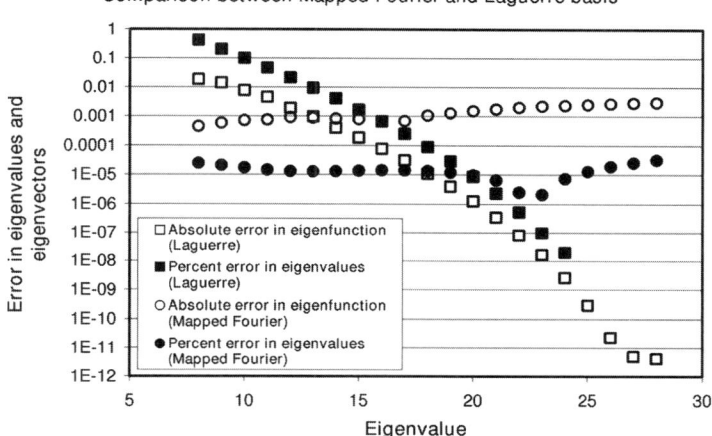

Figure 4. Comparison between the use of a Laguerre basis and a mapped-Fourier basis for the calculation of energy eigenvectors and eigenvalues of the hydrogen atom.

The effect of choosing a symmetry-suited basis is stunning in the accuracy of the calculation,[26] as shown in Fig. 4. One sees that the accuracy of the mapped-Fourier basis (although nominally exponential) reduces to polynomial because of the Coulomb singularity and the artificial method used for domain matching. On the other hand, the error is fairly constant over all the eigenmodes. The Laguerre basis calculation provides extremely accurate eigenvalues (up to machine precision). However, this accuracy is only for a range of eigenmodes.

Thus, to model experiments with spectroscopic accuracy, it might be better to use the Laguerre basis, but to model experiments that need rough estimates of many energy eigenvalues, it might be better to use the mapped-Fourier basis. The limitation of the pseudospectral method is that it is very time consuming to calculate dynamics, especially when multiple angular momenta are involved.

3. Finite-Difference Method

The modelling of atoms in external fields can be effectively accomplished using the finite-difference method. In a finite-difference method, derivatives are approximated by differences between values at infinitesimally close points. Thus, this method lends itself to representation of solutions of differential equations on a grid.[27] The power of this method lies in the ability to calculate the initial state wavefunctions with reasonable accuracy, propagate them in the presence of external fields, and use them to calculate experimentally measurable quantities both for bound states (transition probabilities) and for continuum states (ionization probabilities).

The solution of the field-free eigenproblem is found by expanding the wavefunction (solution) in a mixed basis of discretized radial functions times spherical harmonics, while retaining a finite number of spherical harmonics:

$$\Psi(r_j, \theta, \phi) = \sum_{\ell=0}^{\ell_{max}} \phi_\ell(r_j) Y_\ell^m(\theta, \phi), \qquad (9)$$

where j is an index corresponding to a radial grid point. Thus, the truncation in the number of spherical basis functions (which gives an exponentially small error) is challenged by the truncation in the radial grid extent (which gives a polynomial error). The Coulomb singularity is avoided by using a nonuniform radial grid[28] (see Fig. 5) and a softening of the potential at the origin. The resulting differential equation for $\phi_\ell(r_j)$ is then discretized using second-order approximations for the derivatives and discretization yields an eigenvalue equation of a symmetric tridiagonal matrix, which is then solved.[29]

The unperturbed eigenstates $|k_{n\ell}(r_j)\rangle$ are eigenstates of a real, symmetric, tridiagonal matrix. A diagonalization can yield N eigenstates and eigenvalues (where N is the number of grid points), of which we require only the lowest few. The complexity of this procedure is of $\mathcal{O}(N)$. The grid is chosen to yield eigenvalues with a maximum error of 0.01% by comparing them with known eigenvalues of the hydrogen atom. The radial functions are also in excellent agreement with the analytic solutions to the radial part of the Schrödinger equation for values of the principal quantum number up to $n = 35$.

The time evolution of individual eigenstates can be performed by multiplication with the appropriate phase:

$$|k(t)\rangle = e^{-iE_k t}|k(0)\rangle. \qquad (10)$$

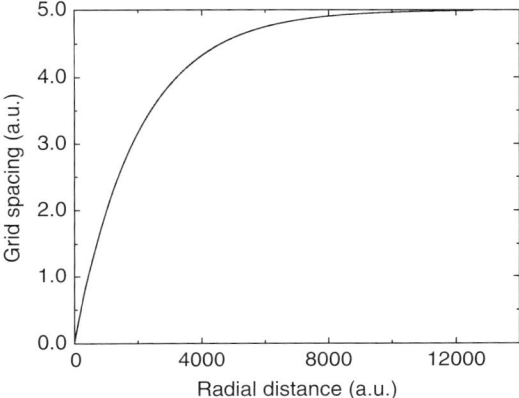

Figure 5. Grid spacing in a nonuniform radial grid used to model the hydrogen atom (spherical quantum dot). The grid points are closely spaced near the origin where the potential has a singularity, and are widely spaced away from the origin where the potential goes asymptotically to 0.

In the presence of an external time-dependent field, the time-dependent Schrodinger equation, a first-order differential equation in time, must be solved. The key problem is that the field also mixes the radial and angular coordinates, making typical implicit methods (that are unconditionally stable and accurate) very resource intensive. Therefore, the Peacemann–Rachford method is recommended.[30] In the total Hamiltonian $H = H_0 + H_1$, H_0 connects adjacent radial points of functions with the same ℓ value, whereas the interaction H_1 couples functions of different ℓ values at the same radial point.

$$\Psi(r, t + \delta t) = \sum_{n=0}^{\infty} \frac{(-i)^n}{n!} \int_0^t dt_1 \int_0^t dt_2 \ldots \int_0^2 dt_n$$
$$T[H(t_1) H(t_2) \ldots H(t_n)] \Psi(r, t), \qquad (11)$$

where T represents the time-ordering operator.
To second order, the short-time propagator is

$$\Psi(r, t + \delta t) = \exp\left[-iH\left(t + \frac{\delta t}{2}\right) \delta t\right] \Psi(r, t). \qquad (12)$$

We expand the propagator as

$$\Psi(r, t + \delta t) = \left(1 + iH_0\frac{\delta t}{2}\right)^{-1}\left(1 + iH_I\frac{\delta t}{2}\right)^{-1}$$
$$\left(1 - iH_I\frac{\delta t}{2}\right)\left(1 - iH_0\frac{\delta t}{2}\right)\Psi(r, t). \quad (13)$$

The Peacemann–Rachford propagator agrees with the full propagator up to the third order in δt. In combining this with the finite-difference method described above, the first two operations on $\Psi(r, t)$ are straightforward, and the next two require finding the solution to five-term and three-term recurrence relations in ℓ and r, respectively. The computational complexity of this operation is of first order in $N_r \times N_\ell$.

The advantage of this method is that since both positive and negative energy states are represented on the same grid, this allows us to study the dynamics of bound states as well as ionization problems. The limitation of this method is that the radial grid is closely spaced near the nucleus, with the spacing increasing to a constant value towards the outer edge. Although this makes it possible to represent field-free wavefunctions accurately, it limits the accurate representation of high-momentum processes away from the nucleus. Thus, the finite-difference method is effective in modelling experiments where the semiconductor quantum dot is probed or manipulated by coherent (laser) light.

IV. SUMMARY

We have presented a variety of methods to model light–matter interactions in nanoelectrochemistry. In particular, the experiments involved include the probing of metal quantum dots using white light (nanoplasmonics), and the manipulation of semiconductor quantum dots using laser light (semiconductor quantum optics). These experiments drive applications in myriad areas such as surface electrochemistry, biosensor development, nanolithography, nanoelectronics and quantum computing. We showed that the symmetry of the problem is an important consideration in choosing the optimal numerical method. This consideration, when carefully applied, can lead to significant cost and time savings in R&D.

ACKNOWLEDGEMENTS

I would like to gratefully acknowledge help from and discussions with Silvia Mittler, Bulent Mutus, Mordechay Schlesinger, Mohammed Hashemi, Jeffrey Rau and Daniel Trojand. Research support by the Natural Sciences and Engineering Research Council of Canada, and Canada Foundation for Innovation is gratefully appreciated. Part of the computations were done on the SharcNet supercomputing network.

REFERENCES

[1] G.A. Somorjai and J.Y. Park, Physics Today, p. 48, Oct. (2007).
[2] C.F. Bohren and D.R. Huffman, "Absorption and scattering of light by small particles", Wiley, New York, (1983).
[3] J.C. Maxwell-Garnett, Philos. Trans. R. Soc. Lond. **203**, 385 (1904).
[4] J.C. Maxwell-Garnett, Philos. Trans. R. Soc. Lond. Ser. A **205**, 237 (1906).
[5] U. Kreibig and M. Vollmer, "Optical properties of metal clusters", Springer, Berlin, (1995).
[6] A.P. Alivisatos, Science **271**, 933 (1996).
[7] V.V. Klimov and D.V. Guzatov, Phys. Rev. B **75**, 024303 (2007).
[8] B.T. Draine and P.J. Flatau, J. Opt. Soc. Am. A **11**, 1491 (1973).
[9] B.T. Draine and J.J. Goodman, Ap. J. **485**, 685 (1993).
[10] J.P. Marton and M. Schlesinger, J. Electrochem. Soc. **115**, 16 (1968).
[11] D. Bedeaux and J. Vlieger, "Optical properties of surfaces", 2nd Edn., World Scientific, Singapore, (2004).
[12] R. Lazzari, I. Simonsen, D. Bedeaux, J. Vlieger and J. Jupille, Eur. Phys. J. B **24**, 267 (2001).
[13] P. Rooney, S. Xu, A. Rezaee, T. Manifar, A. Hassanzadeh, G. Podoprygorina, V. Boehmer, C. Rangan and S. Mittler, Phys. Rev. B **77**, 235446 (2008).
[14] S.M. Hashemi Rafsanjani, C. Rangan and S. Mittler, "A novel measure of refractive index sensitivity in gold nanoparticle biosensors", (submitted for publication).
[15] S. Durocher, A. Rezaee, C. Hamm, C. Rangan, S. Mittler and B. Mutus, 1: J. Am. Chem. Soc. **131**, 2475 (2009).
[16] M.A. Kastner, Ann. Phys. **9**, 885 (2000).
[17] H.A. Bethe and E.E. Salpeter, "Quantum mechanics of one- and two-electron atoms", Springer, New York, (1957).
[18] H. Friedrich, "Theoretical atomic physics", Springer, New York, (1990).
[19] L.I. Schiff, "Quantum mechanics", McGraw-Hill, Singapore, (1968).
[20] U. Fano and A.R.P. Rau, "Symmetries in quantum physics", Academic, New York, (1996).
[21] A.R.P. Rau and M. Inokuti, Am. J. Phys. **65**, 221 (1997).
[22] P. Hawrylak, Phys. Rev. Lett. **71**, 3347 (1993).
[23] U. Woggon, "Optical properties of semiconductor quantum dots," Springer, Berlin, (1997).
[24] T.F. Gallagher, "Rydberg atoms", Cambridge University Press, Cambridge, (1994).

[25] J.P. Boyd, "Chebyshev and Fourier spectral methods", 2nd Edn., Dover, Mineola, New York, (2001).
[26] J.P. Boyd, C. Rangan and P.H. Bucksbaum, J. Comp. Phys., **188**, 56 (2003).
[27] S.E. Koonin, "Computational physics", Benjamin-Cummings, San Fransisco, USA, (1985).
[28] J.L. Krause and K.J. Schafer, J. Phys. Chem. A **103**, 10118 (1999).
[29] C. Rangan, K.J. Schafer and A.R.P. Rau, Phys. Rev. A **61**, 053410 (2000).
[30] R. Varga, "Matrix iterative analysis", Prentice-Hall, Englewood Cliffs, NJ (1963).

4

Applications of Computer Simulations and Statistical Mechanics in Surface Electrochemistry

P. A. Rikvold,[1,2] I. Abou Hamad,[1,3] T. Juwono,[1] D. T. Robb,[4,6] and M. A. Novotny[3,5]

[1]*Center for Materials Research and Technology and Department of Physics, Florida State University, Tallahassee, FL 32306-4350, USA, rikvold@scs.fsu.edu*
[2]*National High Magnetic Field Laboratory, Tallahassee, FL 32310-3706, USA*
[3]*HPC^2, Center for Computational Sciences, Mississippi State University, Mississippi State, MS 39762-5167, USA*
[4]*Department of Physics, Clarkson University, Potsdam, NY 13699, USA*
[5]*Department of Physics and Astronomy, Mississippi State University, Mississippi State, MS 39762-5167, USA*
[6]*Department of Physics, Astronomy, and Geology, Berry College, Mount Berry, GA 30149-5004, USA*

Summary. We present a brief survey of methods that utilize computer simulations and quantum and statistical mechanics in the analysis of electrochemical systems. The methods, molecular dynamics and Monte Carlo simulations and quantum-mechanical density-functional theory, are illustrated with examples from simulations of lithium-battery charging and electrochemical adsorption of bromine on single-crystal silver electrodes.

I. INTRODUCTION

The interface between a solid electrode and a liquid electrolyte is a complicated many-particle system, in which the electrode ions and electrons interact with solute ions and solvent ions or molecules through several channels of interaction, including forces due to quantum-mechanical exchange, electrostatics, hydrodynamics, and elastic deformation of the substrate. Over the last few decades, surface electrochemistry has been revolutionized by new techniques that enable atomic-scale observation and manipulation of solid–liquid interfaces,[1,2] yielding novel methods for materials analysis, synthesis, and modification. This development has been paralleled by equally revolutionary developments in computer hardware and algorithms that by now enable simulations with millions of individual particles,[3] so there is now significant overlap between system sizes that can be treated computationally and experimentally.

In this chapter, we discuss some of the methods available to study the structure and dynamics of electrode–electrolyte interfaces using computers and techniques based on quantum and statistical mechanics. These methods are illustrated by some recent applications. The rest of the chapter is organized as follows. In Sect. II, we present fully three dimensional, simulations in continuous space by molecular dynamics (MD) of ion intercalation during charging of lithium-ion batteries. In Sect. III, we discuss the simplifications that are possible by mapping a chemisorption problem onto an effective lattice-gas (LG) Hamiltonian, and in Sect. IV we demonstrate how input parameters for a statistical-mechanical LG model can be estimated from quantum-mechanical density-functional theory (DFT) calculations. Section V is devoted to a discussion of Monte Carlo (MC) simulations, both for equilibrium problems (Sect. V.1) and for dynamics (Sect. V.2). As an example of the latter, we present in Sect. VI a simulational demonstration of a method to classify surface-phase transitions in adsorbate systems, which is an extension of standard cyclic voltammetry (CV): the electrochemical first-order reversal curve (FORC) method. A concluding summary is given in Sect. VII.

II. MOLECULAR DYNAMICS SIMULATIONS OF ION INTERCALATION IN LITHIUM BATTERIES

The charging process in lithium-ion batteries is marked by the intercalation of lithium ions into the graphite anode material. Here we present MD simulations of this process and suggest a new charging method that has the potential for shorter charging times, as well as the possibility of providing higher power densities.

1. Molecular Dynamics and Model System

MD is based on solving the classical equations of motion for a system of N atoms interacting through forces derived from a potential-energy function.[4–8] From the potential energy E_P, the force on the ith atom, F_i, is calculated. Thus, the equation of motion is

$$F_i(t) = -\frac{\partial E_P}{\partial r_i} = m_i \frac{\partial v_i}{\partial t} = m_i \frac{\partial^2 r_i}{\partial t^2}, \quad (1)$$

where r_i, v_i, and m_i are the position, velocity, and mass of the ith atom, respectively. Consequently, the quality of the simulations strongly depends on the ability of the classical force field to reasonably describe the atomistic behavior.

The newly developed general AMBER force field (GAFF)[9] was used to approximate the bonded interactions of all the simulation molecules, while the simulation package Spartan (Wavefunction, Irvine, CA, USA) was used at the Hartree–Fock/6-31g* level to obtain the necessary point charges for each of the atoms. To simulate a charging field, the charge on the carbon atoms of the graphite sheets was set to $-0.0125e$ per atom. The bonded (first three terms of (2)) and nonbonded (last term) interactions in the AMBER force field are represented by the following potential-energy function:

$$E_P = \sum_{\text{bonds}} K_r (r - r_{\text{eq}})^2 + \sum_{\text{angles}} K_\theta (\theta - \theta_{\text{eq}})^2 \quad (2)$$

$$+ \sum_{\text{dihedrals}} \frac{V_n}{2} [1 + \cos(n\phi - \gamma)]$$

$$+ \sum_{i<j} \left(\frac{A_{ij}}{R_{ij}^{12}} - \frac{B_{ij}}{R_{ij}^6} + \frac{q_i q_j}{\epsilon R_{ij}} \right),$$

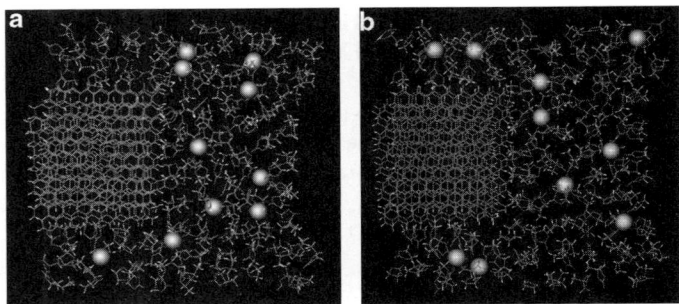

Figure 1. (a) Snapshot of the model system containing four graphite sheets, two PF_6^- ions, and ten Li^+ ions (*spheres*), solvated in 69 propylene carbonate and 87 ethylene carbonate molecules after reaching constant volume in the NPT ensemble. (b) Snapshot after 200 ns molecular dynamics simulation. The ensemble is the NVT ensemble. The system has periodic boundary conditions and is simulated at 1 atm and 300 K. *Top view*, perpendicular to the plane of the graphite sheets.

where K_r, K_θ, and V_n are the bond stretching, bending and torsional constants, respectively, the constants A and B define van der Waals interactions between unbonded atoms, and ϵ is the electrostatic permittivity. The simulation package NAMD[10] was used for the MD simulations, while the graphics package VMD[11] was used for visualization and analysis of the simulation results.

The model system representing the anode half-cell is composed of four graphite sheets (anode) containing 160 carbon atoms each, two PF_6^- ions, and ten Li^+ ions, solvated in an electrolyte made of 69 propylene carbonate and 87 ethylene carbonate molecules (see Fig. 1a). The graphite sheets were fixed from one side by keeping the positions of the edge carbon atoms fixed.

2. Simulations and Results

After energy minimization, the simulations were run at constant pressure using a Langevin piston Nosé–Hoover method[12, 13] as implemented in the NAMD software package until the system had reached its equilibrium volume at a pressure of 1 atm and 300 K in the NPT (constant particle number, pressure, and temperature) ensemble. The system's behavior was then simulated for 200 ns (100 million steps) in the NVT (constant particle number, volume, and temperature) ensemble. Two observations were made: first, the Li$^+$ ions stayed randomly distributed within the electrolyte, and second, none of the Li$^+$ ions had intercalated between the graphite sheets after 200 ns (see Fig. 1b).

While the lithium ions do not intercalate within the simulation time given above, it is expected that given enough time they will move towards the graphite sheets and be intercalated. To test whether intercalation is possible in such a model system, one of the lithium ions was positioned between the graphite sheets at the beginning of a simulation, and we observed whether it diffused out from between the sheets. The lithium ion stayed intercalated, even after 400 ns.

For intercalation to occur, the lithium ion has first to diffuse within the electrolyte until it reaches the graphite electrode. Consequently, faster diffusion would result in faster intercalation and shorter charging time. To increase the diffusion of lithium ions in the electrolyte, we explored a new charging method. In addition to the charging field due to the fixed charge on the graphite carbons, an external oscillating square-wave field (amplitude 5 kcal mol^{-1}, frequency 25 MHz) was applied in the direction perpendicular to the plane of the graphite sheets. Not only does this additional field increase diffusion, but also some of the lithium ions intercalate into the graphite sheets within an average time of about 50 ns. Figure 2

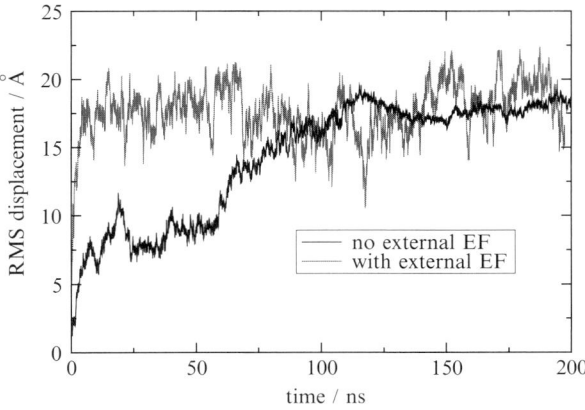

Figure 2. Root-mean-square displacement of lithium ions as a function of time. Diffusion is much faster with the additional oscillating electric field (amplitude 5 kcal mol^{-1}, frequency 25 MHz).

shows a plot of the root-mean-square displacement of lithium ions as a function of time for a system with and without an applied external field. The increased diffusion and intercalation indicate that a charging protocol involving an oscillating field may decrease the charging time and possibly increase the battery's power density.

III. LATTICE-GAS MODELS OF CHEMISORBED SYSTEMS

As mentioned in Sect. I, even the simplest electrosorption systems are extremely complicated. This complexity means that a comprehensive theoretical description that enables predictions for phenomena on macroscopic scales of time and space is still generally impossible with present-day methods and technology. (Note that MD simulations, such as those presented in Sect. II, are only possible up to times of a few hundred nanoseconds.) Therefore, it is necessary to use a variety of analytical and computational methods and to study various simplified models of the solid–liquid interface. One such class of simplified models are LG models, in which chemisorbed particles (solutes or solvents) can only be located at specific adsorption sites, commensurate with the substrate's crystal structure. This can often be a very good approximation, for instance, for halides on the (100) surface of Ag, for which it can be shown that the adsorbates spend the vast majority of their time near the fourfold hollow surface sites.[14] A LG approximation to such a continuum model, appropriate for chemisorption of small molecules or ions,[15–20] is defined by the discrete, effective grand-canonical Hamiltonian,

$$\mathcal{H}_{\mathrm{LG}} = \sum_n \left(-\Phi^{(n)} \sum_{\langle ij \rangle}^{(n)} c_i c_j \right) + \mathcal{H}_3 - \bar{\mu} \sum_i c_i. \qquad (3)$$

Here, the lattice sites i are the preferred adsorption sites (the minima of the continuous corrugation potential), and c_i is a local occupation variable, with 1 corresponding to an adsorbed particle and 0 to a solvated site. The sums $\sum_{\langle ij \rangle}^{(n)}$ and $\sum_i$ run over all nth-neighbor pairs and over all adsorption sites, respectively, $\Phi^{(n)}$ is the effective nth-neighbor pair interaction, and $\sum_n$ runs over the interaction ranges. The term $\mathcal{H}_3$ contains multiparticle interactions.[21–23] The sign convention is such that $\phi < 0$ implies repulsion, and $\bar{\mu} > 0$ favors adsorption. Equation (3) is also easily generalized to multiple species.[24,25]

To connect the electrochemical potentials to the concentrations in bulk solution of species X, [X], and the electrode potential, E, one has (in the dilute-solution approximation)

$$\bar{\mu}_X(T, [X], E) = \bar{\mu}_X^0 + k_B T \ln([X]/[X]^0) - e \int_{E^0}^{E} \gamma_X(E') dE', \quad (4)$$

where k_B is Boltzmann's constant, T the temperature, e the elementary charge, and $\gamma_X(E)$ the electrosorption valency[26–29] of X. The importance of the integral over the potential-dependent electrosorption valency [rather than just the product $e\gamma_X(E)E$ analogous to the case of potential-independent γ_X] was pointed out in Ref. [30]. The quantities superscripted 0 are reference values that include local binding energies. The interaction constants and electrosorption valencies are *effective* parameters influenced by several physical effects, including electronic structure,[21–23] surface deformation, (screened) electrostatic interactions,[31–33] and the fluid electrolyte.[34,35] The density conjugate to $\bar{\mu}_X$ is the coverage relative to the number N of adsorption sites,

$$\Theta_X = N^{-1} \sum_i c_i. \quad (5)$$

IV. CALCULATION OF LATTICE-GAS PARAMETERS BY DENSITY FUNCTIONAL THEORY

There are many methods to estimate LG parameters. One of these is comparison of MC simulations (see Sect. V) of a LG model with experimental adsorption isotherms. For detailed descriptions of this method we refer to Refs. [30,31,36–39]. Here we instead concentrate on the purely theoretical method based on quantum-mechanical DFT calculations.[23]

DFT is the most widely used method to calculate ground-state properties of many-electron systems. It is based on the Hohenberg–Kohn theorem, which states that all properties of the many-particle ground state can be expressed in terms of the ground-state electron charge-density distribution[40] and leads to the Kohn–Sham equations for single-particle wave functions.[41] These are second-order differential equations, which include potential terms due to the ions and the classical Coulomb repulsive energy between the electrons, as well as the electronic exchange–correlation energy, and they are solved self-consistently. For surface structural studies, DFT is usually performed using pseudopotentials with slab models and plane-wave basis sets. The slab consists of a finite number of atomic layers, periodic in the direction parallel to the surface, which can be repeated periodically in the third direction (separated by a vacuum interval), or not. The fluid solvent can be considered either as an effective continuum or by molecular models.

Here we present preliminary results on a DFT calculation of lateral interaction constants pertaining to a LG model for the adsorption of Br on single-crystal Ag(100) surfaces.[29,37–39,42] The LG model is represented by (3) on a square lattice with lattice constant $a = 2.95$ Å, $\mathcal{H}_3 = 0$, infinitely repulsive interactions for adparticles at nearest-neighbor sites, and the long-range repulsion

$$\phi_{ij} = \frac{(\sqrt{2})^3}{r_{ij}^3} \phi_{nnn} \quad \text{for} \quad r_{ij} \geq \sqrt{2} \,, \tag{6}$$

which is compatible with dipole–dipole interactions or elastically mediated interactions. (Here, r_{ij} is given in units of a.) Since the DFT calculations are performed in the canonical ensemble (fixed adsorbate coverage), $\bar{\mu}$ in (3) is replaced by the binding energy of a single adparticle, E_b.

We prepared slabs with seven metal layers, which were placed inside a supercell with periodic boundary conditions. Two different sizes of supercells were used: a 2×2 supercell with size of $2a \times 2a \times 36.95$ Å, and a 3×3 supercell with size of $3a \times 3a \times 36.95$ Å. The vacuum region above the surface was twice the thickness of the slab, and the orientation of the surface normal was in the z direction. One, two, and three Br atoms were placed on the 3×3 surface to represent coverages $\Theta = 1/9, 2/9$, and $1/3$. Two Br atoms were placed on the 2×2 surface to represent $\Theta = 1/2$, and one Br atom was placed on the 2×2 surface to represent $\Theta = 1/4$. Supercells with different coverages of Br are shown in Fig. 3.

The DFT calculations were performed using the Vienna ab initio simulation package (VASP).[43–45] The basis set was plane-wave,

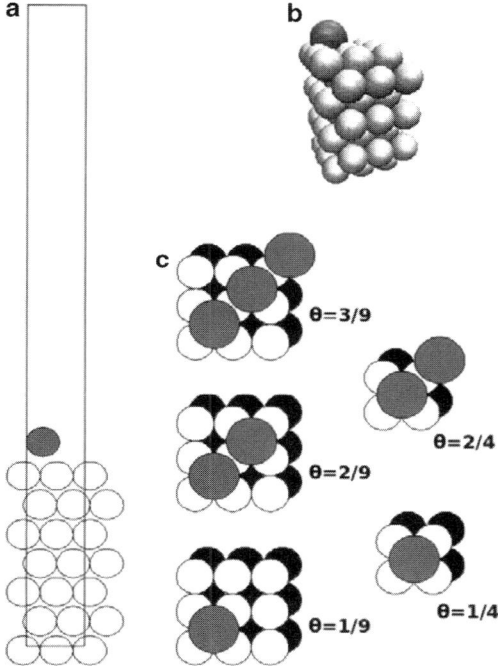

Figure 3. (a) Cross section of a 3 × 3 supercell with $\Theta = 1/9$. (b) Three-dimensional representation of the same cell and coverage. (c) Top view of a 3 × 3 surface and a 2 × 2 surface with various coverages.

with the generalized gradient-corrected exchange–correlation function,[46,47] and Vanderbilt pseudopotentials.[48] The k-point mesh was generated using the Monkhorst method[49] with a 5 × 5 × 1 grid for the 3 × 3 cells and a 7 × 7 × 1 grid for the 2 × 2 cells. All calculations were done on a 54 × 54 × 192 real-space grid.

Individual DFT calculations provide total energies, E, and charge densities, $\rho(x)$. The adsorption energy E_{ads} for a single adatom and the corresponding charge-transfer function $\Delta\rho(x)$ are obtained from calculations of the adsorbed system and isolated slab and atoms as follows:

$$E_{ads} = \left[E_{syst} - E_{slab}\right]/N_{ads} - E_{Br} \tag{7}$$

and[50]

$$\Delta\rho(x) = [\rho(x)_{\text{syst}} - \rho(x)_{\text{slab}}]/N_{\text{ads}} - \rho(x)_{\text{Br}} , \qquad (8)$$

where $N_{\text{ads}} = N\Theta$ is the number of adsorbed Br atoms in the cell, and the quantities subscripted Br refer to a single, isolated Br atom.

Since the system is electrically neutral, the integral over space of $\Delta\rho(x)$ vanishes. The surface dipole moment is defined as

$$p = \int z\Delta\rho(z)\mathrm{d}z . \qquad (9)$$

Kohn and Lau[51] have shown that the nonoscillatory part of the dipole–dipole interaction energy between adsorbates separated by a distance R behaves as

$$\phi_{\text{dip--dip}} = \frac{2p_a p_b}{4\pi\epsilon_0 R^3} \qquad (10)$$

for large R (in our case larger than the nearest-neighbor distance). This result is twice what one might naïvely expect. Thus, the next-nearest-neighbor interaction constant from (6) would be

$$\phi_{\text{dip--dip}_{\text{nnn}}} = \frac{2p^2}{4\pi\epsilon_0 R_{\text{nnn}}^3}, \qquad (11)$$

with p obtained from the DFT by (9). This estimate, which depends on Θ, is included in Fig. 4 as solid circles.

Alternatively, the interaction constant ϕ_{nnn} in the LG Hamiltonian, (3), can be estimated by performing a nonlinear least-squares fit of the Θ-dependent DFT adsorption energy E_{ads} in (7) to

$$E_{\text{ads}} = -\phi_{\text{nnn}}\Sigma_\Theta - E_{\text{b}}\Theta, \qquad (12)$$

with $\phi_{\text{nnn}} = A(1 + B\Theta)^2$, using the three fitting parameters A, B, and E_{b}. This is consistent with the theoretical prediction of (11) with a dipole moment that depends linearly on Θ. The quantity

$$\Sigma_\Theta = \frac{(\sqrt{2})^3}{N} \sum_{i<j} \frac{c_i c_j}{r_{ij}^3} \qquad (13)$$

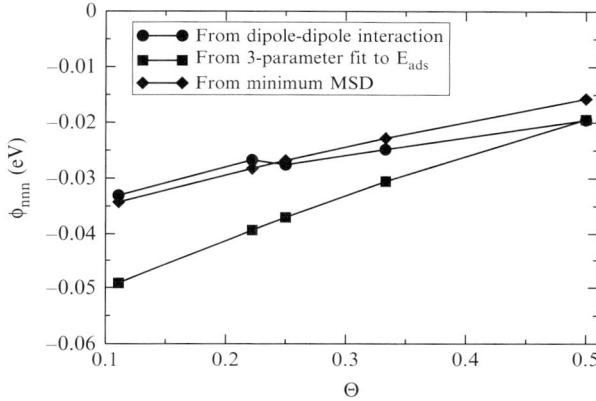

Figure 4. Three different estimates of the lattice-gas interaction constant ϕ_{nnn}. *Circles* based on (10) with the dipole moment p directly obtained from the density functional theory (DFT) calculation. *Squares* based on a three-parameter fit to the DFT adsorption energy E_{ads} as described in (12). *Diamonds* based on minimizing mean-square deviations (MSD) from the estimate based on the DFT dipole moment p, constrained to retain a low value of χ^2 from the fit to E_{ads}. See the discussion in the text.

can be calculated numerically to any given accuracy for a particular coverage and adsorbate configuration. This estimate for ϕ_{nnn} is included in Fig. 4 as solid squares. It does not agree particularly closely with the result obtained from the dipole moments. However, we found that χ^2 of the fit, considered as a function of the fitting parameters, was characterized by an extremely wide and shallow basin surrounding its minimum. We therefore further minimized the mean-square deviation (MSD) between the values of ϕ_{nnn} obtained from this fitting procedure and those obtained directly from (11) with the DFT values for *p within the three-dimensional parameter region for which the original χ^2 was close to its minimum*. This procedure gave significantly improved consistency between the two estimates for ϕ_{nnn}, without a significant increase in χ^2. The final result is shown as solid diamonds in Fig. 4, and the corresponding parameters are listed in Table 1.

The average value of ϕ_{nnn} obtained by this method is consistent with that found by fitting equilibrium MC simulations (see Sect. V.1) to experimental adsorption isotherms in aqueous solution

Table 1.
Results for the fits of the Θ-dependent lattice-gas interaction constant ϕ_{nnn} according to the two methods described in the text. Here, ν is the number of degrees of freedom (number of data points minus number of parameters, here equal to 2) for the initial nonlinear least-squares fit of ϕ_{nnn} to the density functional theory (DFT) adsorption energy E_{ads}, while *MSD* is the mean-square deviation between this estimate and the estimate obtained directly from the DFT dipole moment. Minimizing MSD within the basin of low χ^2 significantly reduces the MSD (see the greatly improved agreement in Fig. 4) without significantly increasing χ^2.

Method	A	B	E_b	χ^2/ν	MSD/ν
Min. χ^2	-6.017×10^{-2}	-0.8632	3.102	2.362×10^{-5}	1.803×10^{-4}
Min. MSD	-4.085×10^{-2}	-0.7595	3.070	2.675×10^{-5}	7.692×10^{-6}

(approximately -21 meV). However, no significant coverage dependence was found in the analysis of the experimental data.[30,39] It is not surprising that results from in situ experiments and in vacuo DFT calculations should show some differences, and we find it encouraging that the average results are consistent. Application of the method described here to Cl/Ag(100) gave less consistent results than for Br, possibly indicating that the effective interactions for Cl are not purely dipole–dipole in nature.[52]

V. MONTE CARLO SIMULATIONS

1. Equilibrium Monte Carlo

As a method to obtain equilibrium properties of a system described by a particular Hamiltonian, MC simulation is more accurate than mean-field approximations, especially for low-dimensional systems near phase transitions.[36,53] This is an effect of fluctuations, which, while ignored or underestimated by mean-field methods, are very important in two-dimensional systems. Given the rapid evolution of computers and the relative ease of programming of MC codes, this is our method of choice for equilibrium and dynamic studies of both LG and continuum models.

The goal of an equilibrium MC code is to bring the system to equilibrium as rapidly as possible, and then sample the equilibrium distribution as efficiently as possible. The only requirement is that the transition rates between two configurations c and c' satisfy *detailed balance*,

$$\mathcal{R}(c' \to c)/\mathcal{R}(c \to c') = \exp\left[-\left(\mathcal{H}(c) - \mathcal{H}(c')\right)/k_B T\right]. \quad (14)$$

This result applies to both continuum and discrete systems, and $\mathcal{H}$ may be a classical potential of predetermined form, or the interaction energies can be calculated "on the fly" by DFT.[54] The sampling can be accomplished with a number of different choices of the transition rates $\mathcal{R}(c' \to c)$,[36,53,55–62] including Metropolis, Glauber, and heatbath algorithms. It is important to note that the stochastic sequence of configurations generated by an equilibrium MC algorithm does *not* generally correspond to the actual dynamics of the system.

2. Kinetic Monte Carlo

To construct a MC algorithm producing a stochastic path through configuration space that is a good approximation to the actual time evolution of the system (in a coarse-grained sense), one can introduce *transition states* between the LG states. Only then can "MC time," measured in MC steps per site in a LG simulation, be considered proportional to "physical time," measured in seconds.[42] In a Butler–Volmer approximation,[26,36] the free energy of the transition state between LG configurations c and c' is given by

$$\mathcal{H}^*(c, c') = \Delta + (1 - \alpha)\mathcal{H}_{LG}(c) + \alpha \mathcal{H}_{LG}(c'), \quad (15)$$

where the symmetry constant $\alpha = 1/2$ for diffusion but may be different for adsorption/desorption.[36] The "bare" barrier Δ must be determined by other methods. These may be ab initio calculations,[35,63–66] MD simulations of the diffusion process on a short time scale as in Sect. II,[4–8] or comparison of dynamic simulations with experiments.[42] The most common choice of transition rate for kinetic MC simulation in chemical applications is the one-step algorithm,[67,68]

$$\mathcal{R}(c \to c') = \nu_0 \exp\left[-\left(\mathcal{H}^*(c, c') - \mathcal{H}_{LG}(c)\right)/k_B T\right], \quad (16)$$

where ν_0 is an attempt frequency [often on the order of a phonon frequency (10^9–10^{13} Hz), but see Ref. [42] for exceptions] that must be determined by other means. As we have shown previously,[55-62] to obtain reliable structural information from a kinetic MC simulation, the transition rates must approximate the real physical dynamics, which includes using transition states with proper energies. While the need for correct transition rates may seem obvious, it is regrettably often ignored in the literature. The most difficult barrier to estimate is that for adsorption/desorption, which requires reorganization of the adparticle's hydration shell.

Since the transition rates used in kinetic MC simulations of activated processes are typically small, simulations that extend to macroscopic times must use a *rejection-free* algorithm, such as the n-fold way[69,70] or one of its generalizations.[67,71-77] These algorithms simulate the *same* Markov process as the "naive" MC approach of proposing and then accepting or rejecting individual moves. Although they require more bookkeeping (see the appendix of Ref. [71] for an example), they avoid the large waste of computer time resulting from rejected moves.

VI. ELECTROCHEMICAL FIRST-ORDER REVERSAL CURVE SIMULATIONS

The FORC method was originally developed to enhance the amount of dynamic information extracted from magnetic hysteresis experiments.[78-81] We recently proposed that the method can be further developed as an extension of traditional CV to study the dynamics of phase transitions in electrochemical adsorption.[82,83]

This electrochemical FORC method consists in saturating the adsorbate coverage Θ in a strong positive electrochemical potential $\bar{\mu}$ and, in each case starting from saturation, decreasing $\bar{\mu}$ at a constant rate to a series of progressively more negative "reversal potentials" $\bar{\mu}_r$ (see Fig. 5a). Subsequently, $\bar{\mu}$ is increased back to the saturating $\bar{\mu}$ at the same rate. (Saturation at negative potentials with reversal potentials in the positive range is also possible.) The method is thus a simple generalization of the standard CV method, in which the negative return potential is decreased for each cycle. This produces a family of FORCs, $\Theta(\bar{\mu}_r, \bar{\mu}_i)$, where $\bar{\mu}_i$ is the instantaneous potential during the increase back toward saturation.

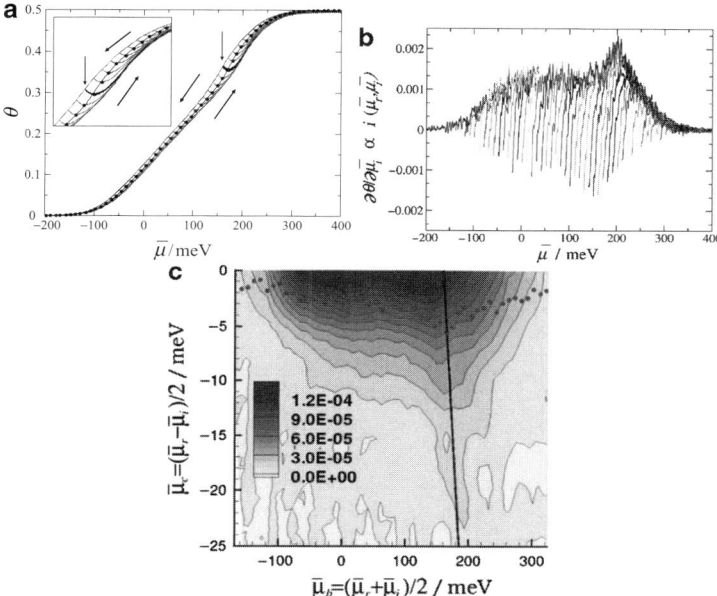

Figure 5. (a) Family of first-order reversal curves (FORCs) for our model of Br/Ag(100), corresponding to potential sweeps back and forth across the continuous phase transition between the disordered and $c(2 \times 2)$ phases. The *bold arrows* show the directions of the potential sweeps, and the *vertical arrow* indicates $\bar{\mu}_r$ for one of the FORCs. The *bold curve* is the FORC whose minimum lies closest to the critical coverage (shown in more detail in the *inset*). The *thin curve* in the middle is the equilibrium isotherm. (b) Voltammetric currents corresponding to the FORCs in (a). (c) Contour plot of the FORC distribution ρ, corresponding to the FORCs in (a). The *jagged curve of dots* in the *upper part* of the diagram corresponds to the minima of the positive-going curves in (a). The area above the curve corresponds to desorption, and the area below it to adsorption. The *slanted, straight line* corresponds to the *bold curve* in (a). After Ref. [82].

In CV experiments, one actually records the corresponding family of voltammetric currents,

$$i(\bar{\mu}_r, \bar{\mu}_i) = -\gamma e \frac{d\bar{\mu}_i}{dt} \frac{\partial \Theta(\bar{\mu}_r, \bar{\mu}_i)}{\partial \bar{\mu}_i} , \quad (17)$$

where γ is the electrosorption valency and e is the elementary charge (see Fig. 5b).

The next step in extracting dynamical information from the FORCs or the corresponding currents is to calculate the *FORC distribution*,

$$\rho = -\frac{1}{2}\frac{\partial^2 \Theta}{\partial \bar{\mu}_r \partial \bar{\mu}_i} = \frac{1}{2\gamma e(\mathrm{d}\bar{\mu}_i/\mathrm{d}t)}\frac{\partial i(\bar{\mu}_r, \bar{\mu}_i)}{\partial \bar{\mu}_r}. \tag{18}$$

This is shown in Fig. 5c as a contour plot commonly known as a *FORC diagram* in terms of the more convenient variables $\bar{\mu}_b = (\bar{\mu}_r + \bar{\mu}_i)/2$ and $\bar{\mu}_c = (\bar{\mu}_r - \bar{\mu}_i)/2$.[79,82] Geometrically, ρ is proportional to the vertical distance between adjacent current traces.

To our knowledge, the data for our model of Br/Ag(100),[29,37–39,42] which are shown in Fig. 5, are the first FORC predictions for a continuous phase transition. The data in all three panels are significantly different from the corresponding data for a discontinuous transition, such as seen in underpotential deposition. In particular, the FORC distribution for a discontinuous transition contains a negative region, while this does not appear for continuous transitions. (See details in Refs. [82, 83].) Closely related to this negative region is an extremum of the current density during the return scan.[84] Electrochemical FORC analysis should be a useful and valuable method to distinguish between continuous and discontinuous phase transitions in experiments.

VII. CONCLUSION

In this chapter we have presented some applications of the statistical-mechanics-based computer-simulation methods of MD and equilibrium and kinetic MC simulations complemented by quantum-mechanical DFT calculations of interaction energies. These include both highly technologically oriented applications to lithium-battery technology, and basic-science investigations into adsorption on single-crystal electrodes. Our hope is that these examples and the list of references will encourage other workers in surface electrochemistry to take advantage of the recent spectacular advances in computational power and algorithmic sophistication to study ever more detailed and accurate models of processes at solid–liquid interfaces.

ACKNOWLEDGEMENTS

This work was supported by US National Science Foundation grants no. DMR-0240078 and DMR-0802288 (Florida State University) and DMR-0509104 (Clarkson University) and by ABSL Power Solutions, award no. W15P7T06CP408.

REFERENCES

[1] D. M. Kolb, Surf. Sci. **500**, 722 (2002).
[2] T. Tansel and O. M. Magnussen, Phys. Rev. Lett. **96**, 026101 (2006).
[3] P. Vashishta, R. K. Kalia, and A. Nakano, J. Phys. Chem. B **110**, 3727 (2006).
[4] M. P. Allen and D. J. Tildesley, *Computer Simulation of Liquids* (Clarendon Press, Oxford, 1992).
[5] J. M. Haile, *Molecular Dynamics Simulation, Elementary Methods* (Wiley, New York, 1992).
[6] D. C. Rapaport, *The Art of Molecular Dynamics Simulation, 2nd ed.* (Cambridge University Press, Cambridge, 2004).
[7] A. F. Voter, Phys. Rev. Lett. **78**, 3908 (1997).
[8] A. F. Voter, J. Chem. Phys. **106**, 4665 (1997).
[9] J. Wang, R. Wolf, J. Caldwell, P. Kollman, and D. Case, J. Comp. Chem. **25**, 1157 (2004).
[10] J. C. Phillips, R. Braun, W. Wang, J. Gumbart, E. Tajkhorshid, E. Villa, C. Chipot, R. D. Skeel, L. Kalé, and K. Schulten, J. Comp. Chem. **26**, 1781 (2005).
[11] A. D. W. Humphrey and K. Schulten, J. Mol. Graphics **14**, 33 (1996).
[12] G. J. Martyna, D. J. Tobias, and M. L. Klein, J. Chem. Phys. **101**, 4177 (1994).
[13] S. E. Feller, Y. Zhang, R. W. Pastor, and B. R. Brooks, J. Chem. Phys. **103**, 4613 (1995).
[14] S. J. Mitchell, S. Wang, and P. A. Rikvold, Faraday Discuss. **121**, 53 (2002).
[15] D. Huckaby and L. Blum, J. Chem. Phys. **92**, 2646 (1990).
[16] L. Blum and D. A. Huckaby, J. Electroanal. Chem. **375**, 69 (1994).
[17] L. Blum, D. A. Huckaby, and M. Legault, Electrochim. Acta **41**, 2207 (1996).
[18] M. Gamboa-Aldeco, P. Mrozek, C. K. Rhee, A. Wieckowski, P. A. Rikvold, and Q. Wang, Surf. Sci. Lett. **297**, L135 (1993).
[19] P. A. Rikvold, M. Gamboa-Aldeco, J. Zhang, M. Han, Q. Wang, H. L. Richards, and A. Wieckowski, Surf. Sci. **335**, 389 (1995).
[20] J. Zhang, Y.-S. Sung, P. A. Rikvold, and A. Wieckowski, J. Chem. Phys. **104**, 5699 (1996).
[21] T. L. Einstein, Langmuir **7**, 2520 (1991).
[22] P. Hyldgaard and T. L. Einstein, J. Cryst. Growth **275**, e1637 (2005).
[23] T. J. Stasevich, T. L. Einstein, and S. Stolbov, Phys. Rev. B **73**, 115426 (2006).
[24] P. A. Rikvold, J. B. Collins, G. D. Hansen, and J. D. Gunton, Surf. Sci. **203**, 500 (1988).
[25] J. B. Collins, P. Sacramento, P. A. Rikvold, and J. D. Gunton, Surf. Sci. **221**, 277 (1989).
[26] W. Schmickler, *Interfacial Electrochemistry* (Oxford University Press, New York, 1996).
[27] K. J. Vetter and J. W. Schultze, Ber. Bunsenges. Phys. Chem. **76**, 920 (1972).

[28] K. J. Vetter and J. W. Schultze, Ber. Bunsenges. Phys. Chem. **76**, 927 (1972).
[29] P. A. Rikvold, Th. Wandlowski, I. Abou Hamad, S. J. Mitchell, and G. Brown, Electrochim. Acta **52**, 1932 (2007).
[30] I. Abou Hamad, S. J. Mitchell, Th. Wandlowski, P. A. Rikvold, and G. Brown, Electrochim. Acta **50**, 5518 (2005).
[31] M. T. M. Koper, J. Electroanal. Chem. **450**, 189 (1998).
[32] J. N. Glosli and M. R. Philpott, in *Microscopic Models of Electrode-Electrolyte Interfaces; Electrochem. Soc. Conf. Proc. Ser.* **93-5**, edited by J. W. Halley and L. Blum (The Electrochemical Society, Pennington, 1993), pp. 80–89.
[33] J. N. Glosli and M. R. Philpott, in *Microscopic Models of Electrode-Electrolyte Interfaces; Electrochem. Soc. Conf. Proc. Ser.* **93-5**, edited by J. W. Halley and L. Blum (The Electrochemical Society, Pennington, 1993), pp. 90–105.
[34] L. Blum, Adv. Chem. Phys. **78**, 171 (1990).
[35] A. Ignaczak, J. A. N. F. Gomes, and S. Romanowski, J. Electroanal. Chem. **450**, 175 (1998).
[36] G. Brown, P. A. Rikvold, S. J. Mitchell, and M. A. Novotny, in *Interfacial Electrochemistry: Theory, Experiment, and Application*, edited by A. Wieckowski (Marcel Dekker, New York, 1999), pp. 47–61.
[37] S. J. Mitchell, G. Brown, and P. A. Rikvold, J. Electroanal. Chem. **493**, 68 (2000).
[38] S. J. Mitchell, G. Brown, and P. A. Rikvold, Surf. Sci. **471**, 125 (2001).
[39] I. Abou Hamad, Th. Wandlowski, G. Brown, and P. A. Rikvold, J. Electroanal. Chem. **554–555**, 211 (2003).
[40] P. Hohenberg and W. Kohn, Phys. Rev. **136**, B864 (1964).
[41] W. Kohn and L. J. Sham, Phys. Rev. **140**, A1133 (1965).
[42] I. Abou Hamad, P. A. Rikvold, and G. Brown, Surf. Sci. **572**, L355 (2004).
[43] G. Kresse and J. Hafner, Phys. Rev. B **47**, 558 (1993).
[44] G. Kresse and J. Furthmüller, Phys. Rev. B **54**, 11169 (1996).
[45] G. Kresse and J. Furthmüller, Comput. Mater. Sci. **6**, 15 (1996).
[46] J. P. Perdew and Y. Wang, Phys. Rev. B **45**, 13244 (1992).
[47] J. P. Perdew, J. A. Chevary, S. A. Vosko, K. A. Jackson, M. R. Pederson, D. J. Singh, and C. Fiolhais, Phys. Rev. B **46**, 6671 (1992).
[48] D. Vanderbilt, Phys. Rev. B **41**, 7892 (1990).
[49] H. J. Monkhorst and J. D. Pack, Phys. Rev. B **13**, 5188 (1976).
[50] S. J. Mitchell and M. T. M. Koper, Surf. Sci. **563**, 169 (2004).
[51] W. Kohn and K.-H. Lau, Solid State Commun. **18**, 553 (1976).
[52] T. Juwono and P. A. Rikvold, unpublished.
[53] D. P. Landau and K. Binder, *A Guide to Monte Carlo Simulations in Statistical Physics, 2nd Ed.* (Cambridge University Press, Cambridge, 2005).
[54] S. Wang, S. J. Mitchell, and P. A. Rikvold, Comp. Mater. Sci. **29**, 145 (2004).
[55] P. A. Rikvold and M. Kolesik, J. Phys. A **35**, L117 (2002).
[56] P. A. Rikvold and M. Kolesik, Phys. Rev. E **66**, 066116 (2002).
[57] P. A. Rikvold and M. Kolesik, Phys. Rev. E **67**, 066113 (2003).
[58] K. Park, P. A. Rikvold, G. M. Buendía, and M. A. Novotny, Phys. Rev. Lett. **92**, 015701 (2004).
[59] G. M. Buendía, P. A. Rikvold, K. Park, and M. A. Novotny, J. Chem. Phys. **121**, 4193 (2004).
[60] G. M. Buendía, P. A. Rikvold, and M. Kolesik, Phys. Rev. B **73**, 045437 (2006).
[61] G. M. Buendía, P. A. Rikvold, and M. Kolesik, J. Mol. Struct.: THEOCHEM **769**, 207 (2006).
[62] G. M. Buendía, P. A. Rikvold, M. Kolesik, K. Park, and M. A. Novotny, Phys. Rev. B **76**, 045422 (2007).

[63] S. Wang, Y. Cao, and P. A. Rikvold, Phys. Rev. B **70**, 205410 (2004).
[64] A. Ignaczak and J. A. N. F. Gomes, J. Electroanal. Chem. **420**, 71 (1997).
[65] S. Wang and P. A. Rikvold, Phys. Rev. B **65**, 155406 (2002).
[66] A. Bogicevic, S. Ovesson, P. Hyldgaard, B. I. Lundquist, H. Brune, and D. R. Jennison, Phys. Rev. Lett. **85**, 1910 (2000).
[67] H. C. Kang and W. H. Weinberg, J. Chem. Phys. **90**, 2824 (1989).
[68] K. A. Fichthorn and W. H. Weinberg, J. Chem. Phys. **95**, 1090 (1991).
[69] A. B. Bortz, M. H. Kalos, and J. L. Lebowitz, J. Comput. Phys. **17**, 10 (1975).
[70] G. H. Gilmer, J. Cryst. Growth **35**, 15 (1976).
[71] S. Frank and P. A. Rikvold, Surf. Sci. **600**, 2470 (2006).
[72] R. J. Gelten, A. P. J. Jansen, R. A. van Santen, J. J. Lukkien, J. P. L. Segers, and P. A. J. Hilbers, J. Chem. Phys. **108**, 5921 (1998).
[73] J. J. Lukkien, J. P. L. Segers, P. A. J. Hilbers, R. J. Gelten, and A. P. J. Jansen, Phys. Rev. E **58**, 2598 (1998).
[74] M. T. M. Koper, A. P. J. Jansen, R. A. van Santen, J. J. Lukkien, and P. A. J. Hilbers, J. Chem. Phys. **109**, 6051 (1998).
[75] F. Nieto, C. Uebing, V. Pereyra, and R. J. Faccio, Vacuum **54**, 119 (1999).
[76] M. A. Novotny, in *Annual Reviews of Computational Physics IX*, edited by D. Stauffer (World Scientific, Singapore, 2001), pp. 153–210.
[77] T. Ala-Nissila, R. Ferrando, and S. C. Ying, Adv. Phys. **51**, 949 (2002).
[78] I. D. Mayergoyz, IEEE Trans. Magn. **22**, 603 (1986).
[79] C. R. Pike, A. P. Roberts, and K. L. Verosub, J. Appl. Phys. **85**, 6660 (1999).
[80] C. R. Pike, Phys. Rev. B **68**, 104424 (2003).
[81] D. T. Robb, M. A. Novotny, and P. A. Rikvold, J. Appl. Phys. **97**, 10E510 (2005).
[82] I. Abou Hamad, D. T. Robb, and P. A. Rikvold, J. Electroanal. Chem. **607**, 61 (2007).
[83] I. Abou Hamad, D. T. Robb, M. A. Novotny, and P. A. Rikvold, ECS Trans. **6** (19), 53 (2008).
[84] S. Fletcher, C. S. Halliday, D. Gates, M. Westcott, T. Lwin, and G. Nelson, J. Electroanal. Chem. **159**, 267 (1983).

5

AC-Electrogravimetry Investigation in Electroactive Thin Films

Claude Gabrielli and Hubert Perrot

UPMC Université Paris 06, UPR 15, LISE, 4 place Jussieu, 75252 Paris Cedex 05, France, hubert.perrot@upmc.fr
CNRS, UPR 15, LISE, 4 place Jussieu, 75252 Paris Cedex 05, France

I. INTRODUCTION

Mixed, i.e. electronic and ionic, conducting materials have been known for a rather long time although their full potential was not always understood. As an example, a conducting polymer such as polyaniline is the oldest conjugated known polymer; it was first reported in 1862 by Letheby.[1] In 1910, polyaniline was described as an octamer existing in four different states[2] and it was later analysed for its electrical properties by Jozefowicz.[3] Later, many works were devoted to conducting polymers and more generally to electroactive materials especially in the form of thin films deposited on a metallic electrode.

A number of useful reviews have been published on the properties of the electroactive materials, especially on conducting polymers. The reviews of Murray[4,5] provide a summary of the early works in this area. The reviews by Albery and Hillman,[6] Hillman,[7] Abruna,[8] Evans,[9] Smyrl and Lien[10] and Lyons[11,12] are also very

M. Schlesinger (ed.), *Modelling and Numerical Simulations II*,
Modern Aspects of Electrochemistry 44, DOI 10.1007/978-0-387-49586-6_5,
© Springer Science+Business Media LLC 2009

useful. More recently, reviews have been provided by Oyama and Ohsaka,[13] Murray,[14] Andrieux and Savéant,[15] Inzelt[16] and Doblhofer,[17] and in the book edited by Lyons.[18] To our knowledge, since the mid 1990s no thorough review has been published. So, this text will mainly focus on the results reported in this field during these last 10 years.

Electroactive materials are often used as thin films coating a metallic electrode. Charge transport in these films and charge transfer reactions at metal/film and film/electrolyte interfaces play key roles in the reduction and oxidation of these electroactive films. As an example, it is commonly accepted that the oxidation process of such a film requires either cation expulsion or anion entry to compensate for the positive charges formed inside the film. However, it has been shown that the redox processes in electroactive films are accompanied not only by the exchange of ions with the electrolyte solution but also by solvent exchanges.[19]

As electroneutrality is demanding, it is generally assumed that the field-assisted transport of charged species is more rapid than the transport of neutral species and, consequently, solvation equilibria can only be established slowly. Therefore, the equilibria associated with electronic, ionic and solvation processes may be established on quite different time scales, but at long enough time scales thermodynamics will prevail and processes will attain a state of global equilibrium. However, the relative rates of all the processes involved in the charge compensation are still an open question.

Electroactive materials have attracted interest in view of their practical applications as electrodes in batteries,[20–22] as gas-separating membranes,[23] in microelectronic devices,[24] for molecular recognition[25] and as sensors for the detection of chemical or biological species,[26] or some inorganic ions in solutions,[27] or even as nanostructured materials.[28] New possibilities have recently been found, e.g. in microwave absorbers for screening external electromagnetic fields.[29] Using these materials, e.g. in the field of electroanalysis,[30] requires a clear understanding of the charge compensation processes following the redox switching of these electroactive materials from reduced to oxidized forms or vice versa.

In this paper, after some generalities, a review of the models and techniques used to investigate ionic and solvent transfer and transport in electroactive materials will be first carried out. As the models employed are largely dependent on the techniques used to test them, two main approaches will be described. The electrochemical investigation of electroactive films, in terms of voltammetry and electrochemical impedances, takes into account the charged species involved in the redox process of the film. The addition of gravimetric investigations thanks to quartz crystal microbalances (QCMs) allows the solvent interaction to be attained. A third approach largely used to investigate these films, as they often have electrochromic properties, is based on optical techniques, but they are out of the scope of this paper. At the end, we shall describe the use of AC electrogravimetry coupled with electrochemical impedance measurements to characterize ions and solvent motion at the film/electrolyte interface during the redox switching of an electroactive material. AC electrogravimetry allows the mass response to a small potential perturbation to be analysed thanks to a fast QCM used in the dynamic regime.[31,32] This technique has already been fruitful in several domains: copper electrodeposition,[33] gold oxidation in an acidic medium,[34] ionic insertion in WO_3[35,36] or passivity of iron in a sulphuric medium.[37] Here, the models proposed in the literature are reviewed for two and three species involved in the oxidation/reduction process taking into account insertion laws based on diffusion and heterogeneous kinetic equations. Then, calculated electrochemical impedances and electrogravimetric transfer functions deduced from these models will be compared with experimental results: the influence of the nature of the ionic species which interact with the films will be discussed.

II. GENERAL CONSIDERATIONS

Before we review ionic and solvent transfer and transport in electroactive materials, some general considerations concerning the thermodynamics, swelling and conductivity of these materials will be given.

1. Thermodynamics

From a thermodynamic point of view, it has been shown that the redox process in an electroactive film is accompanied not only by the exchange of ions with the electrolyte solution but also by solvent exchange,[19,38] e.g. for a polymer P immersed in an aqueous solution of a salt CA, a general redox process where cations, anions and solvent are exchanged with the solution to compensate the electronic charges exchanged with the electrode can be represented by

$$\{[P^{n+}nA^-]\alpha[C^+A^-]\beta[H_2O]\}_P + \nu e^-$$
$$\Leftrightarrow \{[P^{(n-\nu)+}(n-\nu)A^-](\alpha-\delta)[C^+A^-](\beta-\varepsilon)[H_2O]\}_P$$
$$+ \delta C_s^+ + (\delta+\nu)A_s^- + \varepsilon H_2O_s, \qquad (1)$$

where $\{\}_P$ and $\{\}_s$ mean, the species inserted in the polymer and the species in the solution, respectively.

Owing to thermodynamic constraints and electroneutrality requirements, the coefficients, α, β, δ and ε can be of either sign, or equal to 0, and are not necessarily integer numbers.

When an electric field is applied across an ion-containing membrane, ions move through the membrane owing to electromotive forces. This ion transport is accompanied by solvent transport through the membrane. Solvent is transported either by an association with the transported ion, such as a hydration sphere, or by hydrodynamic pumping owing to the movement of the ions and associated solvent molecules. This solvent transport accompanying the ion transport through a membrane is termed "electro-osmosis".[39]

Electroactive polymers are a special case of ion-exchange polymers in that one can control the charge site density; this charge density range is determined by the charge type and volume concentration of the redox sites. For simplicity, for a polymer having cationic sites immersed in a bathing solution containing a single 1:1 electrolyte C^+A^-, the partition of anions and cations (here, counterions and co-ions, respectively) between the bathing solution, C_s^+, A_s^-, and the cationic form of the polymer, C_p^+, A_p^-, is[40]

$$C_s^+ + A_s^- \rightleftarrows C_p^+ + A_p^-. \qquad (2)$$

This process satisfies the activity constraint and is described by the equilibrium constant

$$K_{\text{salt}} = C_p^+ C_p^- \gamma_p^{\pm}/C_s^2 \gamma_s^{\pm 2}, \qquad (3)$$

where K_{salt} is the salt partition coefficient, $\gamma^{\pm}$ denotes the mean activity coefficient in the designated phase C_p^+ and C_p^- are the concentrations of C_p^+ and A_p^- respectively and C_s represents equal concentrations of anion and cation in the bathing solution.

The electroneutrality constraint for the concentration, c_{M^+p}, of the fixed polymer sites, M^+, and the concentrations of mobile ions in the polymer phase, C_p^+, A_p^-, leads to

$$c_{A_p^-} = c_{M_p^+} + c_{C_p^+}. \tag{4}$$

By eliminating $c_{A_p^-}$ between (3) and (4), we obtain a quadratic expression for a "permselectivity index", $R = c_{C_p^+}/c_{M_p^+}$:

$$R^2 + R - S^2 T^2 = 0, \tag{5}$$

where $S = c_s/c_{M_p^+}$ and $T = K_{salt}^{1/2}\left[\gamma_s^\pm/\gamma_p^\pm\right]$.

This expression means that when $ST \ll 1$ (e.g. when $c_s \ll 1$), $R = 0$, which means that the film is ideally permselective (as $c_{C_p^+} = 0$, no co-ions enter the polymer), whereas when ST increases (i.e. when c_s increases), R is different from 0, which shows co-ion ingress in the film. However, it is noticeable that the permselectivity index depends upon the degree of polymer oxidation via $c_{M_p^+}$. The change of the permselectivity index has been thoroughly investigated for various conditions of solvent and salt transfers accompanying complete or partial redox switching.[41,42]

2. Swelling

Among the physicochemical properties of electroactive materials, such as conducting polymers, one which has attracted much attention in recent years is the so-called electrochemomechanical effect, i.e. the expansion and contraction of the sample that arise upon redox switching.[43,44] As an example, for polypyrrole (PPY), it has been shown that a compact structure is attained at cathodic potentials, and only the surface in contact with the electrolyte is electrochemically active, whereas for anodic polarization the structure becomes permeable to ions: every polymeric chain actuates as an electroactive interface.[45–47] For polyaniline, the volume changes can be attributed to the influence of several factors: ion and water exchange with the electrolyte, coulombic repulsion between charged sites in the polymer backbone, anion–polymer interaction and structural changes of the polymer backbone.[48,49] In addition, it has been shown that the anion in the electrolyte has a definite influence on the film volume changes.[50] Finally, the doping degree, the nature of the counterions and the method of preparation (chemical or electrochemical) have a considerable influence upon the polymer structure.[51] In some conditions, the main part of the overall increase of volume (and hence of conformational rearrangements) observed during oxidation has to be attributed to an effect of solvent penetration.[52] For poly (2-methylaniline), the volume changes seem to be due to two different processes, the faster one being proton and anion exchange with the electrolyte, both carrying solvent molecules. The other process is a structural change of the polymer backbone, giving higher volume changes but having a much lower rate during the reduction step.[53] These swelling phenomena have a profound influence on the rate and magnitude of the redox-switching mass response.[54] Finally, this volume change behaviour has allowed arrangements of conducting polymers to be considered as artificial muscles.

These volume and structure changes were studied using various techniques.[55,56] Microscopic observations of a minute drop of polymer[48–50] were very efficient. Quartz crystal admittance measurements (electroacoustic admittance) to follow the departure from rigidity,[57] by following the shear moduli with time,[58] to gain some insight with regard to solvation[59] were also helpful. This technique was also used for identifying the rate-limiting step of the redox switching of some polymers.[60] The authors have shown that the movement of neutral species is often slower than the movement of charged species, and occurs to an extent which depends on the experimental conditions. In particular for polyaniline in HCl, the movement of water is slow and lags behind that of the protons. As a result, at higher sweep rate it has been shown that less water enters the film. It was also shown that the viscoelastic properties are greatly influenced by the anion identity.[61] As examples, perchlorate-doped polyaniline films are compact, whereas sulphate-doped films are more open.

3. Conductivity

The electrical conductivity of conducting polymers is known to be a strong function of their oxidation states (or doping level). It usually increases with the potential applied to an electrode in contact with the film, i.e. when the film is oxidized.[62] This is the case for pure PPY, but for dodecyl sulphate modified PPY the ion conductivity is at its maximum in the reduced state and decreases significantly with increasing potential.[63] For some 3,4-disubstituted

PPYs and polythiophenes, oxidation causes the transition from a low-conductivity to a conductive state as generally found in polyconjugated polymers, but these modified polymers pass to a second low-conductivity state via a maximum.[64,65] The consequence of such a dramatic change in electronic conductivity upon switching might lead to a moving-front phenomenon, which separates a part where the film is in its conductive state from one where it is in its insulating state and propagates from the electrode to the solution.[66,67]

III. ELECTROCHEMICAL APPROACH OF ELECTROACTIVE MATERIALS

Many organic and inorganic materials are electroactive. Among them, electroactive polymers constitute a large family, which can be classified into two major types: redox polymers and electronically conducting polymers.[18] The combination of a deposited polymer film and a supporting electrode constitutes a chemically modified electrode. The mechanism of charge percolation through surface-deposited polymer films is of central importance. Redox polymers are localized-state conductors, containing redox-active groups covalently bound to an electrochemically inactive polymeric backbone. In these materials electron transfer occurs via a process of sequential electron self-exchange between neighbouring redox groups. This process is termed "electron hopping". In contrast with electronically conducting polymers, the polymer backbone is extensively conjugated, which results in considerable charge delocalization. Charge transport (via polarons and bipolarons) along the polymer chain is rapid and interchain charge transfer is rate-limiting. Redox polymers remain conductive over only a limited range of potential. Maximum conductivity is observed when the concentrations of oxidized and reduced sites in the polymer are equal. This occurs at the standard potential of the redox centre in the polymer. In contrast, electronically conducting polymers, such as PPY, display quasi-metallic conductivity and remain conductive over a large potential range. Redox polymers are usually preformed and subsequently deposited onto the support electrode surface via dip or spin coating. In contrast, electronically conducting polymers are usually generated via in situ electrodeposition. In this case there is electropolymerization of the electroactive monomer.

In all cases, the process of redox switching, i.e. the transition from an insulating to a conducting form, is accomplished via an electrochemically induced change in the oxidation state of the layer. Since electroneutrality within the film must be maintained, the oxidation state change is accompanied by the ingress or egress of charge-compensating counterions.

1. Models of the Charge Transport Through the Electroactive Film

Two types of model have been used to describe the charge transport through electroactive films: continuous models, considering ionic transport in a compact film based on the Nernst–Planck equations, and porous models, whose transport is described by transmission line equivalent circuits.

(i) Compact Model (Diffusion–Migration Model)

The geometry of the modified electrode is given in Fig. 1.

By definition, the fluxes of species i, J_i, are positive for outgoing species:

$$J_i(x) > 0 \quad \text{for } x > 0. \tag{6}$$

The global potential across the modified electrode, E (the metal electrode is supposed to be grounded), is the sum of three quantities:

$$E = E_1 + E_2 + E_3, \tag{7}$$

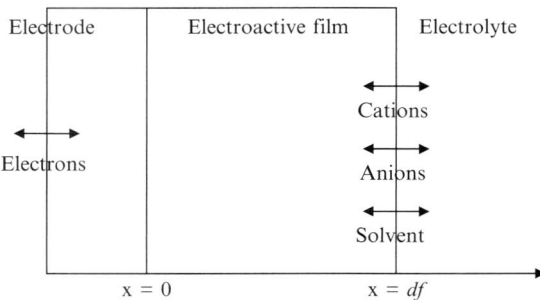

Figure 1. The electrode/film/electrolyte system.

where E_1 and E_3 are the interfacial differences of potential across the metal/film (electron transfer overvoltage) and film/electrolyte (ion transfer overvoltage) interfaces, respectively and E_2 is the potential difference across the film (diffusion + ohmic overvoltages) with thickness d_f.

Generally, the total faradaic electrochemical impedance, $Z_F(\omega)$, is obtained by considering

$$Z_F(\omega) = Z_1(\omega) + Z_2(\omega) + Z_3(\omega), \tag{8}$$

where $Z_1(\omega) = (\Delta E_1/\Delta I_F)$ is relative to the metal/polymer interface, $Z_2(\omega) = (\Delta E_2/\Delta I_F)$ is relative to the bulk polymer and $Z_3(\omega) = (\Delta E_3/\Delta I_F)$ is relative to the polymer/electrolyte interface, to which double-layer capacities have to be added to obtain the measurable electrochemical impedance.[68,69]

As a simplifying assumption, which will be used in the following, the charge transport mechanism, which occurs in the bulk polymer during an oxidation/reduction reaction of the polymer in aqueous media, can be modelled by assuming that the transport of the species in the electrolyte is sufficiently fast and is not a limiting step.

(a) Boundary conditions

The following boundary conditions are supposed to apply. The metal/polymer is an ion blocking interface:

$$\text{i.e. for } x = 0, \quad J_a(0) = J_c(0) = 0. \tag{9}$$

As anions (subscript a), and cations (subscript c), cannot cross the electrode/polymer interface, only the electrons (subscript e), are supposed to cross the interface:

$$J_e(0) = \frac{I_F}{F}, \tag{10}$$

$$J_e(0) = k - k'c_e(0), \tag{11}$$

where I_F is the faradaic current density related to the charge transfer at the metal/film interface, k and k' are the rate constants of the electronic transfer and F is the Faraday number (96,500 C). In addition,

$$k = k_0 \exp bE_1 \text{ and } k' = k'_0 \exp bE_1,$$

where E_1 is the film/electrolyte interfacial potential difference.

On the other hand, as outwards positive fluxes give negative currents for expelled anions and positive currents for expelled cations, by assuming monovalent ions at the film/electrolyte interface, we have

$$\text{for } x = d_f, \quad -J_a(d_f) + J_c(d_f) = \frac{I_F}{F}, \tag{12}$$

where

$$J_i(d_f) = k_i c_i(d_f) - k_i', \quad \text{with } i = a, c \tag{13}$$

and $k_i = k_{i0} \exp b_i E_3$ and $k_i' = k_{i0}' \exp b_i' E_3$ and

$$J_e(d_f) = 0, \tag{14}$$

although for any x in the bulk polymer

$$\frac{I_F}{F} = J_e(x) - J_a(x) + J_c(x), \tag{15}$$

where $J_e(0)$ is the electron flux at the electrode/film interface and $J_a(d_f)$ and $J_c(d_f)$ are the anion and cation fluxes at the film/solution interface, respectively.

(b) Movement of the species

For models considering the electroactive films as a homogeneous medium, the movement of the species is considered to be governed by migration and diffusion. This is easily understandable for conducting polymers but is less obvious for redox polymers where conduction through electron hopping prevails. However, Savéant et al.[70,71] Buck,[72–74] and Albery et al.[75] have shown that electron hopping is not only driven by a concentration gradient but is also field-assisted and then movement of electrons can also be described by Nernst–Planck equations. Then, the flux, J_i, of all the species within the film can be written as

$$J_i(x) = -D_i \frac{\partial c_i}{\partial x} - z_i D_i \frac{F}{RT} c_i \frac{\partial \varphi}{\partial x}, \quad i = e, a, c, \tag{16}$$

where D_i is the diffusion coefficient of species i and $z_e = 1$, $z_a = -1$, and $z_c = 1$.

The concentrations change as

$$\frac{\partial c_i}{\partial t} = -\frac{\partial J_i}{\partial x}. \tag{17}$$

The potential follows the Poisson equation:

$$\varepsilon \frac{\partial^2 \varphi}{\partial x^2} = \sum_i z_i c_i \tag{18}$$

and the total current is equal to

$$I = F \sum_i z_i J_i(x) + \frac{1}{\varepsilon} \frac{\partial^2 \varphi}{\partial x \partial t}. \tag{19}$$

(c) Steady state and quasi steady state

At steady state, $(\partial/\partial t) \equiv 0$ in the movement equations, so, from (17), $J_i(x)$ is a constant and the boundary conditions lead to $J_i(x) = 0$ and $I_F = 0$. In this equilibrium situation, $c_i(x)$ and $E(x)$ are constant throughout the film and are equal to c_i and E, respectively.

The general problem, even without consideration of space-charge effects, is intrinsically non-linear. Only DC solutions for potential, concentrations, fluxes and currents are possible by exact methods in some specific cases. Time-dependence problems, such as the derivation of the impedance, require some simplifications or consideration of special cases. Diffusion–migration models were used to simulate the behaviour of electroactive films both by direct computer integration of the equations to calculate cyclic voltammograms,[76] or potential distributions across the film[77] and by Monte Carlo simulation.[78,79]

(ii) Porous Model (Transmission Line Model)

By considering that polymers have a porous nature,[80–83] Barker,[84] Albery et al.,[85] Buck,[86,87] and Paasch[88–91] have shown that to calculate the impedance of electrode/film/solution systems there is a full equivalence of the transport of species by diffusion–migration and Poisson potential distribution and a transmission line equivalent circuit, like those shown in Fig. 2. Their distributed components are

$$R_i(x) = \frac{RT}{F^2 z_i^2 D_i c_i(x)}, \tag{20}$$

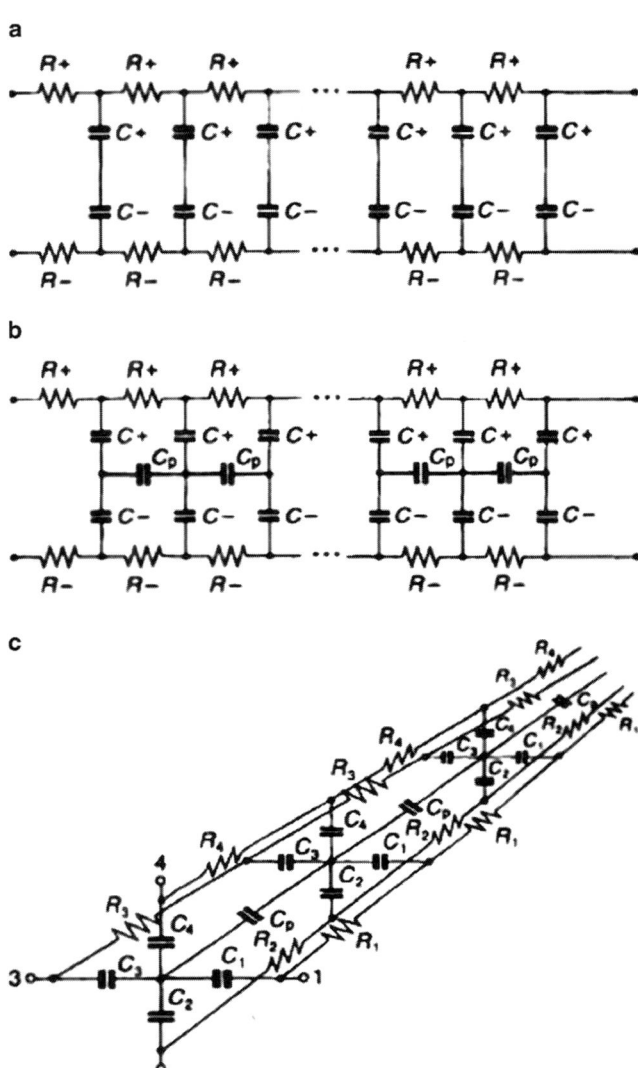

Figure 2. Transmission lines used as equivalent circuits for electroactive films. These aperiodic circuits describe ion and electron transport in conducting polymers with reversible interfacial charge exchange processes. (**a**) Single Warburg impedance, (**b**) circuit describing film bulk and Warburg impedance, capacitors C_P represent the Poisson equation, (**c**) multi-ion and electron exchange processes. From Buck and Mundt.[87]

$$C_i(x) = \frac{z_i^2 F^2 c_i(x)}{RT}, \tag{21}$$

$$C_p(x) = \frac{\varepsilon}{4\pi}. \tag{22}$$

The transmission lines shown in Fig. 2 are such that the distributed capacitance C_p represents the Poisson branch and parallel branches represent each charged species. Figure 2a describes the single Warburg impedance for a single $z_+:z_-$ salt. This leads to an alternative approach of the charged species movement supposed to follow a diffusion–migration process. If the concentrations are supposed to be distance-independent, the impedance is quite easy to obtain. If not, the derivation is more intricate and needs numerical techniques.

The solution of this transport problem in its full generality is difficult; only numerical solutions can be obtained. Using approximations, one can reach analytical solutions in some simple cases. The main hypothesis, often made, is electroneutrality, which means

$$\sum_i z_i c_i = 0 \quad \text{in the transport approach}$$

or

$$C_p = 0 \quad \text{in the transmission line approach.}$$

2. Calculation of the Impedance

At the beginning, the impedance of a polymer film was calculated by taking into account the transport of the species by diffusion alone,[92,93] then both diffusion and migration were considered.[94,95] For the latter, an exact solution can be obtained for two species, whereas only approximate solutions are obtained for three species.

(i) Two-Species Problems

The impedance of an electroactive film where the insertion of one ion (subscript i in the following equations) occurs at the film/electrolyte interface, balanced by the entry of electrons at the metal/film interface has been calculated for compact and porous representations. By assuming local electroneutrality [$c_e(x) = c_i(x)$] in the polymer, Buck[96] and Vorotyntsev[97] followed by others[98–101] found the solution of the Nernst–Planck equations for the following boundary conditions:

$$x = 0, \; J_i = 0 \quad \text{and} \quad J_e = \frac{I}{e} = k_a - k_c c_e(0), \tag{23}$$

$$x = d_f, \; J_e = 0 \quad \text{and} \quad J_i = -\frac{I_F}{e}. \tag{24}$$

At steady state, $J_e = J_i = 0$, $c_e(x) = c_i(x) = c_p$, and $\varphi(x) = \varphi_p$, and then the concentrations of the species and the potential are uniform in the bulk film.

In a low-amplitude sine wave regime, the various impedances were calculated,

$$Z_1(\omega) = R_e + R_p \frac{D_e + D_i}{8 \upsilon D_e D_i} \left[(D_e + D_i) \coth \upsilon - (D_e - D_i) \tanh \upsilon \right], \tag{25}$$

$$Z_2(\omega) = R_p + R_p \frac{(D_e - D_i)^2}{4 \upsilon D_e D_i} \tanh \upsilon, \tag{26}$$

$$Z_3(\omega) = R_i + R_p \frac{D_e + D_i}{8 \upsilon D_e D_i} \left[(D_e + D_i) \coth \upsilon + (D_e - D_i) \tanh \upsilon \right], \tag{27}$$

and then the total faradaic impedance of the electroactive film is

$$Z_F(\omega) = R_e + R_i + R_p + R_p \frac{1}{4 \upsilon D_e D_i} \left[(D_e + D_i)^2 \coth \upsilon + (D_e - D_i)^2 \tanh \upsilon \right], \tag{28}$$

where

$$\upsilon^2 = \frac{j \omega d_f^2 (D_e^{-1} + D_i^{-1})}{8}, \quad R_p = \frac{d_f}{\sigma}, \quad \sigma = (D_e + D_i) c_p \frac{F^2}{RT},$$

$$R_i = \frac{1}{F \left[b_i k_i c_i(d_f) - b_i' k_i' \right]}, \quad R_e = \frac{1}{F \left[b_a k_a - b_c k_c c_e(0) \right]}.$$

R_e and R_i are the electronic and ionic charge transfer resistances at the metal/film and film/electrolyte interfaces, respectively, R_p is the resistance of the bulk film, and σ is the conductivity of the film.

At low frequencies, $\omega \to 0$, we have

$$Z_F(\omega) \sim R_e + R_i + R_p + R_p \frac{1}{4\upsilon D_e D_i} \left[(D_e + D_i)^2 \left(\frac{1}{\upsilon} + \frac{\upsilon}{3} \right) + (D_e - D_i)^2 \upsilon \right]. \tag{29}$$

Then, in the low-frequency range, the impedance is equivalent to a R_{lf}, C_{lf} series circuit, such as

$$R_{lf} = R_e + R_i + R_p + \frac{R_p}{4 D_e D_i} \left((D_e - D_i)^2 + \frac{(D_e + D_i)^2}{3} \right), \tag{30}$$

$$C_{lf} = \frac{d_f^2}{2R_p(D_e + D_i)}, \quad \text{i.e. } C_{lf} = \frac{d_f}{2} c_p \frac{F^2}{RT}. \tag{31}$$

Figure 3 shows the various shapes of the impedance of the thin film calculated for equal coefficients of ions and electrons ($D_e = D_i = D$) but taking into account the double-layer capacities, C_i and C_e, across the film/electrolyte and electrode/film interfaces, respectively.

$$Z(\omega) = \frac{1}{j\omega C_i + \frac{1}{R_i}} + \frac{1}{j\omega C_e + \frac{1}{R_e}} + R_p \left(1 + \frac{\coth \upsilon}{\upsilon} \right), \tag{32}$$

where

$$\upsilon = (j\omega/\omega^*)^{1/2} \text{ and } \omega^* = 4D/d_f^2.$$

For the same geometry, Paasch found for a transmission line like the one in Fig. 2 for the porous model of the conducting polymers[88–90]

$$Z_F(\omega) = \frac{\rho_1^2 + \rho_2^2}{\rho_1 + \rho_2} \frac{\coth(d_f \beta)}{\beta} + \frac{2\rho_1 \rho_2}{\rho_1 + \rho_2} \frac{d_f}{\sinh(d\beta)} + \frac{d_f \rho_1 \rho_2}{\rho_1 + \rho_2}, \tag{33}$$

where

$$\beta^2 = (k + i\omega)CS_c(\rho_1 + \rho_2), \quad k = \frac{g_{ct}}{CS_c},$$

where g_{ct} is the charge transfer conductance which couples electrons and ions, ρ_1 and ρ_2 are the resistivities of the electronically conducting porous material and the pore filled with electrolyte with ionic

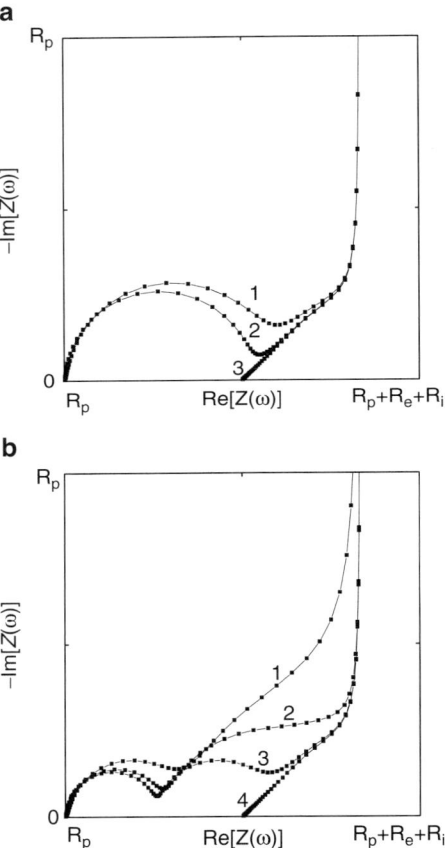

Figure 3. Impedance calculated from (32) plotted in the complex $(-\text{Im}[Z(\omega)], \text{Re}[Z(\omega)])$ for equal diffusion coefficients of ions and electrons $D_e = D_i$ and $\omega^* = 0.4$, $R_e = R_i = R_p/4$, (**a**) $C_e = C_i = 0.01 R_p^{-1}\omega_*^{-1}$ (curve *1*), $0.001 R_p^{-1}\omega_*^{-1}$ (curve *2*), 0 (curve *3*); (**b**) $C_i = 100 C_e$, $C_e = 0.001 R_p^{-1}\omega_*^{-1}$ (curve *1*), $0.0001 R_p^{-1}\omega_*^{-1}$ (curve *2*), $0.00001 R_p^{-1}\omega_*^{-1}$ (curve *3*), 0 (curve *4*). From Buck et al.[96]

conductivity, respectively, C is the capacity of the pore unit area and S_c is the effective pore area per unit volume.

When $k = g_{ct} = 0$, (33) has exactly the same form as (28) when $R_e = R_i = 0$, noting that

$$\coth y = \frac{\coth(y/2) + \tanh(y/2)}{2} \text{ and } \frac{1}{\sinh y} = \frac{\coth(y/2) - \tanh(y/2)}{2}. \quad (34)$$

Then, the two models give equivalent results. This calculation was also given by Buck without the electroneutrality hypothesis (i.e. $C_p \neq 0$). The transmission line approach is often called the "porous model" of a conducting polymer as electrons are supposed to cross the polymer (phase 1) and ions are supposed to move into pores, filled by electrolyte, represented by the second branch of the transmission line. It is noticeable that the transmission line approach allows more complicated kinetics to be tested for a two-species problem, e.g. charge transfer in parallel to the capacity $C_e(x)$ and $C_i(x)$, or diffusion of the ion in the "ionic pores", i.e. to introduce complex impedances instead of the real resistance ρ_1 and/or ρ_2, of the pure capacitances C_1 and/or C_2. It also allows position-dependent parameters to be introduced to mimic concentration gradients in the polymer [$c_i(x) \neq$ constant].

(ii) Three-Species Problems

For three charged species, where electrons crossing the metal/polymer interface and anions and cations crossing the polymer/electrolyte interface are considered, the problem is far more involved and has to be solved by numerical methods.[102] To obtain analytical solutions, questionable assumptions have to be made. Using the alternative model, 3D transmission lines have to be considered and only numerical solutions can be obtained.[86, 103]

Two types of hypothesis were invoked, which lead to position-independent parameters [$c_i(x) =$ constant], to analytically solve the Nernst–Planck equations. Either relationships between the concentrations are supposed in addition to electroneutrality in the bulk film or the migration terms are neglected.

(a) Inzelt model[104]

In this model, in addition to $c_i(x) =$ constant and local steady state electroneutrality ($c_e + c_c - c_a = 0$), a strict relationship between the concentration fluctuations resulting from the potential perturbation was assumed:

$$\Delta c_c = (1 - \gamma)\Delta c_a,$$
$$\Delta c_e = \gamma \Delta c_a. \tag{35}$$

These assumptions lead to the three elementary impedances which form the total impedance of the modified electrode:

$$Z_1(\omega) = \frac{1}{nFk_1} + \frac{k_2^*}{2k_1NF}\left(\frac{a}{P} + \frac{c}{I}\right)\frac{\coth(sd_f/2)}{s}$$
$$+ \frac{k_2^*}{2k_{1NF}}\left(\frac{c}{I} - \frac{a}{P}\right)\frac{\tanh(sd_f/2)}{s}, \tag{36}$$

$$Z_2(\omega) = \frac{d}{NF} + \frac{1}{NF}\left(\frac{c}{I} - \frac{a}{P}\right)U\frac{\tanh(sd_f/2)}{s}, \tag{37}$$

$$Z_3(\omega) = \frac{1}{k_3F} + \frac{k_4^*}{2k_3NF}\left(\frac{c}{I} + \frac{a}{P}\right)\frac{\coth(sd_f/2)}{s}$$
$$+ \frac{k_4^*}{2k_3NF}\left(\frac{a}{P} - \frac{c}{I}\right)\frac{\tanh(sd_f/2)}{s}, \tag{38}$$

where

$$a = \frac{D_e}{RT}C_e, \quad b = \frac{D_c}{RT}C_c, \quad c = \frac{D_a}{RT}C_a$$

and

$$s^2 = j\frac{\omega}{\omega_0},$$

where

$$\omega_0 = \frac{\gamma c D_e + (1-\gamma)cD_c + (a+b)D_a}{N}, \quad N = a+b+c$$

and

$$U = \frac{\gamma n D_e + (1-\gamma)D_c - D_a}{N},$$
$$I = -\frac{(na+b)D_a + \gamma n c D_e + (1-\gamma)cD_c}{N},$$
$$P = \frac{\gamma(b+c)D_e + (1-\gamma)aD_c - aD_a}{N},$$
$$H = \frac{\gamma n b D_e - (1-\gamma)(na+c)D_c - bD_a}{N}.$$

If all the species are supposed to have the same diffusion coefficient, $D_e = D_a = D_c = D$ and $n = 1$, then $U = 0$, $I = -D$, $P = D\gamma((b+c-a)/a+b+c)$ and $H = -D(1-\gamma)$. Then, the impedance is simplified under the form

$$Z_1(\omega) = \frac{1}{nFk_1} + \frac{k_2^*}{2k_1 NFs}\left(\frac{(a+b)(\gamma b - a)}{\gamma b}\coth(sd_f/2)\right.$$
$$\left. + \frac{\gamma b^2 + a(a+b)}{\gamma b}\tanh(sd_f/2)\right), \tag{39}$$

$$Z_2(\omega) = \frac{d_f}{NF}, \tag{40}$$

$$Z_3(\omega) = \frac{1}{Fk_3} + \frac{k_4^*}{2k_3 NFs}\left(\frac{(a+b)(\gamma b - a)}{\gamma b}\coth(sd_f/2)\right.$$
$$\left. + \frac{\gamma b^2 + a(a+b)}{\gamma b}\tanh(sd_f/2)\right). \tag{41}$$

If, in addition, $c_c \ll c_e$ and c_a, then

$$Z_1(\omega) \approx \frac{1}{Fk_1} - \frac{k_2'}{k_1 NFs}\frac{a^2}{\gamma b}\coth(sd_f)$$

and

$$Z_3(\omega) \approx \frac{1}{Fk_3} - \frac{k_4'}{k_3 NFs}\frac{a^2}{\gamma b}\coth(sd_f).$$

The value of the impedance obtained by Inzelt is very close to Vorotyntsev's value[97] when the cation movement is neglected, i.e. when $D_c = 0$, $b = 0$, $\gamma = 1$ and $C_a = C_e$. This is due to the hypothesis made on the concentration fluctuations which limit the two approaches to the same number of degrees of freedom.

(b) Diffusion model

Another possible assumption is to neglect the migration terms in the Nernst–Planck equations so the charged species behave independently. Local electroneutrality is not imposed.[105]

Here, anions, cations and electrons are supposed to be transported through the polymer only by diffusion:

$$\Delta J_i = -D_i \frac{\partial \Delta c_i}{\partial x}, \quad i = \text{e, a, c,} \qquad (42)$$

i.e.

$$j\omega \Delta c_i = -\frac{\partial \Delta J_i}{\partial x} \qquad (43)$$

or

$$j\omega \Delta c_i = \frac{d^2 \Delta c_i}{dx^2}, \quad i = \text{e, a, c.} \qquad (44)$$

At $x = d$, $\quad \Delta J_e(d_f) = 0 \qquad (45)$

and

$$\Delta J_i(d_f) = K_i \Delta c_i(d_f) + G_i \Delta E_3, \quad i = \text{a, c} \qquad (46)$$

At $x = 0$, $\quad \Delta J_e(0) = K_e \Delta c_e(0) + G_e \Delta E_1 \qquad (47)$

and

$$\Delta J_a(0) = \Delta J_c(0) = \Delta J_s(0) = 0. \qquad (48)$$

This leads to

$$\frac{\Delta J_i(d)}{\Delta E_3} = \frac{G_i}{1 + K_i \left((\coth d_f \sqrt{j\omega/D_i})/\sqrt{j\omega D_i} \right)}, \quad i = \text{a, c} \qquad (49)$$

and

$$\frac{\Delta J_e(0)}{\Delta E_1} = \frac{G_e}{1 - K_e \left[\left(\coth d_f \sqrt{j\omega/D_e} \right) \big/ \sqrt{j\omega D_e} \right]}. \qquad (50)$$

Concerning the bulk film, the Poisson equation leads to

$$\frac{d^2 \Delta E}{dx^2} = \frac{4\pi}{\varepsilon \varepsilon_o} \left[\Delta c_e(x) + \Delta c_c(x) - \Delta c_a(x) \right], \qquad (51)$$

where

$$\Delta c_i(x) = \frac{G_i \cosh x \sqrt{j\omega/D_i}}{\sqrt{j\omega D_i} \sinh d_f \sqrt{j\omega/D_i} + K_i \cosh d_f \sqrt{j\omega/D_i}} \Delta E_3, \quad i = \text{a, c} \qquad (52)$$

and
$$\Delta c_e(x) = \frac{G_e \cosh(x-d_f)\sqrt{j\omega/D_e}}{\sqrt{j\omega D_e}\sinh(d_f\sqrt{j\omega/D_e}) - K_e\cosh(d_f\sqrt{j\omega/D_e})}\Delta E_1, \tag{53}$$

so as
$$\Delta E_2 = \Delta E(0) - \Delta E(d_f), \tag{54}$$

then
$$\Delta E_2 = \left\{ \begin{aligned} & \frac{D_c}{j\omega}\frac{G_c(1-\cosh d_f\sqrt{j\omega/D_c})}{\sqrt{j\omega D_c}\sinh d_f\sqrt{j\omega/D_c}-K_c\cosh d_f\sqrt{j\omega/D_c}}\Delta E_3 \\ & -\frac{D_a}{j\omega}\frac{G_a(1-\cosh d_f\sqrt{j\omega/D_a})}{\sqrt{j\omega D_a}\sinh d_f\sqrt{j\omega/D_a}-K_a\cosh d_f\sqrt{j\omega/D_a}}\Delta E_3 \\ & +\frac{D_e}{j\omega}\frac{G_e(\cosh d_f\sqrt{j\omega/D_e}-1)}{\sqrt{j\omega D_e}\sinh d_f\sqrt{j\omega/D_e}-K_e\cosh d_f\sqrt{j\omega/D_e}}\Delta E_1 \end{aligned} \right\}. \tag{55}$$

So
$$Z_1(\omega) = \frac{1}{FG_e}\left(1 - K_e\frac{\coth d_f\sqrt{j\omega/D_e}}{\sqrt{j\omega D_e}}\right), \tag{56}$$

$$Z_3^{-1}(\omega) = F\left(\frac{G_c}{1+K_c\left[(\coth d_f\sqrt{j\omega/D_c})/\sqrt{j\omega D_c}\right]} - \frac{G_a}{1+K_a\left[(\coth d_f\sqrt{j\omega/D_a})/\sqrt{j\omega D_a}\right]}\right), \tag{57}$$

$$Z_2(\omega) = \frac{d_f}{\sigma} + \frac{4\pi}{j\omega\varepsilon\varepsilon_0} \tag{58}$$
$$\times \left\{ \left(\begin{aligned} & \frac{D_c G_c(1-\cosh d_f\sqrt{j\omega/D_c})}{\sqrt{j\omega D_c}\sinh d_f\sqrt{j\omega/D_c}+K_c\cosh d_f\sqrt{j\omega/D_c}} \\ & -\frac{D_a G_a(1-\cosh d_f\sqrt{j\omega/D_a})}{\sqrt{j\omega D_a}\sinh d_f\sqrt{j\omega/D_a}+K_a\cosh d_f\sqrt{j\omega/D_a}} \end{aligned} \right)Z_3(\omega) \\ + \frac{D_e(\cosh d_f\sqrt{j\omega/D_e}-1)}{F\sqrt{j\omega D_e}\sinh d_f\sqrt{j\omega/D_e}} \right\}$$

(c) Limiting cases

If the electron transfer is supposed to be very fast and ΔE_2 is negligible, the total impedance is reduced to $Z_3(\omega)$.[106] On the other hand, if the cation movement is neglected ($G_c = 0$),

$$Z_3(\omega) = -\frac{1}{FG_a}\left(1 + K_a\frac{\coth d_f\sqrt{j\omega/D_a}}{\sqrt{j\omega D_a}}\right) \qquad (59)$$

and

$$Z_2(\omega) = \frac{d_f}{\sigma} + \frac{4\pi}{j\omega\varepsilon\varepsilon_0 F}\left(-\frac{D_a(1 - \cosh d_f\sqrt{j\omega/D_a})}{\sqrt{j\omega D_a}\sinh d_f\sqrt{j\omega/D_a}} + \frac{D_e(\cosh d_f\sqrt{j\omega/D_e} - 1)}{\sqrt{j\omega D_e}\sinh d_f\sqrt{j\omega/D_e}}\right). \qquad (60)$$

Equation (60) has the same general form as (28), which gives the impedance for a permselective film, taking into account (34):

$$Z_2(\omega) = \frac{d_f}{\sigma} + \frac{4\pi}{j\omega\varepsilon\varepsilon_0 F\sqrt{j\omega}}\left(\sqrt{D_a}\tanh\frac{d_f}{2}\sqrt{j\omega/D_a} - \sqrt{D_e}\tanh\frac{d_f}{2}\sqrt{j\omega/D_e}\right),$$

whose low-frequency limit is equal to

$$Z_2(0) = \frac{d_f}{\sigma} + \frac{\pi d_f^3}{6\varepsilon\varepsilon_0 F}\frac{D_a - D_e}{D_a D_e}.$$

(iii) Applications of Impedance Analysis

Impedance techniques were largely applied to investigate the behaviour of electroactive materials. Many conducting polymers, among them polyaniline,[107–110] PPY,[111,112] poly(*o*-toluidine),[113] poly(*o*-aminophenol),[114–116] polythiophene[117] and poly(3-methylthiophene)[118,119] were investigated. It has even possible to separate the transport of two different species in PPY/

polystyrenesulphonate.[120] For PPY, it has been proposed that a fast and a slow charge transport may occur in the polymer. The fast faradaic process is assumed to arise from ionic motion in the bulk of the film and the slow one to arise from ionic motion in the double layer.[121] The study of the transport of species in free-standing polymer membranes was also fruitful.[122] The behaviour of inorganic electroactive materials, among them potassium (Prussian blue),[123–125] cobalt,[126,127] indium,[128] nickel,[129] chromium,[130] and platinum[131] hexacyanoferrates, was also examined by impedance techniques.

For electroactive films, the electrochemical impedance technique alone is not able to discriminate among the various models which have been proposed in the literature because the experimental plots always have the same shape. The scanning electrochemical microscope (SECM) has been used successfully to study the various ion fluxes from and towards the polymer during a redox process and to identify them. The movements of Cl^-[132] and protons[133] have actually been proved by using the SECM for polyaniline in HCl medium. A quantitative approach to the rate of counterion ejection in a PPY film leading to a discussion of the nature (porous or compact) of the film has been reported.[134] SECM investigations have been carried out on many other polymers, such as poly(vinylferrocene),[135] poly(4-vinylpyridine),[136] poly(benzobisimidazobenzophenanthroline),[137] poly(3,4-ethylenedioxythiophene) (PEDOT)[138] and Nafion.[139]

As the electrochemical techniques are sensitive only to charged species, other techniques have been used additionally to provide novel insights into the composition and structure of polymer films. Radiotracer study,[140] surface-enhanced Raman scattering,[141] neutron reflectivity[142–144] and Kelvin probe measurements[145] have been employed. Optical beam deflection is a powerful technique to study ionic movements. The principle of mirage spectroscopy involves the measurement of laser beam deviation provoked by a refractive index gradient at the film/solution interface.[146–149] This technique was developed in the early 1980s by Boccara et al.[150] Coupled with voltammetry, it allows the movements of cations and anions to be easily distinguished, especially if there is no solvent exchange between the polymer and the bathing solution.

IV. COUPLED ELECTROCHEMICAL AND GRAVIMETRIC APPROACH FOR ELECTROACTIVE MATERIALS

Numerous studies of electroactive thin films were conducted with cyclic voltammetry (see the references given in the reviews listed at the beginning of this text) but the performances of this technique were largely improved by using fast scan rates at ultramicroelectrodes.[151,152] However, to gain real insight into the movements of species during the oxidation and reduction of electroactive thin films, particularly concerning the solvent, the QCM has been employed with great success. From the $i(E)$ and $m(E)$ recorded experimental raw data of the current and mass changes with respect to potential, the authors have processed these data as efficiently as possible to extract information on the redox behaviour of the electroactive films, in particular concerning the ingress and egress of neutral species and charged species occurring during the redox switching.

1. Cyclic Voltammetry and Quartz Crystal Microbalance

Since the pioneering works in the 1980s,[153–159] Bruckenstein and Hillman have thoroughly investigated the species involved in the charge compensation occurring mainly in conducting polymers when they are reduced or oxidized by coupling cyclic voltammetry and gravimetry by means of a QCM.

The simplest data processing has been based on the calculation of the charge, $q(t)$, with respect to time by integrating the current $i(t)$. Then, the change of mass per charge unit, $F\Delta m/\Delta q$, or the slope of the $F\Delta m(q)$ function, leads to the apparent molar mass of the species exchanged between the electroactive film and the bathing solution.[160,161] However, this apparent molar mass is the mass of the species if only one species is involved, but can be very different when more than one species is involved in the charge compensation process. So, functions Φ_j, such as $\Phi_j = \Delta m + q(m_j/z_j F)$ for an ion j, where m_j and $z_j F$ are the molar mass and the charge carried per mole of species j, have been considered to eliminate the contribution of ion j from the measured mass. Simple plots involving q, Δm, and Φ (or their time derivatives) as one of the variables permit an unequivocal test for the existence of a global equilibrium. In the absence of a global equilibrium, the nature of the rate-limiting step – motion of electrons, specified ions or neutral species – can be established. Once the identity of the slow species has been established, quantitative interpretation becomes possible using q data (in the case of electrons) or Φ and Δm data (in the case of heavy species).[162,163] In the late 1990s, these authors proposed using a scheme of cubes to visualize complicated electroactive film redox switching mechanisms where multiple redox, solvation and configuration processes might accompany oxidation or reduction of the film.[164–166] As an example, Fig. 4 shows redox switching of polythionine in aqueous acetic acid.

This cube was based on a previous scheme of squares visual approach.[161] Here, the axes x, y and z, respectively, represent coupled electron/proton transfer, solvent transfer and acetic acid coordination. Four equilibrium constants describe the coordination reactions for the four pairs of species on the left and right faces of the cube. The authors interpreted their data on partial redox switching of poly(vinylferrocene) films under permselective conditions in aqueous perchlorate bathing electrolytes which produce films that reach

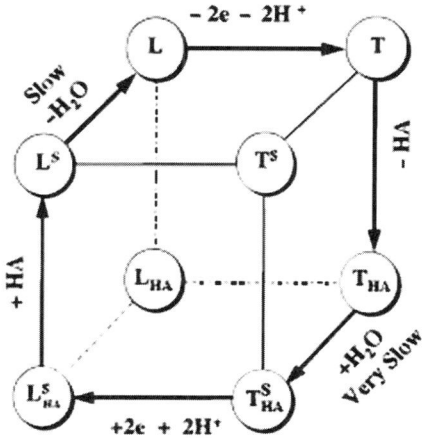

Figure 4. Cube representation for redox switching of polythionine in aqueous acetic acid. L and T denote the reduced (leucothionine) and oxidized (thionine) states, respectively. *Superscript a* denotes the acetic acid coordinated state. The mechanistic pathway for the redox cycle is illustrated by the *heavy arrows*. From Bruckenstein and Hillman.[165]

equilibrium very slowly. The relaxation kinetics involves polymer reconfiguration and solvent (water) transfer. Even after 1 h at an open circuit, the film still exchanges small amounts of solvent with the bathing electrolyte. In addition, they showed that the structure of the polymerized film, which is determined by the counterion introduced during the precipitation, exhibits tortuous pores of various size and voids of molecular dimensions within the film. Then, since a smaller anion meets less resistance than a larger anion during transport within the film, the total amount of counteranion that can enter the film increases with a decrease in anion size and in scan rate. They suggested that a free volume constraint within the film determines the amount of water that transfers between the film and the bathing electrolyte, and concluded that the amount of water transferred decreases with increased size of the counteranion.[167]

For PPY, in general when the anion is small and then mobile, the anion transfer will be dominant, and when the anion is very large (immobile), the cation transfer will be dominant on the time scale of most electrochemical measurements. When PPY is exposed to aqueous tosylate solutions, the authors explored the time scale effects on the competing ion transfers closely associated with solvent transfer. By using the scheme of cubes approach, they showed that on short time scales during reduction, cation entry competes effectively with anion ejection as a means of satisfying film electroneutrality. On longer time scales, the thermodynamically favoured anion mechanism prevails.[168]

To quantify the relative quantity of solvent and ion for a permselective film, the ratio ρ of the flux of water $[f_w = (dm_w/dt)/A]$ divided by the flux of counterions $[f_{ion} = (dm_{ion}/dt)/A]$, at any time, potential E or charge level q, is calculated from experimental data. m_{ion} is obtained from q using Faraday's law ($m_{ion} = M_{ion}q/zF$, where M_{ion} is the molar mass of the ion and zF is its charge) and $m_w = m_{total} - m_{ion}$:

$$\rho = \frac{f_w}{f_{ion}}, \quad \text{i.e.} \quad \rho = \frac{dm_w}{dm_{ion}}. \tag{61}$$

ρ can be used as a diagnostic tool to show whether the redox process is thermodynamically reversible or kinetically controlled. In the latter situation, it allows the slower step, coupled electron/ion or neutral species (water) transfer, to be determined.[169] For Prussian blue in K_2SO_4 as the apparent molar mass of K^+, which is the counterion required for maintaining electroneutrality, is lower than 39 g, Bruckenstein and Hillman concluded that water transfers in the opposite direction to it during redox cycling. In other solutions, the change in molar mass indicates that anion transfer competes with K^+ transfer.[170] The analysis of the mechanism through the ratio of solvent and counterion fluxes has been widened by using cyclic voltammetry with changing sweep rates and potential jump experiments to characterize the rate-controlling process as a function of the extent of film oxidation. This method has the capability to resolve time scale – and potential (charge) – dependent mechanistic shifts and film relaxation phenomena as they are reflected through the flux ratio.[171]

More recently, Bruckenstein and Hillman have proposed a new model for the population of electroactive film mobile species (ion and solvent) under a range of thermodynamically and kinetically controlled conditions that allows the film state to be visualized in three dimensions, including $E(V)$, q (C cm^{-2}), and Λ-space, where Λ represents the film composition and contains the concentrations of the individual mobile species populations, Γ_i (mol cm^{-2}).[172] For a permselective film undergoing a redox process,

$$\text{Red} \Leftrightarrow \text{Ox} + e^-, \tag{62}$$

they have supposed a classical kinetic law for the change of the coupled electron/ion population:[173]

$$\frac{d\Gamma_{\text{Ox}}}{dt} = k_e \left(\Gamma_{\text{Red}} \eta^{(1-\alpha)} - \Gamma_{\text{Ox}} \eta^{(-\alpha)} \right), \tag{63}$$

which can be written

$$\frac{d\Gamma_{\text{Ox}}}{dt} = k_e \left[(\Gamma_T - \Gamma_{\text{Ox}}) \eta^{(1-\alpha)} - \Gamma_{\text{Ox}} \eta^{(-\alpha)} \right], \tag{64}$$

where k_e(s^{-1}) is the rate constant for coupled electron/ion transfer into/out of the film, $\Gamma_T = \Gamma_{\text{Ox}} + \Gamma_{\text{Red}}$ and $\eta = \exp\left[nF(E - E^0)/RT\right]$.

For the solvent transfer, the following solvation model has been considered:

$$\text{Ox} + x\text{S} \underset{k_b}{\overset{k_f'}{\rightleftarrows}} \text{Ox}_S, \tag{65}$$

where Ox is an oxidized site in the polymer, S is the solvent in the solution and Ox_S is a solvated oxidized site in the polymer. The rate equation for the solvent population in the film is

$$\frac{d\Gamma_S}{dt} = k'_f \Gamma_{Ox}[S]^x - k_b \Gamma_{Ox,S}, \qquad (66)$$

where k'_f and k_b are the heterogeneous rate constants for solvent transfer into and out of the film defined by reaction (65). Since a relatively small absolute amount of solvent transfers across the film/solution interface (as compared with the vast excess in bulk solution), the change in solvent concentration in the bulk solution is essentially 0, i.e. [S] $\approx$ constant. Therefore, (66) simplifies to

$$\frac{d\Gamma_S}{dt} = k_f \Gamma_{Ox} - k_b \Gamma_{Ox,S}, \qquad (67)$$

where

$$k_f = k'_f [S]^x.$$

The population (Γ_{Ox}) of unsolvated oxidized sites can be calculated from the total number (solvated and non-solvated) of oxidized sites ($\Gamma_{Ox,T}$):

$$\frac{d\Gamma_S}{dt} = k_f \left(\Gamma_{Ox,T} - \Gamma_{Ox,S} \right) - k_b \Gamma_{Ox,S}. \qquad (68)$$

The total solvent population is the mean number of solvent molecules associated with all oxidized sites and their local environment without distinction as to whether the solvent is "bound" or "free". Thus, $\Gamma_S = x \Gamma_{Ox,S}$, where x is the number of solvent molecules (whether "bound" or "free") per oxidized site.

$$\frac{d\Gamma_S}{dt} = k_f \left(\Gamma_{Ox,T} - \frac{\Gamma_S}{x} \right) - k_b \left(\frac{\Gamma_S}{x} \right). \qquad (69)$$

The film compositional signature in E, q, and Λ-space allows visual diagnosis of thermodynamic compared with kinetic control and the identification of various possible phenomena; these include film reconfiguration, ion and solvent trapping, relative rates of ion and solvent transfer, and relative rates of solvent entry and exit.

Figure 5 shows the 3D (E, q, Γ_S) compositional space representation of the behaviour of the electroactive film mobile species (ion and solvent). This representation has been extended to combine thermodynamic non-ideality (attractive or repulsive interaction

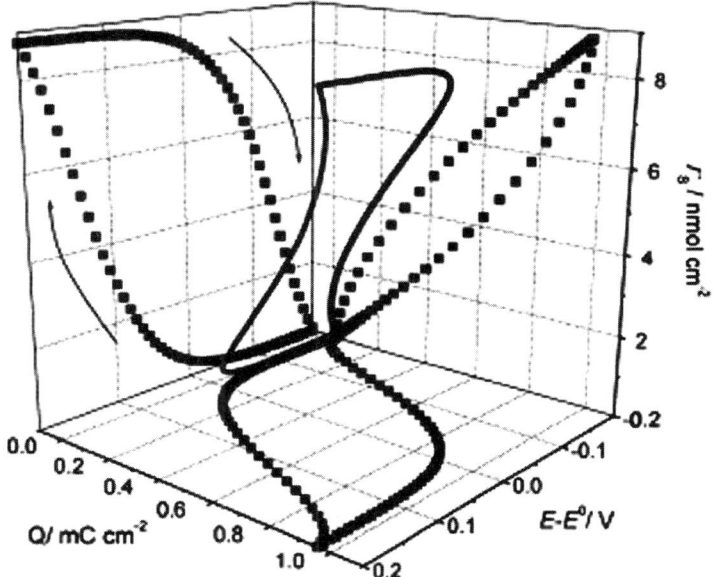

Figure 5. 3D representation in the (Q, E, Γ_s) space for slow electron/ion and slow solvent transfers (equal forward and reverse solvent transfer rates), where Q represents the film ion population and Γ_s is the film solvent population. *Arrows* indicate the potential scan direction. From Jackson et al.[172]

between solvent and redox sites) with slow electron/counterion and solvent transfer kinetics.[174] For polyaniline in perchloric acid, this approach, combined with the analysis of the ion flux to solvent flux ratio, has shown that the early stages of the film oxidation are associated with proton transfer (exit) and the latter stages with perchlorate transfer (entry) to satisfy electroneutrality. By a change of the scan rate, it has been also demonstrated that the film solvent population is in equilibrium on the time scale of slow scan voltammetry, but shows thermodynamic non-idealities.[175] For poly(vinylferrocene) in perchlorate solutions of various cations, it has been shown that the cation as well as perchlorate and water transport participated in the redox switching process. By a change of the scan rate and the use of potential steps, it has been demonstrated that the films exhibit transient non-permselectivity during redox switching. The flux of water per anion was determined in all the media studied. Finally, it has been recognized that the cation and/or water leave/leaves the polymer in an amount that depends on the cations in the solution.[176]

In their recent publication, for permselective films, these authors reported the use of current and microbalance frequency responses to generate time-resolved ion and solvent flux data as functions of potential. The total mass change, $m_T(t)$, is, at any stage of the redox process, the sum of the contributions from anion and solvent transfers, $m_A(t) + m_S(t)$. Then, application of the electroneutrality condition and the Faraday law to the current data yields the ion flux:

$$j_A = \frac{1}{M_A} \frac{dm_A}{dt} = \frac{i}{z_A F}, \tag{70}$$

where M_A and z_A are the molar mass and charge number of the anion.

The total mass flux is defined by

$$j_T = \frac{1}{M_A} \frac{dm_T}{dt} \tag{71}$$

and the solvent flux is calculated from the total mass flux and the ion flux:

$$j_S = \frac{1}{M_S} \frac{dm_S}{dt} = \frac{1}{M_S} \left(\frac{dm_T}{dt} - \frac{i}{z_A F} \right). \tag{72}$$

As $m_T(E)$ and $i(E)$ are recorded experimentally at various scan rates of the voltammetry, j_A, j_T and j_S can be calculated. For PEDOT in tetraethylammonium tetrafluoroborate, acetonitrile or dichloromethane solutions under permselective conditions, it has been shown that the rate of solvent expulsion (during doping) and entry (during dedoping) are key determinants of the switching mechanism, which changes between kinetically limited transfer and rapid solvent transfer, which depend upon the identity of the solvent.[177,178]

Other groups have also proposed other data processing to extract information on the switching mechanism from the raw experimental mass and current changes with respect to the potential. Torresi and colleagues used the following two fundamental equations giving the mass and charge changes:[179]

$$\Delta m = m_c \xi_c + m_a \xi_a + m_s \xi_s, \tag{73}$$
$$q = -F (\xi_c - \xi_a), \tag{74}$$

where m_i and ξ_i are the molar mass and the number of moles of species i exchanged, respectively (where species i is the cation, anion, or solvent). They obtained the flux of cations and anions, each with a contribution from solvent, as a function of the current density and mass flux:

$$\frac{d\left(\xi_c + \frac{m_s}{m_{ca}}\xi_s\right)}{dt} = \frac{1}{W_{ca}}\frac{d\Delta m}{dt} - \frac{m_a}{m_{ca}}\frac{j}{F}, \tag{75}$$

$$\frac{d\left(\xi_a + \frac{m_s}{m_{ca}}\xi_s\right)}{dt} = \frac{1}{W_{ca}}\frac{d\Delta m}{dt} + \frac{m_c}{m_{ca}}\frac{j}{F}, \tag{76}$$

where m_{ca} is the molar mass of the salt.

They were not been able to quantitatively determine the solvent contribution as it is not possible to evaluate the individual amount of each species involved in the charge compensation process. However, with some hypotheses (e.g. assuming that solvent molecules are associated with the hydration layer of the cations, $\xi_s = h\xi_c$) they were able to extract information on the transport number of cations with potential and on the relation between the participation of solvent and the oxidation state of the polymer matrix.[180,181]

By simultaneously measuring the current and mass changes during voltammetric experiments and calculating the instantaneous mass to electrical charge ratio, $F\, dm/dq$, at each potential, one can have access to the atomic mass of the inserted anion or cation if only one species is involved or to the difference between the atomic masses of the two species if two species are involved.[182] By considering the same fundamental equations (73) and (74) (where $\vec{n}_a = \xi_a$ and $\vec{n}_c = -\xi_c$), Ivaska and colleagues have used a quantity similar to $F\, dm/dq$:

$$\bar{m}_r = \frac{\vec{n}_a m_a + \vec{n}_c m_c}{\vec{n}_c - \vec{n}_a}. \tag{77}$$

This quantity can have two extreme values, i.e. either when only cations, i.e. $|\vec{n}_c| \gg |\vec{n}_a|$, or only anions, i.e. $|\vec{n}_a| \gg |\vec{n}_c|$, are contributing to the ionic transfer. Then,

$$m_c \geq \bar{m}_r \geq -m_a.$$

When the observed value of $\bar{m}_r$ is larger than the limits, m_c and $-m_a$, it can be concluded that some solvent molecules are simultaneously transferred with anions or cations. This method has been applied to PEDOT during doping and dedoping processes.[183] The authors have also shown that this polymer has rather slow redox behaviour in water because of its hydrophobic surface properties.[184]

Bund and Neubeck preferred analysing the dm/dt function on a permselective polymer and have shown that the solvent-exchange behaviour depends not only on the dimensions but also on the charge density of the anion.[185] By comparing also dm/dt (called "mass change rate") with the current, Kwak et al. have studied the influence of aprotic solvents (acetonitrile and propylene carbonate) on the charge compensation process during the redox switching of PPY and modified PPY.[186,187]

As shown previously, the use of electrochemistry with either QCM or probe beam deflection (PBD) techniques can only provide two pieces of information. Interpretation of the data then necessitates assumptions to be made, commonly permselectivity (no co-ion transfer) and/or no solvent transfer. By using simultaneously the three techniques, one can achieve a complete description of the charge, mass change and ion flux of all the mobile species (cations, anions and solvent). However, to obtain the relative rates of the species transfers, it has been necessary to vary the experimental time scale during the experiment, either by changing the sweeping rate of the voltammetry[188] or by analysing the responses to potential steps.[189] A comparison of the predicted PBD profiles calculated from the current and mass responses with the profile of the measured PBD signal, by temporal convolution analysis, enabled the contribution of cation (H^+), anion (ClO_4^-),] and solvent (H_2O) transfer at the first redox step to be discriminated quantitatively. A further separation of the slow-moving solvent counter flux from fast ion exchange was possible by shortening the experimental time scale.

Similarly, gravimetric measurements associated with some other techniques can be very fruitful. The use of the SECM to study poly(o-phenylenediamine) has allowed the unusual responses observed at negative potentials to be explained by a precipitation–dissolution of phenazinehydrine charge transfer complexes developed via redox switching of the oligomers.[190] Direct evidence of the exchange capability of $Fe(CN)_6^{4-/3-}$ anions entrapped in PEDOT film with Cl^- anion from the solution was found.[191] By use of contact electric resistance measurements, the doping levels of the synthesized PPY films and the mobility of charge carriers in KCl aqueous solutions were determined.[192] For indium hexacyanoferrate, the use of radiotracers has allowed the movements of anions and cations to be investigated.[193] The use of impedances was also a breakthrough in understanding the movement of neutral and charged species.[194] Also, spectroelectrochemical techniques were used with the QCM with great success.

2. AC Electrogravimetry

For electroactive thin films, the electrochemical impedance technique alone is not able to discriminate among the various models which have been proposed in the literature. So the measurement of an electrogravimetric transfer function, $\Delta m/\Delta E$, by using a fast QCM, in parallel to measurement of the impedance has been proposed to test these various models and to obtain a complete description of the ionic and solvent exchanges between the film and the solution. So far, two groups, one in Korea and the other one in France, have used this technique.

Kwak et al.[195–197] have considered that only two types of species may be involved in the charge compensation process: cations and anions, which can be solvated or not. They considered the electrochemical capacitance, $\Delta q/\Delta E$, and the electrogravimetric transfer function (which they called "electrogravimetric capacitance"), which can be separated into two parts:

$$\frac{\Delta q}{\Delta E} = \frac{\Delta q_+}{\Delta E} + \frac{\Delta q_-}{\Delta E} \tag{78}$$

and

$$\frac{\Delta m}{\Delta E} = \frac{\Delta m_+}{\Delta E} + \frac{\Delta m_-}{\Delta E} = -\frac{m'_{i+}}{z_+ F}\frac{\Delta q_+}{\Delta E} - \frac{m'_{i-}}{z_- F}\frac{\Delta q_-}{\Delta E}, \tag{79}$$

$$m'_i = m_i + Y m_s, \tag{80}$$

where the subscripts $+$ and $-$ represent cation and anion, respectively, m'_i is the molar mass of an ion and its accompanying solvent molecules, z is the electric charge of an ion, m_i is the molar mass of an ion, Y is the number of accompanying solvent molecules per ion and m_s is the molar mass of the solvent. From these equations, $\Delta m_\pm/\Delta E$ and $\Delta q_\pm/\Delta E$ can be obtained separately.

They considered equivalent circuits of the electrochemical impedance of the electrode/polymer film/electrolyte solution system (Fig. 6). If only one kind of ion transport takes part in the charge

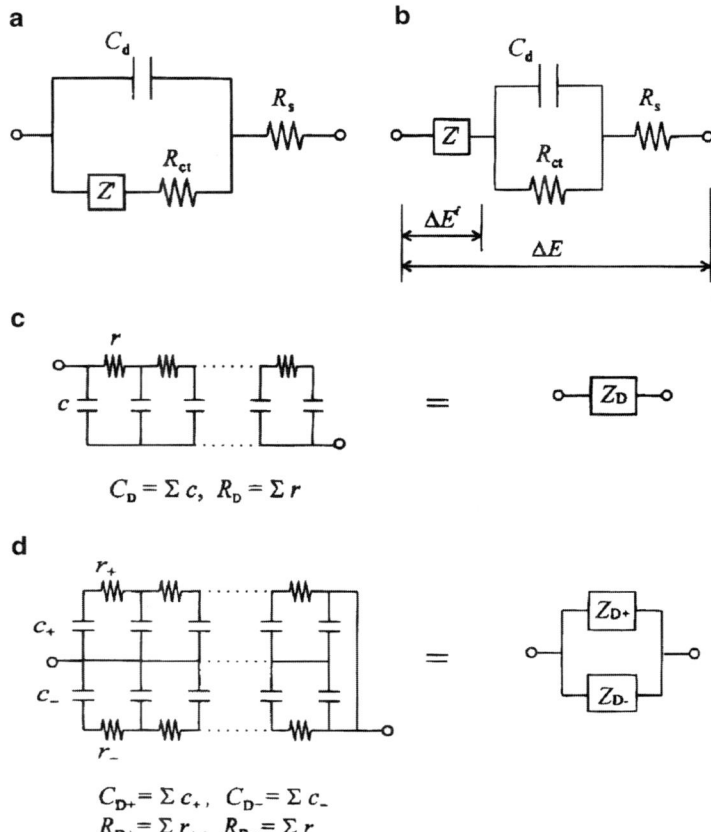

Figure 6. Equivalent circuits for a electrode/film/electrolyte solution system. (a) Classical Randles circuit, where R_s is the solution resistance, R_{ct} is the charge transfer resistance, C_d is the double-layer capacitance, and Z' is the charge transport impedance in the film. (b) Simplified circuit valid when the time scale of the ion diffusion differs from that of the charge transfer phenomena. (c) Z' in the case in which one kind of ion occurs. (d) Z' in the case in which both cation transport and anion transport occurs. Z_{D+} represents Z_D for cation transport and Z_{D-} represents Z_D for anion transport. From Yang et al.[199]

compensation, which is much slower than electron transport, Z', an impedance related to the charge transport in the film, is simplified to Z_D:

$$Z_D = R_D \frac{\coth \sqrt{j\omega R_D C_D}}{\sqrt{j\omega R_D C_D}}, \tag{81}$$

where R_D is the ionic resistance and C_D is the redox capacitance of the film.

Then, when only one ion is involved in the ion transport, the electrogravimetric transfer function is

$$\frac{\Delta m}{\Delta E} = -\frac{m'_i}{z_i F} \frac{\Delta q}{\Delta E} \tag{82}$$

and as

$$\frac{\Delta q}{\Delta E} = \frac{1}{j\omega} \frac{\Delta I}{\Delta E}, \tag{83}$$

$$\frac{\Delta m}{\Delta E} = -\frac{m'_i}{z_i F} \frac{1}{j\omega} \frac{\Delta I}{\Delta E} \tag{84}$$

and

$$\frac{\Delta m}{\Delta E} = -\frac{m'_i}{z_i F} \frac{1}{j\omega} \frac{1}{Z'}. \tag{85}$$

Therefore, when only one ion is involved in the charge transport, the electrogravimetric transfer function is

$$\frac{\Delta m}{\Delta E} = -\frac{m'_i}{z_i F} C_D \frac{\tanh \sqrt{j\omega R_D C_D}}{\sqrt{j\omega R_D C_D}}, \tag{86}$$

which is plotted in Fig. 7. As can be seen in Fig. 7, the electrogravimetric transfer function appears in the third or the first quadrant when ion transport is cation-specific or anion-specific, respectively. If dual transport takes place, Z' is given as the parallel combination of Z_{D+} and Z_{D-}, where Z_{D+} represents Z_D for cation transport and Z_{D-} represents Z_D for anion transport. Z_{D+} and Z_{D-} can be considered as transmission lines (Fig. 6c, d). The electrogravimetric transfer function is then supposed to be a combination of elementary transfer functions, which depend on R_D and C_D, as no free solvent is considered, which depends on the relative rates of anions and cations and on the nature of the dominant ion (Fig. 7c–f).

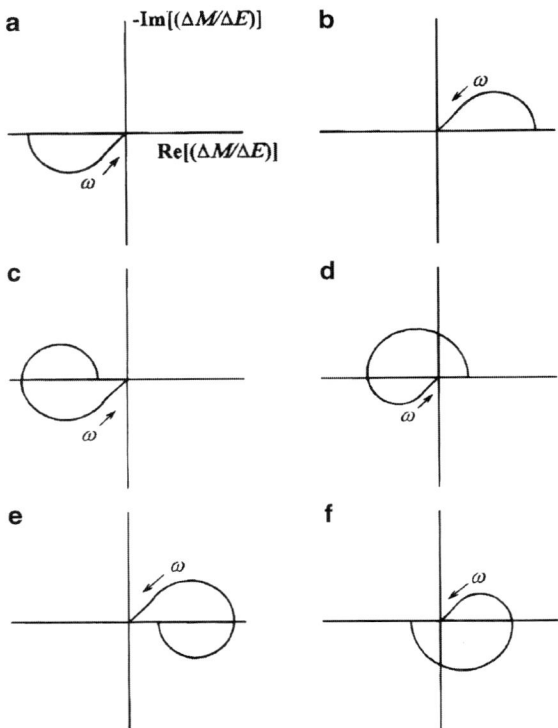

Figure 7. Electrogravimetric transfer function $(\Delta M/\Delta E)$ plots for Z_D in the case in which (**a**) ion transport is cation-specific, $C_{D+} \neq 0$, $C_{D-} = 0$, (**b**) ion transport is anion-specific, $C_{D+} = 0$, $C_{D-} \neq 0$, (**c**) and (**d**) cation transport is faster than anion transport, (**e**) and (**f**) anion transport is faster than cation transport. From Yang et al.[199]

For PPY doped with a large anion such as copper phthalocyanine tetrasulphonate, cation transport is dominant. The authors have shown that the number, Y_i, of accompanying water molecules depends on the nature and concentration of an electrolyte solution as well as on the redox state of the film. They have shown that Y_i increases as the electrolyte concentration decreases. As the decrease in electrolyte concentration causes an increase in dielectric constant of the solution, this results in an increase in the hydration number of the ions. The amount of water inserted into the film during the first cathodic scan is much larger than that during the second scan. As after deposition the film is rather compact, and has few free spaces, when a cation inserts itself into the film during the first cathodic scan, a very strong cathodic peak current appears at more negative potential than during subsequent scans and a substantial amount of water moves with a cation. In addition, within a conducting polymer film, the ionic concentration is high, but the amount of existing water is small, leading to strong ion–ion interaction inside the film.[195]

For PPY doped with an anion such as polystyrenesulphonate, cation transport is dominant, but anions are largely involved as well. By studying the influence of the film thickness, these authors suggested that for films whose thickness varies from 0.2 to 1.5 µm, cation transport would be dominant for thick films, whereas anion transport would be considerable for thin films during the redox reaction. This surprising finding could be due to the distortion of the mass measurements related to the viscoelastic behaviour of the thicker films. Besides, the measured impedance shows large Warburg impedances for the thicker films, showing that mass transport is a rate-limiting step. The electrochemical and electrogravimetric capacitances show two loops which demonstrate the presence of two types of electron transport, one is fast, the second one is slow. The semicircle in the higher-frequency region is related mainly to the cation transport, whereas that in the lower-frequency region is related mainly to anion transport.[197]

For PPY with NO_3^-, they have shown that the apparent diffusion coefficient of an ion in the fast charge transport process is governed by the diffusion coefficient of an ion, whereas the apparent diffusion coefficient of an ion in the slow charge transport process is governed by the diffusion coefficient of an electron. It has been shown that in PPY/NO_3 films anion transport is dominant in the fast charge transport process, whereas cation transport is considerable in the slow one.[198]

Using this technique, Kwak et al. have studied the possible involvement of anion transport in Prussian blue in acetonitrile and propylene carbonate solutions. It is known that cation transport is dominant during the redox reaction of Prussian blue films and that the electroactivity and ion transport behaviour of Prussian blue films are highly limited by the pore size because Prussian blue films have a rigid zeolitic structure. The electroactivity in aqueous solutions containing Na^+ or Li^+ is small, while the electroactivity in solutions containing K^+, Rb^+, Cs^+ or NH_4^+ is large. This results from the larger hydration radius of Na^+ and Li^+ compared with the pore size of Prussian blue films. In the two aprotic solvents, the electrogravimetric transfer function shows that cation transport prevails. However, in polypropylene carbonate solutions, in addition to cation transport, anion transport takes place. Anion transport occurs in the higher-frequency region, whereas cation transport is considerable in the lower-frequency region.[199] By studying ion and water transport in Prussian blue in aqueous solutions, they have shown that 56% (w/w) of the fresh Prussian blue film is water, which shows that this electrodeposited material absorbs substantial amounts of water. After a structural reorganization, ion transport during the redox reaction of the Prussian blue film is cation-dominant, with a small fraction of accompanying water transport.[200]

Our group has based the modelling on the classical redox mechanisms involving cations, anions and solvent, which lead to a generalized partition rate for species i ($J_i < 0$ for inserting species):

$$<P> + e^- + M^+ \underset{k'_c}{\overset{k_c}{\rightleftarrows}} <P, M^+>, \tag{87}$$

$$<P> + A^- \underset{k_a}{\overset{k'_a}{\rightleftarrows}} <P, A^-> + e^-, \tag{88}$$

where $<P>$ is the host film and $<P, M^+>$ and $<P, A^->$ are the inserted cations and inserted anions, whose concentrations are c_c and c_a and whose diffusion coefficients are D_a and D_c, in the film. The insertion/expulsion rates can be taken to be equal to

$$J_a(d) = k_a(c_a - c_{amin}) - k'_a(c_{amax} - c_a)c_{asol}, \tag{89}$$
$$J_c(d) = k'_c(c_c - c_{cmin}) - k_c(c_{c\,max} - c_c)c_{csol}, \tag{90}$$

where $c_{i\,max}$, is the maximum concentration of sites available for ion insertion for species $i = $ a or c, $c_{i\,min}$ is the minimum concentration of sites occupied by ions i in the host film, c_{isol} is the concentration of species i in the solution and

$$k'_c = k'_{c0} \exp b'_c E_3, \quad k_c = k_{c0} \exp b_c E_3, \tag{91}$$
$$k'_a = k'_{a0} \exp b'_a E_3, \quad k_a = k_{a0} \exp b_a E_3. \tag{92}$$

As for thin films $J_i = -d_f (dc_i/dt)$, (89) and (90) have the same form as the Bruckenstein and Hillman equation (64) when $c_{i\,min}$ is considered to be negligible. For the solvent, the same model as used by Bruckenstein and Hillman[172] (68) is adopted and which has formally the same form as the anion and cation dynamic laws:

$$J_s(d) = k_s (c_s - c_{s\,min}) - k'_s (c_{s\,max} - c_s), \qquad (93)$$

where

$$k'_s = k'_{s0} \exp b'_s E_3, \quad k_s = k_{s0} \exp b_s E_3. \qquad (94)$$

If cations, anions and solvent are involved in the redox reaction, the associated mass change and the electric charge passed through the electrode/film interface, per unit surface, are equal to, from (73) and (74),

$$\Delta m = m_c \Delta \xi_c + m_a \Delta \xi_a + m_s \Delta \xi_s, \qquad (95)$$
$$\Delta q = -F (\Delta \xi_c - \Delta \xi_a), \qquad (96)$$

where m_c, m_a and m_s are the molar masses of the cation, anion and solvent, respectively (if ions are solvated, an extra mass nm_s, where n is the number of solvent molecules, is added to their molar mass) and $\Delta \xi_c$, $\Delta \xi_a$ and $\Delta \xi_s$ are the number of moles exchanged for cationic species, anionic species and solvent per surface unit, respectively.

The net instantaneous molar flux of species i (c, a or s) is $J_i = d\xi_i/dt$ $(mol\,cm^{-2}\,s^{-1})$. It can also be expressed in terms of concentration as $c_i = \xi_i/d_f$, where d_f is the film thickness. Therefore, as $J_i(d_f) = -(d\xi_i/dt) = -d_f(dc_i/dt)$,

$$d_f \frac{dc_a}{dt} = -k_a (c_a - c_{a\,max}) + k'_a (c_{a\,min} - c_a) c_{asol},$$
$$d_f \frac{dc_c}{dt} = -k'_c (c_c - c_{c\,min}) + k_c (c_{c\,max} - c_c) c_{csol}, \qquad (97)$$
$$d_f \frac{dc_s}{dt} = -k_s (c_s - c_{s\,min}) + k'_s (c_{s\,max} - c_s),$$

where $dc_a/dt > 0$ for inserted anions. Similar expressions arise for cations and solvent.

The changes of the faradaic current and mass with respect to time are

$$i(t) = FAd_f \left(\frac{dc_c}{dt} - \frac{dc_a}{dt} \right), \quad (98)$$

$$m(t) = A \left[m_c c_c(t) + m_a c_a(t) + m_s c_s(t) \right], \quad (99)$$

where A is the area of the electrode.

(i) Steady State

From (97) and similar equations for cations and solvent, the steady-state concentration of species i at potential E_3, obtained for $(d/dx) \equiv 0$, which gives the insertion isotherm of species i, is

$$c_i(E_3) = \frac{c_{i\max} \exp\left[(b'_i - b_i)(E_3 - E_i)\right] + c_{i\min}}{1 + \exp\left[(b'_i - b_i)(E_3 - E_i)\right]}, \quad i = a, c, s, \quad (100)$$

where E_i is such that

$$c_{isol} \frac{k'_{i0}}{k_{i0}} = \exp\left[-(b'_i - b_i) E_i\right]. \quad (101)$$

Its derivative, i.e. the slope of the insertion or expulsion isotherm, is

$$\frac{dc_i}{dE_3} = \frac{b_i - b'_i}{4} \frac{c_{i\max} - c_{i\min}}{\cosh^2\left[(b'_i - b_i)(E_3 - E_i)/2\right]}, \quad (102)$$

which is bell-shaped and has a maximum for $E_3 = E_i$ equal to

$$\left(\frac{dc_i}{dE_3}\right)_{\max} = \frac{(c_{i\max} - c_{i\min})}{4} (b'_i - b_i). \quad (103)$$

(ii) Dynamic Regime

From the kinetic insertion/expulsion equations, (89), (90) and (93), if small-amplitude potential perturbations, ΔE_3, are considered, the concentration responses, Δc_i, in the frequency domain, are such that

$$\Delta J_i(d) = -j\omega d_f \Delta c_i = K_i \Delta c_i + G_i \Delta E_3, \quad i = a, c, s, \quad (104)$$

where

$$G_i = \left(\frac{\partial J_i}{\partial E_3}\right)_{c_i} = b_i k_i (c_i - c_{i\min}) - b'_i k'_i (c_{i\max} - c_i) c_{i\text{sol}}, \quad i = \text{a, s,}$$

$$K_i = \left(\frac{\partial J_i}{\partial c_i}\right)_{E_3} = k_i + k'_i c_{i\text{sol}} \qquad (105)$$

and

$$G_c = \left(\frac{\partial J_c}{\partial E_3}\right)_{c_c} = b'_c k'_c (c_c - c_{c\min}) - b_c k_c (c_{c\max} - c_c) c_{c\text{sol}},$$

$$K_c = \left(\frac{\partial J_c}{\partial c_c}\right)_{E_3} = k'_c + k_c c_{c\text{sol}}, \qquad (106)$$

where $G_i < 0$ for inserting species and $G_i > 0$ for expelling species.

Then, according to (104), the change of the concentration of species i with potential is

$$\frac{\Delta c_i}{\Delta E_3}(\omega) = \frac{-G_i}{j\omega d_f + K_i} \quad i = \text{a,c,s.} \qquad (107)$$

(iii) Electrochemical Impedance

Indeed, the faradaic current density, ΔI_F, is related to the charge, Δq, by $\Delta I_F = j\omega \Delta q$, and from (96)

$$\frac{\Delta E_3}{\Delta I_F}(\omega) = \frac{1}{j\omega A d_f F \left(-\frac{\Delta c_c}{\Delta E_3}(\omega) + \frac{\Delta c_a}{\Delta E_3}(\omega)\right)} \qquad (108)$$

and the total impedance of the film is equal to

$$\frac{\Delta E_3}{\Delta I}(\omega) = \frac{1}{j\omega C_{\text{int}} + j\omega A d_f F \left(-\frac{\Delta c_c}{\Delta E_3}(\omega) + \frac{\Delta c_a}{\Delta E_3}(\omega)\right)}, \qquad (109)$$

where C_{int} is the polymer/solution interface capacitance, the electrolyte resistance being neglected here. Then,

$$\frac{\Delta E_3}{\Delta I_F}(\omega) = \frac{1}{j\omega A d_f F \left(\frac{G_c}{j\omega d_f + K_c} - \frac{G_a}{j\omega d_f + K_a}\right)}. \qquad (110)$$

At high frequency, a charge transfer resistance is defined such as

$$R_{ct} = \frac{1}{FA(G_c - G_a)} \quad (111)$$

and the measurable impedance is

$$\frac{\Delta E_3}{\Delta I}(\omega) = \frac{1}{j\omega C_{int} + j\omega A d_f F \left(\frac{G_c}{j\omega d_f + K_c} - \frac{G_a}{j\omega d_f + K_a}\right)}. \quad (112)$$

Another interesting quantity can be calculated to discriminate between the influences of the various species: the electric charge/potential transfer function, which is also called the "electrochemical capacitance", $\Delta q / \Delta E_3(\omega)$,

$$\frac{\Delta q}{\Delta E_3}(\omega) = \frac{1}{j\omega} \frac{\Delta I}{\Delta E_3}(\omega) = C_{int} + A d_f F \left(\frac{G_c}{j\omega d_f + K_c} - \frac{G_a}{j\omega d_f + K_a}\right). \quad (113)$$

(a) Comparison with impedance models

The investigation of electroactive materials under the form of thin films deposited on one of the electrodes of a quartz crystal resonator necessitates the use of very thin films as will be explained in Sect. V. Then all the quantities calculated in the previous sections will be examined when $d_f \to 0$ and compared with the proposed model.

For a permselective electroactive film and considering insertion of a counterion from the solution, the impedance of thin films is calculated from (25) and (26) and using the same notation as the proposed model (i.e. (104))

$$Z_3(\omega) = \frac{1}{FG_i} - \frac{d_f K_i}{8FG_i \upsilon D_e D_i} [(D_e + D_i) \coth \upsilon \\ + (D_e - D_i) \tanh \upsilon]. \quad (114)$$

For thin films, as $\upsilon = \alpha d_f$, $\coth \upsilon \sim (1/\upsilon) + (\upsilon/3)$ and $\tanh \upsilon \sim \upsilon$. Hence, when $d_f \to 0$, (25), (26) and (114) become

$$Z_1(\omega) \sim R_e + \frac{1}{\sigma}\frac{D_e + D_i}{8\alpha D_e D_i}\left[(D_e + D_i)\left(\frac{1}{\alpha d_f} + \frac{\alpha d_f}{3}\right)\right.$$
$$\left. - (D_e - D_i)\alpha d_f\right], \tag{115}$$

$$Z_2(\omega) \sim \frac{d_f}{\sigma} + \frac{1}{\sigma}\frac{(D_e - D_i)^2}{4\alpha D_e D_i}\alpha d_f, \tag{116}$$

$$Z_3(\omega) \sim -\frac{1}{FG_i} - \frac{K_i}{G_i}\frac{1}{8F\alpha D_e D_i}\left[(D_e + D_i)\left(\frac{1}{\alpha d_f} + \frac{\alpha d_f}{3}\right)\right.$$
$$\left. + (D_e - D_i)\alpha d_f\right]. \tag{117}$$

Then, the total faradaic impedance is

$$Z_F(\omega) \sim R_e + \frac{1}{\sigma}\frac{(D_e + D_i)^2}{8\alpha^2 d_f D_e D_i} - \frac{1}{FG_i} - \frac{K_i}{G_i}\frac{(D_e + D_i)}{8F\alpha^2 d_f D_e D_i}, \tag{118}$$

i.e.

$$Z_F(\omega) \sim R_e + \frac{RT}{c_p F^2}\frac{1}{j\omega d_f} - \frac{1}{FG_i} - \frac{K_i}{FG_i}\frac{1}{j\omega d_f}, \tag{119}$$

which can be represented by a R_{lf}, C_{lf} series circuit such as

$$R_{lf} = R_e - \frac{1}{FG_i} \text{ and } C_{lf} = d_f\left(\frac{c_p F^2}{RT} - \frac{FG_i}{K_i}\right). \tag{120}$$

To obtain the total impedance of the electrode/film/solution system, it is necessary to take into account the double-layer capacitances at each interface. Therefore, the impedance of thin permselective films depends on both the electron and the ion transfer elementary impedances because the bulk film impedance, $Z_2(\omega)$, is negligible.

It is equivalent to anion insertion without cation movement of the proposed model (112) if R_e and the electron contributions are neglected (for fast electron transfer):

$$\frac{\Delta E_3}{\Delta I_F}(\omega) = -\frac{j\omega d_f + K_a}{j\omega d_f F G_a}. \tag{121}$$

If permselectivity is not supposed, i.e. when anions, cations and solvent are considered to be involved in the switching process of the electroactive film, in addition to the electrons, the full analytical calculation is not possible and the approximate diffusion model described previously can be used. So, the charged species behave independently and local electroneutrality is not imposed.

Here, anions and cations are supposed to be transported through the polymer only by diffusion. From (56)–(58), when $d_f \to 0$, one has

$$Z_1(\omega) \sim \frac{1}{FG_e}\left(1 - K_e \frac{(1/d_f\sqrt{j\omega/D_e}) + (d_f\sqrt{j\omega/D_e}/3)}{\sqrt{j\omega D_e}}\right), \tag{122}$$

$$Z_1(\omega) \sim \frac{1}{FG_e}\left(1 - K_e \frac{1}{j\omega d_f}\right), \tag{123}$$

$$Z_3^{-1}(\omega) \sim F\left(\frac{G_c}{1 + K_c\left((1/d_f\sqrt{j\omega/D_c}) + (d_f\sqrt{j\omega/D_c}/3)\right)/\sqrt{j\omega D_c}}\right.$$
$$\left. - \frac{G_a}{1 + K_a\left((1/d_f\sqrt{j\omega/D_a}) + (d_f\sqrt{j\omega/D_a}/3)\right)/\sqrt{j\omega D_a}}\right), \tag{124}$$

$$Z_3(\omega) \sim \frac{1}{j\omega d_f F\left[(G_c/(j\omega d_f + K_c)) - (G_a/(j\omega d_f + K_a))\right]}, \tag{125}$$

$$Z_2(\omega) \sim \frac{2\pi d_f^2}{\varepsilon\varepsilon_0}\left\{\left[\frac{G_c}{j\omega d_f + K_c} - \frac{G_a}{j\omega d_f + K_a}\right] Z_3(\omega) + \frac{1}{j\omega F d_f}\right\} \tag{126}$$

and then

$$Z_F(\omega) \sim \frac{1}{FG_e}\left(1 - K_e \frac{1}{j\omega d_f}\right)$$
$$+ \frac{1}{j\omega d_f F\left[(G_c/(j\omega d_f + K_c)) - (G_a/(j\omega d_f + K_a))\right]}, \tag{127}$$

which gives the same expression as the impedance calculated from the proposed model if the electron contribution is neglected. This means that the proposed model is coherent with the classical migration–diffusion model but it takes into account only the exchange of species between the film and the solution as the film is supposed to be very thin and the electron transfer very fast.

(iv) Mass/Potential (Electrogravimetric) Transfer Function

Generally, the change of the mass of the film is

$$\Delta m = -A \sum_{i=a,c,s} m_i \int_0^{d_f} \Delta c_i \mathrm{d}x. \tag{128}$$

Hence, for thin films

$$\Delta m = -A d_f \sum_{i=a,c,s} m_i \Delta c_i. \tag{129}$$

The mass/potential transfer function is then defined by

$$\frac{\Delta m}{\Delta E_3} = -A d_f \sum_{i=a,c,s} m_i \frac{\Delta c_i}{\Delta E_3}, \tag{130}$$

i.e.

$$\frac{\Delta m}{\Delta E_3}(\omega) = -A d_f \left(m_c \frac{G_c}{\mathrm{j}\omega d_f + K_c} + m_a \frac{G_a}{\mathrm{j}\omega d_f + K_a} + m_s \frac{G_s}{\mathrm{j}\omega d_f + K_s} \right) \tag{131}$$

and

$$\frac{\Delta m}{\Delta I_F}(\omega) = Z_3(\omega) \frac{\Delta m}{\Delta E_3}, \tag{132}$$

where m_a, m_c and m_s are the molar masses of anions, cations and solvent (the mass of the electron is supposed to be 0). Of course, only the charged species, anions and cations, are involved in the electrochemical impedance, $\Delta E_3 / \Delta I_F(\omega)$, whereas the electrogravimetric transfer function, $\Delta m / \Delta E_3(\omega)$, depends on anions, cations, and solvent all together.

Another interesting quantity is the change of mass per unit charge, $F (\Delta m/\Delta q)$, such as

$$\frac{\Delta m}{\Delta q}(\omega) = j\omega \frac{\Delta m}{\Delta I_F} = j\omega Z_3(\omega) \frac{\Delta m}{\Delta E_3} \quad (133)$$

and then the mass/electric charge transfer function is

$$\frac{\Delta m}{\Delta q}(\omega)$$
$$= \frac{-[m_c (G_c/(j\omega d_f + K_c)) + m_a (G_a/(j\omega d_f + K_a)) + m_s (G_s/(j\omega d_f + K_s))]}{F[(G_c/(j\omega d_f + K_c)) - (G_a/(j\omega d_f + K_a))]}. \quad (134)$$

When only one charged species is involved in the redox switching, the latter function has the particular form

$$F\frac{\Delta m}{\Delta q} = m_i \quad (135)$$

which gives the molar mass of the species involved in the charge compensation process occurring during the redox switching of the electroactive film.

(v) Diagnostic Criterion

As the model takes into account two charged species (cations and anions) plus the solvent, there are, a priori, two degrees of freedom for the impedance, which is concerned with the two charged species, and three degrees of freedom for the electrogravimetric transfer function, which is concerned with the anions, cations and solvent. Generally, three degrees of freedom lead to three time constants and only two quantities $\Delta E/\Delta I$ and $\Delta m/\Delta E$ are experimentally available. However, calculated partial electrogravimetric transfer functions will allow the three species to be distinguished.

By using (130) and (108), one can eliminate the anion contribution in the total mass change, $\Delta m/\Delta E_3$, by considering

$$\frac{\Delta m}{\Delta E_3} - m_a A d_f \frac{\Delta c_a}{\Delta E_3} = A d_f \left(m_c \frac{\Delta c_c}{\Delta E_3} + m_s \frac{\Delta c_s}{\Delta E_3} \right), \quad (136)$$

$$\frac{\Delta m}{\Delta E_3} - m_a A d_f \frac{\Delta c_a}{\Delta E_3} = \frac{\Delta m}{\Delta E_3} - m_a A d_f \left(\frac{\Delta c_c}{\Delta E_3} + \frac{\Delta I_F}{j\omega d_f F \Delta E_3} \right). \quad (137)$$

Therefore, by equating the left-had side of (136) and (137), the quantity

$$\frac{\Delta m_{cs}}{\Delta E_3} = \frac{\Delta m}{\Delta E_3} - \frac{m_a}{j\omega}\frac{A}{FZ_3} \quad (138)$$

takes into account only the mass change of cations and solvent as this partial electrogravimetric transfer function is equal to

$$\frac{\Delta m_{cs}}{\Delta E_3} = Ad_f\left((m_a + m_c)\frac{\Delta c_c}{\Delta E_3} + m_s\frac{\Delta c_s}{\Delta E_3}\right), \quad (139)$$

i.e.

$$\frac{\Delta m_{cs}}{\Delta E_3} = -Ad_f\left((m_a + m_c)\frac{G_c}{j\omega d_f + K_c} + m_s\frac{G_s}{j\omega d_f + K_s}\right). \quad (140)$$

In the same way, if the cation contribution is eliminated in the mass change, $\Delta m/\Delta E_3$, by using (130) and (108), the second partial electrogravimetric transfer function, $\Delta m_{as}/\Delta E_3$, which takes into account only the mass change of anions and solvent, is

$$\frac{\Delta m_{as}}{\Delta E_3} = \frac{\Delta m}{\Delta E_3} + \frac{m_c}{j\omega}\frac{1}{FZ_3}, \quad (141)$$

i.e.

$$\frac{\Delta m_{as}}{\Delta E_3} = Ad_f\left((m_a + m_c)\frac{\Delta c_a}{\Delta E_3} + m_s\frac{\Delta c_s}{\Delta E_3}\right), \quad (142)$$

i.e.

$$\frac{\Delta m_{as}}{\Delta E_3} = -Ad_f\left((m_a + m_c)\frac{G_a}{j\omega d_f + K_a} + m_s\frac{G_s}{j\omega d_f + K_s}\right). \quad (143)$$

From the plot of the cation–solvent, $\Delta m_{cs}/\Delta E_3$, and anion–solvent, $\Delta m_{as}/\Delta E_3$, partial electrogravimetric transfer functions, a diagnostic criterion can be proposed.

In the complex plane, as $\Delta c_s/\Delta E_3$ is the common relaxation loop in $\Delta m_{cs}/\Delta E_3$ and $\Delta m_{as}/\Delta E_3$, if both $\Delta m_{cs}/\Delta E_3$ and $\Delta m_{as}/\Delta E_3$ show only one loop, $\Delta c_s/\Delta E_3 = 0$ and no solvent crosses the polymer/solution interface, and only anions and cations enter or are expelled. However, if $\Delta m_{as}/\Delta E_3$ shows two loops and $\Delta m_{cs}/\Delta E_3$ shows only one loop, then the cations do not interfere in the charge compensation process.

(vi) Simulations

The various time-varying and frequency-varying quantities which characterize the behaviour of the model are calculated for various insertion/expulsion mechanisms. Redox switching of an electroactive film involving a cation alone, then an anion and solvent, and finally a cation, an anion and solvent will be successively examined. First of all, the change of the concentrations with respect to potential are calculated by integrating the differential equations (97) for cations, anions and solvent by assuming a linear change of potential with time. Then the current and the change of mass are calculated by using (98) and (99), respectively. Finally, the quantity $F(dm/dq)$ is calculated through $F(dm/I\,dt)$. Concerning the frequency-varying transfer functions, the impedance $\Delta E/\Delta I\,(\omega)$ is calculated from (109), the electrochemical capacitance $\Delta q/\Delta E\,(\omega)$ is calculated from (113), the electrogravimetric transfer function, $\Delta m/\Delta E\,(\omega)$, is calculated from (131), $F\,\Delta m/\Delta q\,(\omega)$ is calculated from (133), and the partial electrogravimetric transfer functions $\Delta m_{cs}/\Delta E\,(\omega)$ and $\Delta m_{as}/\Delta E\,(\omega)$ are calculated from (138) and (141), respectively. Notice that in the following, as the potential differences across the electrode/film interface and the bulk film will be neglected, the difference of potential E_3 across the film/electrolyte interface will be assimilated to E. All the calculations were carried out using Mathcad 13 (Mathsoft).

Figure 8 shows the various time-varying quantities for a redox switching involving only one cation for a $20\,\text{mV}\,\text{s}^{-1}$ potential scan rate. Figure 8a shows the variation of the concentrations of the cations, $c_c(E)$, from c_{max} to c_{min} for increasing potentials and from c_{min} to c_{max} for decreasing potentials. The difference between the up and down plots is due to kinetics as the system is slow compared with the potential scan rate. The relative change of the mass, $\Delta m\,(E)$, which is arbitrarily taken to be equal to 0 for anodic potentials, is strictly related to the concentrations of the cations as they are the only ions involved in the insertion mechanism (Fig. 8c). Figure 8b shows the cyclic voltammogram, $i\,(E)$; the anodic and cathodic current peaks are not at the same potential as the system is not reversible. Finally, Fig. 8d gives the quantity $F(dm/dq)$, which is a potential-independent constant equal to $-23\,\text{g}$ according to (135). This value is coherent with the Na^+ ion used for the simulation.

Figure 9 shows the various transfer functions related to the insertion/expulsion of cations alone. Figure 9a shows the

AC-Electrogravimetry Investigation in Electroactive Thin Films

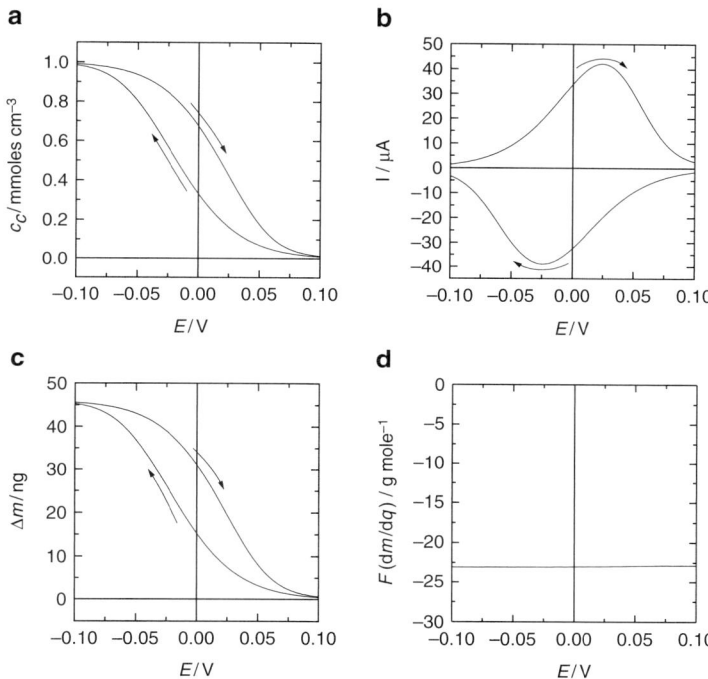

Figure 8. Calculated parameters vs. a voltage scan: (**a**) c_c, (**b**) I, (**c**) Δm, (**d**) $F(\mathrm{d}m/\mathrm{d}q)$. The parameters used for the calculation were as follows: $v = 20\,\mathrm{mV\,s^{-1}}$, $d_f = 0.1\,\mu\mathrm{m}$, $k_{c0} = 5.0 \times 10^{-2}\,\mathrm{cm\,s^{-1}}$, $k'_{c0} = 5.0 \times 10^{-6}\,\mathrm{cm\,s^{-1}}$, $b_c = -20\,\mathrm{V^{-1}}$, $b'_c = 25\,\mathrm{V^{-1}}$, $c_{\mathrm{cmin}} = 10^{-4}\,\mathrm{M}$, $c_{\mathrm{cmax}} = 1\,\mathrm{M}$ and $m_c = 23.0\,\mathrm{g\,mol^{-1}}$.

electrochemical impedance in the $(-\mathrm{Im}\,[Z(\omega)],\ \mathrm{Re}\,[Z(\omega)])$ complex plane calculated from

$$\frac{\Delta E}{\Delta I}(\omega) = \frac{1}{\mathrm{j}\omega C_{\mathrm{int}} + \mathrm{j}\omega d_f\, F\, \frac{G_c}{\mathrm{j}\omega d_f + K_c}}, \tag{144}$$

which is a particular case of (112) for cations alone ($G_a = 0$). At low frequency, a vertical line is found related to the R_{lf}, C_{lf} series circuit (118). At higher frequencies, the semicircle is due to the charge transfer resistance (111) in parallel to the double-layer capacitance across the film/solution interface. Figure 9b shows the

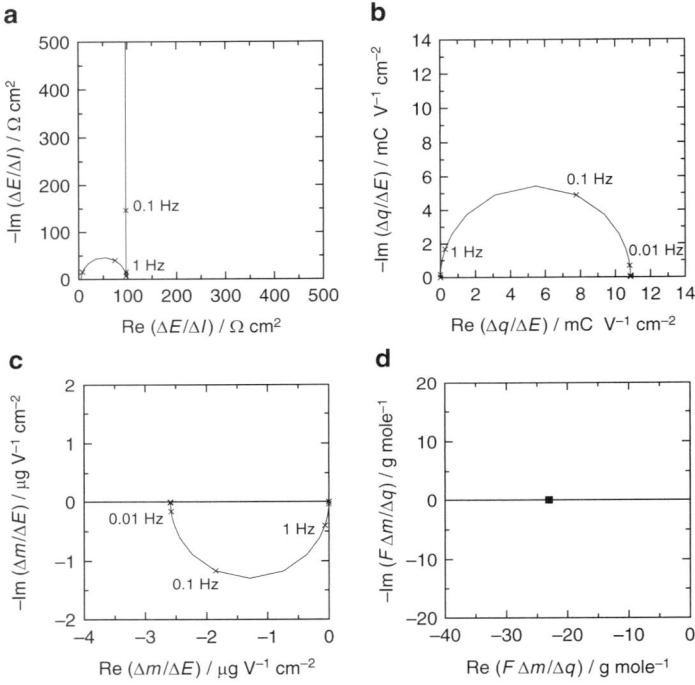

Figure 9. Calculated transfer functions at 0 V: (a) $\Delta E/\Delta I(\omega)$, (b) $\Delta q/\Delta E(\omega)$, (c) $\Delta m/\Delta E(\omega)$, (d) $F\Delta m/\Delta q(\omega)$. The parameters used for the calculation were as follows: $d_f = 0.1\,\mu\text{m}$, $k_{c0} = 5.0 \times 10^{-2}\,\text{cm s}^{-1}$, $k'_{c0} = 5.0 \times 10^{-6}\,\text{cm s}^{-1}$, $b_c = -20\,\text{V}^{-1}$, $b'_c = 25\,\text{V}^{-1}$, $c_{\text{cmin}} = 10^{-4}\text{M}$, $c_{\text{cmax}} = 1\,\text{M}$ and $m_c = 23.0\,\text{g mol}^{-1}$.

electrochemical capacitance function $(\Delta q/\Delta E)$; it is a semicircle whose equation is a particular case of (113) when $G_a = 0$,

$$\frac{\Delta m}{\Delta q}(\omega) = \frac{1}{j\omega}\frac{\Delta I}{\Delta E}(\omega) = C_{\text{int}} + d_f F\left(\frac{G_c}{j\omega d_f + K_c}\right), \quad (145)$$

and which goes from $(\Delta m/\Delta q)(\omega \to \infty) = C_{\text{int}}$ at high frequency to $(\Delta m/\Delta q)(\omega \to 0) = C_{\text{int}} + d_f F(G_c/K_c)$ at low frequency. Figure 9c gives the electrogravimetric transfer function. In agreement with (131) when $G_a = G_s = 0$, it is a semicircle located in the third quadrant of the complex plane as it is a cation which is involved in the charge compensation process. Figure 9d gives the mass change per charge unit ($F\Delta m/\Delta q$) function in the complex plane; it is reduced to one point on the abscissa at -23 g mol^{-1} as only one sodium ion is involved.

For a permselective film involving an anion and solvent, the time-varying quantities are quite similar to those given in Fig. 8 for a cation alone. Hence, only the frequency functions are given in Fig. 10 for this mechanism. The functions shown in Fig. 10a, and b are similar to those shown in Fig. 9a and b as process analysed involved only one charged ion as well. Figure 10c shows the electrogravimetric transfer function which exhibits two loops, one at high frequency, related to the anion movement, is located in the first quadrant, whereas the low-frequency loop, related to the solvent movement, is located in the fourth quadrant. Besides, the partial cation–solvent electrogravimetric transfer function, $\Delta m_{cs}/\Delta E$, where the anion participation has been eliminated, exhibits only a semicircle related to the solvent. Figure 10d shows the $F(\Delta m/\Delta q)$ transfer function. From (134) it is

$$F\frac{\Delta m}{\Delta q}(\omega) = m_a + m_s \frac{G_s\,(j\omega d_f + K_a)}{G_a\,(j\omega d_f + K_s)}. \tag{146}$$

Figures 11 and 12 are relative to an insertion/expulsion mechanisms involving cations, anions and solvent. Figure 11a–c gives the variations of the concentrations of cations, solvent and anions. As the rate of the anion movement is rather low, the kinetic hysteresis between the concentration curves up and down is very large. Figure 11d shows the voltammogram, which is very similar to Fig. 8b as the cation is the dominant species. This observation is also valid for the mass change (Fig. 11e). Figure 11f gives the quantity $F(dm/dq)$; it is rather difficult to interpret this function for the mechanism analysed, although for cathodic potentials, it is close to -23 g mol^{-1}, which is the molar mass assumed for the cation in this case.

Figure 12a gives the impedance plot; as usual it has the shape of a semicircle with a low-frequency capacitive behaviour. The electrochemical capacitance is more attractive as it exhibits two semicircular loops, one for the cation in the high-frequency range, the other one for the anion in the low-frequency range (Fig. 12b). Figure 12c seems to be more interesting as the electrogravimetric transfer function shows three loops, one for each species. In the third quadrant, two loops are found; the higher-frequency one is due to the cation, whereas the other one is related to the solvent. In the second

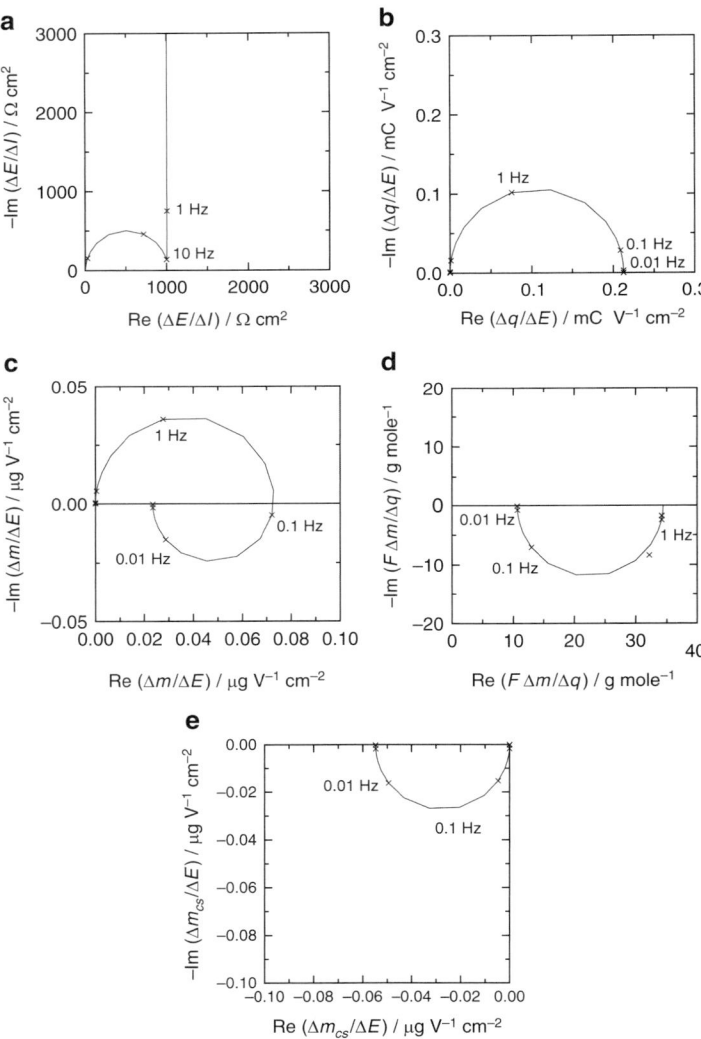

Figure 10. Calculated transfer function for cation-solvent mechanism at 0.05 V: (**a**) $\Delta E/\Delta I(\omega)$, (**b**) $\Delta q/\Delta E(\omega)$, (**c**) $\Delta m/\Delta E(\omega)$, (**d**) $F\Delta m/\Delta q(\omega)$, (**e**) $\Delta m_{cs}/\Delta E(\omega)$. The parameters used for the calculation were as follows: $d_f = 0.1\,\mu m$, $k_{a0} = 8.0 \times 10^{-6}\,cm\,s^{-1}$, $k'_a0 = 1.0 \times 10^{-1}\,cm\,s^{-1}$, $b_a = -30\,V^{-1}$, $b'_a = 30\,V^{-1}$, $c_{amin} = 3 \times 10^{-4}\,M$, $c_{amax} = 10^{-1}\,M$, $m_a = 35.5\,g\,mol^{-1}$, $k_{s0} = 5.0 \times 10^{-3}\,cm\,s^{-1}$, $k'_{s0} = 5.0 \times 10^{-7}\,cm\,s^{-1}$, $b_s = -30\,V^{-1}$, $b'_s = 20\,V^{-1}$, $c_{smin} = 5 \times 10^{-4}\,M$, $c_{smax} = 3 \times 10^{-2}\,M$ and $m_s = 18.0\,g\,mol^{-1}$.

AC-Electrogravimetry Investigation in Electroactive Thin Films

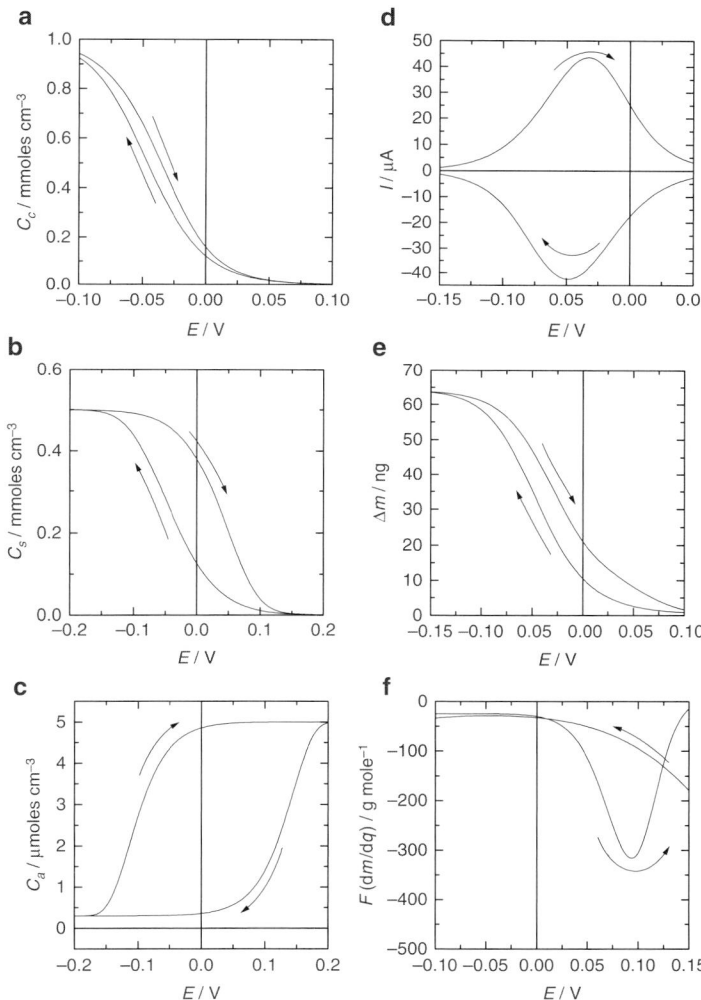

Figure 11. Calculated parameters for a cation-anion-solvent mechanism vs. a voltage scan: (a) c_c, (b) c_s, (c) c_a, (d) I, (e) Δm, (f) $F(dm/dq)$. The parameters used for the calculation were as follows: $v = 20\,\text{mV}\,\text{s}^{-1}$, $d_f = 0.1\,\mu\text{m}$, $k_{c0} = 8.0 \times 10^{-2}\,\text{cm}\,\text{s}^{-1}$, $k'_{c0} = 5.0 \times 10^{-5}\,\text{cm}\,\text{s}^{-1}$, $b_c = -20\,\text{V}^{-1}$, $b'_c = 25\,\text{V}^{-1}$, $c_{c\min} = 10^{-4}\,\text{M}$, $c_{c\max} = 1\,\text{M}$, $m_c = 23.0\,\text{g}\,\text{mol}^{-1}$, $k_{a0} = 2.0 \times 10^{-7}\,\text{cm}\,\text{s}^{-1}$, $k'_{a0} = 8.0 \times 10^{-4}\,\text{cm}\,\text{s}^{-1}$, $b_a = -30\,\text{V}^{-1}$, $b'_a = 30\,\text{V}^{-1}$, $c_{a\min} = 3 \times 10^{-4}\,\text{M}$, $c_{a\max} = 5 \times 10^{-3}\,\text{M}$, $m_a = 35.5\,\text{g}\,\text{mol}^{-1}$, $k_{s0} = 2.0 \times 10^{-2}\,\text{cm}\,\text{s}^{-1}$, $k'_{s0} = 2.0 \times 10^{-6}\,\text{cm}\,\text{s}^{-1}$, $b_s = -15\,\text{V}^{-1}$, $b'_s = 20\,\text{V}^{-1}$, $c_{s\min} = 2 \times 10^{-4}\,\text{M}$, $c_{s\max} = 5 \times 10^{-1}\,\text{M}$ and $m_s = 18.0\,\text{g}\,\text{mol}^{-1}$.

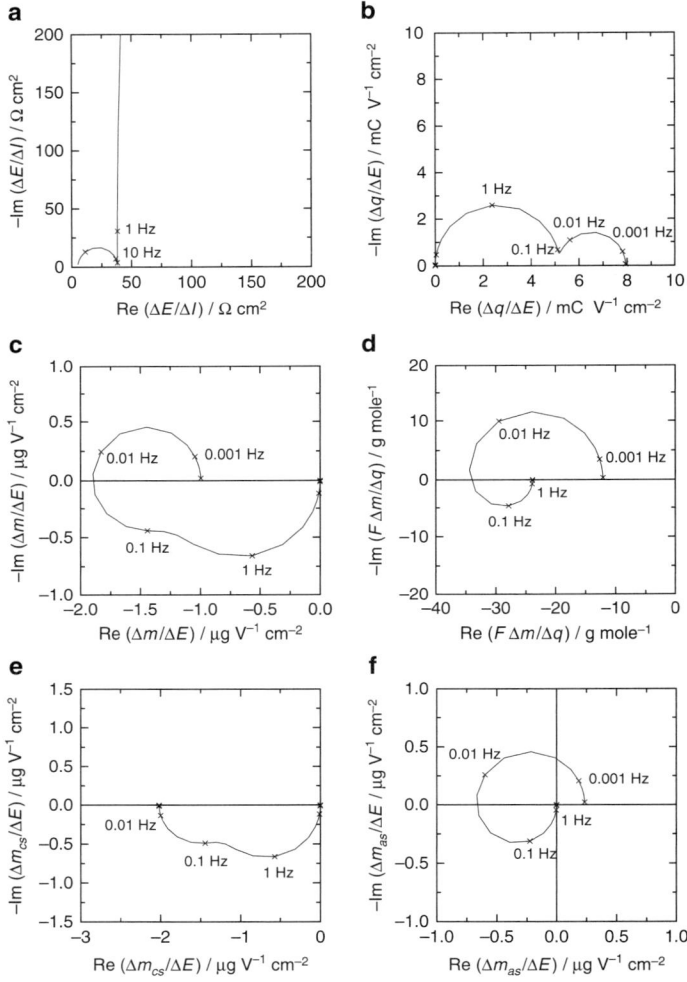

Figure 12. Calculated transfer functions for a cation-anion-solvent mechanism at 0 V: (a) $\Delta E/\Delta I(\omega)$, (b) $\Delta q/\Delta E(\omega)$, (c) $\Delta m/\Delta E(\omega)$, (d) $F\Delta m/\Delta q(\omega)$, (e) $\Delta m_{cs}/\Delta E(\omega)$ and (f) $\Delta m_{as}/\Delta E(\omega)$. The parameters used for the calculation were as follows: $d_f = 0.1\,\mu\text{m}$, $k_{c0} = 8.0 \times 10^{-2}\,\text{cm s}^{-1}$, $k'_{c0} = 5.0 \times 10^{-5}\,\text{cm s}^{-1}$, $b_c = -20\,\text{V}^{-1}$, $b'_c = 25\,\text{V}^{-1}$, $c_{cmin} = 10^{-4}\,\text{M}$, $c_{cmax} = 1\,\text{M}$, $m_c = 23.0\,\text{g mol}^{-1}$, $k_{a0} = 2.0 \times 10^{-7}\,\text{cm s}^{-1}$, $k'_{a0} = 8.0 \times 10^{-4}\,\text{cm s}^{-1}$, $b_a = -30\,\text{V}^{-1}$, $b'_a = 30\,\text{V}^{-1}$, $c_{amin} = 3 \times 10^{-4}\,\text{M}$, $c_{amax} = 5 \times 10^{-3}\,\text{M}$, $m_a = 35.5\,\text{g mol}^{-1}$, $k_{s0} = 2.0 \times 10^{-2}\,\text{cm s}^{-1}$, $k'_{s0} = 2.0 \times 10^{-6}\,\text{cm s}^{-1}$, $b_s = -15\,\text{V}^{-1}$, $b'_s = 20\,\text{V}^{-1}$, $c_{smin} = 2 \times 10^{-4}\,\text{M}$, $c_{smax} = 5 \times 10^{-1}\,\text{M}$ and $m_s = 18.0\,\text{g mol}^{-1}$.

quadrant, in the lowest-frequency range the loop is related to the anion movement. Besides, the partial electrogravimetric transfer functions help in identifying the loops. The cation–solvent electrogravimetric transfer function, $\Delta m_{cs}/\Delta E$ (Fig. 12e), shows the two loops in the third quadrant, characteristic of the cation and the solvent, as the anion loop has disappeared. The anion–solvent electrogravimetric transfer function, $\Delta m_{as}/\Delta E$ (Fig. 12f), shows one loop in the third quadrant, characteristic of the solvent, and one loop in the fourth quadrant, characteristic of the anion, as the cation loop has disappeared.

V. EXPERIMENTAL

AC electrogravimetry consists in the simultaneous measurements of the electrochemical impedance and the mass/potential, or electrogravimetric, transfer function. Impedance measurements are already well documented in the literature;[201,202] hence, in this text, only the electrogravimetric transfer function measurement will be described. It uses a fast QCM whose general concepts are first given below.

1. Basic Microbalance Concepts

Electrochemical and electrogravimetric experiments were carried out by means of a typical three-electrode cell polarized by using a potentiostat (SOTELEM-PGSTAT). Electroactive films were deposited onto one of the gold electrodes of the quartz crystal used as the grounded working electrode, a large-area platinum grid was used as the counter electrode and a saturated calomel electrode (SCE) was the reference electrode.

The electrochemical QCM takes advantage of the change of the resonance frequency of a 6 or 9 MHz "AT-cut" quartz crystal (Matel-Fordhal, France) due to a minute mass change of one of its electrodes exposed to the solution, the other one being kept in air. The two gold electrodes, deposited on the opposite faces of the quartz crystal, allowed the resonator to be electrically connected to an oscillator circuit.[203] The microbalance frequency change, Δf_m, of the oscillator where the quartz crystal is loaded by some material of mass, Δm, is usually interpreted in terms of mass change thanks to the Sauerbrey relationship:[204,205]

$$\Delta f_m = -K_{th}^S \Delta m, \qquad (147)$$

where the theoretical value of the coefficient, K_{th}, is $18.3 \times 10^7 \mathrm{Hz\,g^{-1}\,cm^2}$ for a 9 MHz resonance frequency quartz resonator. This relationship is valid for a small mass increase or decrease.

This relation is only valid for acoustically thin films. Concretely, it means that the relation is not respected when thick layers are used because the influence of the viscoelastic properties of the layer may appear in the frequency change.[206] This effect is, in general, amplified when polymer films are used. To determine the maximum useful thickness, electroacoustic measurements allowed a pertinent value to be evaluated. In a first step, a classical Butterworth–Van Dyke equivalent circuit of the loaded quartz (Fig. 13a) is extracted

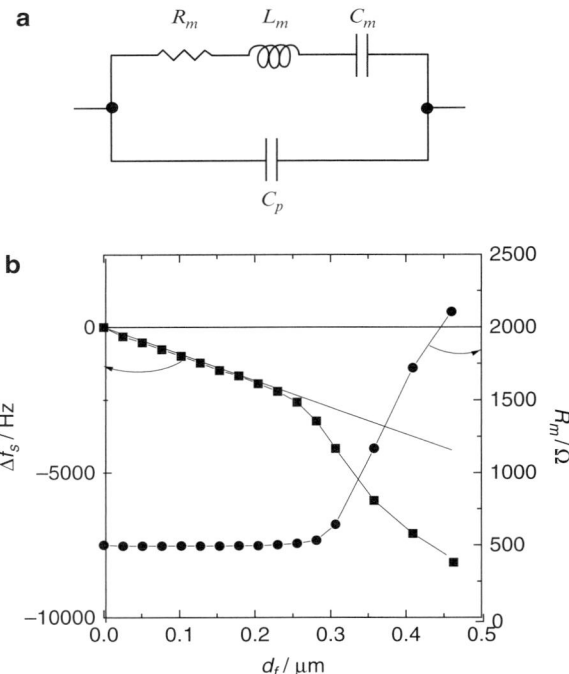

Figure 13. Electroacoustic characterization of the quartz crystal loaded by a polypyrrole film. (**a**) Butterworth–Van Dyke equivalent circuit. (**b**) Change of the frequency, $f_s = 1/(2\pi\sqrt{L_m C_m})$, and motional resistance, R_m, with respect to the polypyrrole film thickness. From Al-Sana et al.[207]

from the electroacoustic measurements performed around the resonance frequency of the quartz resonator with a network analyser. Figure 13b shows the change of two parameters related to the equivalent circuit, $f_s = \left(1/2\pi\sqrt{L_m C_m}\right)$ and R_m, which characterize the gravimetric and the mechanical film properties, with respect to the film thickness. Figure 13b shows that for PPY, up to a film thickness of 0.25 µm, f_s decreases linearly when the thickness (i.e. mass) increases and the motional resistance, R_m, is practically constant.[207] Therefore, the Sauerbrey equation is respected, which is confirmed by a small change of the motional resistance R_m, which means that the mechanical properties of the film do not affect the gravimetric response. After this maximum thickness, f_s does not follow the ideal deviation and R_m increases drastically. Beyond this thickness, the gravimetric regime is not respected and the response given by the microbalance has to be corrected to consider only the mass contribution.

2. AC-Electrogravimetry Aspects

The principle of AC-electrogravimetry measurements is given in Fig. 14. The modified QCM working electrode was polarized at a chosen potential V' and a sinusoidal small-amplitude potential perturbation, $\Delta V'$, was superimposed. The microbalance frequency change, Δf_m, corresponding to the mass response of the film, Δm, of the modified working electrode was detected by means of a frequency/voltage converter. Different systems, a home-made device based on an integrated circuit (VFC110) or a commercial device incorporated in a universal counter (Yokogawa TC110), can be used.[195,208,209]

The resulting signal, ΔV_f, was simultaneously sent with the current response, ΔI, of the electrode to a four-channel frequency response analyser (FRA-Solartron 1254), which allowed the two transfer functions, $\Delta V_f / \Delta V(\omega)$ and $\Delta V / R_{st} \Delta I(\omega)$, to be determined, where ΔV is the raw potential response, which takes into account the electrolyte resistance, and R_{st} is the resistance in the counter electrode circuit, which allows the current ΔI to be measured. Usually, a 10^{-3} Hz to 60 kHz frequency range is analysed to measure the whole impedance. For the electrogravimetric transfer function, the 10^{-3}–100 Hz range is enough. A Fast Fourier Transform analysis was also proposed, but the frequency range analysed is rather narrow.[195]

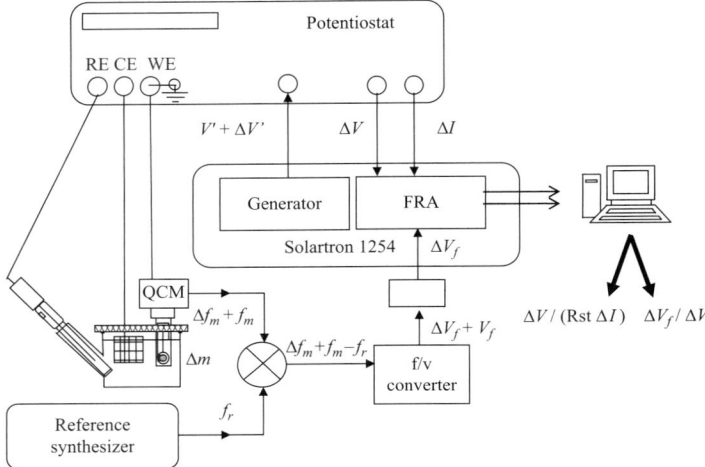

Figure 14. The AC-electrogravimetry and electrochemical impedance set-up. From Gabrielli et al.[209]

As shown previously, the model produces two different transfer functions: $\Delta m/\Delta E(\omega)$ and $\Delta E/\Delta I(\omega)$. So, a numerical treatment is necessary to obtain the final experimental transfer functions comparable with these theoretical quantities from the raw experimental transfer functions, $\Delta V_f/\Delta V(\omega)$ and $\Delta V/R_{st}\Delta I(\omega)$. The most important part is to convert the response ΔV_f from the microbalance frequency changes, Δf_m.

As mentioned previously, the raw transfer function given by the frequency response analyser, $\Delta V_f/\Delta V(\omega)$, must be corrected to obtain the final mass/potential transfer function, $\Delta m/\Delta E(\omega)$.

$$\frac{\Delta m}{\Delta E}(\omega) = \frac{\Delta m}{\Delta f_m}(\omega)\frac{\Delta f_m}{\Delta V_f}(\omega)\frac{\Delta V_f}{\Delta V}(\omega)\frac{\Delta V}{\Delta E}(\omega), \qquad (148)$$

where the four different transfer functions involved in the correction procedure are:

- $\Delta m/\Delta f_m(\omega)$, related to the Sauerbrey relation (147), is equal to $-1/K_{th}^S$ or $-1/K_{exp}^S$, where K_{exp}^S is the experimental mass/microbalance frequency coefficient. At 9 MHz, K_{exp}^S is equal to $16.3 \times 10^7 \text{Hz g}^{-1}\text{cm}^2$ as demonstrated previously.

This quantity is usually evaluated by using copper or silver electrodeposition.[210]

- $\Delta f_m/\Delta V_f(\omega)$ is the inverse of the frequency/voltage converter sensitivity and can be estimated by using the following equation:

$$\frac{\Delta f_m}{\Delta V_f}(\omega) = \frac{\Delta f_m}{\Delta e}(\omega)\frac{\Delta e}{\Delta V_f}(\omega), \qquad (149)$$

where $\Delta f_m/\Delta e(\omega) = \Delta f_{synt}/\Delta e(\omega)$ is the sensitivity of the reference frequency synthesizer by assuming $\Delta f_m = \Delta f_{synt}$ and $\Delta e/\Delta V_f(\omega)$ is the inverse of the global calibration transfer function, $\Delta V_f/\Delta e(\omega)$, obtained by replacing the QCM by a second synthesizer modulated in frequency. Δe is the sinusoidal potential input of the synthesizer used to measure the sensitivity. For an HP 8647A synthesizer, $\Delta f_{synt}/\Delta e$ was determined experimentally and a typical value of 97.75 Hz V^{-1} was obtained.

- $\Delta V_f/\Delta V(\omega)$ is the raw experimental transfer function obtained during the AC-electrogravimetry measurements.
- $\Delta V/\Delta E(\omega)$ allows the ohmic drop correction to be carried out by taking into account the electrolyte resistance, R_e. The following relation is used by incorporating the experimental electrochemical impedance, $\Delta E/\Delta I(\omega)$:

$$\frac{\Delta V}{\Delta E}(\omega) = \frac{R_e}{\Delta E/\Delta I(\omega)} + 1. \qquad (150)$$

3. Dynamic Characterization of the Frequency/Voltage Converter

The frequency/voltage converter has a limited bandwidth owing to the distortions of the electronic components. A global calibration transfer function, $\Delta V_f/\Delta e(\omega)$, has to be determined to evaluate the sensitivity of the frequency/voltage converter and the useful frequency range where the response is linear.

Figure 15 shows the modulus and the phase of the global calibration transfer function, $\Delta V_f/\Delta e(\omega)$, for an home-made configuration. At low frequencies, the phase shift of $(\Delta V_f/\Delta e)$ is close to 0 and the modulus, $|\Delta V_f/\Delta e|$, is equal to 1.02. This result means that the dynamic sensitivity $(\Delta V_f/\Delta f_s)_{dynamic}$ of the system is

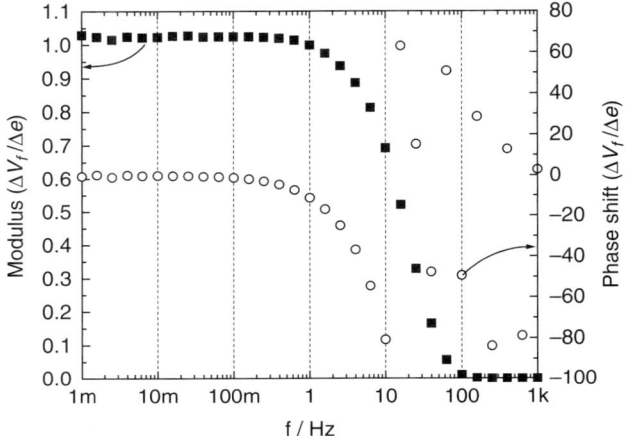

Figure 15. Modulus and phase of the global calibration of a home-made frequency/voltage converter. From Gabrielli et al.[209]

$10.2 \, \text{mV} \, \text{Hz}^{-1}$. The useful frequencies where the system can be used without any correction ranges from low frequencies to 1 Hz for this particular configuration. If measurements are expected above 1 Hz, a modulus and phase shift correction is necessary by using calibration curves such as those given in Fig. 15.[209]

VI. EXAMPLES OF APPLICATIONS OF AC ELECTROGRAVIMETRY

Some examples of the applications of AC electrogravimetry to the study of the redox switching of inorganic and organic films are now reported. First, details about Prussian blue, where mainly cation movement can be detected, will be given. Then, the behaviour of PPY, where cations, anions and solvent participate in the charge compensation process, will be described. Finally, a system of sandwiched polymers will be analysed.

1. Prussian Blue

Prussian blue can be obtained in different ways; however, when Prussian blue films are produced galvanostatically following the procedure described by Itaya et al.[211] it is possible to control in a very accurate way the quality and the amount of Prussian blue deposited. The freshly deposited Prussian blue films are in the "insoluble" form, $Fe_4[Fe(CN)_6]_3$, and after several voltammetric cycles in KCl solution, they are converted into the "soluble" structure, which is characterized by the presence of potassium substituting a quarter of the high-spin iron sites in the "insoluble" Prussian blue structure – $KFeFe(CN)_6$.

Prussian blue films can be reduced to the colourless form, called "Everitt's salt" $\left(KFe_4[Fe(CN)_6]_3 \text{ or } K_2FeFe(CN)_6\right)$, or oxidized to the yellow form called "Prussian yellow" $\left(Fe_4[Fe(CN)_6]_3Cl \text{ or } KFeFe(CN)_6Cl\right)$. These electrochemical processes can be easily detected by cyclic voltammetry of Prussian blue films in KCl solutions. The reduction process for the "soluble" Prussian blue structure has been described as[211]

$$KFe^{III}Fe^{II}(CN)_6 + e^- + K^+ \underset{k'_c}{\overset{k_c}{\rightleftarrows}} K_2Fe^{II}Fe^{II}(CN)_6, \qquad (151)$$

where Fe^{III} and Fe^{II} refer to the different oxidation states of iron atoms in the Prussian blue structure.

The impedance spectra and voltammetric response of Prussian blue films in KCl solutions have been the subject of many papers in the last few years.[212–215] However, many aspects of the electrochemical behaviour of these films have not yet been clarified, especially concerning the entry and exit of ions and solvent into/from the Prussian blue film during redox switching. According to the reaction described in (151), the reduction of Prussian blue to the Everitt's salt form is accompanied by the entrance of potassium ions, which are the counterions here, to maintain film electroneutrality. However, the insertion or expulsion of ions from the film could also be accompanied by an exchange of solvent molecules between the film and the outer solution. To examine the electrochemical subtleties, AC electrogravimetry, i.e. simultaneous coupling of electrochemical impedance and mass/potential transfer function measurements, was used to characterize ion and solvent motions at the Prussian blue film/electrolyte interface.[216–218]

(i) Film Preparation

FeCl$_3$ (chemically pure), K$_3$Fe (CN)$_6$, KCl and HCl (p.a.) (A.R. Merck) were used for the synthesis of Prussian blue films. Water was deionized and distilled. One of the gold electrodes of the quartz crystal was immersed into 0.02 M K$_3$(Fe (CN)$_6$), 0.02 M FeCl$_3$ and 0.01 M HCl aqueous solutions. Electrodeposits of Prussian blue were galvanostatically carried out by applying a controlled cathodic current of 11 µA for 210 s. The estimated film thickness for these experiments was $d_f = 0.14$µm, which is adequate not to have viscoelastic complications.

In a first step, current and mass responses to a potential scan were simultaneously measured in KCl aqueous solutions. Then, measurements of the electrochemical impedance and the electrogravimetric transfer function were carried out.[219]

(ii) Voltammetric and Mass/Potential Curves

During the reduction of Prussian blue films to the colourless Everitt's salt form in KCl solutions, the charge compensation is commonly supposed to occur by the entrance of potassium ions within the film. Figure 16 shows a voltammogram of a film of Prussian blue in a KCl solution and the change of mass measured by means of the QCM accompanying the current variation. The hysteresis observed in the mass/potential curve shows that the electrochemical process is kinetically limited. However, this approach does not allow the determination of the rate-limiting step, either mass transport or mass transfer, at the film/solution interface which is responsible for kinetic limitations. The mass decrease, Δm, between the reduced form (Everitt's salt) and the oxidized form (Prussian blue) was about 1.2 µg cm^{-2}. Meanwhile, the expected mass change evaluated from the electric charge enclosed within the voltammogram, by considering that all the charge compensation involved only the participation of potassium countercations, is about 1.6 µg cm^{-2}. This discrepancy might show that lighter species were involved in the charge compensation process or that there was a flux of water molecules in the opposite sense to the flux cations or that anions were involved. However, it was not possible to discern between these three possibilities from the global information provided by the QCM operating in the quasi steady state.

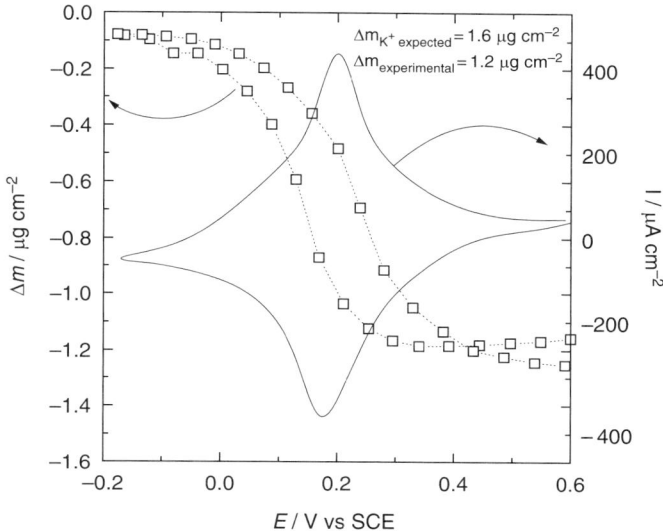

Figure 16. Current and mass with respect to potential for a Prussian blue film in a KCl solution during a linear potential scan. Experimental conditions were as follows: pH 2.8, KCl concentration 0.5 M, scan rate 20 mV s^{-1} and reference saturated calomel electrode (SCE) (potential limits from $+0.6$ to -0.2 V and from -0.2 to $+0.6$ V). From Gabrielli et al.[219]

(iii) Electrogravimetric Transfer Function and Electrochemical Impedance

The current, ΔI, and mass, Δm, responses of Prussian blue films to a small-amplitude sinusoidal potential perturbation, ΔE, were studied at several imposed potentials, with respect to the frequency, to obtain extra information about the role of protons and potassium ions during the electrochemical redox reaction of Prussian blue.[220] Figure 17 shows the transfer functions characteristic of a Prussian blue film at $0.375\,V$ vs. SCE. In the same graphs, the theoretical curves are also represented. Good agreement between the theoretical and the experimental data was obtained for each of the five transfer functions not only regarding shape but also regarding frequencies.

The electrochemical impedance, $\Delta E/\Delta I(\omega)$, is shown in Fig. 17a. $\Delta q/\Delta E(\omega)$ was calculated by using $\Delta E/\Delta I(\omega)$ experimental data and is plotted in Fig. 17b. A single loop was obtained

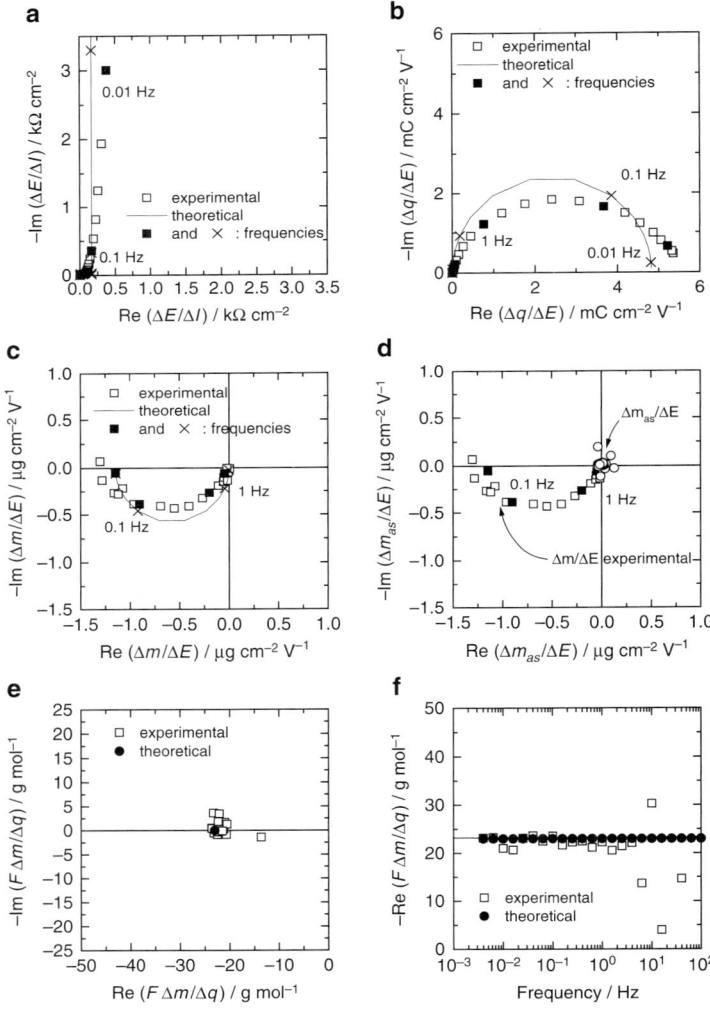

Figure 17. Experimental and theoretical transfer functions for Prussian blue polarized at 0.375 V vs. SCE in 0.5 M KCl. (a) $\Delta E/\Delta I(\omega)$, (b) $\Delta q/\Delta E(\omega)$, (c) $\Delta m/\Delta E(\omega)$, (d) $\Delta m_{as}/\Delta E(\omega)$, (e) $F\Delta m/\Delta q(\omega)$ and (f) $-\mathrm{Re}[F\Delta m/\Delta q(\omega)]$. The parameters used for the calculation were as follows: $d_f = 0.1\,\mu\mathrm{m}$, $C_{int} = 8.0\,\mu\mathrm{F\,cm^{-2}}$, $K_c = 1.2 \times 10^{-5}\,\mathrm{cm\,s^{-1}}$, $G_c = 6.3 \times 10^{-8}\,\mathrm{mol\,s^{-1}\,cm^{-2}\,V^{-1}}$ and $m_c = 23\,\mathrm{g\,mol^{-1}}$. From Gabrielli et al.[219]

for this potential. This means that only one type of charged species, anions or cations, was involved in the redox reaction, but the charge of the ion was not determined. Moreover, the role of the solvent was not clarified at this stage. The experimental electrogravimetric transfer function, $\Delta m/\Delta E(\omega)$, is presented in Fig. 17c. Only one semicircle was obtained and its location in the third quadrant indicates that the ionic species involved here were cations. This single loop confirms that anions do not contribute to the redox switching for this material.

Only the partial electrogravimetric transfer function allows identification of the ions. From (141), $\Delta m_{as}/\Delta E(\omega)$ was calculated and is presented in Fig. 17d, which corresponds to the elimination of a cation having a mass of 23 g. Here, this partial electrogravimetric function, $\Delta m_{as}/\Delta E(\omega)$, is very small, showing that this is mainly a cation with an apparent molar mass of 23 g which is involved at 0.375 V. This result implies that, according to the theory, the contributions of anions and water molecules are always negligible in this potential range.

However, the chemical identification of the species is not available at this step. $F\Delta m/\Delta q(\omega)$ was calculated according to (133). This transfer function is an attractive parameter for identification of the nature of the ion that was involved in the charge compensation process, especially when only one ion acts as a counterion. Therefore, when only one species is involved in the compensation process, $F\Delta m/\Delta q(\omega)$ is frequency-independent and directly gives its molar mass as is plotted in Fig. 17e. At $E = 0.375$ V vs. SCE, $F\Delta m/\Delta q(\omega) = 23$ g mol^{-1} over a wide range of frequencies (Fig. 17f), showing that both K$^+$ and H$_3$O$^+$ participate in the charge compensation process. This value, not so far from the solvated proton mass, 19 g mol^{-1}, indicates that at this potential it is mainly the hydronium ion which participates in the charge compensation process as a countercation. The behaviour of Prussian blue is in very good agreement with the quantities plotted in Fig. 9. It is noticeable that the discrepancy between the experimental quantity and quantiles calculated from the model can be cancelled by considering constant phase elements

$$\frac{\Delta c_c}{\Delta E} = \frac{-G_c}{(j\omega d_f)^\alpha + K_c}, \quad 0 < \alpha < 1 \qquad (152)$$

in the model instead of (107), where $\alpha = 1$.

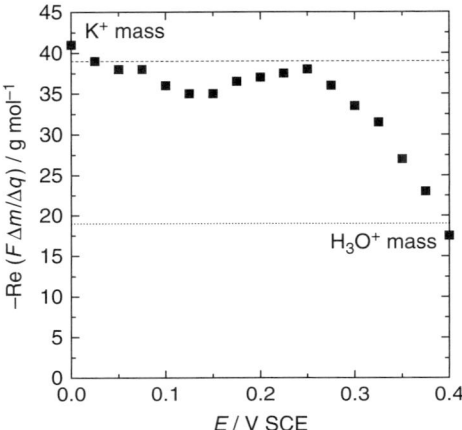

Figure 18. Dependence of the low-frequency limit of the $-\text{Re}[F\Delta m/\Delta q(\omega)]$ function on the potential. Experimental conditions were as follows: pH 2.2 and 0.5 M KCl. From Gabrielli et al.[219]

Figure 18 shows the dependence of the low-frequency limit of $\text{Re}[F\Delta m/\Delta q(\omega)]$, which gives an apparent molar mass, m_c, of the cations expelled from Prussian blue, with respect to the potential. It is observed that over a wide range of potentials, between 0 and 0.25 V, this function reaches values near the potassium molar mass, showing that K^+ is the dominant cation in this potential range. Otherwise, at potentials more anodic than 0.35 V, the molar mass obtained approximated the mass of the hydrated proton. This result explains the smaller values of the apparent molar mass reported in the literature.[221] Between these two potentials, intermediate values of molar mass were obtained. This shows that the two cations are simultaneously involved in the charge compensation process.

Figure 19a shows the variation of the low-frequency limit of the electrogravimetric transfer function $\Delta m/\Delta E$ ($\omega \to 0$), which is equal to the slope of the insertion law. Knowing the apparent mass of the inserted cation allows one to determine the percentage of potassium and hydronium ions and hence the relative participation of these ions in the redox process. As a result of integration of the curves given in Fig. 19a, Fig. 19b shows the change of the concentrations of K^+ and H_3O^+ in the Prussian blue film when the

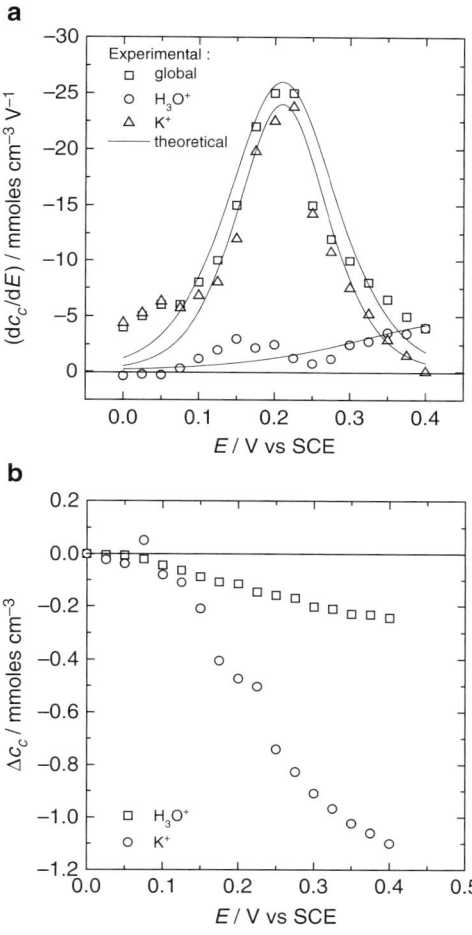

Figure 19. Insertion laws for K^+ and H_3O^+ for Prussian blue in 0.5 M KCl and pH 2.2. (**a**) Variation of the derivative of the insertion law (dc_e/dE); theoretical values were calculated from (102), where $\beta = -0.026$, $\gamma = 10.5$ and $\eta = -2.2$ for the global curve, $\beta = -0.0045$, $\gamma = 5$ and $\eta = -2.2$ for H_3O^+, and $\beta = -0.024$, $\gamma = 12.5$ and $\eta = -2.6$ for K^+, where $\beta = \left[(b_i - b'_i)/4\right](c_{i\,\max} - c_{i\,\min})$, $\gamma = (b'_i - b_i)/2$ and $\eta = \left[(b'_i - b_i)/2\right](E_3 - E_i)$. (**b**) Insertion law for K^+ and H_3O^+ ions with respect to the potential, $\Delta c_c(E_3)$. From Gabrielli et al.[219]

potential changes. It shows that the K^+ concentration decreases by 1.1×10^{-3} mol cm^{-3}, whereas the H_3O^+ concentration decreases by 0.2×10^{-3} mol cm^{-3}, when the potential increases from 0 to 0.4 V vs. SCE.

2. Polypyrrole

In contrast to electroactive films where only one species is involved in the charge compensation process, e.g. Prussian blue, the results for which have already been reported, in the following three species are often involved during the redox switching of PPY. Therefore, it is much more difficult to understand the compensation mechanism is. Indeed, the solution of the problem necessitates the analysis of three unknown processes, namely related to anions, cations and solvent transfers, with only two known quantities as the usual quantities measured being current and mass by means of a potentiostatic arrangement coupled with a QCM.[167] The first investigations analysed the current and mass responses to a potential scan. However, these quasi-steady-state techniques were not able to separate anion and cation transfers and especially solvent interference. So, only qualitative results were gained by means of these techniques. Two ways have been explored to properly analyse the three species transfers. PPY films were chosen as an illustration of the benefits of AC electrogravimetry to gain thorough information on the kinetics of the redox processes of electroactive polymers.[222] To satisfy the QCM requirements, very thin films were used, where charge transport (migration or diffusion) in the polymer is not limiting. The rate-limiting steps are now the ionic and solvent transfers at the polymer/electrolyte interface.

(i) Film Preparation

PPY films can be prepared using different methods described in the literature.[223] The PPY films were prepared from a 0.1 M pyrrole +0.05 M sodium dodecyl sulphate. This low sodium dodecyl sulphate concentration was chosen to avoid surfactant aggregates and to obtain films of good quality in terms of mechanical properties.[224]

From these solutions, a cyclic potential sweep between 0 and 0.675 V vs. SCE at a 10 mV s^{-1} sweep rate was applied twice. These films were sufficiently thin to allow the Sauerbrey equation to be used without limitation from viscoelastic properties of the polymer film.[168] Electrochemical studies were carried out mainly in 0.25 M NaCl to study the influence of the ions present in the bathing solution.

(ii) Voltammetric and Mass/Potential Curves

PPY films were firstly studied by cyclic voltammetry and the mass response to a potential scan was recorded by means of the QCM. Figure 20a shows the current and mass responses to a potential scan of a PPY film in a 0.25 M NaCl solution. The mass decrease, determined through microbalance measurements, between the reduced and oxidized forms was about $2.05\,\mu g\,cm^{-2}$. Meanwhile, the expected mass change evaluated from the electric charge enclosed within the voltammogram was about $2.15\,\mu g\,cm^{-2}$ if only Na^+ was supposed to be involved in the charge compensation process. However, the reduction process of the film between -0.2 and -0.6 V seemed to demonstrate cation ingress as the film gained mass and the oxidation process seemed to demonstrate cation expulsion as the film lost mass. However, when oxidation went beyond $+0.3$ V, the film gained mass again, showing anion insertion. Therefore, the QCM in the quasi steady state led to ambiguous results, as the charge compensation seems to occur not only by cation ingress but also by anion expulsion. Even when the $\Delta m - \Delta q$ curve is plotted (Fig. 20b), the conclusions remain ambiguous: by considering the global mass and charge change during the process, the equivalent molar mass obtained is $-28\,g\,mol^{-1}$ and by taking into account the slope in the linear range, one obtains a value of $-90\,g\,mol^{-1}$.

These results clearly show that the charge compensation process occurring during the oxidation or reduction of the PPY film involves insertion and expulsion of cations and anions. However, it is difficult to separate the various processes. The coupling of the measurement of the electrochemical impedance and the electrogravimetric transfer function will enlighten these points.

(iii) Electrogravimetric Transfer Function and Electrochemical Impedance

Figure 21 shows the various transfer functions measured on PPY films immersed in a 0.25 M NaCl solution at potentials

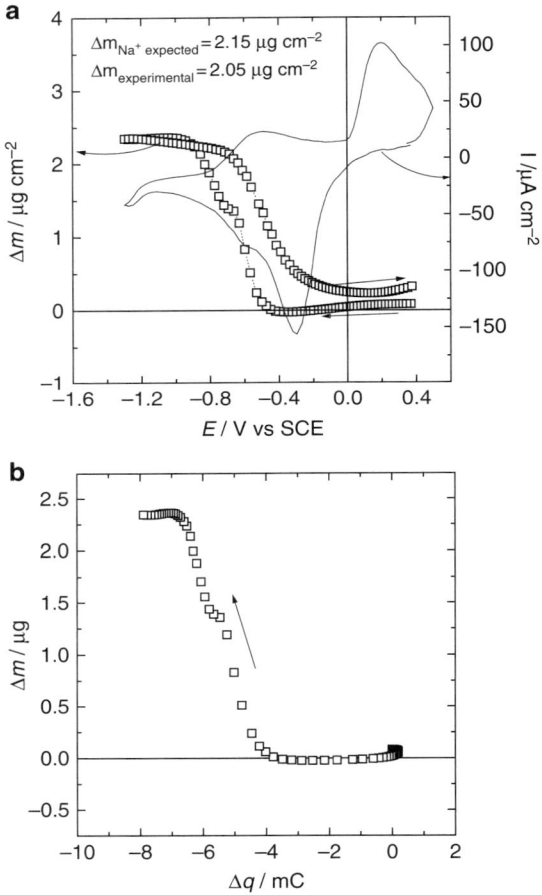

Figure 20. (**a**) Current and mass responses to forward and backward potential scans for a polypyrrole film in 0.25 M NaCl; (**b**) curve mass vs. charge for a cathodic sweep, same conditions as in panel **a**.[216]

-0.55 V vs. SCE. In the same graphs the theoretical curves are also represented. Good agreement between the theoretical and the experimental data was obtained for each transfer function, not only regarding shape but also regarding frequencies.

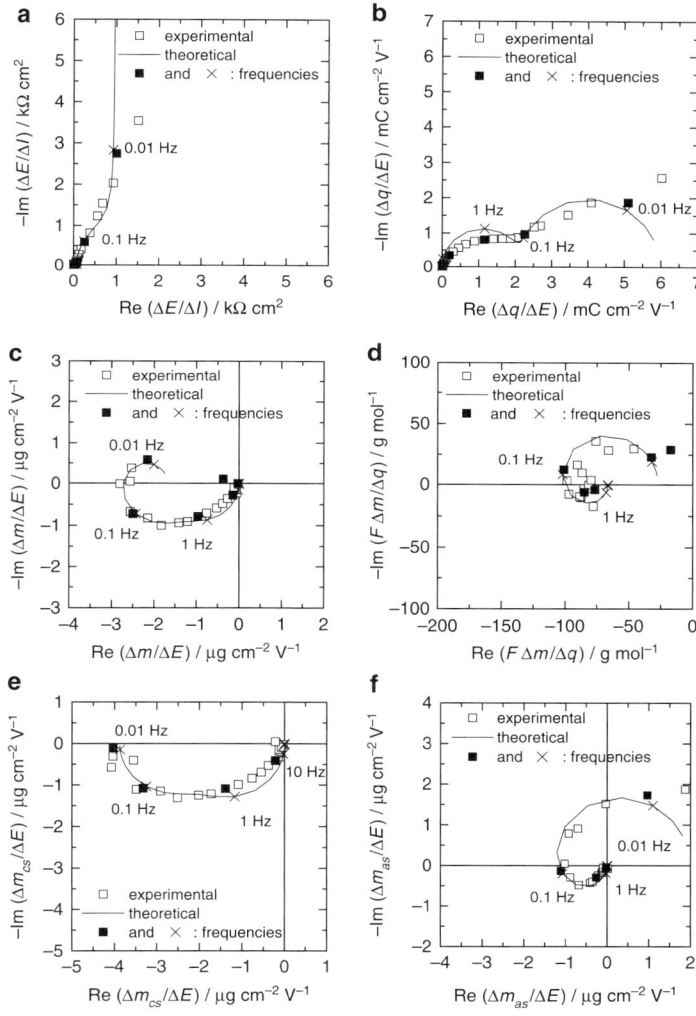

Figure 21. Measured and calculated transfer function quantities for a polypyrrole film in a 0.25 M NaCl solution at -0.55 V vs. SCE. (a) $\Delta E/\Delta I(\omega)$, (b) $\Delta q/\Delta E(\omega)$, (c) $\Delta m/\Delta E(\omega)$, (d) $F\Delta m/\Delta q(\omega)$, (e) $\Delta m_{cs}/\Delta E(\omega)$ and (f) $\Delta m_{as}/\Delta E(\omega)$. The parameters used for the calculation were as follows: $d_f = 0.1\mu m$, $C_{int} = 6.99\mu F\,cm^{-2}$, $K_c = 6.9 \times 10^{-5}\,cm\,s^{-1}$, $K_a = 1.13 \times 10^{-6}\,cm\,s^{-1}$, $K_s = 8.8 \times 10^{-6}\,cm\,s^{-1}$, $G_c = 1.5 \times 10^{-7}\,mol\,s^{-1}\,cm^{-2}\,V^{-1}$, $G_a = -4.5 \times 10^{-9}\,mol\,s^{-1}\,cm^{-2}\,V^{-1}$, $G_s = 8.8 \times 10^{-9}\,mol\,s^{-1}\,cm^{-2}\,V^{-1}$, $m_a = 35.5\,g\,mol^{-1}$, $m_s = 18\,g\,mol^{-1}$, and $m_c = (23+36)\,g\,mol^{-1}$. From Gabrielli et al.[216]

The electrochemical impedance (Fig. 21a) has the usual shape when dealing with this kind of ion blocking electrode, and it is difficult to extract information. It should be noted as there is no part with a slope equal to 45° in the electrochemical impedance response that the rate limiting step is not mass transport in the film but rather ionic transfer between the solution and the film. The electrogravimetric transfer function, $\Delta m/\Delta E(\omega)$, shows (Fig. 21c) at least two loops at -0.55 V. The high-frequency loop, in the third quadrant of the complex plane, seems to be related to cation expulsion from the film, whereas the low-frequency loop, in the second quadrant, seems to be related to anion ingress in the film. The low-frequency limit of this loop gives a negative value for $(\Delta m/\Delta E)$ ($\omega \to 0$); this is in agreement with the slope of the $m(E)$ plot of the mass response to a potential scan given in Fig. 20 in this potential range.

The model was tested by taking into account that the possible species involved in the compensation process could be H_2O for the solvent, Cl^- for the anions and H^+ or Na^+, hydrated or not, for the cations. Below, examples are detailed for 0.25 M NaCl at $E = -0.55$ V vs. SCE.

For this potential, the plot of the electrochemical capacitance $\Delta q/\Delta E(\omega)$ (Fig. 21b) in the complex plane shows two loops. This demonstrates that two charged species are involved in the charge compensation process. For $E = -0.55$ V vs. SCE, the plot of $\Delta m/\Delta q(\omega)$ also shows two loops like in the simulation part (see Fig. 12). This demonstrates that the solvent is involved in the redox reaction in addition to anions and cations. Now, the plots of the partial electrogravimetric transfer function will help to identify the loop related to each species.

Figure 21e shows the quantity $\Delta m_{cs}/\Delta E(\omega)$, obtained by eliminating the anion contribution, which means that this partial electrogravimetric transfer function is only related to the contribution of the cation and solvent and remains in the third quadrant. By taking $m_a = 35.5 \text{ g mol}^{-1}$, i.e. the molar mass of Cl^- ions, the low-frequency loop disappears. This demonstrates that Cl^- is the anion involved in the charge compensation and that it has the lowest time constant.

Finally, the plot of the partial electrogravimetric transfer function, $\Delta m_{as}/\Delta E(\omega)$, obtained by eliminating the cation contribution, is related to the anion and solvent alone (see Fig. 13). By taking m_c as the atomic mass of $Na^+ + nH_2O$, where n is the number of

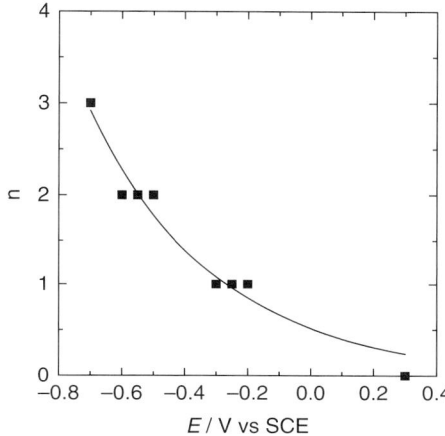

Figure 22. Number of water molecules attached to the Na^+ cations during their insertion/expulsion from polypyrrole immersed in 0.25 M NaCl. From Gabrielli et al.[216]

solvent molecules, the high-frequency loop disappears for $n = 2$. This demonstrates that the cation involved in the charge compensation process is Na^+ hydrated with two molecules of water at this potential and that it has the lowest time constant. If the sodium ions were taken without the solvation shell, the calculations in the cathodic range would show bad agreement between the experimental data and the theoretical simulation. Figure 22 shows the number of water molecules that hydrate the sodium cations. From three water molecules at -0.7 V, this number decreases to none at 0.3 V vs. SCE.

Therefore, the free solvent was related to the intermediate frequency loop and was expelled when PPY was oxidized. To fit the $\Delta m/\Delta E(\omega)$ data, the solvent molar mass has to be taken as $m_s = 18$ g mol^{-1} showing that free water was transferred during the redox process.

By identification of the parameters of the model from the experimental data, it was also possible to determine G_i and K_i for the three species involved in the charge compensation process. Then, the slope of the insertion laws, $\Delta c_i/\Delta E(E) = -(G_i/K_i)$, can be plotted with respect to the potential (Fig. 23a) for the cations, anions and

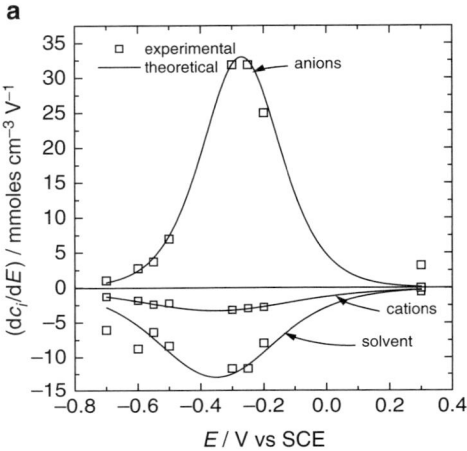

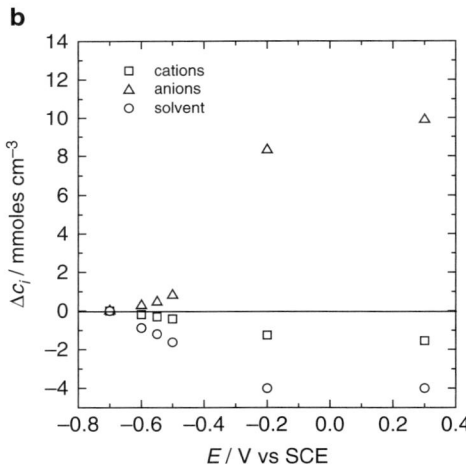

Figure 23. Insertion laws for cations, anions and solvent for polypyrrole in 0.25 M NaCl. (**a**) Variation of the derivative of the insertion law, (dc_e/dE); theoretical values were calculated from (102), where $\beta = -0.0033$, $\gamma = 3$ and $\eta = 1.05$ for cations, $\beta = 0.033$, $\gamma = -6$ and $\eta = -1.6$ for anions, and $\beta = -0.013$, $\gamma = 4$ and $\eta = 1.4$ for solvent, where $\beta = \left[(b_i - b'_i)/4\right](c_{i\,max} - c_{i\,min})$, $\gamma = (b'_i - b_i)/2$, and $\eta = \left[(b'_i - b_i)/2\right](E_3 - E_i)$. (**b**) Insertion law for cations, anions and solvent with respect to the potential, $\Delta c_i (E_3)$. From Gabrielli et al.[216]

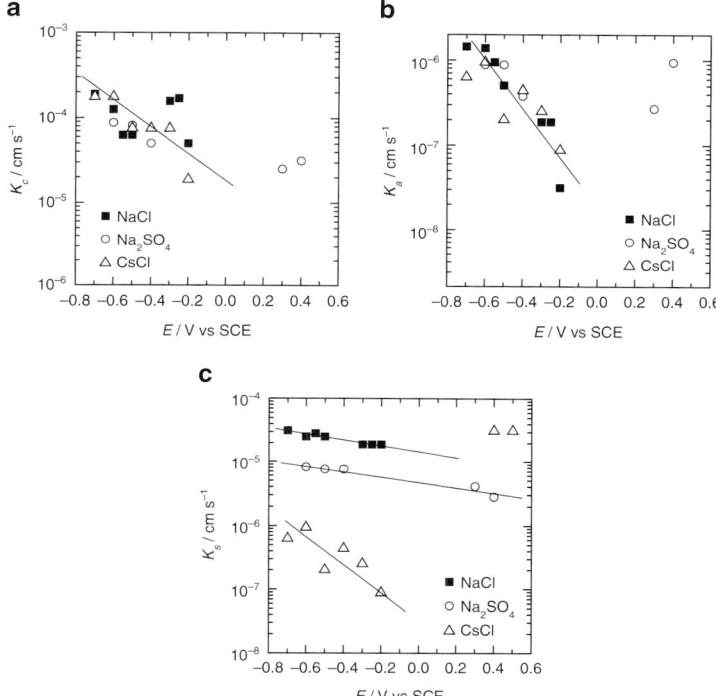

Figure 24. Variations of the global rate constants, K_i, with respect to the potential of the insertion/expulsion mechanism of species during redox switching of polypyrrole immersed in 0.25 M NaCl, 0.1 MNa$_2$SO$_4$ and 0.1 M CsCl. (**a**) cations, (**b**) anions, (**c**) solvent. From Gabrielli et al.[225]

solvent and then through an integration the insertion law, $c_i(E)$, can be determined for each species (Fig. 23b). On the other hand, Fig. 24 gives the plot of the time constants K_i for cations, anions and solvent for various salts, NaCl, CsCl and Na$_2$SO$_4$. It shows that the time constants of cations and anions are practically independent of the salt, whereas the time constant of free solvent depends very much on the salt diluted in the solution.[225] Finally, by determining G_i and K_i for various concentrations of the salt in the aqueous solution, one can evaluate the elementary rate constants of insertion/expulsion, k_i and k'_i (see (93), (105), (106)).

3. Complex Polymeric Structures

Ion-selective electrodes (ISEs) are used in clinical, pharmaceutical or environmental analysis for the detection not only of inorganic ions but also of some organic species such as anionic or cationic surfactants. However, the need for an internal filling solution causes many problems, such as fragility, and it is obstructive for miniaturization. The need for microstructures or even nanostructures leads to the concept of an all-solid-state ISE.

Ivaska et al.[226] proposed the incorporation of a conducting polymer, e.g. PPY, between the membrane, which is only an ionic conductor, and the metallic surface, which is only an electronic conductor, and demonstrated the feasibility of such ISEs.

Since than many authors have tried to develop the two-layer type of sensors by using PPY or other conducting polymers, e.g. polythiophenes,[227] as the internal solid contact. Some authors have even tried to simplify the electrode fabrication by incorporating the conducting polymer into the membrane body.[228] Results showed, however, that the latest systems still need improvement.

PPY is a very well known conducting polymer used in numerous works as the electroactive component of an all-solid-state ISE. Most of the papers dealt with a PPY-coated PVC electrode where PPY is doped with different anions, inorganic such as chloride or organic such as dodecyl sulphate. Several techniques were used to characterize these devices. Potentiometric measurements represent a method allowing thermodynamic characterization. AC electrogravimetry was also used to characterize ion and solvent motions at the PVC/electrolyte interface to understand how the electroactive film (Prussian blue, conducting polymer, etc.) ensures the mediation between the membrane and the electrode.[229,230]

(i) Electrode Preparation

The working electrode was a gold electrode deposited on the quartz crystal. PPY was deposited under a galvanostatic regime through a SOTELEM galvanostat, from an aqueous solution containing 0.3 M pyrrole and 0.1 M KCl. The current applied was 200 µA for 60 s. The film thickness was about 0.2 µm, as measured by scanning electron microscopy. The electrode was then rinsed with distilled water and left in air to dry.

A very thin film is prepared from a PVC solution (20% of high molecular weight PVC, 80% of a plasticizer (dionyl phthalate) dissolved in tetrahydrofuran), an ionophore, valinomycin, was added by means of 1% and a lipophilic salt, potassium tetraphenylboride (KBΦ_4), weighting a third of the ionophore mass was spin-coated on the modified electrode surface by PPY by applying a spinning rate of 10,000 rpm for 10 s after pipetting 10 μL of the above-mentioned solution. The film thickness was estimated by scanning electron microscopy to be 0.2 μm. All electrodes were conditioned in 0.1 M KCl solution for at least 24 h prior to use.

(ii) Electrogravimetric Transfer Function and Electrochemical Impedance

Figure 25 shows the five main transfer functions obtained from electrogravimetric measurements at −0.45 V vs. SCE in a 0.1 M KCl solution. In the same graphs, the theoretical curves are also represented. Good agreement between the theoretical and the experimental data was obtained for each of the five transfer functions not only regarding shape but also regarding frequencies.

From the electrochemical impedance spectra, $\Delta E/\Delta I(\omega)$, little information can be obtained concerning the different species participating in the insertion/expulsion phenomenon as it is presented Fig. 25a. However, the shape of the spectra confirms that there is an ionic transfer between the solution and the film and there is no diffusion (mass transport) usually expressed by a slope equal to 45° appearing at lower frequencies. Moreover, from the impedance data the electrochemical capacitance, $\Delta q/\Delta E(\omega)$ (Fig. 25b), can be calculated. This gives information on the different charged species involved in the charge compensation process. Two loops were obtained corresponding to two different ionic species, maybe a cation and an anion. Additionally, these two transfer functions allow the two constants K_i and G_i to be determined for each of the ions, which will help us to calculate the mass of each ion and so to identify them.

This identification is obtained by using the electrogravimetric transfer function, $\Delta m/\Delta E(\omega)$ (Fig. 25c). The high-frequency loop in the third quadrant is related to the potassium ion. Indeed, by keeping the constants K_i and G_i calculated before, we can estimate the molecular mass of the ionic species and for the cation a mass of 39 g mol^{-1} was found: most of the time the cation was attached to some water molecules. To our knowledge, this technique is the only way to estimate the number n of water molecules attached to the

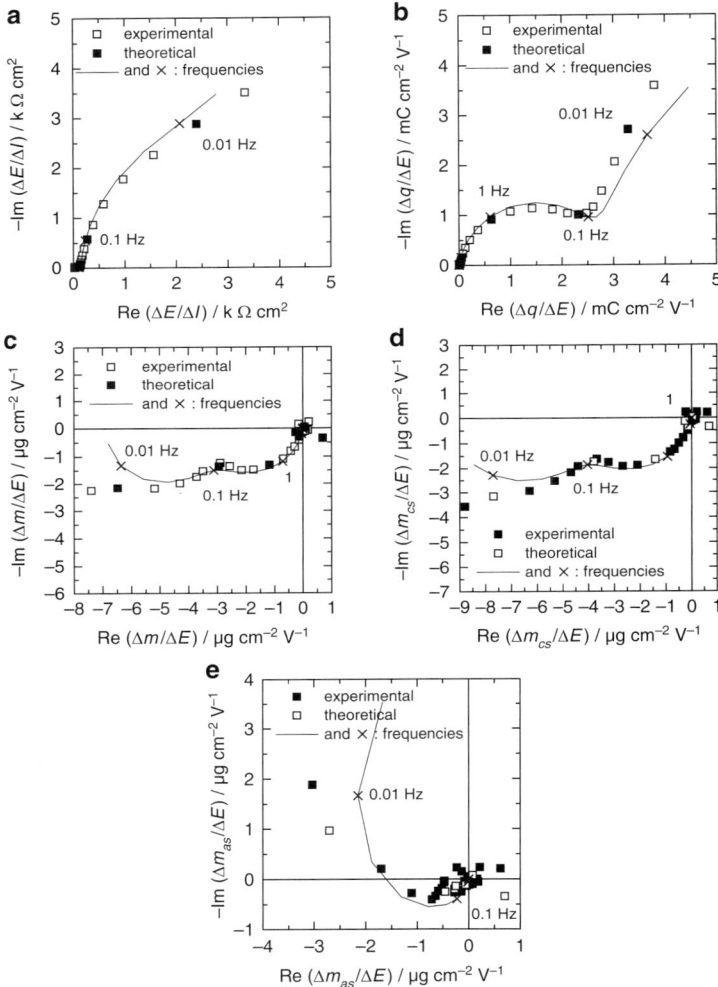

Figure 25. Measured and calculated transfer function quantities for polypyrrole/Prussian blue structures in 0.1 M KCl solution at −0.45 V vs. SCE. (a) $\Delta E/\Delta I(\omega)$, (b) $\Delta q/\Delta E(\omega)$, (c) $\Delta m/\Delta E(\omega)$, (d) $\Delta m_{cs}/\Delta E(\omega)$ and (e) $\Delta m_{as}/\Delta E(\omega)$. The parameters used for the calculation were as follows: $d_f = 0.4\,\mu m$, $C_{int} = 6.99\,\mu F\,cm^{-2}$, $K_c = 9.4 \times 10^{-5}\,cm\,s^{-1}$, $K_a = 1.2 \times 10^{-6}\,cm\,s^{-1}$, $K_s = 5.0 \times 10^{-5}\,cm\,s^{-1}$, $G_c = 5.3 \times 10^{-8}\,mol\,s^{-1}\,cm^{-2}\,V^{-1}$, $G_a = -1.1 \times 10^{-9}\,mol\,s^{-1} = cm^{-2}\,V^{-1}$, $G_s = 3.3 \times 10^{-8}\,mol\,s^{-1}\,cm^{-2}\,V^{-1}$, $m_a = 35.5\,g\,mol^{-1}$, $m_s = 18\,g\,mol^{-1}$, and $m_c = (23 + 3 \times 18)\,g\,mol^{-1}$. From Gabrielli et al.[230]

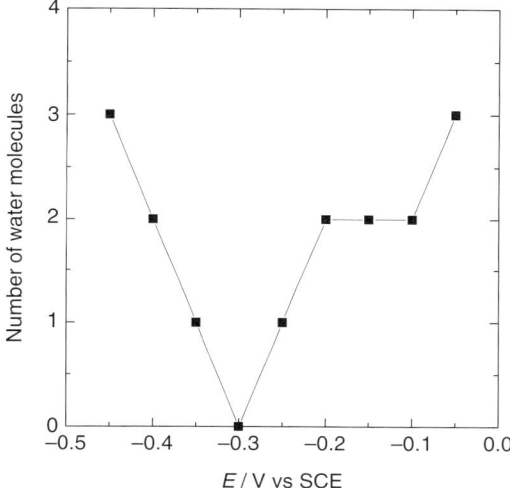

Figure 26. Number of water molecules n attached to the inserted cation against potential. From Gabrielli et al.[230]

cation inserted in the PVC film. Figure 26 shows the evolution of the number of water molecules n with the potential. The minimum corresponds to a potential close to the equilibrium potential (vs. SCE) in 0.1 M KCl . Furthermore, at -0.45 V a second well-defined loop also appears, related to the free solvent as shown in Fig. 25d for the partial electrogravimetric transfer function $\Delta m_{cs}/\Delta E(\omega)$. Anions are also involved in the phenomenon as indicated by the partial electrogravimetric transfer function $\Delta m_{as}/\Delta E(\omega)$ (Fig. 25e) where the cation contribution was eliminated. At this potential the anion was involved in the insertion/expulsion process. Several questions arise from these results concerning anion and solvent motions into the PVC film. In the first place the difficulty of carrying out accurate measurements at low frequencies (less than 10 mHz) might be the reason for the not so well defined loop for each of the species, especially for the anion. Microbalance frequency instabilities can explain this behaviour and could be correlated to the swelling effect. The fact that the solvent has an active role in the kinetics indicates that the membrane is not completely hydrophobic and, consequently, the influence of the soaking time of the electrode before measurements has to be considered. Different parameters can be determined according

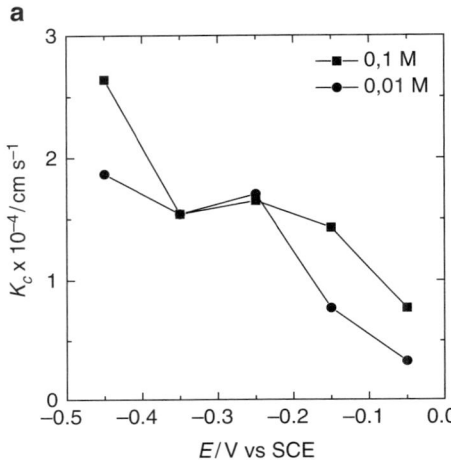

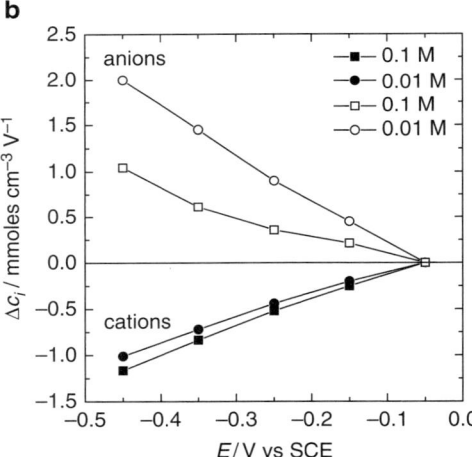

Figure 27. (**a**) Variation of the global kinetic constant K_c for cation insertion in the PVC membrane with potential in 0.01 and 0.1 M KCl solutions; (**b**) isotherms of the insertion law with respect to the potential for cations and for anions in 0.1 and 0.01 M KCl. From Gabrielli et al.[230]

to the potential. In Fig. 27a, the global kinetic parameter K_c was determined at each potential value. A constant decrease is observed when the potential goes towards the anodic region. Two KCl concentrations were examined and a small difference is observed, which indicates poor influence of the c_{csol} values on the K_c evolution. Another key point was the determination of the relative concentrations of the cations and the anions, respectively, Δc_c and Δc_a, over the potential changes. In Fig. 27b, the estimated response is given: when the potential goes towards the cathodic regime, c_c increases and c_a decreases. This result is in good agreement with the charge compensation process shown previously.

VII. CONCLUSION

First of all, the use of the QCM to study the behaviour of electroactive thin films was a real breakthrough in understanding the charge compensation process occurring during the redox switching, above all for simple processes where only one counterion is involved. AC electrogravimetry, which consists in combining impedance and mass/potential transfer function measurements, allows a better understanding of more complex insertion/expulsion mechanisms. Different key points can be explored by using this dynamic approach, such as:

- Determination of the interfacial transfer kinetics of the charged or non-charged species
- Quantification of the electroactive sites in the host film
- Identification of the species involved in the electrochemical process
- Quantification of the charged or non-charged species present in the film

Here, where diffusion of the species in the polymer is not limiting, ionic transfer at the polymer/solution interface is the rate-limiting step of the charge transfer related to the charge compensation. More precisely, anions, cations and solvent exchanges between the electroactive film and the solution due to the charge compensation process occurring during a redox switching can be separated. The ionic and free solvent transfer rates can be determined and the concentration changes of the species in the electroactive film can be estimated at each potential value. Finally, the molar mass of the species involved in the charge transfer can be obtained, which allows the species to be clearly identified. Sometimes, and in addition, the solvent bound to the ions can be quantified according to the polarization value applied to the electroactive film.

Finally, it should be noted that this technique is non-destructive and is a very good way to clarify the insertion/expulsion mechanism of free solvent which could occur during electrochemical redox processes of electroactive materials.

REFERENCES

[1] H. Letheby, *J. Chem. Soc.*, **15** (1862) 161.
[2] A. G. Green, A. E. Woodhead, *J. Chem. Soc.*, **97** (1910) 2388.0
[3] R. de Surville, M. Jozefowicz, L. T. Yu, J. Perichon, R. Buvet, *Electrochim. Acta*, **13** (1968) 1451.
[4] R. C. Murray, *Acc. Chem. Res.*, **13** (1980) 135.
[5] R. C. Murray, in *Electroanalytical Chemistry*, vol. 13, Ed. by A. J. Bard, Marcel Dekker, New York, 1984.
[6] W. J. Albery, A. R. Hillman, *Ann. Rep. Prog. Chem., Sect. C*, **78** (1981) 277.
[7] A. R. Hillman, in *Electrochemical Science and Technology of Polymers*, vol. 1, Ed. by R. G. Lindfors, Elsevier, Amsterdam, 1987.
[8] H. D. Abruna, in *Electroresponsive Molecular and Polymeric Systems*, vol. 1, Ed. by T. A. Stockheim, Marcel Dekker, New York, 1988.
[9] G. P. Evans, in *Advances in Electrochemical Science and Engineering*, vol. 1, Ed. by H. Gerisher, C. W. Tobias, VCH Verlag, Weinheim, Germany, 1990.
[10] W. H. Smyrl, M. Lien, in *Applications of Electroactive Polymers*, Ed. by B. Scrosati, Chapman and Hall, London, 1993.
[11] M. E. G. Lyons, *Ann. Rep. C. R. Soc. Chem.*, **87** (1990) 119.
[12] M. E. G. Lyons, in *Advances in Chemical Physics*, vol. XCIV, Ed. by I. Prigogine, S. A. Rice, Wiley, New York, 1996.
[13] N. Oyama, T. Ohsaka, in *Molecular Design of Electrode Surfaces*, Ed. by R. W. Murray, Techniques of Chemistry Series, vol. 22, Wiley Interscience, New York, 1992.
[14] R. W. Murray, in *Molecular Design of Electrode Surfaces*, Ed. by R. W. Murray, Techniques of Chemistry Series, vol. 22, Wiley Interscience, New York, 1992.
[15] C. Andrieux, J. M. Savéant, in *Molecular Design of Electrode Surfaces*, Ed. by R. W. Murray, Techniques of Chemistry Series, vol. 22, Wiley Interscience, New York, 1992.
[16] G. Inzelt, in *Electroanalytical Chemistry*, vol. 18, Ed. by A. J. Bard, Marcel Dekker, New York, 1994.
[17] K. Doblhofer, in *Frontiers in Electrochemistry*, Ed. by J. Lipkowski, P. N. Ross, Electrochemistry in Novel Materials, vol. 3, VCH Publishers, New York, 1994.
[18] M. E. G. Lyons, in *Electroactive Polymer Electrochemistry*, Ed. by M. E. G. Lyons, Plenum, New York, 1994.
[19] S. Bruckenstein, A. R. Hillman, *J. Phys. Chem.*, **92** (1988) 4837.
[20] J.-F. Fauvarque, *J. Chim. Phys.*, **86** (1989) 5.
[21] B. Scrosati, *Prog. Solid State Chem.*, **18** (1988) 1.

[22] T. Tatsunada, H. Daifuku, T. Nakajima, T. Kawagoe, *Polym. Adv. Technol.*, **1** (1990) 33.
[23] M. R. Anderson, B. R. Mattes, H. Reiss, R. B. Kaner, *Synth. Met.*, **41** (1991) 11.
[24] M. Angelopoulos, J. M. Shaw, R. D. Kaplan, S. Perreault, *J. Vac. Sci. Technol. B*, **7** (1989) 1519.
[25] H. Shigi, M. Kishimoto, H. Yakabe, B. Deore, T. Nagaoka, *Anal. Sci.*, **18** (2002) 41.
[26] A. Boyle, E. M. Genies, M. Lapkowski, *Synth. Met.*, **28** (1989) C769.
[27] Y. Ikarijima, W. R. Heineman, *Anal. Chem.*, **56** (1986) 1803.
[28] A. Malinauskas, J. Malinauskiene, A. Ramanavicius, *Nanotechnology*, **16** (2005) R51.
[29] C. Y. Lee, H. M. Kim, J. W. Park, Y. S. Gal, J. I. Jin, J. Joo, *Synth. Met.*, **117** (2001) 109.
[30] M. Jiang, J. Wang, *J. Electroanal. Chem.*, **500** (2001) 584.
[31] S. Bourkane, C. Gabrielli, M. Keddam, *J. Electroanal. Chem.*, **256** (1988) 471.
[32] S. Cordoba-Torresi, C. Gabrielli, M. Keddam, H. Takenouti, R. Torresi, *J. Electroanal. Chem.*, **290** (1990) 261.
[33] S. Bourkane, C. Gabrielli, M. Keddam, *Electrochim. Acta*, **34** (1989) 1081.
[34] S. Bourkane, C. Gabrielli, F. Huet, M. Keddam, *Electrochim. Acta*, **38** (1993) 1023, 1827.
[35] C. Gabrielli, M. Keddam, H. Perrot, R. Torresi, *J. Electroanal. Chem.*, **378** (1994) 85.
[36] O. Bohnke, B. Vuillemin, C. Gabrielli, M. Keddam, H. Perrot, H. Takenouti, R. Torresi, *Electrochim. Acta*, **40** (1995) 2755, 2765.
[37] C. Gabrielli, M. Keddam, F. Minouflet, H. Perrot, *Electrochim. Acta*, **41** (1996) 1217.
[38] A. R. Hillman, D. C. Loveday, S. Bruckenstein, C. P. Wilde, *J. Chem. Soc. Faraday Trans.*, **86** (1990) 437.
[39] B. S. Pivovar, *Polymer*, **47** (2006) 4194.
[40] S. Bruckenstein, A. R. Hillman, *J. Phys. Chem.*, **95** (1991) 10748.
[41] E. Pater, S. Bruckenstein, A. R. Hillman, *J. Chem. Soc. Faraday Trans.*, **92** (1996) 4087.
[42] E. Pater, S. Bruckenstein, A. R. Hillman, *J. Phys. Chem. B*, **110** (2006) 14761.
[43] E. M. Andrade, F. V. Molina, M. I. Flotit, D. Posadas, *Electrochem. Solid-State Lett.*, **3** (2000) 504.
[44] M. Kaneko, K. Kaneko, *Synth. Met.*, **76** (1999) 1350.
[45] T. F. Otero, H. Grande, J. Rodríguez, *Electrochim. Acta*, **41** (1996) 1863.
[46] T. F. Otero, H.-J. Grande, *Colloids Surf. A*, **134** (1998) 85.
[47] H. Grande, T. F. Otero, *Electrochim. Acta*, **44** (1999) 1893.
[48] L. Lizarraga, E. M. Andrade, F. V. Molina, *J. Electroanal. Chem.*, **561** (2004) 127.
[49] L. Lizarraga, E. M. Andrade, M. I. Florit, F. V. Molina, *J. Phys. Chem. B*, **109** (2005) 18815.
[50] L. Lizarraga, E. M. Andrade, F. V. Molina, *Electrochim. Acta*, **53** (2007) 538.
[51] Y. M. Volfkovich, V. S. Bagotzky, T. K. Zolotova, E. Y. Pisarevskaya, *Electrochim. Acta*, **41** (1996) 1905.
[52] H. Grande, T. F. Otero, I. Cantaro, *J. Non-Crystalline Solids*, **235** (1998) 619.
[53] E. M. Andrade, F. V. Molina, M. I. Florit, D. Posadas, *Electrochem. Solid-State Lett.*, **3** (2000) 504.
[54] D. M. Kelly, J. G. Voss, A. R. Hillman, *J. Chem. Soc. Faraday Trans.*, **92** (1996) 4121.

[55] R. Batra, A. L. Sharma, M. K. Pandey, K. K. Saini, B. D. Malhotra, *Curr. Appl. Phys.*, **3** (2003) 317.
[56] E. Smela, N. Gadegaard, *J. Phys. Chem. B*, **105** (2001) 9395.
[57] H. L. Bendey, M. Gonsalves, A. R. Hillman, A. Glidle, S. Bruckenstein, *J. Electroanal. Chem.*, **410** (1996) 219.
[58] C. M. Lagier, I. Efimov, A. R. Hillman, *Anal. Chem.*, **77** (2005) 335.
[59] I. Efimov, A. R. Hillman, *Anal. Chem.*, **78** (2006) 3616.
[60] L. P. Bauermann, P. N. Bartlett, *Electrochim. Acta*, **50** (2005) 1537.
[61] M. A. Mohamoud, A. R. Hillman, *Electrochim. Acta*, **53** (2007) 1206.
[62] B. J. Feldman, P. Burgmayer, R. W. Murray, *J. Am. Chem. Soc.*, **107** (1985) 872.
[63] C. Ehrenbeck, K. Juttner, *Electrochim. Acta*, **41** (1996) 1815.
[64] G. Zotti, S. Zecchin, G. Schiavon, L. Groenendaal, *Chem. Mater.*, **12** (2000) 2996.
[65] G. Zotti, S. Zecchin, G. Schiavon, B. Vercelli, A. Berlin, E. Dalcanale, L. Groenendaal, *Chem. Mater.*, **15** (2003) 4642.
[66] J. C. Lacroix, K. Fraoua, P. C. Lacaze, *J. Electroanal. Chem.*, **444** (1998) 83.
[67] T. Johansson, N. K. Persson, O. Inganas, *J. Electrochem. Soc.*, **151** (2004) E119.
[68] M. A. Vorotyntsev, J. Heinze, Electrochim. Acta, **46** (2001) 3309.
[69] V. V. Malev, O. V. Levin, M. A. Vorotyntsev, *Electrochim. Acta*, **52** (2006) 133.
[70] J. M. Savéant, *J. Electroanal. Chem.*, **201** (1986) 211.
[71] C. P. Andrieux, J. M. Savéant, *J. Phys. Chem.*, **92** (1988) 6761.
[72] R. P. Buck, *J. Electroanal. Chem.*, **219** (1987) 23.
[73] R. P. Buck, *J. Phys. Chem.*, **92** (1988) 4196.
[74] R. P. Buck, *J. Phys. Chem.*, **92** (1988) 6445.
[75] W. J. Albery, Z. Chen, B. R. Horrocks, A. R. Mount, P. J. Wilson, D. Bloor, A. T. Monkman, C. M. Elliott, *Faraday Discuss. Chem.* Soc., **88** (1989) 247.
[76] T. Yeu, K.-M. Yin, J. Carbajal, R. E. White, *J. Electrochem. Soc.*, **138** (1991) 2869.
[77] M. A. Vorotyntsev, A. A. Rubashkin, J. P. Badiali, *Electrochim. Acta*, **41** (1996) 2313.
[78] F. Miomandre, M. N. Bussac, E. Vieil, L. Zuppiroli, *Electrochim. Acta*, **44** (1999) 2019.
[79] F. Miomandre, M. N. Bussac, E. Vieil, L. Zuppiroli, *J. Chem. Phys.*, **255** (2000) 291.
[80] I. Rubinstein, E. Sabatini, J. Rishpon, *J. Electrochem. Soc.*, **134** (1987) 3079.
[81] C. D. Paulse, P. G. Pickup, *J. Phys. Chem.*, **92** (1988) 7002.
[82] S. Fletcher, *J. Electroanal. Chem.*, **337** (1992) 127.
[83] G. Lang, M. Ujvari, G. Inzelt, *Electrochim. Acta*, **46** (2001) 4159.
[84] G. C. Barker, *J. Electroanal. Chem.*, **41** (1973) 201.
[85] W. J. Albery, C. M. Elliott, A. R. Mount, *J. Electroanal. Chem.*, **288** (1990) 15.
[86] T. R. Brumleve, R. P. Buck, *J. Electroanal. Chem.*, **126** (1981) 73.
[87] R. P. Buck, C. Mundt, *J. Chem. Soc. Faraday Trans.*, **92** (1996) 3947.
[88] K. Rosberg, G. Paasch, L. Dunsch, S. Luswig, *J. Electroanal. Chem.*, **443** (1998) 49.
[89] P. H. Nguyen, G. Paasch, J. Electroanal. Chem., **460** (1999) 63.
[90] C. Ehrenbeck, K. Juttner, S. Ludwig, G. Paasch, *Electrochim. Acta*, **43** (1998) 2781.
[91] G. Paasch, *J. Electroanal. Chem.*, **600** (2007) 131.
[92] C. Gabrielli, O. Haas, H. Takenouti, *J. Appl. Electrochem.*, **17** (1987) **82**.
[93] C. Gabrielli, H. Takenouti, O. Haas, A. Tsukada, *J. Electroanal. Chem.*, **302** (1991) 59.
[94] M. F. Mathias, O. Haas, *J. Phys. Chem.*, **96** (1992) 3174.
[95] M. F. Mathias, O. Haas, *J. Phys. Chem.*, **97** (1993) 9217.
[96] R. P. Buck, M. B. Madaras, R. Mackel, *J. Electroanal. Chem.*, **362** (1993) 33.
[97] M. A. Vorotyntsev, L. I. Daikhin, M. D. Levi, *J. Electroanal. Chem.*, **364** (1994) 37.

98. C. Deslouis, M. M. Musiani, B. Tribollet, M. A. Vorotyntsev, *J. Electrochem. Soc.*, **142** (1995) 1902.
99. M. A. Vorotyntsev, J. P. Badiali, E. Vieil, *Electrochim. Acta*, **41** (1996) 1375.
100. M. A. Vorotyntsev, J. P. Badiali, G. Inzelt, *J. Electroanal. Chem.*, **472** (1999) 7.
101. M. A. Vorotyntsev, C. Deslouis, M. M. Musiani, B. Tribollet, K. Aoki, *Electrochim. Acta*, **44** (1999) 2105.
102. T. R. Brumleve, R. P. Buck, *J. Electroanal. Chem.*, **90**(1978) 1.
103. R. P. Buck, C. Mundt, *Electrochim. Acta*, **44**(1999) 1999.
104. G. Lang, G. Inzelt, *Electrochim. Acta*, **44**(1999) 2037.
105. C. Gabrielli, J. Garcia-Jareno, H. Perrot, *Models Chem.*, **137**(2000) 269.
106. C. Gabrielli, M. Keddam, N. Nadi, H. Perrot, *J. Electroanal. Chem.*, **485** (2000) 101.
107. S. H. Glarum, *J. Electrochem. Soc.*, **134**(1987) 142.
108. C. Deslouis, M. M. Musiani, B. Tribollet, *J. Phys. Chem.*, **98**(1994) 2936.
109. H. N. Dinh, P. Vanysek, V. I. Birss, *J. Electrochem. Soc.*, **146**(1999) 3324.
110. A. Benyaich, C. Deslouis, T. El Moustafid, M. M. Musiani, B. Tribollet, *Electrochim. Acta*, **41**(1996) 1781.
111. M. Grzeszczuk, G. Zabinska-Olszak, *J. Electroanal. Chem.*, **427**(1997) 169.
112. A. Kepas, M. Grzeszczuk, *J. Electroanal. Chem.*, **582**(2005) 209.
113. M. J. Rodriguez Presa, H. L. Bandey, R. I. Tucceri, M. I. Florit, D. Posadas, A. R. Hillman, *Electrochim. Acta*, **44**(1999) 2073.
114. F. J. Rodriguez Nieto, D. Posadas, R. I. Tucceri, *J. Electroanal. Chem.*, **434**(1997) 83.
115. F. J. Rodriguez Nieto, R. I. Tucceri, *J. Electroanal. Chem.*, **416**(1996) 1.
116. T. Komura, Y. Ito, T. Yamaguti, K. Takahasi, *Electrochim. Acta*, **43** (1998) 723.
117. H. Ding, Z. Pan, L. Pigani, R. Seeber, C. Zanardi, *Electrochim. Acta*, **46** (2001) 2721.
118. K. Juttner, R. H. J. Schmitz, A. Hudson, *Electrochim. Acta*, **44**(1999) 4177.
119. K. Juttner, R. H. J. Schmitz, A. Hudson, *Synth. Met.*, **101** (1999) 579.
120. X. Ren, P. G. Pickup, *Electrochim. Acta*, **41** (1996) 1877.
121. T. Amemiya, K. Hashimoto, A. Fujishima, *J. Phys. Chem.*, **97** (1993) 4187.
122. C. Deslouis, M. M. Musiani, B. Tribollet, *J. Phys. Chem.*, **100** (1996) 8994.
123. J. J. Garcia-Jareno, J. J. Navarro-Laboulais, A. F. Roig, H. Scholll, F. Vicente, *Electrochim. Acta*, **40** (1995) 1113.
124. J. J. Garcia-Jareno, J. J. Navarro-Laboulais, F. Vicente, *Electrochim. Acta*, **41** (1996) 835.
125. J. J. Garcia-Jareno, A. Sanmatias, J. J. Navarro-Laboulais, D. Benito, F. Vicente, *Electrochim. Acta*, **43** (1998) 235.
126. Z. Gao, J. Bobacka, A. Ivaska, *Electrochim. Acta*, **38** (1993) 379.
127. N. R. de Tacconi, R. Rajeshwar, R. D. Lezna, *J. Electroanal. Chem.*, **587** (2006) 42.
128. V. Malev, V. Kurdakova, V. Kondratiev, V. Zigel, *Solid State Ionics*, **169** (2004) 95.
129. U. Retter, A. Widmann, K. Siegler, K. Hahlert, *J. Electroanal. Chem.*, **546** (2003) 87.
130. Z. Gao, *J. Electroanal. Chem.*, **370** (1994) 95.
131. H. Durliat, P. Delorme, M. Comtat, *Electrochim. Acta*, **30** (1985) 1071.
132. G. Denuault, M. H. Troise Frank, L. M. Peter, *Faraday Discuss.*, **94** (1992) 23.
133. M. H. Troise Frank, G. Denuault, *J. Electroanal. Chem.*, **354** (1993) 331.
134. M. Arca, M. V. Mirkin, A. J. Bard, *J. Phys. Chem.*, **99** (1995) 5040.
135. F. R. F. Fan, M. V. Mirkin, A. J. Bard, *J. Phys. Chem.*, **98** (1994) 1475.
136. J. Kwak, F. C. Anson, *Anal. Chem.*, **64** (1992) 250.
137. M. Quinto, S. A. Jenekhe, A. J. Bard, *Chem. Mater.*, **13** (2001) 2824.
138. N. Yang, C. G. Zoski, *Langmuir*, **22** (2006) 10338.

[139] C. Lee, F. C. Anson, *Anal. Chem.*, **64** (1992) 528.
[140] G. Inzelt, G. Horanyi, *J. Electrochem. Soc.*, **136** (1989) 1747.
[141] Y.-C. Liu, K.-H. Yang, C.-C. Wang, *J. Electroanal. Chem.*, **549** (2003) 151.
[142] A. Glidle, L. Bailey, C. S. Hadyoon, A. R. Hillman, A. Jackson, K. S. Ryder, P. M. Saville, M. J. Swann, J. P. R. Webster, R. W. Wilson, J. M. Cooper, *Anal. Chem.*, **73** (2001) 5596.
[143] A. Glidle, J. Cooper, A. R. Hillman, L. Bailey, A. Jackson, J. P. R. Webster, *Langmuir*, **19** (2003) 7746.
[144] A. R. Hillman, L. Bailey, A. Glidle, J. M. Cooper, N. Gadegaard, J. R. P. Webster, *J. Electroanal. Chem.*, **532** (2002) 269.
[145] K. Wapner, M. Stratmann, G. Grundmeier, *Electrochim. Acta*, **51** (2005) 3303.
[146] T. Matencio, E. Vieil, *Synth. Met.*, **44** (1991) 349.
[147] T. Matencio, M. A. De Paoli, R. C. D. Peres, R. M. Torresi, S. I. Cordoba de Torresi, *Synth. Met.*, **72** (1995) 59.
[148] E. Vieil, K. Meerholz, T. Matencio, J. Heinze, *J. Electroanal. Chem.*, **368** (1994) 183.
[149] C. Barbero, M. C. Miras, E. J. Calvo, R. Kotz, O. Haas, *Langmuir*, **18** (2002) 2756.
[150] A. C. Boccara, D. Fournier, J. Badoz, *Appl. Phys. Lett.*, **36** (1980) 130.
[151] C. P. Andrieux, P. Audebert, P. Hapiot, M. Nechtschein, C. Odin, *J. Electroanal. Chem.*, **305** (1991) 153.
[152] L. M. Abrantes, J. C. Mesquita, M. Kalaji, L. M. Meter, *J. Electroanal. Chem.*, **307** (1991) 275.
[153] S. Bruckenstein, M. Shay, *Electrochim. Acta*, **30** (1985) 1295.
[154] S. Bruckenstein, C. P. Wilde, M. Shay, A. R. Hillman, D. C. Loveday, *J. Electroanal. Chem.*, **258** (1989) 457.
[155] S. Bruckenstein, C. P. Wilde, A. R. Hillman, *J. Phys. Chem.*, **94** (1990) 6458.
[156] S. Bruckenstein, C. P. Wilde, M. Shay, A. R. Hillman, *J. Phys. Chem.*, **94** (1990) 787.
[157] D. Orata, D. Buttry, *J. Am. Chem. Soc.*, **109** (1987) 3574.
[158] K. Naoi, M. M. Lien, W. H. Smyrl, *J. Electroanal. Chem.*, **272** (1989) 273.
[159] K. K. Kanazawa, J. G. Gordon, *Anal. Chim. Acta*, **175** (1985) 99.
[160] S. Bruckenstein, C. P. Wilde, A. R. Hillman, *J. Phys. Chem.*, **94** (1990) 6458.
[161] S. Bruckenstein, C. P. Wilde, A. R. Hillman, *J. Phys. Chem.*, **97** (1993) 6853.
[162] A. R. Hillman, M. J. Swann, S. Bruckenstein, *J. Phys. Chem.*, **95** (1991) 3271.
[163] A. R. Hillman, N. A. Hughes, S. Bruckenstein, *J. Electrochem. Soc.*, **139** (1992) 74.
[164] S. Bruckenstein, P. Krtil, A. R. Hillman, *J. Phys. Chem. B*, **102** (1998) 4994.
[165] S. Bruckenstein, A. R. Hillman, *J. Phys. Chem. B*, **102** (1998) 10826.
[166] S. Bruckenstein, A. R. Hillman, H. L. Bandey, *Compr. Chem. Kinet.*, **37** (1999) 489.
[167] I. Jureviciute, S. Bruckenstein, A. R. Hillman, *J. Electroanal. Chem.*, **488** (2000) 73.
[168] S. Bruckenstein, K. Brzezinska, A. R. Hillman, *Electrochim. Acta*, **45** (2000) 3801.
[169] I. Jureviciute, S. Bruckenstein, A. R. Hillman, A. Jackson, *Phys. Chem. Chem. Phys.*, **2** (2000) 4193.
[170] K. Kim, I. Jureviciute, S. Bruckenstein, *Electrochim. Acta*, **46** (2001) 4133.
[171] S. Bruckenstein, I. Jureviciute, A. R. Hillman, *J. Electrochem. Soc.*, **150** (2003) E285.
[172] A. Jackson, A. R. Hillman, S. Bruckenstein, I. Jureviciute, *J. Electroanal. Chem.*, **524** (2002) 90.
[173] E. Laviron, L. Roullier, *J. Electroanal. Chem.*, **115** (1980) 1.

[174] A. R. Hillman, M. A. Mohamoud, S. Bruckenstein, *Electroanalysis*, **17** (2005) 1421.
[175] A. R. Hillman, M. A. Mohamoud, *Electrochim. Acta*, **51** (2006) 6018.
[176] I. Jureviciute, S. Bruckenstein, A. R. Hillman, *Electrochim. Acta*, **51** (2006) 2351.
[177] A. R. Hillman, S. J. Daisley, S. Bruckenstein, *Electrochem. Commun.*, **9** (2007) 1316.
[178] A. R. Hillman, S. J. Daisley, S. Bruckenstein, *Phys. Chem. Chem. Phys.*, **9** (2007) 2379.
[179] G. Maia, R. M. Torresi, E. A. Ticianelli, F. C. Nart, *J. Phys. Chem.*, **100** (1996) 15910.
[180] H. Varela, R. M. Torresi, *J. Electrochem. Soc.*, **147** (2000) 665.
[181] H. Varela, R. M. Torresi, D. A. Buttry, *J. Electrochem. Soc.*, **147** (2000) 4217.
[182] J. Agrisuelas, C. Gabrielli, J. J. Garcia-Jareno, D. Giménez-Romero, J. Gregory, H. Perrot, F. Vicente, *J. Electrochem. Soc.*, **154** (2007) F134.
[183] L. Niu, C. Kvarnstrom, A. Ivaska, *J. Electroanal. Chem.*, **569** (2004) 151.
[184] J. Bobacka, A. Lewenstam, A. Ivaska, *Talanta*, **40** (1993) 1437.
[185] A. Bund, S. Neubeck, *J. Phys. Chem. B*, **108** (2004) 17845.
[186] H. Lee, H. Yang, J. Kwak, *J. Phys. Chem. B*, **103** (1999) 6030.
[187] H. Lee, H. Yang, J. Kwak, *Electrochem. Commun.*, 4 (2002) 128.
[188] M. J. Henderson, A. R. Hillman, E. Vieil, *J. Phys. Chem. B*, **103** (1999) 8899.
[189] M. J. Henderson, A. R. Hillman, E. Vieil, *Electrochim. Acta*, **45** (2000) 3885.
[190] X. Tu, Q. Xie, C. Xiang, Y. Zhang, S. Yao, *J. Phys. Chem. B*, **109** (2005) 4053.
[191] V. Syritski, R. E. Gyurcsanyi, A. Opik, K. Toth, *Synth. Met.*, **152** (2005) 133.
[192] V. Syritski, A. Opik, O. Forsen, *Electrochim. Acta*, **48** (2003) 1409.
[193] M. A. Malik, G. Horanyi, P. J. Kulesza, G. Inzelt, V. Kertesz, R. Schmidt, E. Czirok, *J. Electroanal. Chem.*, **452** (1998) 57.
[194] W. Plieth, A. Bund, U. Rammelt, S. Neudeck, L. Duc, *Electrochim. Acta*, **51** (2006) 2366.
[195] H. Yang, J. Kwak, *J. Phys. Chem. B*, **101** (1997) 774.
[196] H. Yang, J. Kwak, *J. Phys. Chem. B*, **101** (1997) 4656.
[197] H. Yang, J. Kwak, *J. Phys. Chem. B*, **102** (1998) 1982.
[198] H. Yang, H. Lee, Y. T. Kim, J. Kwak, *J. Electrochem. Soc.*, **147** (2000) 4239.
[199] H. Yang, H. Lee, Y. T. Kim, J. Kwak, *J. Electrochem. Soc.*, **147** (2000) 3801.
[200] I. Oh, H. Lee, H. Yang, J. Kwak, *Electrochem. Commun.*, **3** (2001) 274.
[201] C. Gabrielli, *Identification of electrochemical processes by frequency response analysis*, Solartron Instruments, Farnborough, UK, 1981.
[202] M. Orazem, B. Tribollet, *Electrochemical impedance analysis*, Wiley, New York, 2008.
[203] H. Ehahoun, C. Gabrielli, M. Keddam, H. Perrot, P. Rousseau, *Anal. Chem.*, **74** (2002) 1119.
[204] G. Sauerbrey, *Z. Phys.*, **155** (1950) 206.
[205] K. Bizet, C. Gabrielli, H. Perrot, *Appl. Biochem. Biotechnol.*, **89** (2000) 139.
[206] J. J. Garcia-Jareno, C. Gabrielli, H. Perrot, *Electrochem. Commun.*, **2** (2000) 195.
[207] S. Al-Sana, C. Gabrielli, H. Perrot, *Russ. J. Electrochem.*, **40** (2004) 275.
[208] R. Torres, A. Arnau, H. Perrot, J. Garcia-Jareno, C. Gabrielli, *Electron. Lett.*, **42** (2006) 1272.
[209] C. Gabrielli, H. Perrot, D. Rose, A. Rubin, J. P. Toqué, M. C. Pham, B. Piro, *Rev. Sci Instrum.*, 78, **074103** (2007) 1.
[210] C. Gabrielli, M. Keddam, R. Torresi, *J. Electrochem. Soc.*, **138** (1991) 2657.
[211] K. Itaya, T. Ataka, S. Toshima, *J. Am. Chem. Soc.*, **104** (1982) 4767.
[212] B. J. Feldman, O. Melroy, *J. Electroanal. Chem.*, **234** (1987) 213.
[213] B. J. Feldman, R. W. Murray, *Inorg. Chem.*, **26** (1987) 1702.

[214] J. J. Garcia-Jareño, J. Navarro, A. F. Roig, H. Scholl, F. Vicente, *Electrochim. Acta*, **40** (1995) 1113.

[215] P. J. Kulesza, *J. Electroanal. Chem.*, **289** (1990) 103.

[216] C. Gabrielli, J. J. García-Jareño, M. Keddam, H. Perrot, F. Vicente, *J. Phys. Chem. B*, **106** (2002) 3192.

[217] D. Gimenez-Romero, P. R. Bueno, C. Gabrielli, J. J. Garcia-Jareno, H. Perrot, F. Vicente, *J. Phys. Chem. B*, **110** (2006) 19352.

[218] P. R. Bueno, D. Gimenez-Romero, C. Gabrielli, J. J. Garcia-Jareno, H. Perrot, F. Vicente, *J. Amer. Chem. Soc.*, **128** (2006) 17146.

[219] C. Gabrielli, J. J. García-Jareño, M. Keddam, H. Perrot, F. Vicente, *J. Phys. Chem. B*, **106** (2002) 3182.

[220] D. Gimenez-Romero, P. R. Bueno, J. J. Garcia-Jareno, C. Gabrielli, H. Perrot, F. Vicente, *J. Phys. Chem. B*, **110** (2006) 2715.

[221] M. Zadronecki, P. K. Wrona, Z. Galus, *J. Electrochem. Soc.*, **146** (1999) 620.

[222] C. Gabrielli, H. Perrot, A. Rubin, M. C. Pham, B. Piro, *Electrochem. Commun.*, **9** (2007) 2197.

[223] A. F. Diaz, J. F. Rubinson, H. B. Mark, *Adv. Polym. Sci.*, **84** (1988) 113.

[224] R. C. D. Peres, J. M. Pernault, M. A. De Paoli, *J. Polym. Sci.*, **29** (1991) 225.

[225] C. Gabrielli, J. J. Garcia-Jareno, H. Perrot, *Electrochim. Acta*, **46** (2001) 4095.

[226] A. Cadogan, Z. Gao, A. Lewenstam, A. Ivaska, *Anal.Chem.*, **64** (1992) 2496.

[227] J. Bobacka, *Anal. Chem.*, **71** (1999) 4932.

[228] T. Lindorfs, A. Ivaska, *Anal. Chim. Acta*, **404** (2002) 101.

[229] C. Gabrielli, P. Hémery, P. Liatsi, M. Masure, H. Perrot, *J. Electrochem. Soc.*, **152** (2005) 219.

[230] C. Gabrielli, P. Hémery, P. Liatsi, M. Masure, H. Perrot, *Electrochim. Acta*, **51** (2006) 1704.

6

Monte Carlo Simulations of the Underpotential Deposition of Metal Layers on Metallic Substrates: Phase Transitions and Critical Phenomena

M. Cecilia Giménez[1], Ezequiel P. M. Leiva[2], and Ezequiel Albano[3]

[1] *Physics Department, San Luis University, Chacabuco 917, C.P. 5700, San Luis, Argentina, cecigime@unsl.edu.ar*
[2] *INFIQC, Chemical Sciences Faculty, Córdoba National University, Haya de La Torre and Medina Allende, C.P. 5000, Córdoba, Argentina, eleiva@mail.fcq.unc.edu.ar*
[3] *INIFTA, CCT-La Plata, CONICET, UNLP, Casilla de Correo 16, Sucursal 4, 1900 La Plata, Argentina, ealbano@inifta.unlp.edu.ar*

Summary. The underpotential deposition (UPD) of metal submonolayers and monolayers on metal substrates for the systems Ag/Au(100), Au/Ag(100), Ag/Pt(100), Pt/Ag(100), Au/Pt(100), Pt/Au(100), Au/Pd(100), and Pd/Au(100) is studied by means of lattice Monte Carlo simulations. Interaction energies among different metal atoms are evaluated by using the embedded-atom method. A wide variety of physical situations are found and discussed, including systems exhibiting the sequential adsorption of atoms on kink and step sites, prior to the completion of the monolayer. On the other hand, for other systems, we observe the formation of 2D alloys between substrate and adsorbate atoms, and our predictions are compared with available experimental data. The adsorption isotherms determined for most of the systems studied exhibit sharp transitions in the coverage when the chemical potential is finely tuned. In particular, on the basis of the fact that the UPD of Ag atoms on the Au(100) surface exhibits a sharp first-order phase transition, at a well-defined value of the (coexistence) chemical potential, we also performed extensive simulations aimed at investigating the hysteretic dynamic behavior of the system close to coexistence upon the application of a periodic potential signal.

I. INTRODUCTION: SOME BASIC ASPECTS OF THE UNDERPOTENTIAL DEPOSITION PHENOMENON

Among electrochemical surface phenomena, the deposition of a metal onto a foreign metal surface opens the way to a wealth of possibilities for preparing and designing surfaces with a variety of electrocatalytic activities.[41] Furthermore, since the growth of a new phase is involved, the problem is by itself of fundamental importance for understanding a number of processes involved in the formation of the new phase.

Let us first briefly discuss the advantages of electrochemical deposition of a metal with respect to the same processes achieved from the gas phase. Figure 1 schematically illustrates the binding (free) energy of an atom binding to a surface as a function of the distance from it. In the case of metallic systems, the binding-energy curve typically exhibits a minimum with values in the range 3–5 eV, which at room temperature amount to between $120kT$ and $200kT$, kT being the thermal energy. An estimation of the desorption rate can be made in terms of the equation

$$\nu = \nu_0 \exp\left(-\frac{E^{\#}}{kT}\right), \tag{1}$$

where the preexponential factor ν_0 contains entropic contributions and shows a weak dependence on the temperature, and $E^{\#}$ is the activation energy for the desorption process. Using an attempt frequency ν_0 of 1 THz, the resulting frequencies range is between 10^{-40} and 10^{-75} s^{-1} for the desorption of a single adatom. Even considering

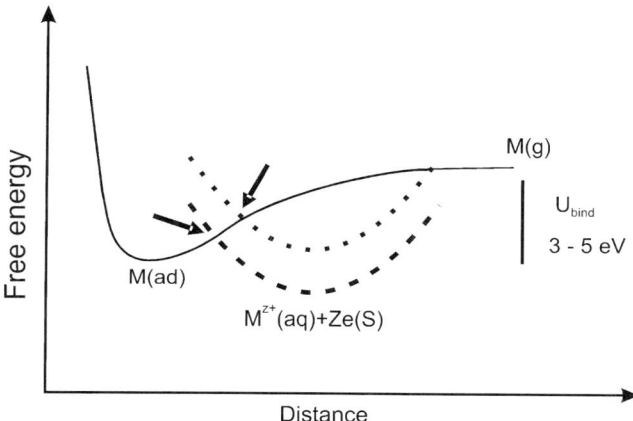

Figure 1. The binding (free) energy of a metal atom (M) to a metal surface as a function of the distance from it (*full line*) and the modification in the case of the electrochemical situation. Here *(g)* and *(ad)* refer to the gas and adsorbed phases, respectively. Also, M^{z+}(aq) represents a metal ion in an aqueous solution. The *dotted* and the *broken lines* represent states of the system where the desorbed state corresponds to an electron in the metal and a solvated ion in the solution. The *broken line* corresponds to a more positive potential applied to the metal surface than that shown by the *dotted line*.

an ensemble of the order of 10^{23} atoms, the times resulting for the desorption of a single adatom are in the range of 10^{17} and 10^{52} s, which are extraordinarily long times (the age of the universe is of the order of 4×10^{17} s). These estimations show that any attempt at attaining an adsorption–desorption equilibrium would be unsuccessful for metallic atoms at room temperature. Once a free atom arrives at the surface, it can be practically stated that it will never leave it at 300 K. However, when electrochemistry is put to work, the situation changes drastically, as illustrated in Fig. 1. The dotted curve introduces an alternative state to that of the desorbed adatom. This new state corresponds to an ion core solvated in solution and an electron (or the number of electrons corresponding to the valence) in the electrode. The energetic situation of the latter can be changed by the application of a potential difference, as we shall show below, so that one can have a state with an even lower energy, as marked by the dashed line. Thus, the activation energy for adatom desorption can be practically changed by the application of a suitable potential to the substrate. It must be pointed out that although we have used so far the term energy, the proper thermodynamic quantity to consider in the case of the present system is the Gibbs free energy, G, since the usual experimental conditions are those where the pressure and the temperature of the system are given.

The state of the art of several underpotential deposition (UPD) systems and a thorough compilation of them prior to 1996 have been given in the book by Budevski et al.[8] More recent reviews have been given by Aramata[6] and Herzog and Arrigan,[18] the latter in an analytical context. Although the evidence of UPD phenomena dates back to the 1930s (see Chap. 3 in Ref. [8] for a historical review), experimental studies on UPD on polycrystalline surfaces bloomed in the 1960s and the 1970s. At the end of the latter decade, Kolb[23] reviewed the work on UPD on polycrystalline surfaces. At this stage, modeling on this phenomenon was phenomenological, being related to a thermodynamic description of the problem. In this respect, Kolb and coworkers[14,24] found a linear relationship between the underpotential shift and the work function that was interpreted in terms of the differences of the electronegativities of the metals participating in the binding. Following this line of modeling, Trasatti[46] proposed an ionic bond model to explain the correlation between underpotential shift and work function differences. A quantum mechanical description of UPD in terms of the jellium model that revealed the role played by the work function and the surface energies of the metals involved in the UPD problem was introduced by Leiva and Schmickler.[27–29] The use of single-crystal surfaces to study the UPD phenomenon started at the end of the 1970s, providing a new insight into this problem.

Figure 2 shows typical examples of UPD on single-crystal surfaces, that is, potentiodynamic runs obtained for the deposition of Tl on Ag(111) and Ag(100) single-crystal surfaces. The technique employed to get these results involves the application of a linear potential sweep to the working electrode, with the measurement of the resulting current I. This is usually referred to the electrode area, A, so the most commonly reported quantity is the current density $i = I/A$. The latter can be written as

$$i = -Q\frac{d\theta}{dt}, \qquad (2)$$

where Q is the charge flowing for the desorption of a complete monolayer, and θ is the coverage degree. If the potential sweep rate

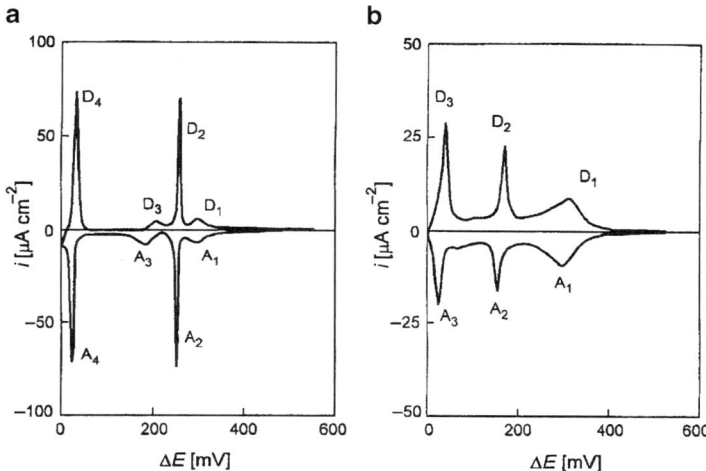

Figure 2. Cyclic voltammograms corresponding to the electrodesorption/adsorption of Tl^+ ions on a Ag(111) surface (**a**) and on a Ag(100) surface (**b**). The electrolyte was 1.5×10^{-3} M $TlClO_4^+$, 0.5 M $NaClO_4^+$, 10^{-2} M $HClO_4$. The potential sweep rate was 10 mV/s and the temperature was 298 K. (Reprinted from Ref. [45], with permission from Elsevier.)

is low enough so that equilibrium conditions are approached, (2) shows us that a voltammogram such as that shown in Fig. 2 should give us the derivative of the adsorption isotherm. Furthermore, the hysteresis observed between the positive and negative sweep is a measure of the deviation from equilibrium, so the peak shift between the upper and lower parts of the figure tells us something about the nature of the different processes involved. Also, several processes are observed on the different single-crystal faces, which involve the formation of different phases on the surface. Their nature was discussed by Lorenz and coworkers.[44] The fact that even a simple system shows the existence of several phases (and, thus, phase transitions) on a well-defined single crystal, as well as hysteresis, indicates that UPD systems may be a rich area for experimental and theoretical work.

Let us now analyze UPD in terms of the classical approach employed to study the deposition of a metal on a foreign substrate,[8] as summarized in Fig. 3. According to this view, the deposition mechanisms are considered taking into account two main parameters. One

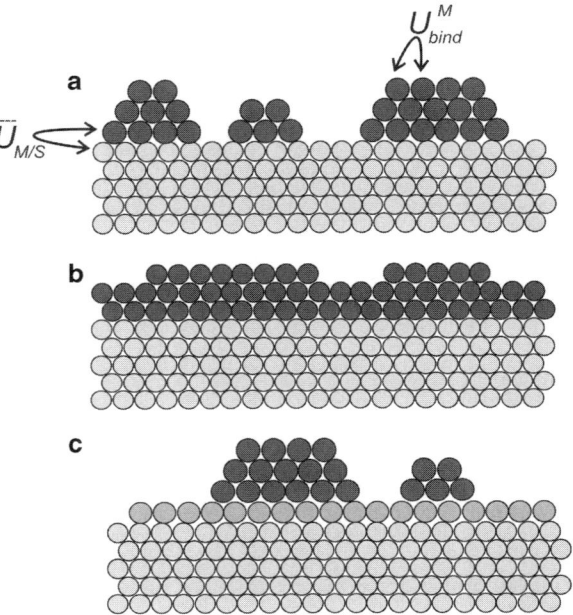

Figure 3. Different metal growth types depending on the metal adsorbate–metal substrate interaction and the lattice misfit. (**a**) Adsorbate–adsorbate interaction, U_{bind}^{M}, is stronger than adsorbate—substrate interaction, $\overline{U}_{M/S}$. Only overpotential deposition is observed, with the formation of adsorbate metal clusters on the surface (Volmer–Weber mechanism). (**b**) Substrate–adsorbate interaction, $\overline{U}_{M/S}$, is stronger than adsorbate–adsorbate interaction, U_{bind}^{M}. Lattice misfit, $\varepsilon_{mf}^{\%}$, is negligible. A pseudomorphic layer is observed at underpotential. Overpotential growth takes place in a layer-by-layer fashion (Frank–van der Merwe mechanism). (**c**) Substrate–adsorbate interaction, $\overline{U}_{M/S}$, is stronger than adsorbate–adsorbate interaction, U_{bind}^{M}. Lattice misfit, $\varepsilon_{mf}^{\%}$, is important. A first underpotential-deposited layer is formed, which exhibits a significant strain and is different from its bulk three-dimensional phase. Further overpotential growth involves the formation of unstrained three-dimensional clusters that resemble the bulk structure of the deposit (Stranski–Krastanov mechanism).

of them is the interaction of the adsorbate with the substrate, $\overline{U}_{M/S}$, with respect to the bulk interactions of metal being deposited, U_{bind}^{M}, and the other is the lattice misfit, which can be defined according to

$$\varepsilon_{\text{mf}}^{\%} = \frac{a_{\text{S}}^{\text{bulk}} - a_{\text{M}}^{\text{bulk}}}{a_{\text{S}}^{\text{bulk}}} \times 100, \qquad (3)$$

where $a_{\text{S}}^{\text{bulk}}$ and $a_{\text{M}}^{\text{bulk}}$ denote the substrate and the adsorbate bulk lattice parameters, respectively.

When $U_{\text{bind}}^{\text{M}}$ is lower than $\overline{U}_{\text{M/S}}$, a more negative potential than the reversible one is necessary to generate the new phase. Metal deposition under these conditions is denominated "overpotential deposition" (OPD). Since the homoatomic binding is stronger than the heteroatomic one, it is expected that the deposited atoms will try to join each other instead of wetting the surface, with the concomitant formation of adsorbate islands on the substrate surface (Fig. 3a). This is the so-called Volmer–Webe deposition mechanism. The opposite case, $U_{\text{bind}}^{\text{M}} > \overline{U}_{\text{M/S}}$, means UPD. We will address the thermodynamics of UPD and OPD in more detail below. Depending on $\varepsilon_{\text{mf}}^{\%}$, two different UPD mechanisms may result. If $\varepsilon_{\text{mf}}^{\%}$ is negligible, the UPD growth will closely follow the morphology of the substrate, yielding the so-called pseudomorphic growth. The usual situation is that one or two monolayers can be deposited onto the substrate in the UPD region without appreciable strain. The application of a negative overpotential would lead this system to the so-called Frank–van der Merwe layer-by-layer growth (see Fig. 3b). On the other hand, if $\varepsilon_{\text{mf}}^{\%}$ is important, UPD is still possible, but the growth may no longer be pseudomorphic. In any case, a strained monolayer results, with characteristics different from those of the bulk metal. A negative overpotential would lead this system into the so-called Stranski–Krastanov deposition mechanism, where metal islands will occur on the surface of the strained monolayer (Fig. 3c). This third case, where $\varepsilon_{\text{mf}}^{\%}$ is relatively large, may in turn be subdivided into two types of growth. On the one hand, if $\varepsilon_{\text{mf}}^{\%}$ is positive and the adsorbate tries to follow the substrate structure, an expanded structure will result with respect to bulk metal. This will imply a work function of the system lowered not only with respect to bulk substrate but also with respect to bulk metal, providing a favorable scenario for anion adsorption due to the shift of the potential of zero charge of the surface. On the other hand, if $\varepsilon_{\text{mf}}^{\%}$ is negative and large, repulsion between metal adatoms will prevent them from following the structure of the substrate, with the occurrence of expanded structures with respect to the substrate, which may be higher-order commensurate or incommensurate. Rojas[38] has analyzed the different types of atomic packing

Table 1.

Crystallographic misfits as defined according to (3) (*first line*) and the status of the monolayer (*second line*) according to the results of an off-lattice Monte Carlo simulation.

Substrate/adsorbate	Cu	Pd	Pt	Au	Ag
Cu	0	−7.6	−8.4	−12.9	−13.2
	p	p	p	e	e
Pd	7.0	0	−0.7	−4.9	−5.3
	p	p	p	p	p
Pt	7.7	0.7	0	−4.1	−4.5
	p	p	p	p	p
Au	11.4	4.7	4.0	0	−0.3
	c	p	p	p	p
Ag	11.7	5.0	4.3	0.3	0
	c	p	p	p	p

Pseudomorphic systems are reasonable candidates for lattice model simulations. Taken from Ref. [39].
p pseudomorphic, e expanded with respect to a 1×1 layer on the substrate, c compressed with respect to a 1×1 layer on the substrate.

at the surface for binary systems involving Ag, Au, Pd, Pt, and Cu, by means of an off-lattice Monte Carlo simulation technique. These simulations allowed the positions of adsorbate atoms to relax freely to their equilibrium values, without enforcement of any underlying lattice positions. The results are summarized in Table 1. Those systems remaining commensurate are good candidates for lattice model simulations, such as those that are the subject of the present work.

II. SOME THERMODYNAMICS ON THE UPD PHENOMENON

We discuss now the electrochemical problem in some more detail. To do this, it is useful to draw a thermodynamic cycle, as shown in Fig. 4. Electrochemical measurement systems usually consist of three electrodes: the so-called working electrode, where the processes under study take place, the reference electrode, with respect to which the potential applied to the working electrode is measured

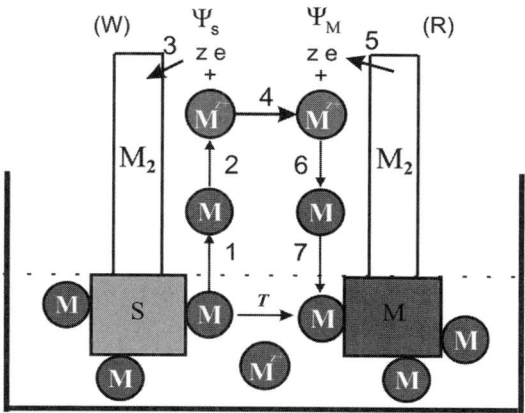

Figure 4. Thermodynamic cycle representing the transfer of a core ion of a metal adsorbed on a substrate to the bulk metal in an electrochemical system.

(this remains unpolarized), and the counter electrode, which provides a circuit for electron flow, acting as an electron source or sink, depending on the reactions taking place at the working electrode. For the sake of simplicity, in this scheme we consider the working and reference electrodes only. Alternatively, our reference electrode could be thought to be infinitely large, playing at the same time the role of the counter electrode, providing the current necessary to compensate the reactions at the working electrode without changing its potential. Furthermore, we will assume that the reference electrode is made of the same metal as that being deposited on the working electrode and that both the reference and the working electrodes are immersed in a solution containing the core ions M^{z+} of the metal being deposited. Let us assume first that the potential difference is measured under zero current flow conditions, and that equilibrium has been established. Let us further assume that a certain number of atoms have been deposited on the working electrode, which is made of the substrate material. Thus, the ion cores M^{z+} of the deposit on the substrate are in equilibrium with the ion cores in the bulk metal of the reference electrode. Thus, the free-energy change for the transfer of an ion from the deposit to the bulk metal, say, ΔG_T, must be equal to zero:

$$\Delta G_T = 0. \tag{4}$$

An alternative pathway for the same ion is also shown in Fig. 4, involving the corresponding free-energy changes from 1 to 7, i.e., the free energy change required to take the core ion M^{z+} out from the surface of the substrate to a point in the metal. This can be summarized in the following set of equations:

$$\Delta G_1 = -\mu_{M/S}, \tag{5}$$

$$\Delta G_2 = I_{+z}, \tag{6}$$

$$\Delta G_3 = -ze\Phi_{M_2}, \tag{7}$$

$$\Delta G_4 = ze\left(\Psi_M - \Psi_S\right), \tag{8}$$

$$\Delta G_5 = ze\Phi_{M_1}, \tag{9}$$

$$\Delta G_6 = -I_{+z}, \tag{10}$$

$$\Delta G_7 = \mu_M, \tag{11}$$

where $\mu_{M/S}$ is the chemical potential of the metal atoms adsorbed on the substrate, I_{+z} is the energy required to remove z electrons from the metal atom, Ψ_S and Ψ_M are the outer potentials of the working and the reference electrodes, respectively, and μ_M is the chemical potential of the metal atoms in bulk metal. Use of (5)–(11) along with the condition $\Delta G_T = \sum_{i=1}^{7} \Delta G_i$ leads to the equality

$$\Delta\phi = (\Psi_S - \Psi_M) = \frac{\mu_M - \mu_{M/S}}{ze}, \tag{12}$$

where $\Delta\phi$ is the potential difference measured between the working and the reference electrodes. If for some type of atomic arrangement of the metal atoms on the substrate surface $\Delta\phi > 0$, we are in the presence of the so-called UPD phenomenon. That is, the adsorption of the first monolayer of adsorbate atoms may occur at potentials that are positive with respect to the Nernst equilibrium potential. On the other hand, if $\Delta\phi < 0$, this means that the atomic arrangement with the chemical potential is less stable than the bulk metal and could only be observed at overpotentials (OPD), constituting a metastable state. In the case of UPD, (12) defines the so-called *underpotential shift*,[23] say, $\Delta\phi_{UPD}$, which is a measure of the stability of the adatoms in a monolayer with respect to the bulk. To discuss it in energetics terms, let us neglect for the moment entropic and volume effects, considering the case $T = 0$ and $P = 0$. This is not a severe approximation in the case of the formation of a pure monolayer, as we have discussed previously.[34] In such a case, the configurational entropy is zero, and the vibrational entropic and volume changes were found to provide a very small contribution to the underpotential shift. In this limit $\Delta\phi_{\text{UPD}}$ can be calculated from the binding energy of the adatoms on the monolayer and the bulk binding energy of the metal, $U_{\text{bind}}^{\text{M}}$,

$$\Delta\phi_{\text{UPD}} \approx \frac{U_{\text{bind}}^{\text{M}} - (U_{\text{M/S}} - U_{\text{S}})/N_{\text{M/S}}}{ze}, \qquad (13)$$

where $U_{\text{M/S}}$ is the energy of the adatoms plus the substrate system, U_{S} is the energy of the naked substrate, and $N_{\text{M/S}}$ is the number of adsorbate atoms. Note that we have replaced the differential quantity $\mu_{\text{M/S}}$ by the term $(U_{\text{M/S}} - U_{\text{S}})/N_{\text{M/S}}$. In fact, by neglecting entropic effects, $\mu_{\text{M/S}}$ should be given by the derivative of $U_{\text{S/M}}$ with respect to the number of metal atoms in the system, say, N_{M},

$$\mu_{\text{M/S}} \approx \left(\frac{\partial U_{\text{M/S}}}{\partial N_{\text{M/S}}}\right)_{P,T}, \qquad (14)$$

which in principle could be different from $(U_{\text{M/S}} - U_{\text{S}})/N_{\text{M/S}}$. However, if we model the UPD monolayer as an infinite system, the binding energy of the adatoms, namely, $U_{\text{M/S}} - U_{\text{S}}$, can be considered to be an *extensive* quantity and can be written as

$$U_{\text{M/S}} - U_{\text{S}} = N_{\text{M}}\overline{U}_{\text{M/S}}, \qquad (15)$$

where $\overline{U}_{\text{M/S}}$ is the binding energy per atom, a quantity that, in principle, is independent of the system size owing to the absence of border effects. Thus, differentiation of (15) yields

$$\mu_{\text{M/S}} \approx \left(\frac{\partial U_{\text{M/S}}}{\partial N_{\text{M/S}}}\right)_{P,T} = \overline{U}_{\text{M/S}} = \frac{(U_{\text{M/S}} - U_{\text{S}})}{N_{\text{M/S}}}. \qquad (16)$$

The physical meaning of (16) is that if we assembled a monolayer atom by atom, most of the time we would obtain the same energy as the binding energy of an adatom in the monolayer. This is analogous to the considerations that are made concerning the disassembly of a bulk metal by removing atoms in a kink position.[8] This equation is a consequence of the extensivity assumption made in (15) and would be not be valid for a nanometric system or for a metal deposit where border effects are not negligible.

Note that in the derivation of (12) no current flow through an external circuit was considered. In fact, in the third step the electrons of the adatom were brought back to the working electrode and only the transfer of an ion was taken into account. In other words, we considered what we usually denominate an "N, P, T system." An alternative derivation can be made taking into account electron exchange between the system and the environment. Let us now assume that the working and the reference electrodes are connected to two infinite electronic reservoirs with electrochemical potentials $\widetilde{\mu}_e^w$ and $\widetilde{\mu}_e^r$, respectively, thus providing a Volta potential difference $\Psi_S - \Psi_M = \widetilde{\mu}_e^r - \widetilde{\mu}_e^w$ (the contacts are made of the same material, so the chemical part of the electrochemical potential is the same). These reservoirs may remove electrons from (or inject electrons into) the system by keeping $\Psi_S - \Psi_M$ constant. With these boundary conditions, the proper thermodynamic function to study the system is

$$\widetilde{G} = U + PV - TS - N_e^w \widetilde{\mu}_e^w - N_e^r \widetilde{\mu}_e^r \qquad (17)$$

where N_e^w and N_e^r are the number of electrons in the W and the R electrodes respectively.

On the other hand, the energy of the system can be written as

$$U = TS - PV + N_e^w \widetilde{\mu}_e^w + N_e^r \widetilde{\mu}_e^r + N_M \widetilde{\mu}_{M^{+z}} + N_{M/S} \widetilde{\mu}_{M^{+z}/S} \quad (18)$$

$N_{M/S}$ is, as before, the number of adsorbate atoms on the substrate S, N_M is the number of M bulk atoms making the R electrode, $\widetilde{\mu}_{M^{+z}}$ and $\widetilde{\mu}_{M^{+z}/S}$ are the electrochemical potentials of the M^{+z} core ions in the reference and working electrode respectively.

Let us now consider the transfer of a single M atom from the W to the R electrode, assuming that the atom is transferred as a neutral species. Thus we have that $\Delta N_e^w = -z = -\Delta N_e^r$, $\Delta N_{M/S} = -\Delta N_M = -1$, so the two previous equations become:

$$\Delta \widetilde{G} = \Delta U + \Delta(PV) - \Delta(TS) + z\widetilde{\mu}_e^w - z\widetilde{\mu}_e^r \qquad (19)$$

and

$$\Delta U = \Delta(TS) - \Delta(PV) - z\widetilde{\mu}_e^w + z\widetilde{\mu}_e^r + \widetilde{\mu}_{M^{+z}} - \widetilde{\mu}_{M^{+z}/S} \quad (20)$$

Taking into account that the chemical potentials of the M atoms in the reference and working electrodes are $\mu_M = \widetilde{\mu}_{M^{+z}} + z\widetilde{\mu}_e^r$ and $\mu_{M/S} = \widetilde{\mu}_{M^{+z}/S} + z\widetilde{\mu}_e^w$ and combining eqs. (19) and (20) we get:

$$\Delta \widetilde{G} = \mu_M - \mu_{M/S} + z\widetilde{\mu}_e^w - z\widetilde{\mu}_e^r = \mu_M - \mu_{M/S} - ze\eta \quad (21)$$

The equilibrium condition $\Delta \tilde{G} = 0$ leads, as before, to Eq. (12), which provides the basis for computer simulations of the present system. In fact, this equation defines the relationship between a parameter that is experimentally controlled, that is, the overpotential η, and its counterpart in computer simulations, which is the chemical potential of the atoms being deposited μ_M. In principle, since we are dealing with macroscopic systems, simulations within different ensembles should lead to the same conclusions, i.e., canonical (NVT) as well as grand canonical simulations could be employed to study UPD systems. However, the grand canonical conditions are closer to the experimental ones, and the control of the chemical potential in many cases provides a straightforward emulation of the experimental conditions, so this is the technique more frequently applied. Furthermore, if nanosystems are under consideration, this is the only valid alternative, since the equivalence between ensembles is no longer valid for them.[19]

III. DESCRIPTION OF THE MONTE CARLO SIMULATION METHOD AND THE MODEL FOR METAL DEPOSITION

In this section we will describe the model and simulation method employed for the results presented in this chapter.

1. The Lattice Model

Lattice models are widely used in computer simulations to study the adsorption of atoms and molecules on surfaces, because they allow us to deal with a large number of particles at a relatively low computational cost. Furthermore, the assumption that particle adsorption can only occur at definite sites is a good approximation for some systems. Such is the case of the adsorption of Ag on Au, where there is no crystallographic misfit. For some systems, one also has strong evidence, obtained by performing continuum computer simulations using the canonical Monte Carlo method, that indicates that at least one of the phases present during the UPD on (100) surfaces possesses a pseudomorphic structure.[39] Continuum Monte Carlo simulations have shown that an adsorbate monolayer spontaneously acquires a (1×1) coincidence cell, in agreement with the experimental finding at low underpotentials.[13]

However, it should be recognized that continuum Hamiltonians, where particles are allowed to take any position in space, are much more realistic in cases where the epitaxial growth of an adsorbate leads to incommensurate adsorbed phases[40] or to adsorbates with large coincidence cells.

Throughout this chapter, we report results obtained by using the grand canonical Monte Carlo simulation applied to a lattice model that represents the square (100) surface lattice of a metal substrate. Solvent effects are neglected, but this should not be a major problem for the metal couples considered, since the partial charge on the adatoms is expected to be small, thus minimizing the metal–water dipole interactions. The model also neglects all kinds of anion effects that may coadsorb during the metal deposition process. This approximation may lead to some underestimation or overestimation of the underpotential shift,[24] depending on whether anions adsorb more strongly or more weakly on the adsorbate than on the substrate, respectively.[42]

Square lattices with periodic boundary conditions are used to represent the surface. Each lattice node represents an adsorption site for an atom. The adsorbate may adsorb, desorb, or hop between neighboring sites. Also, atoms of the same nature as the substrate may only move on the surface. In this way, the model represents an open system for one of its components (the adsorbate), as in the case of adatom deposition on a foreign surface.

2. The Grand Canonical Monte Carlo Method

One of the most appealing characteristics of the grand canonical Monte Carlo method is that, as in many experimental situations, the chemical potential μ is one of the independent variables. This is the case of low-sweep-rate voltammetry, an electrochemical technique where the electrode potential can be used to control the chemical potential of species at the metal–solution interface. This technique offers a straightforward way of obtaining the adsorption isotherms provided the sweep rate is slow enough to ensure equilibrium for the particular system considered. At the solid–vacuum interface, the chemical potential is related to the vapor pressure of the gas in equilibrium with the surface. However, as we discussed in Sect. I, chemical potential sweeps under equilibrium conditions cannot be achieved at the solid–vacuum interface because the desorption process is too slow owing to the high energy barrier that the metal adatoms have to surmount (typically of the order of some electronvolts).

The 2D adsorption system is taken as a square lattice with $M = L \times L$ sites (in the present simulations, we will take $L = 100$). Each adsorption site is labeled by index 0, 1, or 2, depending on whether it is empty, occupied by one atom of the same kind as those of the substrate, or occupied by one adsorbate atom, respectively.

Following the procedure proposed by Metropolis and coworkers,[5] the acceptance probability for a transition from state $\vec{n}$ to $\vec{n}'$ is defined as

$$W_{\vec{n} \to \vec{n}'} = \min(1, \frac{P_{\vec{n}'}}{P_{\vec{n}}}), \qquad (22)$$

where $P_{\vec{n}}$ is the probability of finding the system in state $\vec{n}$, and $P_{\vec{n}'}$ is the probability of finding the system in state $\vec{n}'$, so that detailed balance is granted.

In the grand canonical Monte Carlo simulation, three different types of events are allowed for, as follows:

1. Adsorption of an adsorbate atom onto a randomly selected lattice site
2. Desorption of an adsorbate atom from an occupied lattice site selected at random
3. Motion of an atom from the lattice site where it is adsorbed to one of its four nearest neighboring sites.

Let us now describe the Monte Carlo implementation of these events in detail.

(i) Change of Occupation

This process corresponds to the first two items enumerated in the previous section, which can be considered as a Monte Carlo trial consisting of an attempt to change the occupation state of a randomly selected site.

Let us call p_{io} (probability of input/output) the probability of making this kind of trial. Then, if a random number within the interval [0, 1) is smaller than p_{io}, a change of occupation is attempted. In this way, one of the M adsorption sites is selected at random with probability $1/M$. By defining the matrix element surface(i, j), such that it contains the occupation state of the site (i, j), one has that:

- If surface$(i, j) = 0$, the site was empty and an attempt to adsorb a particle on it is made [the state would change to surface$(i, j) = 2$]. In this case, the change in the number of particles will be $\Delta N = 1$ and the change in energy will be $\Delta E = E_{\text{ads}}$, where E_{ads} is the adsorption energy associated with the adsorbate particle coming to position (i, j), according to the corresponding environment.
- If surface$(i, j) = 1$, it means that the site is already occupied by a substrate particle, so the trial ends.
- If surface$(i, j) = 2$, one has that the site is occupied by an adsorbate particle. Then, an attempt to desorb it is made [the state would change to surface$(i, j) = 0$]. In this case, the change in the number of particles will be $\Delta N = -1$, and the change in energy will be $\Delta E = -E_{\text{ads}}$.

The acceptance probability of the trial for the first or the third case is

$$p = \min\left[1, \exp\left(-\frac{\Delta E - \mu \Delta N}{k_{\text{b}} T}\right)\right]. \qquad (23)$$

(ii) Diffusion

The probability of occurrence of a diffusion event is complementary to that corresponding to the change of occupation, namely,

$$p_{\text{dif}} = 1 - p_{\text{io}}. \qquad (24)$$

To allow the diffusion process on the surface, the particles are labeled from 1 to N (total number of particles) and an attempt to move all particles, each one toward a neighboring site, is made. For each particle, one of the four first neighboring sites is selected with probability $1/4$. If the site is occupied, the trial ends. However, if the site is empty, an attempt to move the particle toward the selected neighboring site is made with probability

$$p = \min\left[1, \exp\left(-\frac{\Delta E}{k_b T}\right)\right], \qquad (25)$$

with

$$\Delta E = E_{\text{ads}}^{\text{final}} - E_{\text{ads}}^{\text{initial}}, \qquad (26)$$

where $E_{\text{ads}}^{\text{initial}}$ is the adsorption energy of the particle on the site initially occupied by it, and $E_{\text{ads}}^{\text{final}}$ is the adsorption energy on the final site where diffusion is attempted.

(iii) Calculation of the Coverage

During the dynamic evolution of the system, the relevant thermodynamic properties are calculated, after disregarding a large enough number of Monte Carlo steps to allow for equilibration. In this way, average values of instantaneous magnitudes are stored during each simulation run. For example, the average coverage of the adsorbate atoms $\langle\theta\rangle_{\text{Ads}}$, at a given chemical potential μ, is calculated as the average of the instantaneous value, $\theta(\mu)_{\text{Ads},i}$, which is defined as follows:

$$\theta(\mu)_{\text{Ads},i} = \frac{N_{\text{Ads},i}}{M - N_{\text{Su}}}, \qquad (27)$$

where $N_{\text{Ads},i}$ is the number of adsorbate atoms, M is the total number of sites, and N_{Su} is the number of substrate-type atoms present on the surface at the time step i.

3. Interatomic Potential: The Embedded-Atom Method

A very important feature to be taken into account when comparing the results of a simulation with experiments is the quality of the interatomic potentials used to perform the simulations. In this work, the embedded-atom method (EAM)[11] is used, because it is able to reproduce important characteristics of the metallic binding, such as the equilibrium lattice constants and heats of solution of the binary alloys.

The EAM considers that the total energy U_{tot} of an arrangement of N particles may be calculated as the sum of energies U_i, corresponding to individual particles, that is,

$$U_{\text{tot}} = \sum_{i=1}^{N} U_i, \tag{28}$$

where U_i is given by

$$U_i = F_i(\rho_{h,i}) + \frac{1}{2} \sum_{j \neq i} V_{ij}(r_{ij}). \tag{29}$$

Here, F_i is the embedding function and represents the energy necessary to embed an atom i in the electronic density $\rho_{h,i}$, at the site where this atom is located. $\rho_{h,i}$ is calculated as the superposition of the individual electronic densities $\rho_i(r_{ij})$:

$$\rho_{h,i} = \sum_{j \neq i} \rho_i(r_{ij}). \tag{30}$$

Thus, the attractive contribution to the embedded-atom potential is given by the embedding energy, which accounts for many-body effects.

On the other hand, the repulsion between ion cores is represented through a pair potential $V_{ij}(r_{ij})$, which only depends on the distance between the cores r_{ij}:

$$V_{ij} = \frac{Z_i(r_{ij}) Z_j(r_{ij})}{r_{ij}}, \tag{31}$$

where $Z_i(r_{ij})$ may be envisaged as a sort of effective charge, which depends on the nature of particle i.

It is worth mentioning that to obtain reliable interatomic potentials, the energies resulting from the application of the EAM have been parameterized by fitting available experimental data, such as elastic constants, dissolution enthalpies of binary alloys, bulk lattice constants, and sublimation heats.[11]

4. Surface Defects

Concerning the electrodeposition process and owing to the presence of surface defects, one may expect not only the formation of 2D phases but also of 0D and 1D phases.[8] In fact, Fig. 5 shows different kinds of surface defects. The 0D phases are associated with the adsorption of individual particles on single surface defects, such as vacancies or kink sites (Fig. 5b). On the other hand, the 1D phases are associated with the step decoration (Fig. 5c). Finally, Fig. 5d shows the adsorption of a complete monolayer forming a 2D phase.

The simulation of annealing processes has often been used to obtain minimal-energy structures or to solve ergodicity problems.[35] A suitable way to implement these simulations is by using the canonical Monte Carlo method at different temperatures. In the present case, simulated annealing is used to obtain different surface defects, such as islands of various sizes and shapes. In all cases the initial state corresponds to a coverage degree of $\theta = 0.1$ substrate atoms distributed at random. Simulations start at a very high initial temperature T_0, of the order of 10^4 K, and the system is later cooled following a logarithmic law:

$$T_\mathrm{f} = T_0 K^{N_{\mathrm{cycles}}}, \tag{32}$$

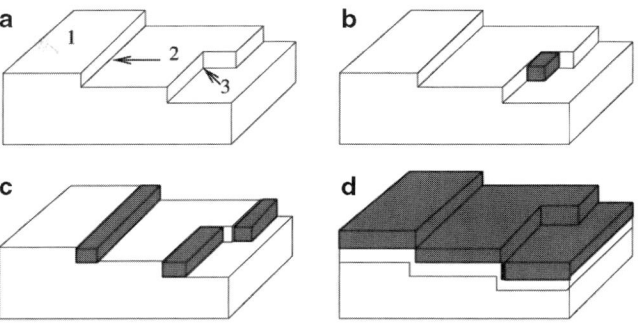

Figure 5. Surface defects. (**a**) *1* terrace sites; *2* step sites; *3* kink sites. (**b**) Adsorption on kink sites. (**c**) Adsorption on steps. (**d**) Adsorption on a complete monolayer. (Reprinted from Ref. [15], with permission from Elsevier.)

where T_f is the final temperature, N_{cycles} is the number of cooling steps, and K is a constant between 0 and 1 (in the present case, $K = 0.9$ is used). Here, a Monte Carlo step consists of the diffusion of all particles present on the surface, as described in Sect. III.2. A certain number of Monte Carlo steps were run at each temperature and the simulation stopped when T_f was reached.

In general, this technique is often used to reach equilibrium configurations. In these cases, a large number of steps are needed for each temperature. However, since our goal is to obtain far-from-equilibrium configurations of substrate islands, a relatively small number of Monte Carlo steps were used. As expected, the use of few Monte Carlo steps, which corresponds to fast cooling, generates a large number of small islands. On the other hand, by using a large number of Monte Carlo steps, which corresponds to slow cooling, one observes the coarsening of the islands, so one generates bigger and more compact aggregates (the limiting case would be to obtain a single island, as in the equilibrium situation).

5. Energy Tables

One of the main advantages of lattice models is their simplicity: since the distances between the adsorption nodes are fixed, the energy values that the system can take become reduced to a discrete set. Furthermore, the potentials used are relatively short ranged, so a very important simplifying assumption can be made to obtain the adsorption energies E_{ads}.

Let us consider the adsorption (desorption) of a particle at a node immersed in a certain environment, as shown in Fig. 6. The adsorption site for the particle is located in the central box, and the calculation of the interactions is limited to a circle of radius R. Then, the adsorption energy for all the possible configurations for the environment of the central atom can be calculated before the simulation. The present results were obtained considering configurations involving first-, second-, and third-nearest neighbors, giving a total of 13 sites, including the central atom (under consideration). Each site has three possible occupation states: 0 (empty), 1 (occupied by a substrate-type atom), or 2 (occupied by an adsorbate-type atom). Therefore, there are $3^{12} = 531,441$ possible configurations, depending on the different environments that an atom can have on the surface. Each possible value of E_{ads} is then calculated as the difference of the total energies of the system with and without the central atom.

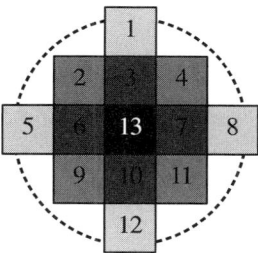

Figure 6. Environment involved in the calculation of the adsorption energy. The sites involved are numbered to distinguish the different configurations. The central site is the darkest and the color of the sites is clearer as we consider first, second, and third neighbors.

One advantage of the Monte Carlo method implemented is that all the adsorption energies of an atom can be tabulated, so during the Monte Carlo simulation the most expensive numerical operations are reduced to the reconstruction of the number I that characterizes the configuration surrounding the particle on the adsorption node. Computationally speaking, I is nothing but the index of the array in which the energy is stored.

IV. ADSORPTION ISOTHERMS

In this section we will show the simulations of adsorption isotherms for several systems. We will study the influence of the temperature and the influence of surface defects on the adsorption of a metal monolayer on a different substrate.

1. Systems Studied and Adsorption Energies

In Fig. 7, 25 relevant adsorption configurations, taken from the 531,441 that are possible for the environments that an atom can have on the surface, are shown. The corresponding adsorption energies are summarized, as illustrative examples, in Table 2 for the

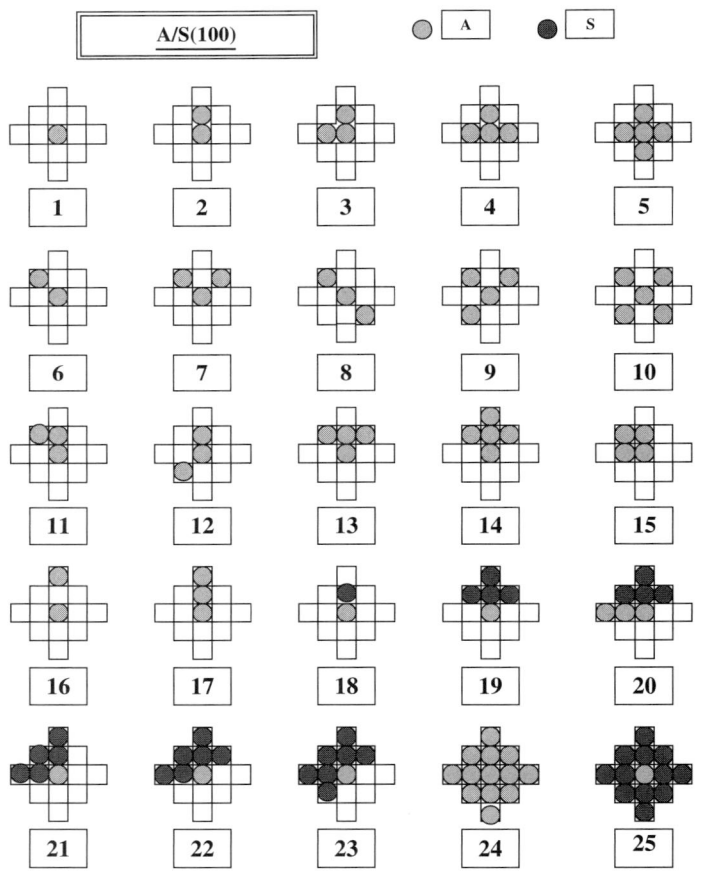

Figure 7. Typical configurations, taken from the 531, 441 possible ones, which an adsorbing atom can have on the surface. *A* adsorbate-type atoms; *S* substrate-type atoms. (Reprinted from Ref. [16], with permission from the American Chemical Society.)

systems studied, namely, Ag on Au(100), Au on Ag(100), Ag on Pt(100), Pt on Ag(100), Au on Pt(100), Pt on Au(100), Pd on Au(100), and Au on Pd(100).

It is interesting to compare different physical situations which can be expected as a consequence of the energies involved in the configurations analyzed. Let us start by considering configurations

Table 2.
Energy differences (eV) associated with the deposition of an atom in the environments represented in Fig. 7 for the adsorption systems considered in this work. (Reprinted from Ref. [16], with permission from the the American Chemical Society.)

Config.	Ag/Au	Au/Ag	Ag/Pt	Pt/Ag	Au/Pt	Pt/Au	Au/Pd	Pd/Au
1	−2.580	−3.106	−3.127	−4.224	−3.672	−4.228	−3.283	−3.117
2	−2.866	−3.566	−3.340	−4.828	−4.160	−4.847	−3.901	−3.502
3	−3.126	−3.987	−3.521	−5.397	−4.585	−5.434	−4.444	−3.862
4	−3.366	−4.380	−3.681	−5.938	−4.965	−5.986	−4.453	−3.863
5	−3.591	−4.744	−3.825	−6.455	−5.300	−6.508	−5.358	−4.535
6	−2.582	−3.142	−3.122	−4.267	−3.705	−4.246	−3.335	−3.137
7	−2.585	−3.177	−3.116	−4.310	−3.738	−4.254	−3.388	−3.157
8	−2.585	−3.177	−3.116	−4.310	−3.738	−4.264	−3.388	−3.157
9	−2.588	−3.212	−3.110	−4.353	−3.770	−4.282	−3.439	−3.178
10	−2.590	−3.247	−3.105	−4.396	−3.802	−4.300	−3.490	−3.198
11	−2.842	−3.563	−3.303	−4.837	−4.130	−4.833	−3.878	−3.497
12	−2.868	−3.599	−3.332	−4.869	−4.188	−4.863	−3.948	−3.521
13	−2.825	−3.567	−3.275	−4.852	−4.112	−4.817	−3.863	−3.500
14	−2.810	−3.541	−3.259	−4.826	−4.071	−4.785	−3.815	−3.485
15	−3.076	−3.947	−3.455	−5.370	−4.495	−5.388	−4.347	−3.833
16	−2.580	−3.106	−3.126	−4.224	−3.671	−4.227	−3.282	−3.116
17	−2.840	−3.529	−3.309	−4.792	−4.103	−4.814	−3.833	−3.477
18	−3.002	−3.470	−3.692	−4.634	−4.336	−4.708	−3.905	−3.497
19	−2.917	−3.473	−3.564	−4.651	−5.139	−4.640	−3.885	−3.450
20	−3.149	−3.859	−3.174	−5.185	−4.626	−5.193	−4.382	−3.777
21	−3.250	−3.760	−4.010	−4.996	−4.758	−5.048	−4.351	−3.758
22	−3.230	−3.770	−3.976	−5.009	−4.747	−5.029	−4.363	−3.747
23	−3.211	−3.780	−3.942	−5.022	−4.736	−5.010	−4.374	−3.736
24	−3.358	−3.498	−3.534	−6.257	−4.808	−6.176	−4.796	−4.384
25	−3.731	−4.287	−4.621	−5.661	−5.505	−5.635	−5.176	−4.223

1–17 and 24, which correspond to environments involving only adsorbate atoms. By comparing configurations 1–5, one can get insight into the influence of the first neighbors, which in all cases have a favorable effect on the adsorption of the central atom. On the other hand, by comparing cases 6–10, one can observe the influence of second neighbors (in the absence of first neighbors) and one may conclude that, except for Ag/Pt, in all the remaining cases they favor the adsorption of an atom at the central site. The direct influence of third neighbors is almost negligible, as follows from the comparison of configurations 1 and 16.

Configurations 19 and 20 are meaningful when considering adsorption at step sites. In these cases, the data in Table 2 indicate that upon adsorption on Pt steps, Ag and Au atoms will tend to avoid the presence of neighboring homoatoms. This fact is no longer unexpected, since both Ag and Au exhibit an important compressive surface stress when adsorbed on Pt. On the other hand, in the remaining systems, the adatoms would prefer to adsorb beside other adatoms.

Configurations 21–23 represent the environment of the three kinds of kink sites that an atom can find on the surface. By comparing configurations 20 and 22 for all systems, one notices that for the systems Ag/Au, Ag/Pt, and Au/Pt adsorption onto kink sites (configuration 22) is favored as compared with step decoration (configuration 20), whereas for the systems Au/Ag, Pt/Ag, Pt/Au, Au/Pd, and Pd/Au the growth of a monodimensional phase at step sites should be preferred (configuration 20).

A similar trend can be noticed when comparing configurations 14 and 19, which correspond to the adsorption of an adatom close to a step of the same (configuration 14) or a different (configuration 19) nature. For the systems Ag/Au, Ag/Pt, Au/Pt and Au/Pd, configuration 19 is more stable than configuration 14, so in these cases the adsorption of a new atom at the step of a substrate island should be more favorable than adsorption on the edge of an adsorbate island. On the other hand, the opposite trend is observed for the systems Au/Ag, Pt/Ag, Pt/Au, and Pd/Au.

2. Evaluation of Adsorption Isotherms for Defect-Free Surfaces

We will first analyze the adsorption isotherms on defect-free surfaces.

(i) UPD Compared with OPD: First-Order Phase Transitions

Adsorption isotherms were calculated for the different systems in the case of defect-free surfaces. For a fixed temperature and chemical potential, a simulation has to be performed to evaluate the average coverage degree corresponding to the equilibrium state of the system. By repeating this procedure for different values of the chemical potential, one obtains the isotherms, such as those

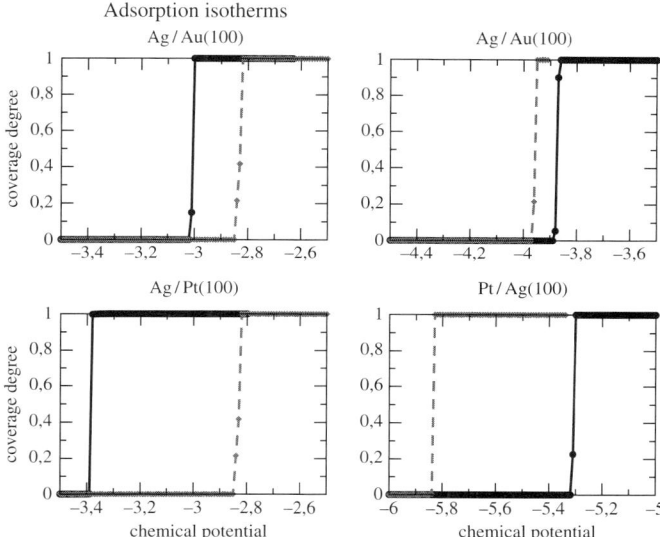

Figure 8. Adsorption isotherms of four model systems at $T = 300$ K. *Full lines* and *circles*, adsorbate/substrate isotherms; *dashed lines* and *diamonds*, adsorbate/adsorbate isotherms.

shown in Fig. 8 for the cases of the systems Ag/Au(100), Ag/Pt(100), Au/Ag(100), and Pt/Ag(100).

In all examples shown in Fig. 8, each isotherm exhibits an abrupt jump in the coverage degree, as expected for the case of a first-order phase transition.

The isotherms were obtained at $T = 300$ K and each graph includes two of them, which can generically be symbolized as A/S(100) and A/A(100), where A and S denote adsorbate and substrate, respectively. Thus, the comparison of the behavior of the heteroatomic A/S(100) system with that of the pure metal A/A(100) system allows us to determine the existence of either UPD or OPD.

The difference between the chemical potentials $\Delta \mu = \mu_{A/S} - \mu_A$, at which the transition occurs for the systems A/S(100) and A/A(100), is then a measure of the underpotential shift, namely,

$$\Delta \phi_{UPD} = -\frac{\Delta \mu}{ze_0}. \tag{33}$$

Table 3.

Chemical potential (eV) at coexistence μ_{coex}, which is identified by the abrupt step observed in the adsorption isotherms. Results corresponding to different systems and obtained at $T = 300$ K. (Reprinted from Ref. [16], with permission from the the American Chemical Society.)

Substrate/adsorbate	Ag	Au	Pt	Pd
Ag	−2.83	−3.87	−5.30	−
Au	−3.00	−3.95	−5.29	−3.78
Pt	−3.38	−4.37	−5.83	−
Pd	−	−4.21	−	−3.92

Table 4.

Excess of chemical potential (eV) calculated as the difference of the values of μ_{coex} in Table 3 for the systems A/S(100) and A/A(100). (Reprinted from Ref. [16], with permission from the the American Chemical Society.)

Substrate/adsorbate	Ag	Au	Pt	Pd
Ag	0.00	0.08	0.53	−
Au	−0.17	0.00	0.54	0.14
Pt	−0.55	−0.42	0.00	−
Pd	−	−0.26	−	0.00

Table 3 summarizes the approximate values for the chemical potential at which transition occurs for each system, while Table 4 shows the excess of chemical potential for the systems A/S(100), calculated as the difference of the values listed in Table 3 for the systems A/S(100) and A/A(100). Negative values indicate UPD, while positive ones correspond to OPD.

According to the results shown in Table 4, UPD is predicted for the systems Ag/Au, Ag/Pt, Au/Pt, and Au/Pd, and OPD is expected for the remaining systems. Although this prediction is in agreement with the experimental results for Ag/Au and Ag/Pt, it does not agree with the experimental finding of UPD for Pd/Au.[22] This discrepancy may be understood since the presence of adsorbed anions may yield part of the energy excess required for the observation of UPD.

(ii) The Influence of Temperature on the Isotherms

Adsorption isotherms have also been calculated, for the systems studied, at different temperatures, as shown in Fig. 9 for four typical cases.

It can be observed that, in general, at low temperatures the isotherms exhibit an abrupt jump at a certain "coexistence" chemical potential, as expected for first-order phase transitions. However, at higher temperatures, the isotherms become smooth, approaching the Langmuir-type isotherm. This behavior is due to the fact that at high temperatures the lateral interaction among the adsorbed particles becomes less important. The limiting case would be a Langmuir isotherm, where interactions among adsorbed particles are no longer present.

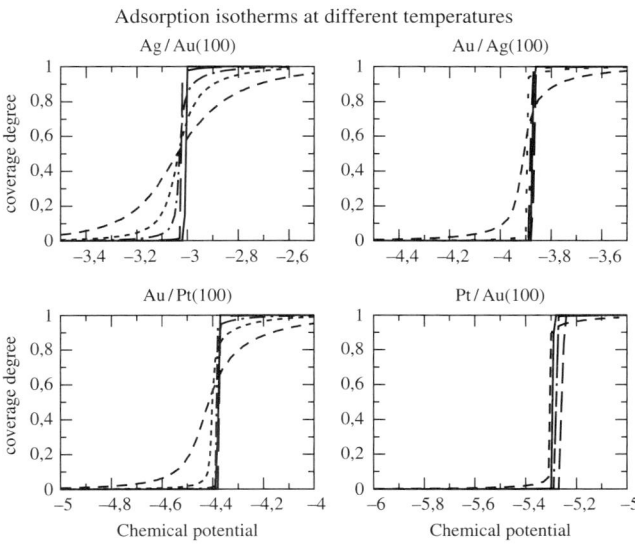

Figure 9. Adsorption isotherms obtained for different temperatures of four model systems. *Full lines* 300 K; *long-dashed lines* 1,000 K; *dot-dashed lines* 1,500 K; *dotted lines* 2,000 K; *dashed lines* 3,000 K.

For each system, there is a critical temperature that separates two well-defined regions: the first one, where first-order phase transitions exist (abrupt jumps, at low temperatures), and the second one, where one observes smooth isotherms (at high temperatures). The exact calculation of that critical temperature is beyond the scope of this work because it would involve a deeper and computationally demanding study with the aid of finite-size scaling techniques.

3. Study of the Influence of Surface Defects

To study the influence of surface defects on the adsorption isotherms, simulations were performed in the presence of substrate atoms in the monolayer, with a coverage degree of 0.1. As described in Sect. III.4, the substrate-type atoms are present in the form of islands, giving place to step and kink sites.

(i) Isotherms Corresponding to UPD Systems: The Effect of Kinks and Steps as Compared with the Complete Monolayer

As already discussed, for $T = 300$ K, the adsorption isotherms on defect-free surfaces show an abrupt change of coverage degree for a certain coexistence chemical potential. This situation turns out to be different in the presence of surface defects, as shown in Fig. 10 for the systems Ag/Au, Ag/Pt, Au/Pd, and Au/Pt. In these cases the coverage degree starts to rise smoothly at chemical potentials more negative than the coexistence chemical potential. Relative coverage degrees θ_k (θ_s) for kink (step) sites may be defined as the number of occupied kink (step) sites divided by the total number of kink (step) sites. Figure 10 shows the dependence of θ_k, θ_s, and the total coverage degree θ on the chemical potential, μ. The behavior of these isotherms indicates that kink sites are occupied first, then adsorption takes place on step sites, and finally the rest of the surface becomes covered. A careful inspection of this figure shows that the chemical potential values at which $\theta_k = 0.5$ are close to the corresponding energy values for adsorption at kink sites (see configurations 21–23 in Fig. 7 and the corresponding values in Table 3). Something similar occurs with the μ values at which $\theta_s = 0.5$, which are close to the energy values of configuration 20 in Fig. 7 (adsorption on step

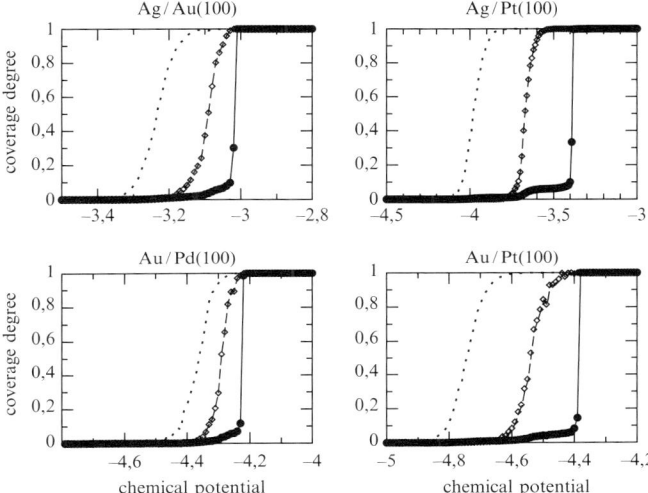

Figure 10. Adsorption isotherms obtained at $T = 300$ K in the presence of surface defects, for the systems that present underpotential deposition. *Full lines* and *filled circles*, adsorption isotherm for the complete monolayer; *dashed lines* and *open diamonds*, adsorption isotherms for the step sites; *dotted lines*, adsorption isotherms for the kink sites.

sites). However, it must be noticed that an abrupt jump is no longer observed in the θ_k and θ_s isotherms, as expected for both 0D and 1D systems.

It is very illustrative to analyze the final state of the surface, for some typical systems, at different chemical potentials. In fact, Fig. 11 shows frames corresponding to the final state of the Ag/Pt(100) system obtained at six different chemical potentials. Since the substrate island remains unchanged during the adsorption process, it is concluded that the diffusion of substrate atoms at the temperature considered ($T = 300$ K) is negligible. It is observed that adsorbate atoms are adsorbed on the free sites of the surface according to the preferences described above, that is, first they fill the kink sites, then steps sites are decorated, and finally the rest of the surface becomes covered. A similar behavior is observed for the systems Ag/Au(100) and Au/Pt(100), which are not shown here for the sake of space. However, a detailed view of the neighborhood of a single Pt island is presented in Fig. 12 for the system Au/Pt, showing the main features discussed above.

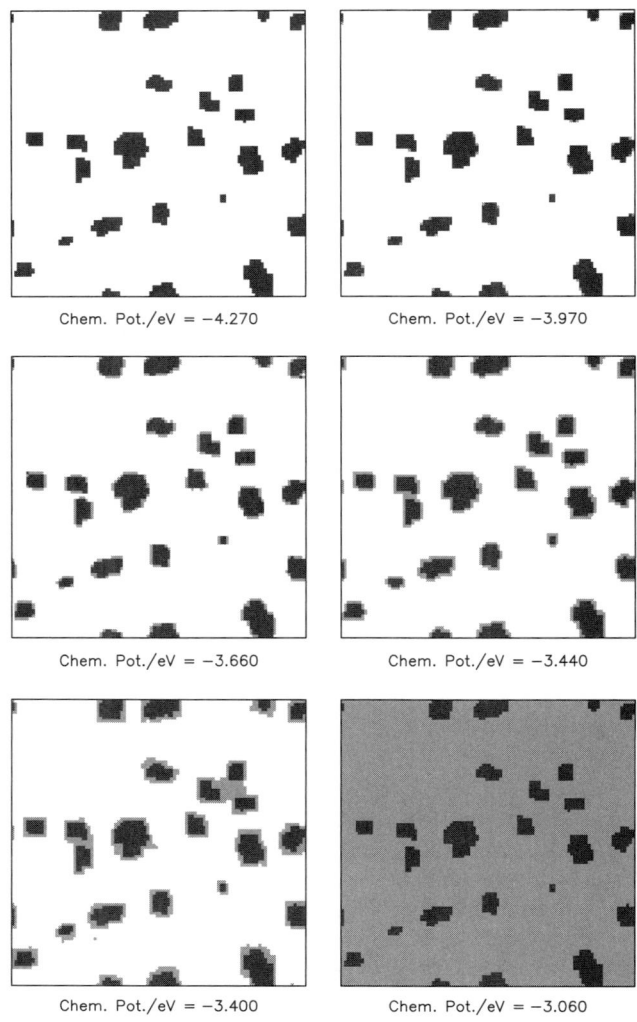

Figure 11. Snapshots of the final state of the surface at six different chemical potentials (-4.27, -3.97, -3.66, -3.44, -3.40, and -3.06 eV) for the system Ag on Pt(100), as obtained for $T = 300$ K. *Light gray*, Ag atoms; *dark gray*, Pt atoms. (Reprinted from Ref. [16], with permission from the American Chemical Society.)

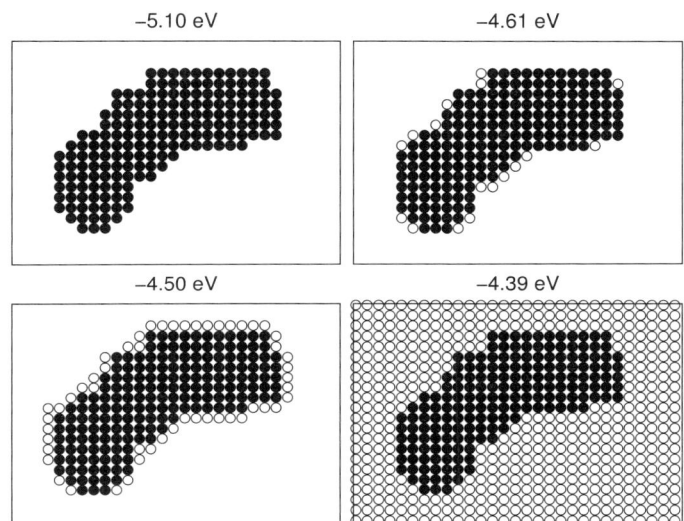

Figure 12. Final state of the surface at four different chemical potentials (−5.10, −4.61, −4.50, and −4.39 eV) for the adsorption of Au on Pt. Defects are those of a single Pt island. *Open circles*, gold atoms; *filled circles*, Pt atoms. (Reprinted from Ref. [16], with permission from the the American Chemical Society.)

(ii) *Isotherms Corresponding to OPD Systems: The Formation of Surface Alloys*

Unlike the systems considered in Fig. 10, the systems Au/Ag, Pt/Ag, Pd/Au, and Pt/Au do not present an appreciable widening of the adsorption isotherms due to the presence of surface defects (Fig. 13). Figure 13 also shows that the partial coverages θ_k and θ_s for this type of system do not exhibit a clear-cut trend as in the case of the systems depicted in Fig. 10. This can be understood through a comparative analysis of configuration 20 against configurations 21–23 (see Fig. 7, Table 3). For these systems there is not a clear preference for adsorption at steps or at kink sites, as compared with adsorption on terraces, in contrast to the predictions for the systems that exhibit UPD and are presented in Fig. 10.

From Fig. 13 it also follows that kink sites are not filled before step decoration, but instead both processes take place simultaneously. Furthermore, as pointed out above, configuration 14 is

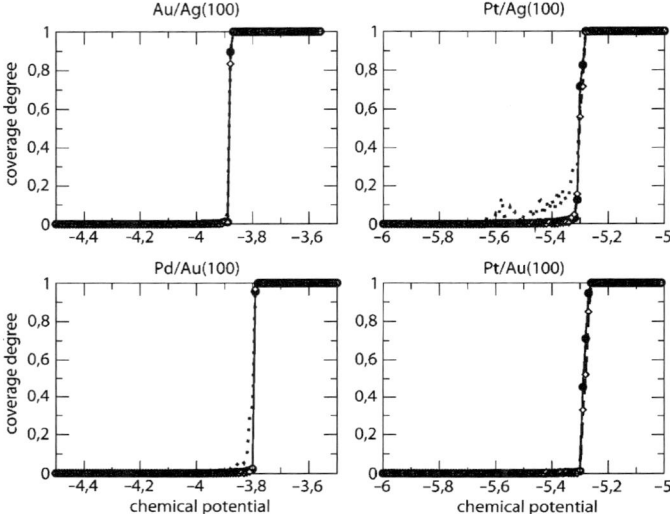

Figure 13. Adsorption isotherms in the presence of surface defects, calculated at $T = 300\,\text{K}$ for the systems that present overpotential deposition. *Full lines* and *circles*, adsorption isotherm for the complete monolayer; *dashed lines* and *diamonds*, adsorption isotherms for the step sites; *dotted lines*, adsorption isotherms for the kink sites.

characterized by lower energy values than configuration 19, so atom deposition at the steps of substrate islands becomes delayed as compared with deposition at sites belonging to islands of the same nature. It is also remarkable that, within the present model, all these systems exhibit a positive excess of binding energy, and that the binding energy of the adsorbate is larger than or similar to that of the substrate.

For the systems Au/Ag, Pt/Ag, Pd/Au, and Pt/Au, adsorbate atoms diffuse into the islands, forming an alloy, while the islands tend to disintegrate. This fact is illustrated in Fig. 14 for Pt deposition on Ag(100) and in Fig. 15 for the system Pt/Au(100). At low coverage degrees, the adatoms start to be embedded into the islands, and even form small nuclei. This process continues and upon completion of the monolayer, the substrate islands have incorporated an important number of adatoms.

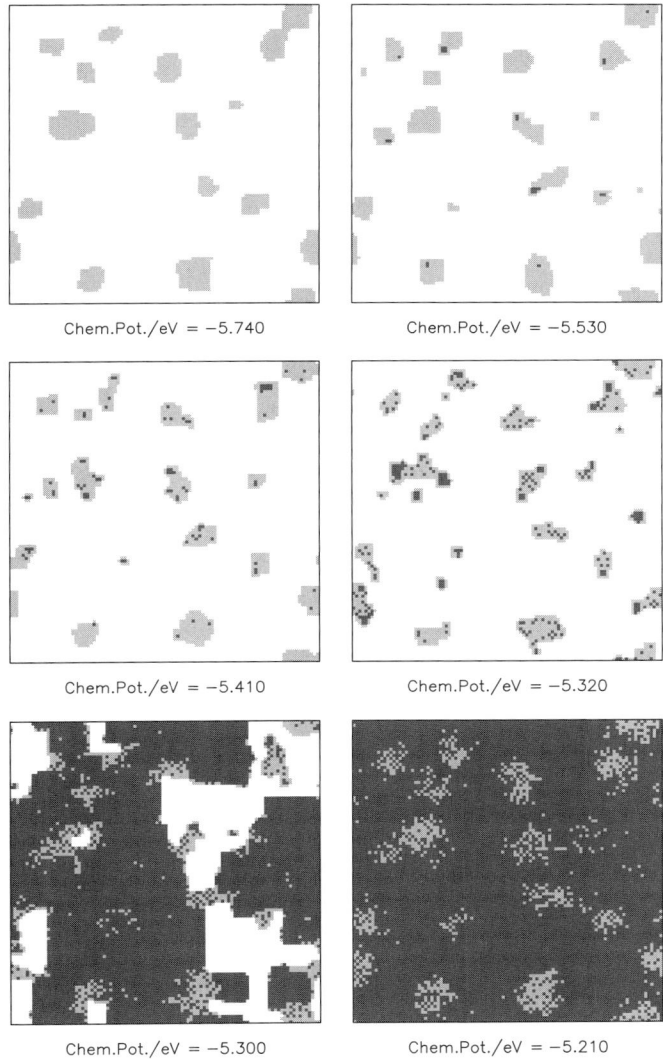

Figure 14. Snapshots of the final state of the surface obtained at six different chemical potentials (−5.74, −5.53, −5.41, −5.32, −5.30, and −5.21 eV), for the adsorption of Pt on Ag, in the presence of Ag islands. Results obtained at $T = 300$ K. *Light gray*, silver atoms; *dark gray*, platinum atoms. (Reprinted from Ref. [16], with permission from the the American Chemical Society.)

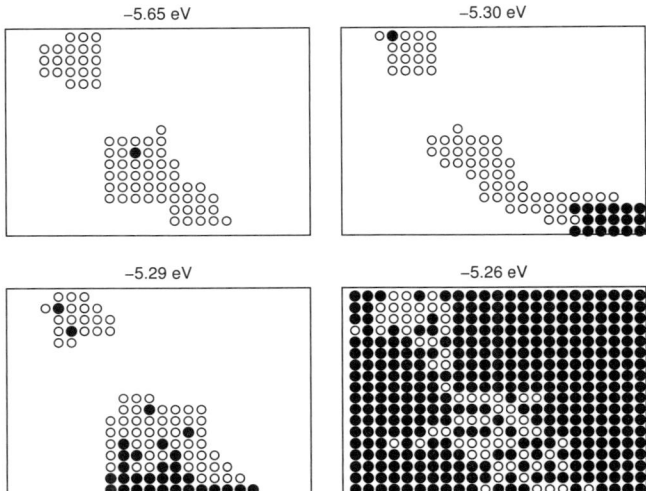

Figure 15. Typical configurations showing final states of the surface obtained for four different chemical potentials (−5.65, −5.30, −5.29, and −5.26 eV) for the adsorption of Pt on Au(100), with Au defects. *White circles*, Au atoms; *black circles*, Pt atoms. Results obtained at $T = 300$ K. (Reprinted from Ref. [16], with permission from the the American Chemical Society.)

4. Comparison with Experiments

While the simplicity of the present model, which, among others, neglects surface reconstruction, anion adsorption, and specific kinetic features, does not allow a quantitative comparison with experiments, qualitative predictions can be made concerning surface alloy formation in the presence of islands.

Table 5 summarizes the predictions of the Monte Carlo simulations for the systems studied in this work, along with some results of experimental observations. Within this context, it is worth mentioning the results reported by Kolb and coworkers for Pd and Pt deposition on Au(100).[22,47] For the former system, the authors proposed that alloying upon Pd deposition should proceed involving Au atoms from islands and step edges. The results of the Monte Carlo simulations strongly support this explanation on thermodynamic grounds, based on the energetics of the Pd/Au system. Interesting results have also been obtained for the system Pt/Au. Waibel et al.[47] studied Pt deposition on Au(100), finding that nucleation of Pt starts mainly at

Table 5.
Comparison between the predictions of Monte Carlo simulations for the formation of surface alloys in the presence of islands of substrate atoms and experimental observations. (Reprinted from Ref. [16], with permission from the the American Chemical Society.)

System	Surface alloy prediction	Experimental observation
Ag/Au(100)	No	No
Ag/Pt(100)	No	No
Au/Pd(100)	Slight alloying	Not available
Au/Pt(100)	No	Not available
Au/Ag(100)	Yes	Nor available
Pt/Ag(100)	Yes	Not available
Pd/Au(100)	Yes	Surface alloying[22]
Pt/Au(100)	Yes	Au islands became stable against dissolution[47]

defects, such as step edges, for low deposition rates. On the other hand, at high deposition rates some nuclei also appear on the terraces, at random sites. Figure 7 in Ref. [47] shows that upon Pt deposition on Au, the shape of the islands becomes progressively blurred as Pt is deposited. According to the Monte Carlo results, at low deposition rates, Pt atoms could be incorporated into the islands, yielding the enhanced stability observed. Also, owing to its high binding energy, Pt is expected to present 3D growth, as pointed out by Waibel et al.,[47] but this feature is not considered in the model employed.

V. DYNAMIC RESPONSE OF AG MONOLAYERS ADSORBED ON AU(100) UPON AN OSCILLATORY VARIATION OF THE CHEMICAL POTENTIAL

1. Dynamic Phase Transitions: Basic Concepts

The term "hysteresis" is used to describe the lagging of an effect behind its cause. Hysteresis is a common phenomenon that is observed in a great variety of physical, chemical, and biological systems. However, the magnetization response of a ferromagnet in the presence of an oscillatory magnetic field is probably the best known example of hysteresis.

Hysteresis in thermodynamic systems is often related to the existence of first-order phase transitions, which are the sources of nonlinearities always associated with hysteretic behavior. Within this context, physical systems exhibiting first-order phase transitions and capable of becoming coupled to an external oscillatory drive, as in the case of metal adsorption on metal surfaces, as described in previous sections, are excellent candidates for the observation of (out-of-equilibrium) dynamic phase transitions (DPTs). DPTs may be observed when a physical system is forced by an oscillatory (external) drive. In this case, one has a symmetry breaking between a dynamically ordered state (DOS), such that the system cannot follow the external oscillatory drive, and a dynamically disordered state (DDS), where the system becomes coupled to the drive. Of course, DPTs involve the competition of two characteristic time scales: the relaxation time of a stationary state of the physical system and the period of the external drive. DPTs have been observed, among others, in the Ising magnet,[1,2,9,12,25,26,43,45] the Heisenberg model,[20,21] the XY model,[48] and the Ziff–Gulari–Barshad (ZGB) model for the catalytic oxidation of CO.[30,32] In the case of the Ising ferromagnet on a 2D square lattice, an oscillatory magnetic field causes the occurrence of a second-order DPT that is believed to belong to the universality class of the standard Ising model in the same dimensionality.[36,37,43] For a recent review, see Ref. [3].

2. Simulation Method

Monte Carlo simulations were performed using the model described in Sect. III.2. Runs started from an empty substrate and subsequently, the chemical potential was swept harmonically according to

$$\mu(t) = \mu_{\text{coex}} + \mu_0 \sin(\omega t), \qquad (34)$$

where μ_0 is the amplitude of the sweep, while $\omega = 2\pi/\tau$ is the pulsation such that τ is the period of the applied perturbation. Also, μ_{coex} is the coexistence chemical potential measured at the first-order phase transition observed upon deposition of Ag on Au(100) (see Sect. 2), given by $\mu_{\text{coex}} = -3.03\,\text{eV}$.

Simulations were performed at $T = 300$ K, so the system is far below its critical temperature. During the simulation, the time dependence of the Ag coverage ($\theta(t)$) was recorded. After several initial sweeps to allow the system to achieve a nonequilibrium stationary state had been neglected, the quantities of interest were evaluated by averaging over many cycles.

3. Dynamic Response of the Coverage Degree

Figure 16 shows two examples of the dynamic behavior of the system as obtained by applying oscillatory sweeps of the same amplitude $\mu_0 = 0.3$ eV but different periods τ.

For the shortest period ($\tau = 10\,\mu$s, see Fig. 16a), the system becomes trapped within a low-coverage regime, with $\theta_{Ag} < 0.5$. Since the highest chemical potential achieved is given by $\mu_{coex} - \mu_0 = -2.73$ eV, the applied signal clearly drives the system well inside the high-coverage regime of the equilibrium isotherm. So, the behavior observed in Fig. 16 for $\tau = 10$ is clear evidence that such a period is much smaller than the relaxation time (τ_{relax}) required by the system to jump from the low-coverage to the high-coverage regimes. On the other hand, for $\tau = 100$ (see Fig. 16b) the system reaches the high-coverage regime ($\theta_{Ag} \approx 1$) during all sweeps, indicating that for this case one has $\tau > \tau_{relax}$. So, from Fig. 16 one concludes that it is possible to identify the dynamic competition between two time scales in the system: the half period of the external drive and the relaxation time (also known as the metastable lifetime of the system in a given state). For large periods, a complete decay of the metastable phase always occurs during each half period. Consequently, the coverage describes a limiting cycle (almost symmetric). In contrast, for short periods, the system does not have enough time to change the coverage from $\theta \approx 0$ to $\theta \approx 1$, and the symmetry of the hysteresis loop is broken. It should be mentioned that the phenomenon of symmetry breaking between limiting cycles in an externally driven system has been the subject of considerable attention. It was first reported in the context of numerical and mean-field studies of the magnetization of a ferromagnet in an oscillating magnetic field[45] and subsequently it has been studied by means of Monte Carlo simulations of the kinetic Ising model.[3,9,36,37,43]

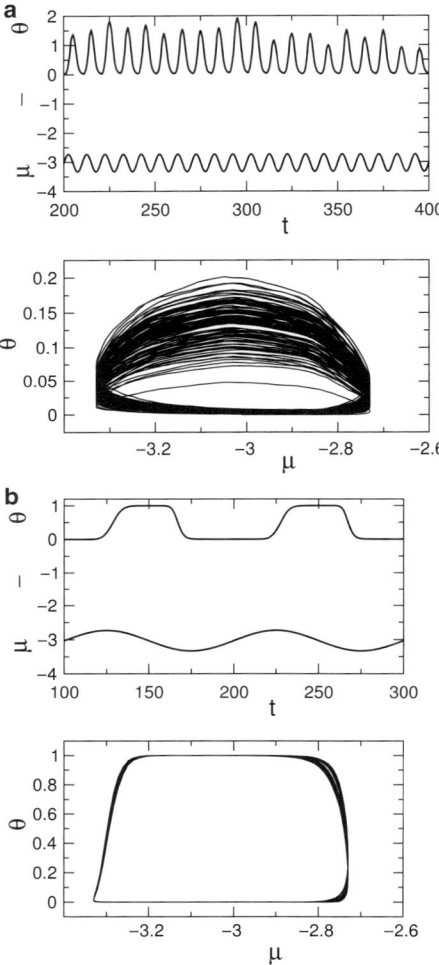

Figure 16. The temporal dependence of the coverage and the chemical potential (*upper panel*), and the coverage–chemical potential loops (*lower panel*). (**a**) Results obtained using lattices of side $L = 100$ and by taking $\mu_0 = 0.3$ eV and $\tau = 10$. (**b**) As in (**a**) but for $\tau = 100$. In (**a**) the coverage is amplified by a factor of 10 for the sake of clarity. So, to properly obtain the actual coverage, the θ-scale has to be divided by a factor of 10. (Reprinted from Ref. [17], with permission from the the American Chemical Society.)

4. Dynamic Phase Transitions

To study the DPTs involved in the already discussed symmetry breaking process, it is useful to define the dynamic order parameter (Q) as the period-averaged surface coverage with Ag atoms, namely,

$$Q = \frac{1}{\tau} \oint (2\theta_{Ag} - 1) dt. \tag{35}$$

Notice that for finite systems, as in the present study, one actually computes $\langle |Q| \rangle$, where $\langle \rangle$ means averages over different cycles of the time series $\theta_{Ag}(t)$. So, if $\tau < \tau_{relax}$, the coverage cannot change from $\theta \approx 0$ to $\theta \approx 1$ (and vice versa) within a single period, and therefore one has $\langle |Q| \rangle > 0$. This situation is regarded as the DOS. In contrast, when $\tau > \tau_{relax}$ and the coverage follows the applied chemical potential, one has $Q \approx 0$ in the so-called DDS. Between these two extreme regimes, one expects the existence of a critical period such as $\langle |Q| \rangle$ should vanish in the thermodynamic limit. This behavior may be the signature of a nonequilibrium DPT.

Further insight into the nature of DPTs can be obtained by measuring the fluctuations of the order parameter (χ), given by

$$\chi = L^2 \text{Var}(|Q|) = L^2 [\langle |Q|^2 \rangle - \langle |Q| \rangle^2], \tag{36}$$

where $\text{Var}(|Q|)$ is the variance of the order parameter. This measurement is motivated by the fact that, as is well known, systems undergoing second-order phase transitions exhibit a divergency of the susceptibility. Of course, for equilibrium systems the fluctuations of the order parameter are related to the susceptibility through the fluctuation-dissipation theorem. It is not obvious if such a theorem would hold for our out-of-equilibrium dynamic system. However, if our system obeyed the fluctuation-dissipation theorem, both quantities would be proportional.

Figures 17 and 18 show the dependence of the order parameter (see (35)) and its fluctuations (see (36)) on the chemical potential and the period of the applied signal, respectively. For a given period and low enough values of μ_0, one has $\langle |Q| \rangle > 0$, although it decreases smoothly and monotonically when μ_0 is increased (see Fig. 17a). On the other hand, if the chemical potential is fixed, the order parameter also decreases when τ is increased (see Fig. 18a). So, the behavior of $\langle |Q| \rangle$ suggests the existence of continuous transitions

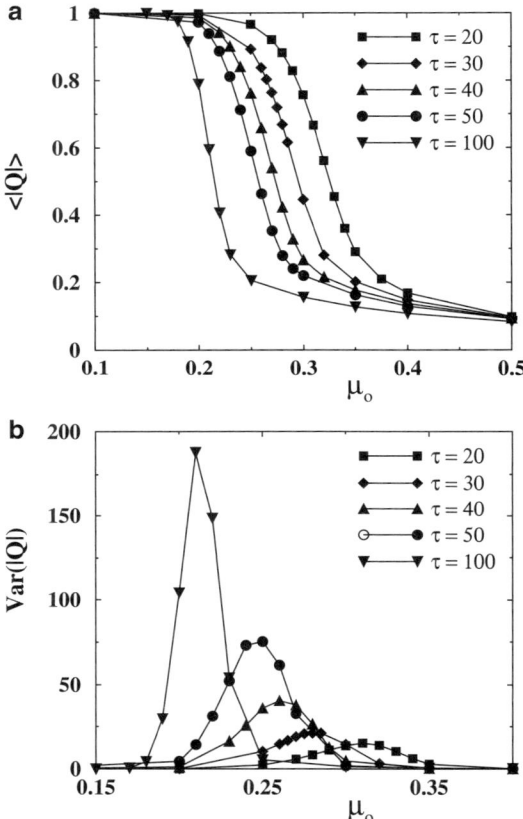

Figure 17. (a) The order parameter and (b) the fluctuations of the order parameter vs. the amplitude of the applied (oscillatory) chemical potential. Results obtained by using lattices of side $L = 100$ and by taking different periods τ of the applied signal, as listed. (Reprinted from Ref. [17], with permission from the the American Chemical Society.)

between a DOS (for $\langle |Q| \rangle > 0$) and a DDS (for $\langle |Q| \rangle \approx 0$). This preliminary conclusion, which has already been anticipated within the context of the discussion of Fig. 16, may also be supported by the characteristic peaks observed by plotting the variance of the order parameter, as shown in Figs. 17b and 18b and c.

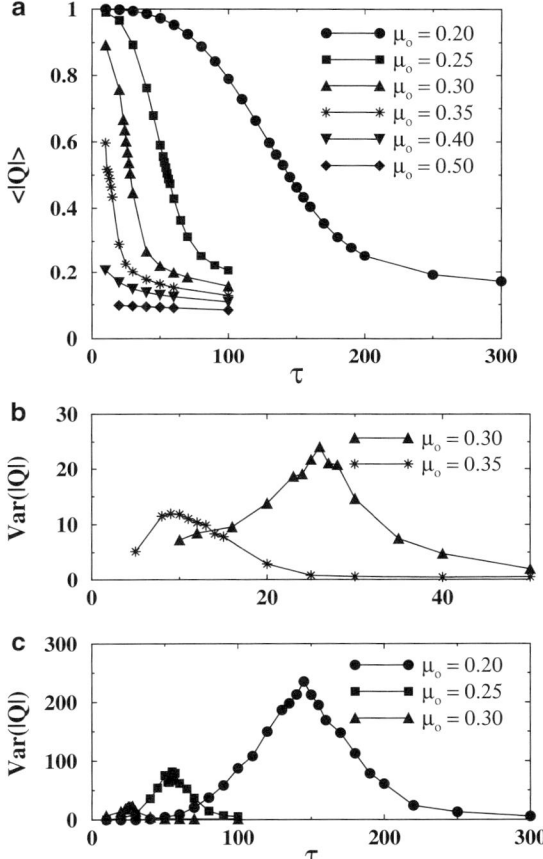

Figure 18. (**a**) The order parameter and (**b**) the fluctuations of the order parameter vs. the period τ of the applied chemical potential. Results obtained by using lattices of side $L = 100$ and by taking different values of the amplitude of the applied signal, as listed. (Reprinted from Ref. [17], with permission from the the American Chemical Society.)

In view of these findings, it would be desirable to perform a systematic study of finite-size effects. In fact, it is well known that continuous (second-order) phase transitions become shifted and rounded owing to the finite size of the samples used in numerical

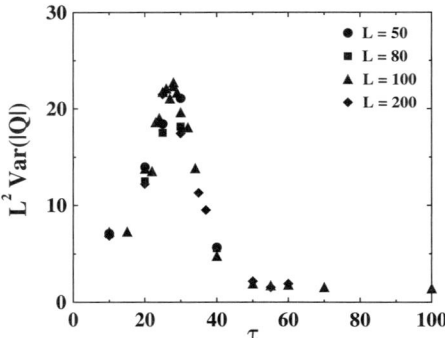

Figure 19. The fluctuations of the order parameter, as defined by (36), vs. the period τ of the applied chemical potential. Results obtained by using lattices of different side as listed. (Reprinted from Ref. [17], with permission from the the American Chemical Society.)

simulations, while actual transitions can only be observed in the thermodynamic limit ($L \to \infty$). This shortcoming can be overcome by applying the finite-size scaling theory to the numerical results obtained by using a wide range of sample sizes.[7] Therefore, we performed simulations up to $L = 1,024$ for a few typical values of the parameters (not shown here for the sake of clarity). Our first finding was that the dependence of $\langle |Q| \rangle$ on μ_0 does not show any appreciable finite-size effect (not shown here for the sake of space). Furthermore, the fluctuations of the order parameter (χ, as measured according to (36)) are independent of the lattice size, as is shown in Fig. 19.

Since one has $\chi = L^2 \text{Var}(Q)$, it follows that the variance of the order parameter actually vanishes in the thermodynamic limit. Owing to this evidence we conclude that for the range of parameters used, rather than undergoing a true phase transition, the Ag/Au(100) system actually exhibits a crossover between the DOS and the DDS. A possible reason for the observation of a crossover instead of a true DPT could be the absence of symmetry between the adsorption and desorption processes. To test this possibility we measured the relaxation times for Ag-covered and uncovered surfaces as a function of the applied overpotential (see Fig. 20).

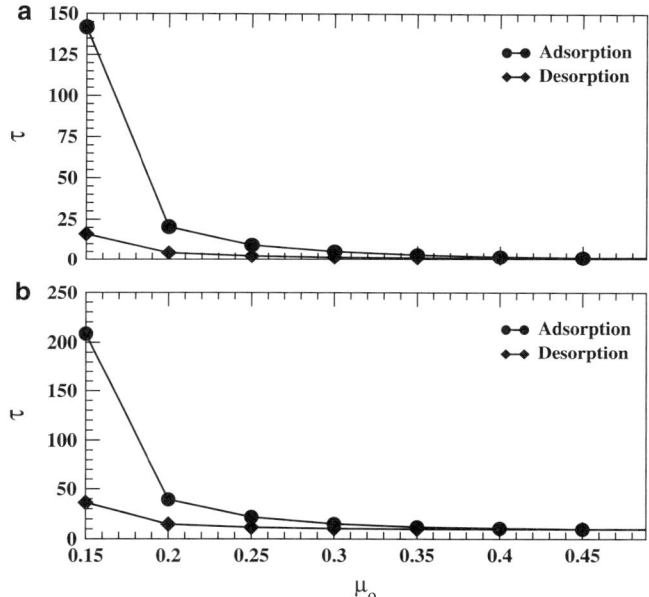

Figure 20. The relaxation times vs. the applied overpotential. $\tau_{des}^{1/2}$ and τ_{des}^{0} are shown by means of *full diamonds* in (**a**) and (**b**), respectively. These data correspond to the desorption processes up to $\theta_{Ag} = 1/2$ and $\theta_{Ag} = 0$, respectively. $\tau_{ads}^{1/2}$ and τ_{ads}^{1} are shown by means of *full circles* in (**a**) and (**b**), respectively. These data corresponds to the adsorption processes up to $\theta_{Ag} = 1/2$ and $\theta_{Ag} = 1$, respectively. Results obtained by averaging over 100 different configurations. More details are provided in the text. (Reprinted from Ref. [17], with permission from the the American Chemical Society.)

Two relaxation times are measured for each process, namely, (1) $\tau_{des}^{1/2}$ and τ_{des}^{0} for the desorption processes up to $\theta_{Ag} = 1/2$ and $\theta_{Ag} = 0$, respectively, and (2) $\tau_{ads}^{1/2}$ and τ_{ads}^{1} for the adsorption processes up to $\theta_{Ag} = 1/2$ and $\theta_{Ag} = 1$, respectively. The results obtained, plotted in Fig. 20, show that for low overpotentials the relaxation times corresponding to both processes are different, quantitatively confirming the asymmetry between them. Also, for large overpotentials the relaxation times tend to be almost the same, suggesting that in this limit the asymmetry may be irrelevant. However, it should also be noticed that in that case neither crossovers nor DPTs are observed owing to the fact that one would need to apply signals with periods shorter than 1 μs (see Fig. 18). Within this context, it is worth mentioning that the response to a pulsed perturbation of the ZGB model[49] for the catalytic oxidation of CO has also been studied very recently.[32] The ZGB model exhibits a first-order irreversible phase transition between an active state with CO_2 production and an absorbing (or poisoned) state with the surface of the catalyst fully covered by CO, such that in this regime the reaction stops irreversibly[49] (for a recent review, see Ref. [31]). To study DPTs in the ZGB model, it is convenient to perform a generalization by introducing a small probability for CO desorption that, on the one hand, preserves the first-order nature of the transition[4] and, on the other hand, prevents the occurrence of CO poisoning. Measurements of the lifetimes associated with the decay of the metastable states of the ZGB model indicate that they depend on the direction of the process, showing a marked asymmetry as in the case of our adsorption–desorption simulations. In fact, the contamination time τ_d (measured when the system is quenched from high to low CO coverage) is different from the poisoning time τ_p (measured from low to high CO coverage).[32,33] Therefore, DPTs are observed by applying a periodic external – asymmetric – signal of period $\tau = \tau_p + \tau_d$.[32] On the basis of this evidence, we conclude that the observation of the crossover in our simulations should – most likely – be related to the fact that we applied a periodic – symmetric – potential, and the adsorption–desorption processes involved are not symmetric. Of course, it could also be possible that at higher temperatures the asymmetry becomes irrelevant. However, in the present work we restricted ourselves to $T = 300$ K since we expect to contribute to the understanding of the dynamic behavior of the Ag/Au(100) system in a standard electrochemical environment.

To further characterize the crossover between different states, we take advantage of the well-defined peak exhibited by χ and Var($|Q|$) (see Figs. 17–19). So, the crossover period (τ^{cross}) and the corresponding crossover chemical potential (μ_0^{cross}) are identified with the location of the above-mentioned peak. The results obtained are displayed in Fig. 21, which shows a logarithmic dependence of μ_0^{cross} on τ^{cross}. In the "state diagram" of Fig. 21, the full line shows the border between DOSs obtained for intermediate values of both μ_0^{cross} and τ^{cross}, and DDSs that are found for large enough values

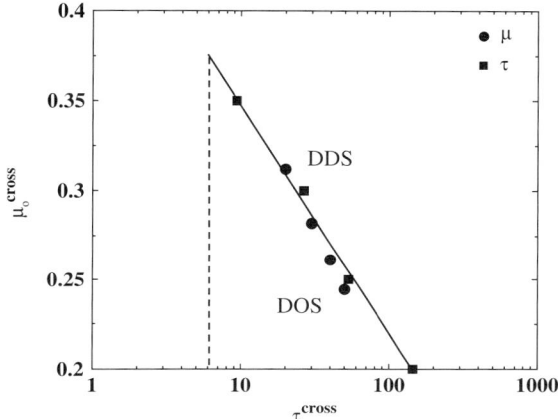

Figure 21. Linear–logarithmic plot of the crossover chemical potential vs. the crossover period. Data obtained from the peaks observed in plots of Var($|Q|$) vs. μ_0 (*full circles*), and Var($|Q|$) vs. τ (*full squares*). The *full line* showing the border between dynamic ordered states and dynamic disordered states has been drawn in order to guide the eye. Also, the *vertical dashed line* corresponds to the minimum period at which dynamic disordered states were observed. More details are provided in the text. (Reprinted from Ref. [17], with permission from the the American Chemical Society.)

of the period. It should be noted that for very low periods (i.e., $\tau^{\text{cross}} < 6$) we were unable to observe DOS (see, e.g., Fig. 18) even after greatly increasing the chemical potential.

VI. CONCLUSIONS

Grand ganonical Monte Carlo simulations using realistic interatomic potentials were performed for a significant number of metallic systems, allowing us to draw a number of interesting conclusions. One of the novel features of the work is the exploration of the electrochemical phenomena of UPD and OPD in terms of lattice models that consider the many-body interactions typical of metallic systems. Thus, without the need to assume a particular type of interaction potential between the particles, phase transition phenomena in metallic monolayers could be studied. These studies comprised the formation of the metal monolayer phase under conditions close to equilibrium, as well as the dynamic perturbation of the interface. In the latter case, a crossover between different behaviors of the systems is predicted, which should stimulate further experimental and theoretical work in the area.

Concerning the energetics of the different systems, it can be stated that the systems Ag/Au(100), Ag/Pt(100), Au/Pt(100), and Au/Pd(100) present a positive excess of binding energies (negative excess of chemical potential) as compared with the homoepitaxial growth of adsorbate-type atoms, indicating that in these systems UPD is expected. On the other hand, for the Au/Ag(100), Pt/Ag(100), Pt/Au(100), and Pd/Au(100) systems, the monolayer adsorption is more favorable on substrates of the same nature than on the substrates considered.

Simulations were also performed in the presence of surface islands made of the same metal as the substrate to emulate surface defects. For the first family of systems mentioned, the islands remained almost unchanged, being decorated by the adatoms before completion of the monolayer. In the case of the systems Au/Pt(100) and Ag/Pt(100), the adsorbate atoms filled first the kink and then the step sites in a clear sequence. These processes are somewhat closer in the case of Ag/Au(100) and very close in Au/Pd(100). For the second family of systems, the substrate islands showed disgregation to form 2D alloys with the adsorbate atoms and there is no differentiation in the filling of kink, step, or terrace sites. The system Au/Pd(100) presents borderline behavior, as a small quantity of Au is embedded into the Pd islands without altering their structure. On the basis of these results, it can be suggested that, in general, the stability of the substrate islands upon deposition of a foreign metal is mainly determined by the difference of the cohesive energies.

Monte Carlo simulation results showed that most metal–metal systems studied exhibit abrupt, first-order transitions when the chemical potential is tuned around a coexistence value. This observation, as well as the detection of hysteretic and relaxation effects, confirms the existence of true first-order transitions upon the UPD of a great variety of metal overlayers on different metal substrates.

On the other hand, on the basis of the fact that many adsorption isotherms of metal/metal systems exhibit an abrupt jump between a low-coverage state and a high-coverage state, for a well-defined (coexistence) chemical potential, the dynamic behavior of one particular system [Ag adsorption on the Au(100) surface], upon the application of an oscillatory chemical potential signal, has been carefully studied. The situation where the chemical potential is varied periodically around μ_{coex} was analyzed and the influence of the period and the amplitude of that variation on the dynamic behavior of the system was discussed.

The numerical results show that a silver layer adsorbed on Au(100) smoothly changes from a DOS to a DDS when the period (amplitude) of the chemical potential is increased by keeping $\mu_0(\tau)$ constant. Since the size dependence of the order parameter is negligible and its fluctuations scale with the system size, it is concluded that the system does not exhibit a true DPT, but rather a crossover between two different dynamic states, namely, DOS and DDS. The absence of true DPTs is due to the asymmetry between the adsorption and desorption processes of Ag atoms. The crossover points are identified by the position of the peaks of the fluctuations of the order parameter. In this way it is possible to draw the corresponding diagram of states characteristic of the system. The amplitude of the chemical potential at the crossover point exhibits a logarithmic dependence on the crossover period. However, for low enough periods, the DOS is no longer observed.

These theoretical results could be verified by performing experiments consisting in the study of the adsorption of Ag on a Au(100) surface, under UPD conditions, and the subsequent analysis of the influence of a periodic variation of the applied potential. One has to recognize that it would be difficult to establish a correlation between the actual time scale of the experiments and the Monte Carlo time step, although qualitative similar observations are expected. Furthermore, by determining the rate constants of the relevant electrochemical processes, one may perform real-time Monte Carlo simulations.[10]

ACKNOWLEDGEMENT

This work was financially supported by CONICET, UNLP, SecyT(UNC), and Agencia Nacional de Promoción Científica y Tecnológica(Argentina), PAE nos. 22711, PICT06-00036, and PICT 06-12485.

REFERENCES

[1] M. Acharyya, *Phys. Rev. E* **56** (1997) 1234.
[2] M. Acharyya, *Phys. Rev. E* **56** (1997) 2407.
[3] M. Acharyya, *Int. J. Mod. Phys. C* **16** (2005) 1631.
[4] E. V. Albano, *Appl. Phys. A* **55** (1992) 226.
[5] M. P. Allen and D. J. Tildesley, *Computer Simulation of Liquids*, Oxford University Press, Oxford, 1987.
[6] A. Aramata, Underpotential deposition on single-crystal metals, in *Modern Aspects of Electrochemistry*, Vol 31, pp 181–250, J. O'M. Bockris, Ralphe E. White and B. E. Conway (eds), Springer,New York, 2002 .
[7] K. Binder and D. W. Heermann, *Monte Carlo Simulation in Statistical Physics: An Introduction*, Springer, Berlin, 1988.
[8] E. Budevski, G. Staikov and W. J. Lorenz, *Electrochemical Phase Formation and Growth*, VCH, Weinheim, 1996.
[9] B. Chakrabarti and M. Acharyya, *Rev. Mod. Phys.* **71** (1999) 847.
[10] K. A. Fichthorn and W. H. Weinberg, *J. Chem. Phys.* **95** (1991) 1090; *Phys. Rev. Lett.* **68** (1992) 604.
[11] S. M. Foiles, M. I. Baskes and M. S. Daw, *Phys. Rev. B* **33** (1986) 7983.
[12] H. Fujisaka, H. Tutu and P. A. Rikvold, *Phys. Rev. E* **63** (2001) 016120; **63** (2001) 059903(E).
[13] S. G. García, D. Salinas, C. Mayer, E. Schmidt, G. Staikov, W. J. Lorenz, *Electrochim. Acta* **43**(1998) 3007.
[14] H. Gerischer, D. M. Kolb and M. Przasnyski, *Surf. Sci.* **43** (1974) 662.
[15] M. C. Giménez, M. G. del Pópolo and E. P. M. Leiva, *Electrochim. Acta* **45** (1999) 699–712.
[16] M. C. Giménez and E. P. M. Leiva, *Langmuir* **19** (2003) 10538–10549.
[17] M. C. Giménez and E. V. Albano, *J. Phys. Chem. C* **111** (2007) 1809–1815.
[18] W. Herzog and D. W. M. Arrigan, *Trends Anal. Chem.* **24**(3) (2005) 208–217.
[19] T. Hill, *Nano Lett.* **1** (2001) 273–275.
[20] H. Jang and J. Grimson, *Phys. Rev. E* **63** (2001) 066119.
[21] H. Jang, J. Grimson and C. K. Hall, *Phys. Rev. B* **67** (2003) 094411; *Phys. Rev. E* **68** (2003) 046115.
[22] L. A. Kibler, M. Kleinert and D. M. Kolb, *Surf. Sci.* **461** (2000) 155.
[23] D. M. Kolb in *Advances in Electrochemistry and Electrochemical Engineering*, Vol 11, p. 125, H. Gerischer and C. W. Tobias (eds), Wiley, New York, 1978.
[24] D. M. Kolb, M. Przasnyski and H. Gerischer, *J. Electroanal. Chem.* **54** (1974) 25.
[25] G. Korniss, P. A. Rikvold and M. A. Novotny, *Phys. Rev. E* **66** (2002) 056127.
[26] G. Korniss, C. J. White, P. A. Rikvold and M. A. Novotny, *Phys. Rev. E* **63** (2000) 016120.
[27] E. Leiva, *J. Electroanal. Chem.* **350** (1993) 1.
[28] E. Leiva and W. Schmickler, *Chem. Phys. Lett.* **160** (1989) 75.
[29] E. Leiva and W. Schmickler, *Electrochim. Acta* **39** (1994) 1015; *Electrochim. Acta* **40** (1995) 37.
[30] A. C. Lopez and E. V. Albano, *J. Chem. Phys.* **112** (2000) 3890.
[31] E. Loscar and E. V. Albano, *Rep. Prog. Phys.* **66** (2003) 1343.
[32] E. Machado, G. Buendía, P. Rikvold and R. Ziff, *Phys. Rev. E* **71** (2005) 016120.
[33] E. Machado, G. Buendía and P. Rikvold, *Phys. Rev. E* **71** (2005) 031603.
[34] O. A. Oviedo, E. P. M. Leiva and M. I. Rojas, *Electrochim. Acta* **51** (2006) 3526–3536.

[35] W. H. Press, S. A. Teukolsky, W. T. Vetterling and B. P. Flannery, *Numerical Recipes in FORTRAN, The Art of Scientific Computing*, Second Edition, Cambridge University Press, Cambridge, 1992.

[36] P. A. Rikvold and B. M. Gorman, in *Annual Reviews of Computational Physics I*, D. Stauffer (ed), World Scientific, Singapore, 1994.

[37] P. A. Rikvold, H. Tomita, S. Miyashita and S. W. Sides, *Phys. Rev. E* **49** (1994) 5080.

[38] M. Rojas, *Surf. Sci.* **569** (2004) 76.

[39] M. I. Rojas, M. G. Del Pópolo and E. P. M. Leiva, *Langmuir* **16** (2000) 9539–9546.

[40] M. G. Samant, M. F. Toney, G. L. Borges, L. Blum and O. R. Melroy, *J. Phys. Chem.* **92** (1998) 220.

[41] C. Sánchez and E. P. M. Leiva, Handbook of fuel cell technology, in *Catalysis by upd Metals*, Chapter 5, Vol 2, Part 1, pp 47–61, W. Vielstich, A. Lamm and H. Gasteiger (eds), Wiley, Chichester, 2003.

[42] C. G. Sánchez, E. P. M. Leiva and J. Kohanoff, *Langmuir* **17** (2001) 2219–2227.

[43] S. W. Sides, P. A. Rikvold and M. A. Novotny, *Phys. Rev. Lett.* **81** (1998) 834; *Phys. Rev. E* **59** (1999) 2710.

[44] H. Siegenthaler, K. Jüttner, E. Schmidt and W. J. Lorenz, *Electrochim. Acta* **23** (1978) 1009.

[45] T. Tomé and M. J. de Oliveira, *Phys. Rev. A* **41** (1990) 4251.

[46] S. Trasatti, *Z. Phys. Chem. N. F.* **98** (1975).

[47] H. F. Waibel, M. Kleinert, L. A. Kibler and D. M. Kolb, *Electrochim. Acta* **47** (2002) 1461.

[48] T. Yasui, H. Tutu, M. Yamamoto and H. Fujisaka, *Phys. Rev. E* **66** (2002) 036123; **67** (2003) 019901(E).

[49] R. Ziff, E. Gulari and Y. Barshad, *Phys. Rev. Lett.* **56** (1986) 2553.

7

Topics in the Mathematical Modeling of Localized Corrosion

Kurt R. Hebert and Bernard Tribollet

Department of Chemical and Biological Engineering, Iowa State University, Ames, IA 50014, USA
Laboratoire Interfaces et Systèmes Electrochimiques, UPR 15 du CNRS, Université Pierre et Marie Curie, 75252 Paris Cedex 05, France

I. GENERAL INTRODUCTION

Localized corrosion describes dissolution processes concentrated at specific areas on the surfaces of metals. In some types of localized corrosion, enhanced dissolution rates arise from partial or complete destruction of the protection normally afforded by the passive oxide film covering the metal surface. Oxide breakdown can be due to mechanical rupture (stress corrosion cracking), the chemical action of aggressive anions such as chloride (pitting corrosion), the impaction of solid particles on the surface (erosion corrosion), or the concentration of corrosion products within small solution-filled gaps (crevice corrosion). Other localized corrosion processes are initiated at metal compositional inhomogeneities such as grain boundaries in alloys (intergranular corrosion), or interfaces between dissimilar metals (galvanic corrosion). The economic impact of all forms of localized corrosion is severe. For example, pitting and stress corrosion cracking together account for about one fourth of equipment failures in the chemical process industries.[1]

Metal dissolution rates during localized corrosion are high enough so that large concentration or potential gradients are typically found near the dissolving metal surface. Characterization of these gradients is a necessary precursor for understanding the mechanisms controlling the corrosion rate. Thus, experimental research on localized corrosion has always been closely coupled to quantitative analysis of mass transport processes by mathematical modeling. In this chapter, three examples are presented which illustrate the range of models applied to localized corrosion processes, reflecting the particular interests of the authors. Section II, written by Hebert, is a review of recent work on the modeling of pitting corrosion. The remainder of the chapter communicates results of recent work by Tribollet on galvanic corrosion (Sect. III) and on the simulation of the impedance in crevice-type geometries.

II. PITTING CORROSION

1. Introduction

Pitting corrosion refers to the localized dissolution of metals in small cavities called "pits," which appear spontaneously when the metal is in contact with aqueous solutions containing certain "aggressive" anions, such as the chloride ion. Pitting occurs on passive metals, such as iron, nickel, chromium, aluminum, and titanium, as well as most industrially relevant alloys, including stainless steel and aluminum alloys. The initiation of a pit apparently involves the localized breakdown of the surface oxide film to expose the reactive underlying metal, by means of a mechanism which is not yet well understood. Once exposed, the metal corrodes very rapidly, at equivalent current densities which exceed those on the surrounding passive surface by factors of 10^5 or more. While as-initiated pits may be as small as 0.1 μm in width, they can rapidly grow to sizes which can penetrate walls of metal structures, or act as initiation sites for stress corrosion cracks.

The solution composition and potential inside pits differ dramatically from those in the bulk solution. Large concentration gradients develop within the pit and in the neighboring solution, to support the diffusion of metal ions away from the corroding surface. The high metal ion concentration in the pit also produces large concentrations of chloride ions, which migrate into the pit to preserve electrical neutrality, and the solution in the pit is acidified by hydrolysis of the metal ions. Large anodic current densities flow from pits, which under natural corrosion conditions are discharged by cathodic reactions occurring on the outside metal surface. The electrostatic potential gradients required to drive these currents result in electrode potentials in the pit which are displaced cathodically relative to the outside surface. The metal corrosion rate in the pit is determined by the response of the dissolution kinetics to the different solution composition and potential in the cavity.

The fundamental electrochemical reactions and transport processes occurring during pit growth can be formulated as equations, which taken together constitute a mathematical model for the pit. Modeling of pit growth has been actively pursued for the past several decades. The solution of the model equations yields the distribution of the corrosion rate along the surface of the pit, and in principle can be used to predict the evolution of the pit shape with time. Models of pitting have been pursued from both engineering and mechanistic points of view. The former category includes stochastic and deterministic models of pitting implemented in predictions of long-term corrosion failure probability of existing structures.[2-5] Other models have been developed which delineate ranges of solution composition and potential for which active pits will be repassivated, to identify materials suitable for particular process environments.[6-8] Mechanistic modeling studies, on the other hand, usually incorporate hypotheses about processes controlling the metal dissolution rate in pits, and critical conditions of pit stability, which are evaluated by comparing the simulation output with experimental measurements. The present review focuses on this latter type of model because of the author's perspective that the efficacy of engineering models depends ultimately on the resolution of mechanistic issues regarding pit growth and stability. Because of the focus on individual pits, the review will not discuss another area of significant activity, the modeling of pit–pit interactions.[9,10]

The following section describes the general framework of pitting models, in terms of the theory of coupled transport and reaction in electrochemical systems.[11] Previously, the application of these equations to pitting and crevice corrosion was reviewed 15–20 years ago by Sharland[12] and Turnbull.[13] Section 3 covers progress in the modeling of pitting, focusing on the period since these reviews. Section 4 discusses an important remaining challenge, the mathematical description of transport in concentrated solutions found in pits.

2. General Structure of Pitting Models

The fundamental equations of a pit model are the differential mole balances for chemical species dissolving in the pit. The mole balance for species i is

$$\frac{\partial c_i}{\partial t} = -\nabla \cdot \mathbf{N}_i + R_{vi}. \quad (1)$$

R_{vi} represents the rates of homogeneous reactions in the pit, expressed as moles of i generated per time per volume of the solution. If the solution is very dilute, the species flux $\mathbf{N}_i$ is

$$\mathbf{N}_i = -D_i \nabla c_i - z_i F u_i c_i \nabla \phi + c_i \mathbf{v}. \quad (2)$$

In practice, (2) is usually applied even when relatively concentrated solutions in pits are expected. Possible errors due to the dilute solution approximation are examined below (Section 4). Poisson's equation, when applied to electrolyte solutions, suggests that the length scale of regions of ionic space charge (the Debye length) is on the order to 1 nm.[11] Thus, on the much larger scale of pits and crevices, the solution composition must be electrically neutral,

$$\sum_i z_i c_i = 0. \quad (3)$$

Equations (1) and (3) provide $n + 1$ model equations to be solved for the n chemical species concentrations c_i and the electrostatic potential ϕ. The species in the model include the metal ions formed by corrosion, and the anion (most frequently Cl^-) and cation of the external electrolyte solution. Models also incorporate balances for species formed by homogeneous reactions in the cavity, such as hydrolysis reactions generating H^+ ions and hydrolyzed metal ions, and complexation reactions producing metal chloride complexes. Many models of pits contain at least five chemical species.

Boundary conditions on reactive surfaces are based on continuity of transport flux and reaction rates. If species 1 is produced by dissolution at current density i_d, the boundary condition would be

$$\frac{i_d}{z_i F} = \mathbf{N}_1 \cdot \mathbf{n}, \quad (4)$$

where $\mathbf{n}$ is the outward unit normal vector at the surface. The reaction current density i_d is coupled to the local potential and concentration through a kinetic rate expression. The fluxes of non-reactive species at the metal surface are zero. If the pit's length-to-width aspect ratio is large, reactions on the cavity sidewalls can then be modeled as pseudohomogeneous reactions in the volume of the cavity solution. The wall reactions are then represented as R_{vi} terms in species balances (1) where R_{vi} represents the rate of the surface reaction multiplied by the surface area per unit volume. The model pit geometry should resemble that of the actual pit or crevice, if possible. Many approximations to the pit geometry have been employed in models, which will be discussed in detail in the next section.

3. Review of Recent Models

This section summarizes modeling studies reported since the reviews of Sharland and Turnbull.[12,13] First, models are discussed which considered the pit surface to be actively dissolving, and did not take into account repassivation of surfaces in the pit. Then, the results of two modeling studies are described which included both dissolution and passivation. We choose to highlight repassivation in this way, since it underlies the important issue of pit stability. All models were based on the dilute solution transport equations in the previous section, except as indicated.

(i) Models of Pit Growth

Sharland simulated one-dimensional crevices or pits in steel, with either passive or active sidewalls.[14,15] The model included hydrolysis chemistry and formation of corrosion products, but did not consider obstructed transport due to the solid material in the pit. Metal dissolution kinetics were taken from earlier measurements on macroscopic electrodes.[16] The model predictions were compared with the findings of previous experiments on large-scale artificial pits with lengths and widths of order 1 and 0.1 cm, respectively. Because these dimensions are at least an order of magnitude larger than those of actual pits, the simulations may have been representative of crevice corrosion rather than pitting; however, encouraging comparisons were reported with some pH and potential measurements. In other model predictions, the quality of agreement with experiment was unclear owing to uncertain values of transport parameters in the concentrated solutions present in the pit.

Harb and Alkire reported comparisons of experimental measurements and finite-element simulations of pit growth on stainless steel and nickel in the presence of flow.[17,18] The simulation domain was a two-dimensional analog of the pit geometry, consisting of a cylindrical trench oriented transverse to the flow direction. Experimental evidence presented by the authors indicated that, for the conditions studied, pit growth on stainless steel was controlled by mass transport, while that on nickel was potential-dependent and therefore determined by electrode kinetics. Thus, the model for stainless steel used a constant concentration boundary condition based on the solubility of a precipitated salt film. The predicted pit currents were about 3 times smaller than those measured experimentally, possibly owing to the simplified geometry or the neglect of migration mass transport. The electrode kinetics used in the model for nickel were taken from an earlier study using macroscopic electrodes.[19] Experiments on both stainless steel and nickel indicated that pit stability in the presence of flow is governed by a critical Péclet number, i.e. $Pe = u_o r_o / D$, where u_o is the characteristic velocity in the pit, r_o is the pit radius, and D is the metal ion diffusivity. In the case of nickel, pits were repassivated when Pe exceeded 1,000, while on steel the critical Pe was about 10. The simulations of nickel found that the pit solution was significantly diluted by flow for Pe in the range 100–1,000, as shown in Fig. 1. The results therefore suggested that a concentrated pit solution may be necessary to sustain the kinetically controlled dissolution of nickel in pits.

The same authors also applied the model of nickel to simulate pit growth in quiescent solutions.[20] The predictions yielded potential-dependent pit growth currents in reasonable agreement with experimental findings. The same model equations and parameters were later solved by Verbrugge et al. using the quasipotential method.[21] This technique is a dependent-variable transformation through which the simulation is carried out in two steps: the solution of the Laplace equation in the geometric domain of the pit for the quasipotential; and the integration of a nonlinear ordinary differential equation initial value problem to obtain the species concentrations and potential as functions of the quasipotential. This method is computationally more efficient than the finite-element method, as the nonlinear species balance differential equations are replaced by the much simpler Laplace equation. The currents predicted by this simulation were again in reasonable agreement with experimental

Topics in the Mathematical Modeling of Localized Corrosion

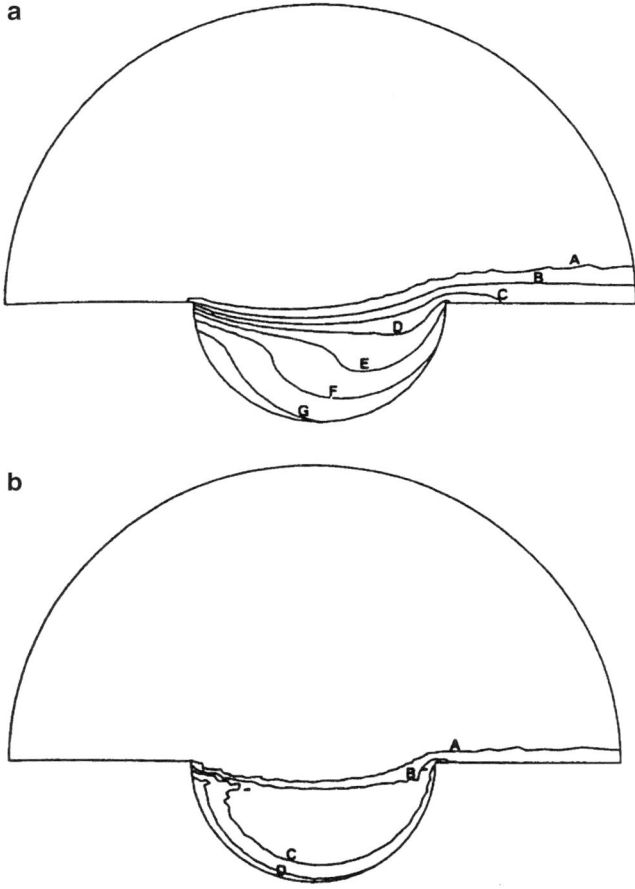

Figure 1. Lines of constant Ni^{2+} concentration at (**a**) Pe = 100 and (**b**) Pe = 1,000.[17] A–G correspond to concentrations of 0.01 (A), 0.04 (B), 0.07 (C), 0.10 (D), 0.13 (E), 0.16 (F), and (G) 0.19 mol/l.

findings. However, the metal ion concentrations at the dissolving surface were found to exceed saturation, which would be inconsistent with kinetically controlled dissolution. This discrepancy led the authors to speculate that knowledge of concentration-dependent transport coefficients in concentrated solutions would be necessary to obtain quantitatively accurate predictions.

Engelhardt and Strehblow developed a method employing the quasipotential transformation to predict the evolution of the pit shapes during dissolution.[22] The model was highly flexible in that it applied for different initial pit shapes, and accounted for transport by both diffusion and migration, and transport resistance both inside and outside pit. It was apparently the first model to calculate the development of the pit shape. The pit solution consisted of the dissolved metal ion, along with the anion and cation of the bulk electrolyte. The boundary condition at the dissolving surface assumed either Tafel kinetics or a saturated solution composition, or a combination of these conditions applying on different parts of the pit surface. The model explained some widely observed general characteristics of pit growth. For example, it predicted that when pit growth was controlled by mass transport, the pit depth should increase proportionally to the one-third power of time, and that pits should evolve to ellipsoidal and ultimately to hemispherical shapes, independent of their initial geometry (Fig. 2). Figure 2 also demonstrates good agreement with experimentally measured pit geometries.[23] On the

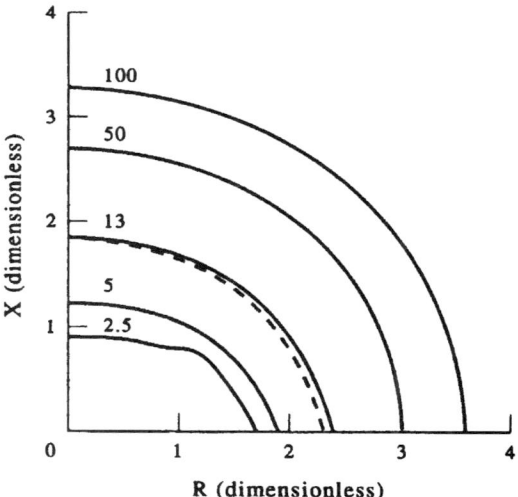

Figure 2. Shapes of axially symmetric pits evolved from dissolution of an initial plane disk, after the dimensionless times indicated.[22] X represents depth and R distance from the pit axis. The *dashed line* is an experimentally measured pit shape.[23]

other hand, pit growth controlled jointly by transport and kinetics (on the pit bottom and sidewalls, respectively) led to shallow-dished shapes, as also found experimentally.

In a later paper, Engelhardt et al. examined the use of approximations to models of one-dimensional pits of the type studied previously by Sharland.[24] Their objective was to determine the minimum level of model complexity required to predict pit growth rates with acceptable accuracy. Simplification was achieved by carefully identifying chemical species which could be eliminated. It was found that the pit solution could frequently be viewed as a binary electrolyte, even when the principle anion in the pit is present in low concentrations in the bulk solution. Inclusion of the potential drop outside the pit was shown to be important when pit sidewalls were actively dissolving. More recently, Engelhardt and Macdonald used a similar modeling approach to analyze the potential drops in pits which grow at low dissolution current densities on the order of $0.1 - 1 \text{ mA/cm}^2$, as expected from penetration rates observed in service conditions.[25] They concluded that practical electrolytes such as seawater give rise to negligible potential drops and hence the pit growth rate should be constant. However, it is not clear how the conditions of pit stability, which usually require concentrated solutions in the pit (see the next section), would be satisfied at such low current densities. In any case, this study indicated that greater attention should be paid to the conditions of slow pit growth during long-term exposures, which may differ appreciably from those of typical laboratory studies.

Verhoff and Alkire simulated the pitting of aluminum in alkaline chloride solutions, comparing their model predictions with experimental measurements of the growth of individual pits.[26,27] Their first paper reported calculations using a model which assumed steady-state transport by diffusion and migration in a domain including both the hemispherical pit and the surrounding solution. The solution chemistry was described at different levels of complexity, incorporating various hydrolysis and complexation reactions. pH measurements were successfully predicted by the model with only one hydrolysis reaction, but not with more complex schemes. Also, the simulations apparently predicted unrealistic metal ion concentrations much higher than saturation. In the next paper, transient model calculations carried out to predict the evolution of pit solution compositions during interruptions of anodic polarization were reported. The results were used to rationalize experimental observations of pit-size-dependent critical interruption times above which pits were passivated during the interruption. While the comparisons with experiment were not definitive, they suggested critical pit stability criteria related to the minimum chloride or metal ion concentrations in the pit.

More recent work by Alkire and coworkers has focused on predictions of critical solution compositions needed to initiate pitting corrosion in preexisting microscopic crevices adjacent to MnS inclusions in stainless steel.[28–30] Webb and Alkire attempted to predict the chloride concentration dependence of the pitting potential, assuming that pit initiation depended on a critical thiosulfate concentration in the crevice.[28] The solution chemistry in these models is highly complex, as they include the species produced by steel and MnS dissolution, as well as the products of their solution-phase reactions. In view of the resulting large number of model parameters, the authors have developed techniques to highlight parameters on which the model behavior critically depends. These techniques should be generally useful in other simulations of electrochemical systems which employ complex speciation schemes.

(ii) Models of Pit Growth and Repassivation

Laycock and coworkers developed simulations of pit growth on stainless steel at constant applied potential.[31,32] The present discussion focuses on their second paper, in which the model was substantially improved by the extension of the simulation domain to include the solution outside the pit cavity. The pit growth model incorporated a hypothesis of the critical condition for passivation within the pit, and the resulting development of the pit shape was evaluated. The metal surface was assumed to be passivated at metal ion concentrations lower than a critical value; the dissolution rate increased with metal ion concentration above the critical value. These concentration-dependent dissolution kinetics were closely related to results of experimental studies of artificial pits,[33–35] and the criterion for passivation was supported by measurements of actual pit growth.[36] It should be mentioned that description of mass transport in the model did not conform to the dilute solution theory as outlined above, since a current continuity equation was used which included contributions from migration but not diffusion; hence, the potential gradient in the pit was probably overestimated. However, inaccuracies in the prediction of potential were likely compensated by adjustment of empirical dissolution kinetic parameters in the model.

The pit shapes and current transients produced by the simulation were strongly influenced by passivation in the pit, and showed unique characteristics which are also found experimentally. At relatively low applied potentials, the surface near the pit mouth tended to be passivated because the concentration was below the critical value. As a result, the dissolution front undercut the passivated portion of the pit sidewall, and then propagated upward until the pit wall was eventually breached. Metal ions near the perforation rapidly diffused out of the pit, causing the neighboring surface to be passivated. This process was then repeated, producing cavities covered by perforated thin metal layers, in which the perforations were spaced at predicable intervals. These features strongly resembled perforated pit covers found experimentally, supporting the general concept of passivation embodied by the model (see Figs. 3, 4). It was shown that each

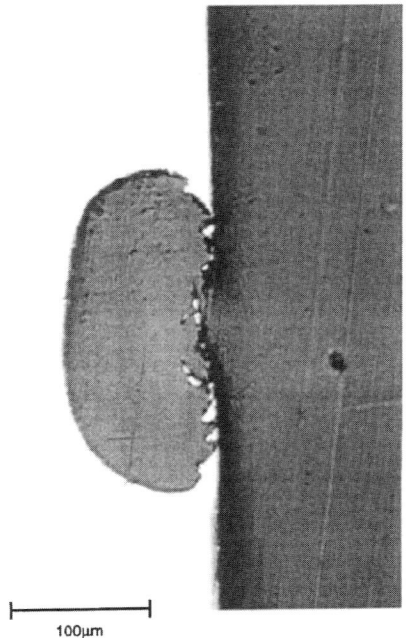

Figure 3. Cross section through a typical pit in 316 stainless steel, grown at −0.23 V in 1 M NaCl.[31]

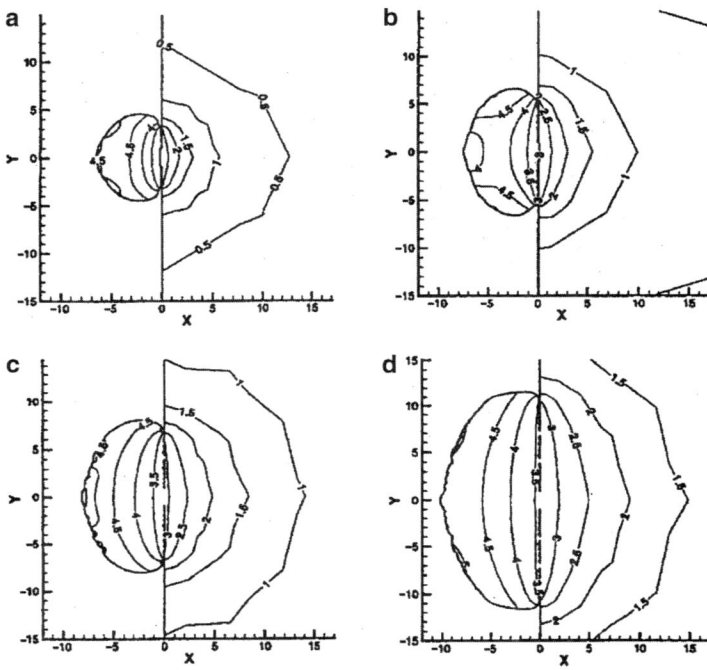

Figure 4. Cross sections through simulated pits grown at 0.19 V vs. the saturated calomel electrode, after (**a**) 0.5 s, (**b**) 1.0 s, (**c**) 1.5 s and (**d**) 2.5 s.[31] Contour lines correspond to the constant metal ion concentrations indicated, in moles per liter, and axis scales are in micrometers. Note the perforated cover over the pit predicted by the simulation.

perforation event decreased the overall mass transfer resistance in the pit, and that the relative decrease of resistance became larger as the pit size increased. Thus, a perforation event would cause small pits to be completely passivated, since the concentration increase outside the pit was generally not sufficient to overcome the sudden decrease of internal mass transfer resistance. The resulting pit current transients of small pits grew with time until they were suddenly terminated, and in this aspect resembled measured transients of "metastable pits" found at low potentials. However, experiments indicate that metastable pits are prevalent in a well-defined potential range, while in the simulations, the existence of such a potential range depended on the details of the assumed initial pit geometry. Therefore, while the modeling results support the existence of a threshold concentration for repassivation, it is not clear that the potential dependence of repassivation was captured. Experimental studies of stainless steel and iron–chromium alloys suggest that the potential as well as the electrolyte concentration may control repassivation.[35,37]

Aluminum etch tunnels are another type of pit which exhibit repassivation on their internal surfaces. Tunnels are formed by anodic etching of aluminum in chloride solutions at temperatures above 60°C.[38] They maintain widths on the order of 1 μm while they penetrate the metal in the crystallographic <100> direction, to depths of tens of micrometers. Figure 5 shows a scanning electron microscope image of an oxide replica of an etched aluminum surface, illustrating the typical morphology of tunnels.[39] The high length-to-width aspect ratio of tunnels is due to continuous oxide passivation on the sidewalls, while the tip surface corrodes at current densities of several amperes per square centimeter. Because of the small width dimension of tunnels, lateral concentration and potential gradients across the tip surface are very small. Thus, both the passive sidewall adjoining the tip and the corroding tip itself are in contact with the same solution composition and potential, which may be viewed as a critical condition for repassivation. Hebert and coworkers have carried out an extended experimental and modeling study of etch tunnels, with the goal of understanding the nature of this critical condition. These studies took advantage of unique opportunities to study repassivation afforded by the geometry of tunnels, such as the amenability of the linear tunnel shape to accurate transport modeling. The outcome of this investigation was a model of tunnel growth incorporating a mathematical description of repassivation, which successfully explained the tunnel shape.[40] In the following paragraphs, the experimental background of this repassivation model is discussed, after which the modeling results are described.

In the experiments of Hebert and coworkers, repassivation was initiated by either step or ramp reductions of the applied current during etching.[41–44] Repassivation in tunnels produces morphological changes of the otherwise flat tip surface, which were clearly apparent in electron microscope images. The current step experiments showed that oxide passivation on the tip occurred within times of 1 ms, and was accompanied by characteristic potential transients

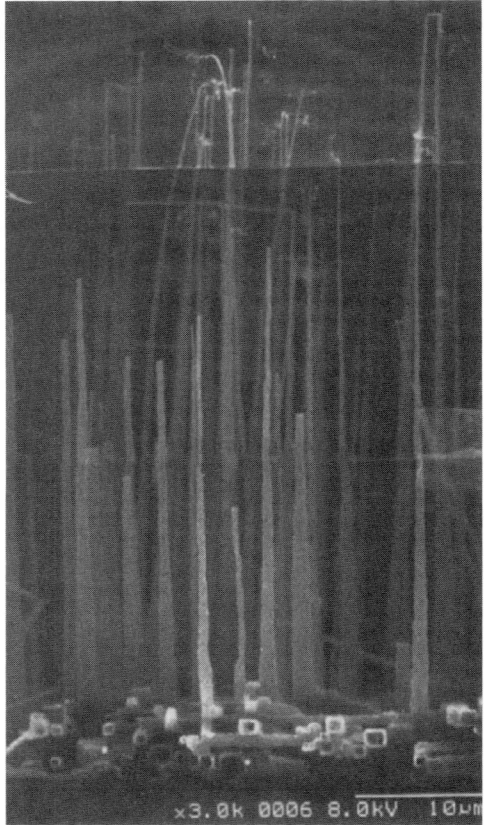

Figure 5. Oxide replica of aluminum etch tunnels grown in 1N HCl+6N H_2SO_4 at 77°C.[39] The replica is formed by anodic oxidation of the etched surface, followed by dissolution of the metal.

over the same range of times.[41,42] The time scale of passivation was much smaller than the time needed for the occurrence of solution composition changes, which is approximately $L^2/D \sim 100$ ms, where L is the tunnel length and D the electrolyte diffusion coefficient. Thus, in these experiments the solution composition at the tip remained constant during repassivation. It is evident that passivation was controlled by a critical potential, and not a critical solution composition, as discussed above with reference to stainless steel pitting. However, both the ohmic potential drop and the electrolyte concentration in tunnels increase significantly during their growth. If passivation were controlled only by the potential, the entire tip surface would be rapidly passivated. Therefore, to explain passivation in the transient experiments and also during steady tunnel growth, it was recognized that the critical potential should decrease with increasing electrolyte concentration.[45]

The critical repassivation potential, as measured using current–potential curves, has this required dependence on chloride concentration, and in fact the tunnel growth is observed at the repassivation potential corresponding to the bulk solution composition. Transport calculations during steady tunnel growth showed that the decrease of the repassivation potential at the tip surface, due to the elevated chloride concentration there, almost exactly compensated for the effect of increasing ohmic drop.[46] Thus, the experiments indicated that the commonly measured repassivation potential, evaluated at the local chloride concentration at the metal surface, is the fundamental criterion for repassivation in tunnels. This finding contrasts with other interpretations stating that the repassivation potential is not fundamentally significant in terms of surface–solution interactions.[47] In further experiments by Hebert and coworkers, the kinetics of repassivation were quantitatively determined with experiments using ramped reductions of applied current during tunnel etching.[43] It was argued that the repassivation potential probably controls chloride adsorption on dissolving metal surface in the pit.

Hebert's model for tunnel growth predicted the tunnel shape on the basis of the repassivation model just described.[40] Starting from the initial condition of a cubic etch pit, the model calculated the evolution of the pit shape resulting from dissolution and sidewall passivation. The dissolution rate was taken directly from experimental measurements.[48] Since the passivation kinetics were potential-dependent, it was necessary to accurately predict the potential at the tunnel tip. This required the use of concentrated solution transport equations, for the first time in a pitting model. All transport and kinetic parameters used in the model were taken from independent sources. The calculations showed that pits growing at the bulk solution repassivation potential spontaneously transformed into tunnels by sidewall passivation (Fig. 6). The tunnels then grew with parallel walls until the concentration at the tip approached saturation, at

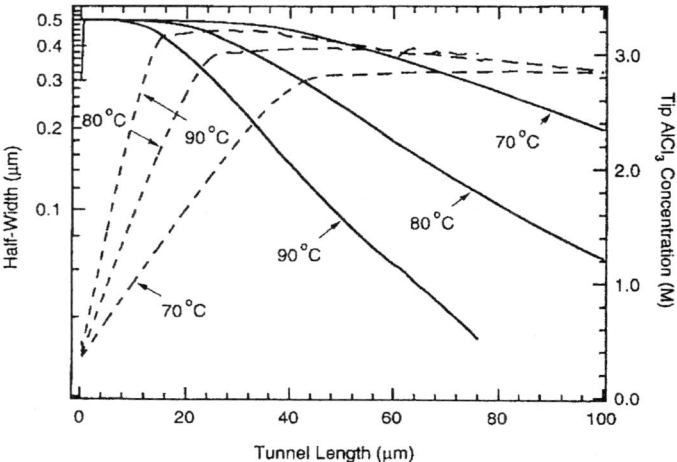

Figure 6. Predicted width profiles of aluminum etch tunnels at 70 – 90°C (*solid lines*), along with AlCl₃ electrolyte concentrations at the dissolving tip surface (*dashed lines*). The attenuation lengths of the predicted exponential width decays were 30, 19, and 11 μm at 70, 80, and 90°C, respectively, while values of 36, 22, and 13 μm were measured experimentally.[39] (Reproduced with permission from Ref. [40] Copyright 2001, The Eletrochemical Society.)

which point the tunnel width began to decrease exponentially with increasing length (the constant-width regime is not clearly apparent in Fig. 5 owing to extensive surface dissolution in this etchant). The exponential decay slopes in the tapering-width regime increased with temperature, and were in quantitative agreement with those observed experimentally (Fig. 6). The ability of the model to predict both the onset of tunneling and the detailed development of the tunnel shape validated the conceptual basis of the repassivation model.

The apparently different repassivation concepts embodied in the models of Laycock and White and Hebert may in fact share important common elements.[31,40] Repassivation occurs in the former model below a critical electrolyte concentration, while in the latter it takes place below a critical potential, the value of which decreases with increasing concentration. Therefore, at a given potential, both concepts indicate that active dissolution occurs only above a minimum concentration. The hypotheses differ in that Hebert additionally accounted for a direct influence of the potential on passivation, which was not done by Laycock and White. As mentioned earlier, there is some evidence that the potential influences repassivation on stainless steel and iron–chromium alloys.[35,37] Future work might explore whether these investigations might lead to a common understanding of pit stability applying to both aluminum and stainless steel.

4. Transport in Concentrated Electrolyte Solutions

Nearly all pit models have been based on transport equations which strictly apply in solutions much more dilute than those usually found in pits, which exceed 1 M and often approach saturation in the metal chloride salt. The fundamental shortcoming of dilute solution transport theory is that it accounts only for interactions between ions and solvent molecules, and not between pairs of ions. Ion–ion interactions are manifested, for example, by deviations of the solution conductivity from values predicted by dilute solution theory, which become appreciable at concentrations as low as 0.01 M.[49] This section will examine specific inaccuracies resulting from the dilute solution approximation, and point out cases where the use of concentrated solution transport models is tractable. Dilute and concentrated solution approaches will be compared in the context of a simple example of a one-dimensional pit with passive sidewalls. The metal and electrolyte solution were taken to be aluminum in 0.1 M NaCl. There are no cathodic reactions or homogeneous reactions in the pit, and the solution composition at the pit mouth is that of the bulk solution. This example was described in more detail in an earlier publication.[50] This example is chosen because of its simplicity and since the behavior of the dilute solution model may be familiar to readers.

The general multicomponent flux equations in concentrated solutions appear in the form[11]

$$c_i \nabla \mu_i = RT \sum_j \frac{c_i c_j}{c_T \bar{D}_{ij}} \left(\mathbf{v}_j - \mathbf{v}_i \right), \quad (5)$$

where μ_i is the electrochemical potential of ionic species i, c_T is the total concentration of all solution species, $\bar{D}_{ij}$ is the diffusivity for the interaction between species i and j, and $\mathbf{v}_i$ is the species velocity of i. The interspecies diffusivities $\bar{D}_{0j}$, which refer to ion–solvent interactions, reduce to the infinite-dilution diffusivities D_j in (2) in the limit as the c_i approach zero. The fluxes of species i, $\mathbf{N}_i = c_i \mathbf{v}_i$, which appear in the species-balance equations (1), are obtained in principle by inverting the system of equations like (5) written for each species. For the simple example system containing water, and Na^+, Cl^-, and Al^{3+} ions, six $\bar{D}_{ij}$ coefficients would describe all possible species interactions. Moreover, since the $\bar{D}_{ij}$ depend on composition, their values would be required over the range of all possible solution compositions which might be found in a pit. These considerations render impractical the direct use of models based on (5).

Considerable simplification results if the electrolyte solution may be approximated as a binary electrolyte. The use of this approximation in pitting models was shown to be widely applicable by Engelhardt et al., in the context of dilute solution transport theory.[24] In the example, it may be recognized that since electrical migration drives Na^+ ions out of the pit, these ions would be present only very close to the pit mouth, and hence the pit solution could be approximated as an $AlCl_3$ electrolyte. The inverted flux equations for water, Al^{3+}, and Cl^- ions may be combined with the species conservation equations, i.e., (1), to obtain a diffusion equation in terms of the electrolyte concentration c,

$$\frac{d}{dx}\left[D\left(1 - \frac{d \ln c_o}{d \ln c}\right)\frac{dc}{dx}\right] - \frac{i_d}{z_+ \nu_+ F}\frac{dt_+}{dx} = 0, \qquad (6)$$

where x is the coordinate along the pit length (the origin is at the pit mouth), c_o is the water concentration, and z_+, ν_+, and t_+ are the metal ion charge, stoichiometric number, and transference number. Given these properties, (6) can be solved using the boundary conditions $c = c_b/z_+$ at $x = 0$, where c_b is bulk NaCl concentration, and at the dissolving surface,

$$\frac{i_d(1 - t_+)}{z_+ \nu_+ F} = D\left(1 - \frac{d \ln c_o}{d \ln c}\right)\frac{dc}{dx}. \qquad (7)$$

The electrolyte concentration profile $c(x)$ is used to determine two types of potential. First, the potential which would be measured using reference electrodes equilibrating with the local solution is calculated. We take the reference electrode to be of Ag/AgCl type, which equilibrates with Cl^- ions in solution according to

$$\mu_{Ag} + \mu_- = \mu_{AgCl} + \mu_{e^-}. \qquad (8)$$

The potential of a Ag/AgCl reference electrode in the bulk solution, relative to a hypothetical electrode of the same type positioned at the dissolving surface ($x = L$) is

$$\Delta E = \frac{1}{F}[\mu_-(x = L) - \mu_-(x = 0)]. \tag{9}$$

The gradient of μ_- is related to the current density and the local composition gradient,

$$\frac{d\mu_-}{dx} = -\frac{F}{\kappa}i_d + \frac{1}{z_+\nu_+}\frac{d\mu_A}{dx}. \tag{10}$$

where κ is the solution conductivity and the AlCl$_3$ electrolyte chemical potential μ_A (where $\mu_A = \nu_+\mu_+ + \nu_-\mu_-$) can be computed at a given concentration with knowledge of the electrolyte activity coefficient. ΔE is found by integrating (10),

$$\Delta E = -i_d \int_0^L \frac{dx}{\kappa} + RT\left(\frac{\nu_+ + \nu_-}{z_+\nu_+ F}\right) \int_{c(0)}^{c(L)} t_+ d\ln(cf_{+-}). \tag{11}$$

where f_{+-} is the electrolyte activity coefficient. If ΔE is added to the potential of the metal with respect to a reference electrode in the bulk, one obtains the potential of the metal with respect to a hypothetical reference electrode of the same type, positioned just outside the electrical double layer on the dissolving surface of the pit. In effect, ΔE is a potential correction which, when added to the measured potential, produces the "true" electrode potential. The first term contributing to ΔE on the right side of (11) may be identified as the ohmic drop in the pit, and the second term represents the concentration overpotential.

The quasielectrostatic potential as defined by Newman was also calculated, for comparison with the electrostatic potential ϕ in the dilute solution model.[11] The quasielectrostatic potential is defined according to the electrochemical potential and concentration of a reference species. Here this species is taken to be the chloride ion,

$$\mu_- = RT \ln c_- - F\Phi. \tag{12}$$

The value of the quasielectrostatic potential Φ depends on the choice of the reference species used in the definition. As the solution becomes very dilute, it becomes the same as the electrostatic potential ϕ in the dilute solution model.

The solutions of (6) and (11) require three concentration-dependent transport properties, D, t_+, and κ, along with the electrolyte activity coefficient in Eq. (11). These properties may be measured using well-known techniques.[51] The three independent transport coefficients D, t_+, and κ may be related to the diffusivities $\bar{D}_{0^-}$, $\bar{D}_{0^+}$, and $\bar{D}_{+-}$ from (5). In the dilute solution model there would be only two diffusivities, D_+ and D_-, equivalent to $\bar{D}_{0^-}$ and $\bar{D}_{0^+}$. There is no analog of $\bar{D}_{+-}$ in dilute solution theory because ion–ion interactions are not considered. When $\bar{D}_{+-}$ is significant, the predictions of the potential in dilute solution models deviate strongly from those in concentrated solution models. These deviations cannot be compensated by adjusting D_+ and D_- from their values at infinite dilution. However, since the properties D and t_+ in (6) and (7) do not depend on $\bar{D}_{+-}$, the predictions of the concentration are not affected by ion–ion interactions.

Calculations were carried out at 70°C because transport properties and activity coefficients of $AlCl_3$ had been assembled at this temperature.[40] Measurements indicated that D and t_+ were approximately constant at concentrations at least up to 1.5 M, with values of 2.1×10^{-5} cm^2/s and 0.21, respectively. Also, since even at saturation the electrolyte concentration was small compared with that of water, the water concentration derivative in (6) could be neglected. Thus, (6) led to a simple linear concentration profile in the pit, as was also predicted by the dilute solution model.

$$c = c(x=0) + \frac{i_d(1-t_+)}{z_+ \nu_+ FD} x. \tag{13}$$

Since both κ and f_{+-} varied strongly at high concentration, it was necessary to explicitly consider their concentration dependences when evaluating the integrals in (11). In Figs. 7–9, respectively, predictions of the chloride concentration at the dissolving surface, the potential correction ΔE, and the quasielectrostatic potential are compared between the concentrated and dilute solution models. According to the latter model, μ_- is determined by the chloride concentration and local electrostatic potential ϕ ((12), with ϕ replacing Φ). Accordingly

$$\Delta E = \frac{RT}{F} \ln \frac{c_-(x=L)}{c_-(x=0)} - [\phi(x=L) - \phi(x=0)]. \tag{14}$$

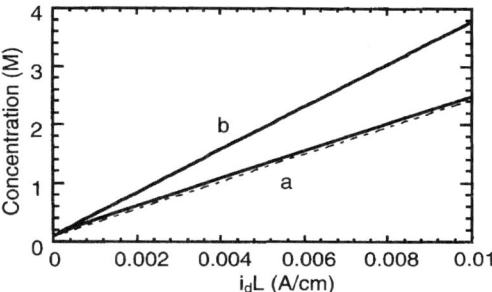

Figure 7. Comparison of chloride concentration at the dissolving surface of a pit, calculated using models based on (**a**) dilute solution and (**b**) concentrated solution transport equations. The *dashed line* is calculated from the dilute solution model assuming a binary electrolyte in the pit. (Reproduced with permission from Ref. [50]. Copyright 1999, The Eletrochemical Society.)

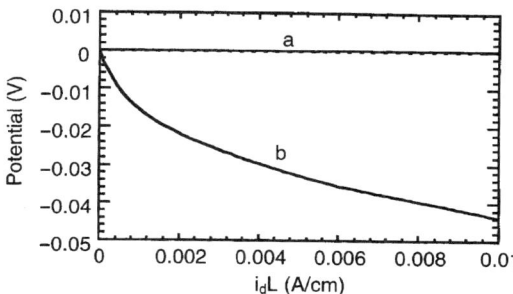

Figure 8. Comparison of reference electrode potential drops in pits, calculated using models based on (**a**) dilute solution and (**b**) concentrated solution transport equations. The *dashed line* is calculated from the dilute solution model assuming a binary electrolyte in the pit. (Reproduced with permission from Ref. [50]. Copyright 1999, The Eletrochemical Society.)

The infinite-dilution diffusivities were temperature-corrected using empirical formulas.[52] Since the predictions of both concentrated and dilute solution models depend only on the product $i_d L$, the results in Figs. 7–9 are plotted against this parameter.

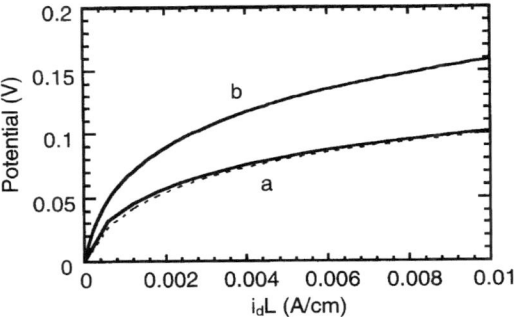

Figure 9. Comparison of quasi-electrostatic potential at a dissolving surface in a pit, calculated using models based on (**a**) dilute solution and (**b**) concentrated solution transport equations.[50] The *dashed line* is calculated from the dilute solution model assuming a binary electrolyte in the pit.

Figure 7 shows that in both models, the Cl^- concentration at the dissolving surface increases linearly with $i_d L$. Both trends are represented by (13), representing the binary electrolyte approximation. The larger slope of the line representing the concentrated solution model is due to the smaller D and larger t_+, respectively, compared with their infinite-dilution values. Aside from the different transport property values, Fig. 7 is consistent with the mathematical equivalence of the concentrated and dilute solution models for predictions of concentration. If one is primarily interested in predictions of concentration, the dilute solution model may suffice, provided that the calculations account for deviations of transport properties from their values at infinite dilution. However, most models require knowledge of the potential to calculate rates of electrochemical reactions.

The reference electrode potential drop ΔE is displayed in Fig. 8. Here, the results of the two models are quite different, in that the dilute solution model predicts zero ΔE at all depths, while negative values are calculated by the concentrated solution model. The behavior in Fig. 2 is due to the dependence of the reference electrode potential on μ_- (8). Since Cl^- ions are not involved in the reaction, the species balance (1) requires that their flux is zero in both models. In the dilute solution model, the Cl^- flux depends only on the gradient of μ_-, and so zero flux requires that μ_- is uniform in the pit. On the other hand, the Cl^- flux in the concentrated solution model depends on the gradients of both μ_- and μ_+, the latter through the $\bar{D}_{+-}$ term in (5). Therefore, the predicted μ_- is nonuniform, leading to a nonzero potential drop. The fundamental reason for the erroneous potential calculations of the dilute solution model is its neglect of effects due to ion–ion interactions, which appear in the concentrated solution model through the dependence of the conductivity on $\bar{D}_{+-}$. A reliable measure of whether ion–ion interactions might be important in a model is the whether the experimental conductivity deviates from its value predicted by dilute solution theory.

Figure 9 compares the quasielectrostatic potential with the electrostatic potential in the dilute solution model. The plotted potentials are those at the dissolving surface relative to the values at the pit mouth. The concentrated solution model predicts a much larger potential drop in the pit, primarily because it depends on the experimentally measured conductivity, which is again is reduced in concentrated solutions relative to the conductivity according to the dilute solution model. By comparison with Fig. 7, it may be seen that significant relative errors in the dilute solution calculation appear at concentrations less than 1 M. Since pit solutions are usually even more concentrated, large relative errors in potential calculations would be expected. This may be a serious concern because the metal dissolution rate increases exponentially with potential.

5. Concluding Remarks

The foregoing review shows that in the past 15–20 years modeling of pitting corrosion has benefited from significant technical advances, including the use of finite-element methods and the quasipotential transformation. Pitting models have enabled predictions of shape evolution and convection effects, and can handle increasingly complex solution-phase chemistry. Models of metastable pits in stainless steel and aluminum etch tunnels were developed incorporating repassivation criteria, which were validated by the successful prediction of the highly unusual morphologies of these pits.[31,40]

The discussion of concentrated solution models has indicated that, while the transport flux equations in their rigorous form (5) may be intractable, the use of the binary electrolyte approximation allows the convenient implementation of concentrated solution theory in pitting corrosion models. Engelhardt et al. have shown that this approximation is valid over a surprisingly wide range of conditions.[24] Some pitting models in recent years have pursued a different approach using the dilute solution model, implementing complex descriptions of the solution composition with a large number of chemical species. The calculations presented in this section suggest that the errors due to the dilute solution equations, particularly regarding potential calculations, may outweigh the benefits of a detailed knowledge of the solution composition.

III. GALVANIC COUPLING AT THE INTERFACE BETWEEN TWO METALS

The galvanic coupling between two metals is a well-known phenomenon. Generally it is studied by using two electrodes of pure metals immersed in the same solution and electrically connected; thus, the current flowing between the two metals can be measured directly. However, while it is frequently assumed that each electrode is homogeneous, experimental observation shows that in some cases a particular behavior is visible near the interface between the two metals. From a qualitative point of view, this heterogeneity can be due to a current or potential distribution. This kind of distribution was well studied in particular by Newman for an interface between a metal and an insulator.[53,54] However, the case of the interface between two conducting materials and in particular two metals was considered only recently.[55] The particular case of aluminum and copper is considered here as an example.

1. Theoretical Description of the Currents and Potentials at the Interface of Two Metals

Numerical simulations were performed to obtain a description of the potential and current distributions on the surface of the disk electrode and in the surrounding electrolytic solution. Figure 10 gives a schematic representation of the disk electrode used in this model. In the absence of concentration gradients, the potential ϕ in the solution surrounding the electrode is governed by the Laplace equation:

$$\nabla^2 \phi = 0. \quad (15)$$

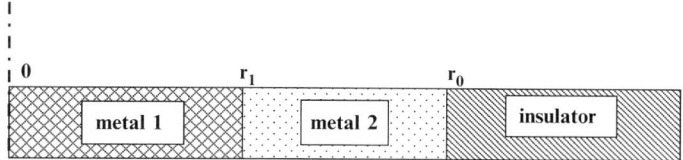

Figure 10. The electrode.

Using cylindrical coordinates (r, θ, z), one can express (15) as

$$\frac{1}{r}\frac{\partial}{\partial r}\left(r\frac{\partial \phi}{\partial r}\right) + \frac{1}{r^2}\frac{\partial^2 \phi}{\partial \theta^2} + \frac{\partial^2 \phi}{\partial z^2} = 0, \tag{16}$$

where z is the normal distance to the electrode surface, r is the radial coordinate, and θ is the azimuth. The cylindrical symmetry condition requires the geometry to be invariant under rotation about the z axis, i.e.,

$$\frac{\partial \phi}{\partial \theta} = 0. \tag{17}$$

The combination of (16) and (17) yields the governing equation in a two-dimensional domain as

$$\frac{\partial^2 \phi}{\partial r^2} + \frac{1}{r}\frac{\partial \phi}{\partial r} + \frac{\partial^2 \phi}{\partial z^2} = 0. \tag{18}$$

On the surrounding insulator and far from the electrode surface, the boundary conditions are given by

$$\left.\frac{\partial \phi}{\partial z}\right|_{z=0} = 0 \text{ at } r > r_0 \tag{19}$$

and

$$\phi \to 0 \text{ as } r^2 + z^2 \to \infty. \tag{20}$$

Under the assumption of a kinetic regime, the current density at the electrode surface can be expressed as

$$i = -\kappa \left.\frac{\partial \phi}{\partial z}\right|_{z=0}, \tag{21}$$

where κ is the electrolyte conductivity.

The kinetic current expressions $f_i(\phi)$ are a function of the potential ϕ, and the corresponding kinetic parameters were determined from experimental measurements performed individually on each material.

$$\text{For metal 1}: 0 \leq r \leq r_1, \quad f_1(\phi) = -\kappa \left.\frac{\partial \phi}{\partial z}\right|_{z=0}. \quad (22)$$

$$\text{For metal 2}: r_1 \leq r \leq r_0, \quad f_2(\phi) = -\kappa \left.\frac{\partial \phi}{\partial z}\right|_{z=0}. \quad (23)$$

$f_1(\phi)$ and $f_2(\phi)$ can correspond to anodic current, to cathodic current, or to the sum of anodic and cathodic current.

2. Application to the Al–Cu Coupling

(i) *Mathematical Model*

Metal 1 is copper (Fig. 11), and according to our knowledge of this system, the copper electrode is under cathodic polarization and the oxygen reduction reaction occurs. The current can then be expressed as

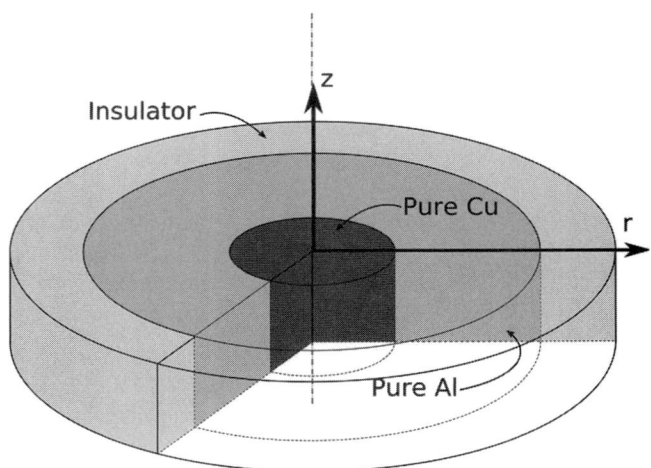

Figure 11. The electrode under investigation.

$$i_{Cu} = f_{Cu}(\phi) = -k_{O_2} \exp\left(-\frac{2.303}{b_{O_2}}\phi\right), \tag{24}$$

where the constant k_{O_2} and b_{O_2} were determined from experimental measurements. The values determined experimentally were $b_{O_2} = 220\,\text{mV/dec}$ and $k_{O_2} = 7.6 \times 10^{-6}\,\text{A cm}^{-2}$. Metal 2 is aluminum and since this metal is passive, the current is approximately uniform on the electrode. Thus, the boundary condition corresponds to an anodic current of $1\,\mu\text{A cm}^{-2}$ independent of the potential:

$$i_{Al} = f_{Al}(\phi) = 1\,\mu\text{A cm}^{-2}. \tag{25}$$

The simulations were performed using the finite-element package FEMLAB with the conductive DC module in a two-dimensional axial symmetry. The mesh size was refined to obtain a numerical error lower than 0.1% evaluated from the net current of the system, which is the sum of the current passing through the copper and aluminum electrodes. At the corrosion potential this net current is zero.

Figure 12 shows the potential distribution calculated on the electrode surface along the electrode radius with the electrolyte conductivity as a parameter. Independent of the electrolyte conductivity,

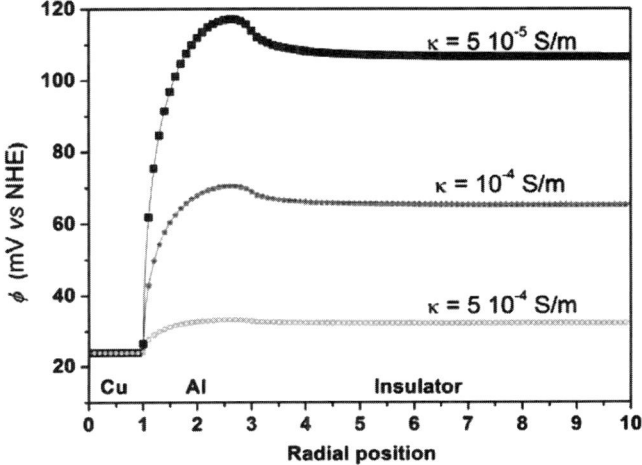

Figure 12. Potential distributions on the surface of the Al/Cu model couple deduced from theoretical calculations for different values of the electrolyte conductivity.

the general shape of the potential distribution remains similar. The potential is seen to be constant over the copper electrode; it strongly increases immediately after the copper/aluminum interface and reaches a maximum value over the aluminum electrode near the aluminum/insulator interface, just before a slight decrease. Such behavior is fully consistent with the boundary conditions determined for the calculation. Figure 12 also shows that the variations of the potential along the disk electrode radius are higher for a low conductivity of the electrolyte, which, from a practical point of view, made these variations easier to detect.

Figure 13 shows the influence of Ag/AgCl probe position (i.e., readings at two distances from the electrode surface). Calculations were performed in the case of a weakly conductive electrolyte to highlight the influence of the position of the probe on the potential distribution. No significant difference was observed between the two measurements: when the probe was withdrawn from the electrode surface, the sudden variation of the potential at the aluminum/copper interface was barely reduced in comparison with measurements performed on the disk electrode surface itself. Moreover, it should be noticed that the amplitude of the potential variations was smaller when the electrode was far from the substrate.

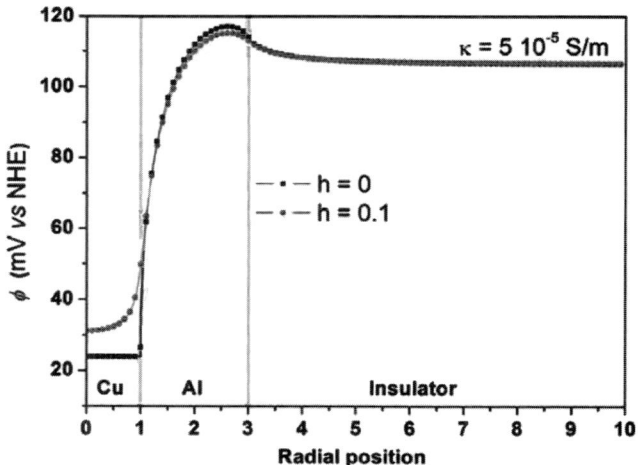

Figure 13. Potential distributions on the surface of the Al/Cu model couple deduced from theoretical calculations for different values of the probe position (electrolyte conductivity equal to 5×10^{-5} S m^{-1}).

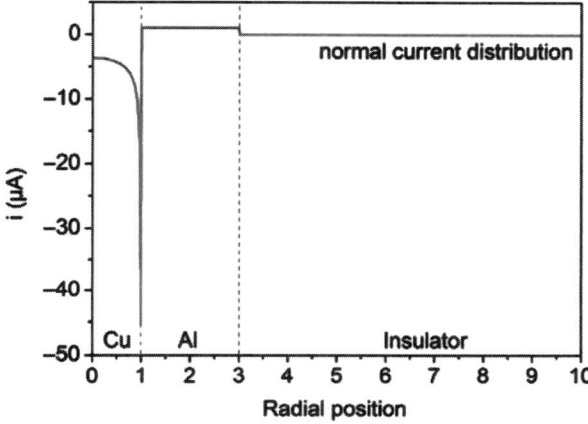

Figure 14. Current distributions on the surface of the Al/Cu model couple deduced from theoretical calculations.

The solution of the Laplace equation also allows the distribution of the normal current on the disk electrode surface to be calculated, as shown in Fig. 14. The normal current on aluminum is constant and corresponds to the boundary condition chosen, while for the copper electrode, the calculations show a cathodic current distributed along the electrode radius. Thus, calculation led to results in good agreement with the initial hypothesis: a passive behavior for aluminum and a cathodic current related to oxygen reduction on copper. Furthermore, potential and current distributions revealed a particular behavior at the aluminum/copper interface. Figure 14 shows a very high cathodic current on copper at the aluminum/copper interface corresponding to an increase of oxygen reduction on copper:

$$O_2 + 2H_2O + 4e^- \rightarrow 4OH^-. \tag{26}$$

This suggests that local variations of the chemical composition of the electrolyte can occur from the beginning of immersion with a local alkalinization of the electrolyte at the aluminum/copper interface. Thus, the consequence is a particular evolution of the interface morphology of the disk electrode after immersion in the electrolyte related to corrosion phenomena restricted to the aluminum/copper interface.

(ii) Experimental

The sample consisted of a pure aluminum/pure copper couple. A cylinder of pure aluminum (99.999 wt%), provided by Alfa/Aesar, was drilled in its center and a cylinder of pure copper (99.9 wt%), provided by Goodfellow, was then introduced by force into the hole. The assembly of the two materials gave a perfectly joined interface avoiding any crevice corrosion due to surface defects. The diameters of the two cylinders were chosen to obtain an aluminum surface area to copper surface area ratio of 10 (external diameters were equal to 10 and 3.15 mm for the aluminum and copper bars, respectively). The electrode was then embedded in an epoxy resin so that a disk electrode was obtained at the extremity. Before immersion in the electrolyte, the disk electrode was mechanically polished with SiC papers up to 4,000 grade, ultrasonically cleaned with ethanol, then with distilled water. The electrolyte was a 10^{-3} M Na_2SO_4 solution prepared with analytical-grade chemicals in contact with an air atmosphere and at room temperature.

Local electrochemical measurements were performed using a homemade device. The local potential variations were measured using a 100 − µm-diameter Ag/AgCl microelectrode. It consisted of a silver wire (100 µm in diameter, Goodfellow) laterally insulated with cataphoretic paint, and sealed in a glass capillary with epoxy resin. The AgCl was deposited in a 2 M KCl solution under potentiostatic oxidation of the silver electrode at 0.4 V vs. the saturated calomel electrode. The microreference electrode was moved with a three-axis positioning system (UTM25, Newport) driven by a motion encoder (MM4005, Newport), allowing a spatial resolution of 0.2 µm in the three directions. A homemade analog differential amplifier with both variable gain and high input impedance was used to record the local potential. All measurements were performed at the corrosion potential E_{corr} using a Keithley 2000 digital multimeter. The experimental setup was controlled with homemade data-acquisition software developed under a LabView® environment.

(iii) Experimental Results and Discussion

The aluminum/copper interface was observed using optical and scanning electron microscopy before and after immersion in the sodium sulfate solution. Figure 15 shows an optical micrograph of the interface before immersion (copper on the right, aluminum on

Figure 15. Optical observation of the Al/Cu interface of the disk electrode before immersion.

the left of the micrograph). The two materials are perfectly joined without any defects observable at the interface. After 24 h of immersion in the sulfate solution (Fig. 16), two main differences were observed in the vicinity of the interface in comparison with the observations performed before immersion:

1. A deep crevice was formed at the aluminum/copper interface. Close to the crevice, aluminum has a bright color, while except in this zone it seems to be covered by an oxide film over its whole surface, which is in good agreement with previous hypotheses concerning a passive state for aluminum in the couple. The bright color suggests that the crevice was at least partially related to the dissolution of the aluminum close to the interface.
2. An orange-brown ring about 50 µm thick occurred all around the aluminum/copper interface. It was located on the aluminum material at a distance of about 150 µm from the interface. The ring was analyzed by energy-dispersive spectroscopy as being copper and is attributed to the deposition of copper coming from the dissolution of the aluminum/copper interface.[56,57] This copper deposition will not be analyzed in the present work.

Figure 16. Optical observations after 24 h of immersion in a 10^{-3} M Na_2SO_4 solution.

The observations also show that the corrosion processes were restricted to the aluminum/copper interface and only concerned the interface or the zone very close to the interface (copper ring at 150 μm from the interface). No other domain of the sample exhibited corrosion damage, whether on aluminum or on copper. Further scanning electron microscope observations allowed the zone close to the interface to be more accurately described. In Fig. 17, the gray zone corresponds to aluminum, while the white zone corresponds to copper. The deep crevice previously observed using optical microscopy was more accurately seen on the scanning electron microscope micrographs. Moreover, aluminum was found to be corroded very close to the aluminum/copper interface with a dissolution depth decreasing from the interface to some micrometers from the interface. Except in this zone, aluminum was not corroded.

(iv) Comparison Between Theoretical Calculations and Experimental Observations

Observations of the morphology of the aluminum/copper interface after immersion in the electrolyte suggest that corrosion

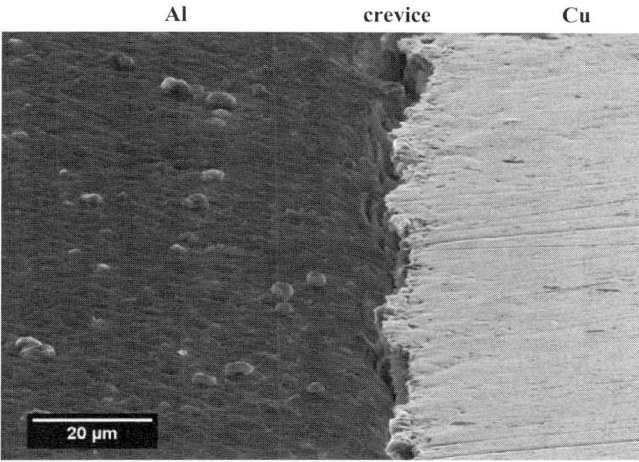

Figure 17. Scanning electron microscope observations of the Al/Cu interface of the disk electrode after 24 h of immersion in a 10^{-3} M Na$_2$SO$_4$ solution.

occurs in different steps. The strong dissolution of aluminum at the aluminum/copper interface partially leads to the formation of the crevice, as observed. Thus, the initial step of the corrosion process appears to be the dissolution of aluminum at the aluminum/copper interface, while aluminum remains passive on the remaining surface of the electrode. This is in good agreement with the current distribution previously shown from the numerical calculations (Fig. 14). A strong cathodic current density, related to increased reduction of the oxygen, was predicted on copper at the aluminum/copper interface. Such a reaction should lead to a local alkalinization of the electrolyte.[57]

Figure 18 shows the potential distribution measured at the beginning of immersion at the surface of the disk electrode using a Ag/AgCl microreference electrode. Two consecutive measurements were performed and both gave the same results. The curves were also in good agreement with the potential distribution deduced from the theoretical calculations ($h = 0.1$), showing that the hypotheses of the calculations were relevant.

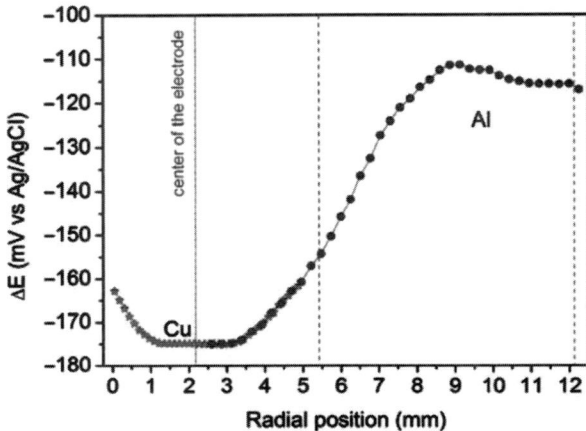

Figure 18. Potential distributions measured in a 10^{-3} M Na_2SO_4 solution on the surface of the pure aluminum/pure copper model couple using a Ag/AgCl microreference electrode. Two consecutive measurements were performed.

Thus, from these results, a two-step mechanism is proposed to explain the corrosion phenomena observed on the aluminum/copper model couple:

- Step 1: During the immersion of the disk electrode in the electrolyte, aluminum is the anode of the system and is in the passive state, while copper is cathodically polarized. The current distribution shows that significant oxygen reduction occurs on the copper electrode close to the aluminum/copper interface, whereas copper dissolution can be ignored during the first moments of immersion. Thus, owing to the strong cathodic reduction, alkalinization of the electrolyte occurs locally.
- Step 2: When the pH reaches a value of 9, the aluminum is depassivated and dissolves, leading to the formation of a crevice. The current distribution shows that the high cathodic current is restricted to the aluminum/copper interface, which explains that the dissolution of aluminum only occurs at the aluminum/copper interface, leading to the specific topography observed with scanning electron microscopy.

Such phenomena are not dependent on the scale of the events and the two-step mechanism could explain the corrosion mechanisms in commercial alloys such as AA2024. Indeed, the results showed that the most significant feature was the current distribution, which provokes the increase of oxygen reduction at the interface and thus induces a strong variation of the pH of the electrolyte. This study shows that a simple model (pure aluminum/pure copper) is relevant for the study of what happens in copper-rich aluminum alloys and also underlines the specific and important role of the aluminum/copper interface. This feature cannot be observed if the galvanic coupling is studied in the usual way with two independent electrodes in the same solution.

3. Conclusions

A simple system with two pure metals as described in Fig. 10 is relevant for understanding some local corrosion phenomena. The potential and current distributions on the surface of the model couple were deduced from theoretical calculations and the potential distribution could be experimentally checked using a Ag/AgCl microreference electrode. In the example of pure aluminum/pure copper, an increased cathodic activity related to oxygen reduction was observed at the aluminum/copper interface from the beginning of immersion. Combination of the measurements and calculations with optical and scanning electron microscopy observations of the model couple after a longer immersion time (24 h) allowed a two-step mechanism to be proposed to explain the corrosion damage. This work underlines that the local alkalinization of the electrolyte at the aluminum/copper interface plays a key role in the degradation of the system.

IV. IMPEDANCE IN A CONFINED MEDIUM

1. Introduction

Thin-layer cells are experimental devices of interest for simulation of various localized corrosion processes. For example, a thin moisture film covers the metal surface during atmospheric corrosion, and the thin-layer geometry is also relevant to crevice corrosion, and corrosion under delaminated protective films.[58–62] Thin-layer cells are usually achieved by confining a thin electrolyte layer (the thickness of which is less than a few hundred micrometers) between the surface of a working electrode and a parallel waterproof wall which is mechanically set to define the desired thickness.[61] Most classical thin-layer cells have a cylindrical geometry and involve large disk electrodes (with radii of several millimeters). The geometrical accuracy is generally limited by the difficulty of achieving strict parallelism between the electrode surface and the wall ensuring electrolyte confinement.[63,64] With use of a positioning control procedure based on the measurement of the electrolyte resistance, parallelism errors could, however, be accurately quantified and then minimized in these cells.[63]

In a thin-layer cell, the residual convection ("natural convection") which always exists in a large volume of solution becomes negligible. When the migration processes of the electroactive species can also be neglected (for instance, in presence of a supporting electrolyte), the mass transport is purely achieved by diffusion and is determined by Fick's laws:

$$N = -D\nabla c, \tag{27}$$

$$\frac{\partial c}{\partial t} = D\nabla^2 c, \tag{28}$$

where N is the flux of the diffusing species $(\text{mol s}^{-1}\,\text{cm}^2)$, and D $(\text{cm}^2\,\text{s}^{-1})$ and c (mol cm^{-3}) are its diffusion coefficient and concentration, respectively.

Electrochemical impedance spectroscopy is a powerful electrochemical technique allowing the mass transport of electroactive species in the vicinity of electrodes to be investigated. In a thin-layer-cell configuration, the corresponding mass transport impedance can be calculated from Fick's equations (27) and (28). With the assumption that diffusion processes occur only along the normal direction of the electrode surface (one-dimensional diffusion), analytical expression of the diffusion impedance can be derived in a thin-layer cell. The diffusion impedance will depend obviously on the boundary conditions considered in the calculations. It could also be modified by an eventual time constant distribution within the layer of the ionic conductor covering the electrode or by the occurrence of homogeneous chemical reactions coupled to the diffusion processes.

From a more general point of view, Keddam et al. also demonstrated that the impedance measured in a thin-layer-cell configuration is significantly modified by the existence of a radial potential distribution within the liquid film confined over the electrode surface.[62] This effect was ascribed to the ohmic drop existing within the liquid film covering the electrode surface.

In this context, the aim of this work was to experimentally investigate mass transport in a cylindrical thin-layer cell involving a large disk electrode, using electrochemical impedance spectroscopy. In addition to this experimental approach, a new transmission line model was proposed to take into account the radial potential drop existing within the cell in the theoretical diffusion impedance calculation. In contrast to the model previously proposed by Keddam et al., the cylindrical geometry of the electrochemical cell was rigorously taken into account in this new mathematical development.[62] With use of this model, modified one-dimensional spatially restricted diffusion impedances were calculated using two methods. First, it was assumed that mass transport proceeds only normally to the electrode surface. Second, both normal and radial mass transport contributions were considered.

2. Experimental

The cylindrical thin-layer cell used in this study is depicted in Fig. 19. To guarantee quasiperfect mechanical stability during measurements, this setup is supported by an antivibration table. The working electrode is the cross section of a platinum cylinder (5 mm in diameter). This electrode is sealed in a cylindrical epoxy resin insulating holder (30 mm in diameter). Prior to use, the electrode was successively polished with a 4,000 grit silicon carbide paper, cleaned with ethanol, rinsed with deionized water, and dried with a N_2 flux. The electrode holder is fixed at the bottom of a glass cell. The glass cell/electrode holder assembly is attached to a micropositioning system made of a bidirectional $x - y$ translation stage mounted on a three-axis rotation platform. Thanks to this configuration, the position and spatial orientation of the electrode surface can be completely defined with an accuracy of $2 \times 10^{-7 \circ}$ in the angular settings, and a precision of 1 µm in the linear positioning. Facing the platinum working electrode, a mobile waterproof polytetrafluoroethylene cylinder (30 mm in diameter) ensures the confinement of the electrolyte, forming a thin layer over the metallic sample. This cylinder is attached to a z-translation stage and can be moved vertically with a precision of 2 µm. The geometry of the confined zone facing the working electrode surface was set according to the

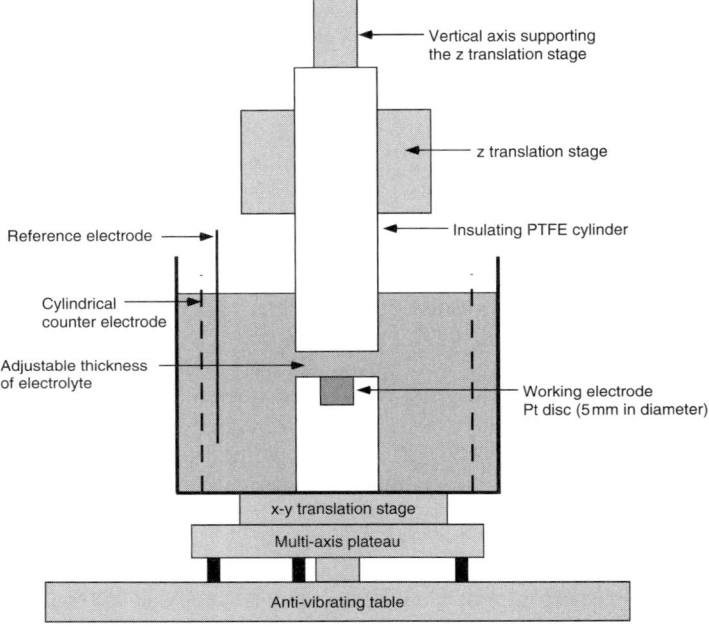

Figure 19. The cylindrical thin-layer cell used for the impedance measurements.

positioning procedure proposed in a previous work.[63] Thanks to this method, the average electrolyte film thickness was set with a precision estimated at ±5 μm.

The impedance measurements were carried out using a classical setup: a frequency response analyzer (Solartron 1250) and an electrochemical interface (Solartron 1286) monitored by a personal computer. A standard sulfate electrode and a large platinum grid were used as reference and counter electrodes, respectively. These two electrodes were located outside the confined zone facing the working electrode surface. This configuration prevents notably the counter electrode reactions from affecting the local chemistry in the vicinity of the working electrode. All experiments were performed in a 0.01 M $Fe(CN)_6^{3-}$ + 0.01 M $Fe(CN)_6^{4-}$ + 0.5 M KCl solution. Solutions were prepared in deionized and doubly distilled water from analytical-grade chemicals (Sigma). All solutions were freshly prepared.

3. Theory

As discussed above, if we assume that diffusion processes occur only normal to the electrode surface (one-dimensional model),[62,65] the theoretical diffusion impedance associated with the transport of an electroactive species can be derived analytically in a thin-layer-cell configuration. For a thin-layer cell of thickness e, the expression of this impedance will depend on the condition considered at the boundary.

In the case of a boundary impermeable to the diffusing species [i.e., $(\partial c/\partial x)_{x=e} = 0$], a one-dimensional spatially restricted diffusion impedance Z_{1D} ($\Omega\,\text{cm}^2$) is obtained at the electrode:

$$Z_{1D}(\omega) = R_d \frac{e}{D} \frac{\coth\left(\sqrt{\frac{e^2}{D}j\omega}\right)}{\sqrt{\frac{e^2}{D}j\omega}}, \qquad (29)$$

where ω is the angular frequency of the sine wave perturbation, R_d is a scaling factor depending on the kinetics of the interfacial reactions, and e is the electrolyte film thickness.[65]

The Nyquist plot of this spatially restricted diffusion impedance, calculated for $R_d = 0.05\,\Omega\,\text{cm}^3\text{s}^{-1}$ and $D = 6.3 \times 10^{-6}\,\text{cm}^2\text{s}^{-1}$, is presented in Fig. 20 for various values of the electrolyte film thickness e. The diagrams exhibit a classical Warburg behavior at high frequency (straight line with a 45° slope) followed by a capacitive behavior at lower frequencies. Electrically, the low-frequency behavior is equivalent to a resistor–capacitor series connection. The value of this low-frequency resistor decreases as the electrolyte film thickness decreases. An increase of the characteristic frequencies (corresponding to the transition frequency between the Warburg and the capacitive behavior, for example) is also observed as the electrolyte film thickness decreases.

As discussed in the previous section, the existence of a radial potential distribution should be taken into account for modeling suitably the impedance response on a large disk electrode located in a cylindrical thin-layer cell. Thus, a transmission line model adapted to the geometry of the cylindrical thin-layer cells was derived according to Fig. 21. Let us assume first that the electrolyte resistivity was homogeneous in the whole liquid film confined at the

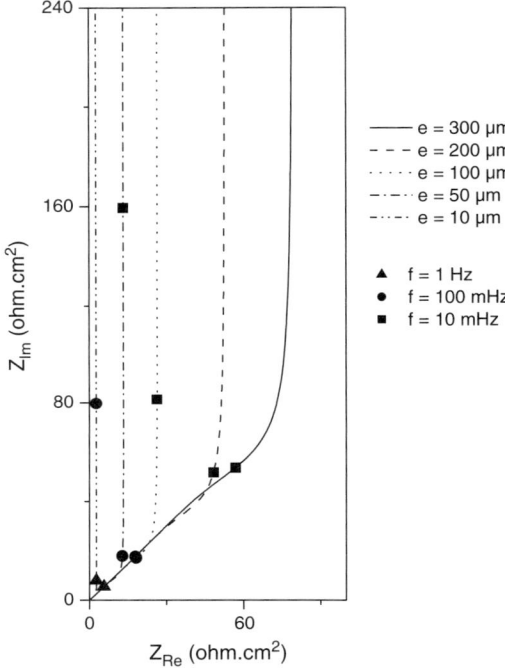

Figure 20. One-dimensional spatially restricted diffusion impedance (calculated according to (28) with $D = 6.3 \times 10^{-6}$ cm^2 s^{-1} and $R_d = 5 \times 10^{-2}\,\Omega\,\text{cm}^3\,\text{s}^{-1}$) with the diffusion layer thickness e as a parameter.

working electrode surface, and second that the interfacial impedance of the electrode Z_{int} $(\Omega\,\text{cm}^2)$ was homogeneous at the electrode surface (i.e., no reactivity or capacitance distribution exists at the electrode surface).

This model is very similar to that of de Levie,[66] which was also used by Keddam et al. for describing the impedance response of electrodes located in a cylindrical thin-layer cell.[62] However, in contrast to the previous work of Keddam et al.,[62] this approach takes into account the cylindrical geometry of the thin-layer cell. This was achieved by introducing the dependence on the radial coordinate of the ring-shaped electrode surface elements $dS_{\text{elec}}(r)$ (30) and on the

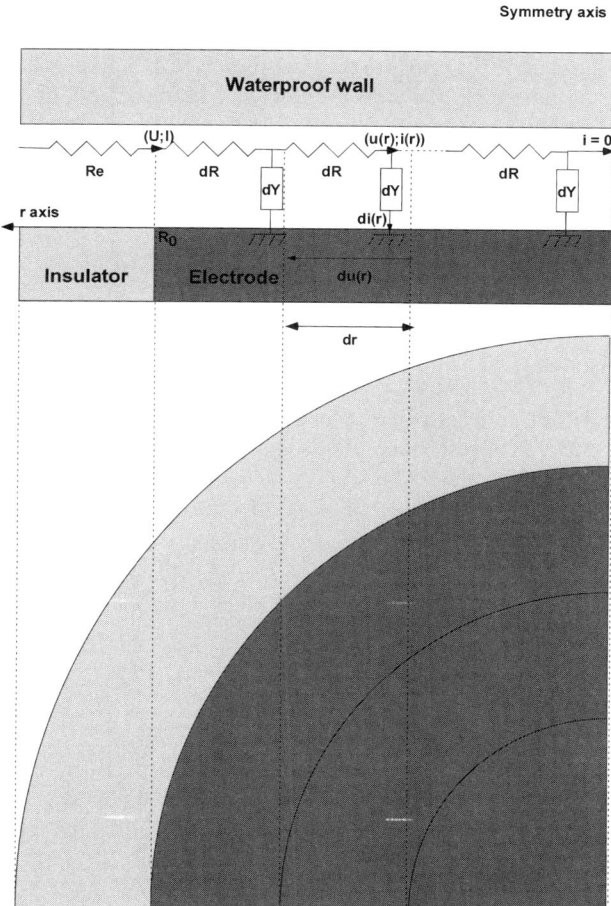

Figure 21. The transmission line model proposed to take into account the potential drop within the thin-layer cell (lateral view).

lateral surface of the cylindrical electrolyte film confined within the cell S_{lat} (31):

$$dS_{\text{elec}}(r) = 2\pi r\, dr, \qquad (30)$$
$$S_{\text{lat}}(r) = 2\pi r e. \qquad (31)$$

where r (cm) is the radial coordinate, $dS_{elec}(r)$ (cm^2) is a ring-shaped electrode surface element located at a distance r from the electrode center, S_{lat} (cm^2) is the lateral surface of a cylindrical electrolyte film element of radius r confined within the cell, dr (cm) is a length element following the radius of the electrode, and e (cm) is the thickness of the thin-layer cell.

Under these assumptions, one can express the elementary resistances and admittances (dR and dY) involved in the model (see Fig. 21) as

$$dR = R_0 \, dr, \tag{32}$$

$$dY = \frac{dr}{Z_0}, \tag{33}$$

where dR is the resistance of a length element of the liquid layer, R_0 is the electrolyte resistance, dY is the admittance of a ring-shaped electrode surface element $dS_{elec}(r)$, and Z_0 is the impedance of the interface for a radial unit length of electrode. Equations (32) and (33) can be rewritten as

$$dR = \frac{\rho}{dS_{lat}(r)} dr = \frac{\rho}{2\pi r e} dr, \tag{34}$$

$$dY = \frac{dS_{elec}(r)}{Z_{int}} = \frac{2\pi r \, dr}{Z_{int}}, \tag{35}$$

where ρ is the electrolyte resistivity.

The local currents I and di (A), flowing respectively along the radial coordinate through the electrolyte layer and normally through a ring-shaped electrode surface element, are linked by the following charge conservation law:

$$di(r) = I(r + dr) - I(r). \tag{36}$$

Equation (36) can be rewritten as

$$\frac{di(r)}{dr} = \frac{\partial}{\partial r}[I(r)]. \tag{37}$$

Otherwise, these currents can also be expressed using Ohm's law according to

$$I = \frac{du}{dR}, \tag{38}$$

$$di = u\,dY, \tag{39}$$

where u and du are the local potential within the thin-layer cell and the local ohmic drop in a length element dr of the liquid layer, respectively.

Using (34) and (35), one can rewrite (38) and (39) as

$$\frac{du}{dr} = \frac{\rho}{2\pi r e} I, \tag{40}$$

$$\frac{di}{dr} = \frac{2\pi r}{Z_{\text{int}}} u. \tag{41}$$

Using relations (37) and (41), the derivative of expression (40) vs. the radial coordinate leads to

$$\frac{d^2 u}{dr^2} - \frac{1}{r}\frac{du}{dr} - \frac{\rho}{e Z_{\text{int}}} u = 0. \tag{42}$$

This differential equation can be solved numerically; however, if the quantity $\frac{1}{r}\frac{du}{dr}$ is negligible, (42) becomes

$$\frac{d^2 u}{dr^2} - \frac{\rho}{e Z_{\text{int}}} u = 0. \tag{43}$$

This last equation can be solved analytically and the general solution is given by

$$u = A \exp(\alpha r) + B \exp(-\alpha r), \tag{44}$$

where A and B are two constants to be determined with the boundary conditions and $\alpha\ (\text{cm}^{-1})$ is equal to $\sqrt{\frac{\rho}{e Z_{\text{int}}}}$.

Otherwise, the cylindrical symmetry of the cell imposes the boundary condition (45):

$$\left[\frac{du}{dr}\right]_{(r=0)} = 0. \tag{45}$$

Thus, starting from (44), the boundary condition (45) leads to

$$A = B. \tag{46}$$

From (44) and (46), the local potential u can then be expressed according to
$$u = 2A \cosh(\alpha r). \tag{47}$$

By defining U as the potential at the edge of the electrode [$U = u(R_0)$] and I_0 as the global current flowing through the interface [$I_0 = I(R_0)$], one can express the impedance of the electrode $Z\,(\Omega\,\text{cm}^2)$ according to

$$Z = \frac{U}{I_0} \pi R_0^2. \tag{48}$$

Using (42), one can express the global current flowing through the interface as

$$I_0 = \int_0^{R_0} \mathrm{d}i = \int_0^{R_0} \frac{2\pi r}{Z_{\text{int}}} u\,\mathrm{d}r. \tag{49}$$

Using (47), one can express I_0 more explicitly as

$$I_0 = \frac{4\pi A}{Z_{\text{int}}} \left(\frac{R_0}{\alpha} \sinh(\alpha R_0) + \frac{1}{\alpha^2} [1 - \cosh(\alpha R_0)] \right). \tag{50}$$

Finally, the impedance of the electrode $Z\,(\Omega\,\text{cm}^2)$ is given by

$$Z = \frac{R_0^2 Z_{\text{int}}}{2\left[\frac{R_0}{\alpha} \tanh(\alpha R_0) + \frac{1}{\alpha^2}\left(\frac{1}{\cosh(\alpha R_0)} - 1\right) \right]}. \tag{51}$$

For e very large, thus in bulk solution, it is easy to verify that Z is equal to Z_{int}; and for e very small, Z is proportional to $\sqrt{Z_{\text{int}}}$, in agreement with the de Levie theory. Knowledge of the interfacial impedance Z_{int} is required for the calculation of the global impedance of the electrode, which is given by (51). In the present work, this interfacial impedance corresponds to a diffusion impedance and was calculated using the classical analytic expression of the one-dimensional spatially restricted diffusion impedance Z_{1D} given by (29).

To know the validity of (51), the differential equation (42) was solved numerically. The results obtained from the direct numerical integration and from (51) are given in Fig. 22, where only the phase shift, as a function of the dimensionless frequency

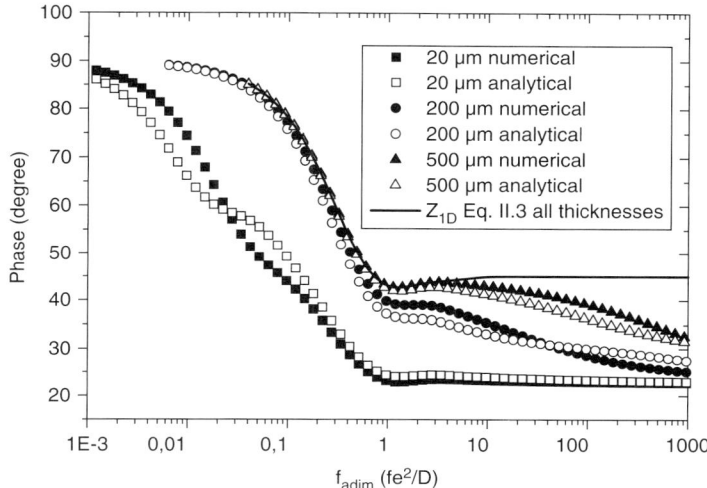

Figure 22. Bode representation of the theoretical impedance diagrams calculated with the transmission line model ((51) with Z_{int} calculated according to (29)). Comparison with the numerical integration of (42) is given. With $R_d = 0.05\,\Omega\,cm^2\,s^{-1}$, $D = 6.3 \times 10^{-6}\,cm^2\,s^{-1}$, $\rho = 18\,\Omega\,cm$, and different values of the thickness e.

$f_{a\,dim} = (fe^2)/D$, is plotted. The comparison with the theoretical one-dimensional spatially restricted diffusion impedance (29) is also given in Fig. 22, and shows that the transmission line model must be applied for each thickness. For small values of e, the phase shift tends towards 22.5°, which corresponds to half of the phase limit (45°) for the diffusion impedance (see (29)). As shown in Fig. 22, the analytical solution given by (51) is a relatively good approximation.

4. Results and Discussion

The experimental impedance diagrams measured in the thin-layer cell for relatively thin electrolyte films (thicknesses greater than 115 µm) and for very thin electrolyte films (thicknesses less than 115 µm) are presented in Nyquist coordinates in Figs. 23 and 24, respectively. In bulk conditions (electrolyte film thicknesses greater than several centimeters), the diagram obtained experimentally consists of a straight line exhibiting a phase angle very close to 45° (Fig. 23). This result is consistent with the assumption of a linear

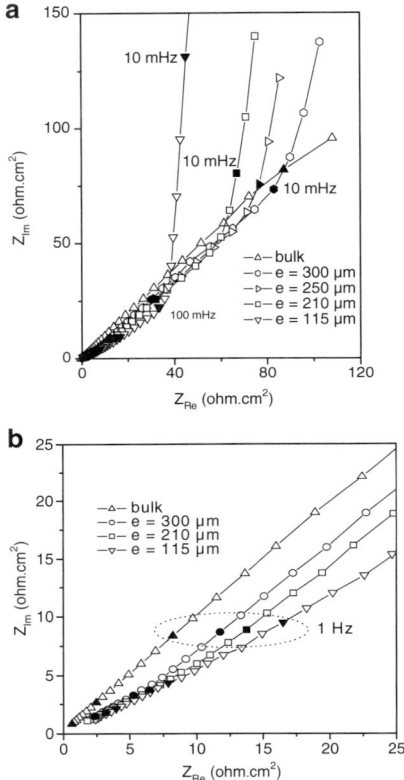

Figure 23. (a) Experimental impedance diagrams as a function of the electrolyte film thickness for $e \geq 115\,\mu m$; (b) magnification of the high-frequency range of the diagram.

mass transport normal to the electrode surface. Indeed, in bulk conditions, such behavior is well predicted by the classical one-dimensional diffusion model in the case of a semi-infinite diffusion process (Warburg impedance corresponding to e infinite in (29)).

In the whole range of thin electrolyte film thicknesses investigated (i.e., thicknesses equal to or less than $300\,\mu m$), the general shape of the experimental impedance diagrams was changed. As a first approximation, they can be described by a combination of a

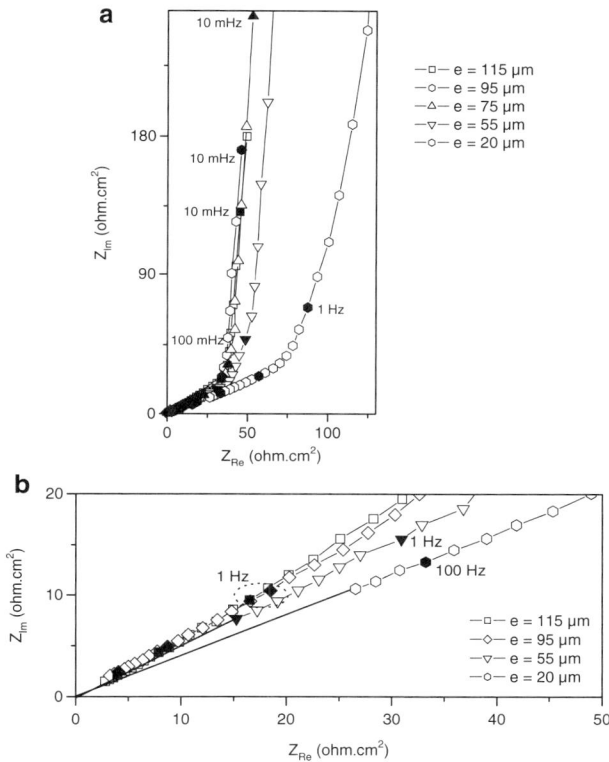

Figure 24. (**a**) Experimental impedance diagrams as a function of the electrolyte film thickness for $e \leq 115\,\mu m$; (**b**) magnification of the high-frequency range of the diagrams.

straight line in the high frequencies followed by a capacitive arc in the low-frequency domain, showing behavior quite similar to that predicted by one-dimensional spatially restricted diffusion models. However, several significant differences exist between the experimental (Figs. 23, 24) and the theoretical (Fig. 20) impedance diagrams calculated from (29). First, whatever the range of electrolyte film thicknesses, the one-dimensional spatially restricted diffusion model predicts a decrease of the low-frequency limit of the real part of the impedance as the electrolyte film thickness decreases (see Fig. 20). As shown in Fig. 23, this prediction was achieved experimentally only for relatively thick electrolyte films (thicknesses ranging from 95 to 300μm), but for smaller film thickness, the low-frequency limit of the real part of the impedance increases as the electrolyte film thickness decreases.

Indeed, as clearly illustrated by the high-frequency magnification of the experimental Nyquist plots in Figs. 23b and 24b, the phase angles decrease monotonically as the electrolyte film thickness decreases. Quantitatively, in the thin electrolyte film conditions investigated in this study ($e \leq 300\,\mu m$), the experimental values of the phase angle are systematically lower than 45° in the high frequencies. This result is in agreement with Fig. 22.

In the whole range of thicknesses investigated ($20\,\mu m \leq e \leq 300\,\mu m$), the theoretical phase angles calculated from (51) decrease in the high-frequency range ($f_{adim} > 1$) as the electrolyte film thickness decreases. This prediction from the transmission line model is also consistent with the experimental measurements. These results show the significant improvement provided by the transmission line model for the description of the diffusion impedance in cylindrical thin-layer cells involving large disk electrodes and demonstrate how this potential distribution could have a noteworthy impact on the impedance response in a thin-layer cell involving a large disk electrode.

The fit between experimental measurements and the one-dimensional spatially restricted modified diffusion impedance calculated from the transmission line model remains, however, imperfect. A possible explanation for the quantitative differences existing between the theoretical diagrams calculated from (51) and the measurements may be suggested by the conclusions of Gabrielli et al.[67] These authors showed that neglecting radial mass transport leads to inaccurate impedance calculations in the case of cylindrical thin-layer cells involving microelectrodes. The discrepancy between the measurements and the predictions from the transmission line model proposed in this work may be consequently explained by the use of a too strong approximation consisting in neglecting the radial diffusion processes in (51). Experimentally, the occurrence of such radial diffusion is, however, not obvious in our case. Indeed, Gabrielli et al. demonstrated that radial diffusion appears in the impedance measured at a microelectrode as an additional low-frequency contribution.[67] In the frequency range investigated in this work (100 kHz − 5 mHz), such a contribution was not clearly evidenced for large disk electrodes. A careful examination of the results reveals, however, that the purely capacitive behavior predicted theoretically at low frequency by both one-dimensional models previously discussed ((29) and (51)) was not achieved in this range of frequency. In fact, the experimental Nyquist diagrams (Figs. 23, 24) do not reach clearly the perfect vertical asymptotic behavior predicted by these models at low frequency (Fig. 20). Similarly, the phase angle of 90° predicted by both models was not observed in the frequency range investigated (Fig. 22).

To check the possible occurrence of a radial contribution in the impedance measured on large disk electrodes, some measurements were performed with a frequency limit as low as 100 µHz (Fig. 25). Moreover, radial mass transport was also taken into account in the transmission line model and theoretical impedance diagrams (Z_{2D}; also plotted in Fig. 25) were calculated numerically. Figure 25 shows an additional time constant on the experimental diagrams in the very low frequency range, which is evidenced by the existence of a maximum of the experimental phase angle in Bode representation. The value of this maximum increases as the electrolyte film thickness decreases, whereas the dimensionless frequency corresponding to this maximum increases as the electrolyte film thickness decreases. This low-frequency behavior remains totally unexplained if only a one-dimensional diffusion path is considered.

Diagrams calculated numerically from the two-dimensional model (Fig. 25) are also in very good accordance with the experimental results. In the low-frequency range, the additional time constant is totally predicted by the model, and for high frequencies, the transmission line model proposed in this paper allows one to explain the small (lower than 45°) and frequently distributed values of the phase angle. Thus, these results are consistent with the findings of previous work by Gabrielli et al., but the use of an electrode in the millimeter or centimeter range requires very low frequency (less than 100 µHz in certain cases) to allow the radial mass transport effect to be observed.[67]

5. Conclusions

The use of an accurate experimental device allowed the measurement of the electrochemical impedance corresponding to the diffusion of ferrocyanide ions on a large disk electrode in a cylindrical

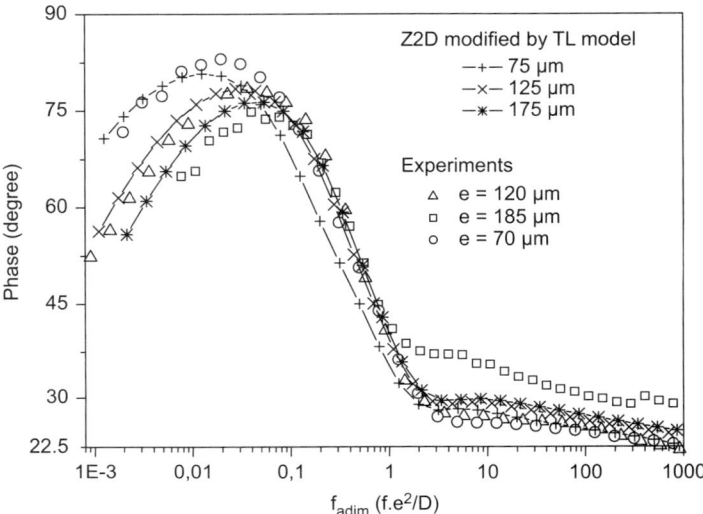

Figure 25. Bode representation of the theoretical impedance diagrams calculated with the transmission line model by taking into account radial mass transport ((50) with Z_{int} calculated numerically using an adaptation of the model of Gabrielli et al.,[67] $D = 6.3 \times 10^{-6}$ cm^2/s and $\rho = 18\,\Omega$ cm) and comparison with experimental measurements.

thin-layer cell configuration. Even if the diagrams obtained experimentally exhibit some strong similarities with the theoretical diagrams predicted by the classical spatially restricted linear diffusion impedance models, some significant differences between theory and experiments were evidenced. These differences consisted notably in a nonmonotonic evolution of the low-frequency limit of the real part of the experimental diagrams as well as small (less than 45°) values of the phase angle in the high frequencies. This behavior was ascribed to the potential distribution in the electrolyte film confined at the electrode surface due to ohmic drop. This effect was taken into account by developing a new transmission line model which accounts for the cylindrical geometry of the electrochemical cell. Moreover, an analytical derivation of the impedance was possible by neglecting the radial mass transport within the cell. Under this linear mass transport assumption, the model allowed qualitative explanation of the experimental results in a large frequency range.

However, to fit quantitatively the diagrams and to further explain the additional time constant evidenced experimentally in a very low frequency range (frequencies lower than 1 mHz), the two-dimensional character of the mass transport should be taken into account.

REFERENCES

[1] J.A. Collins and M.L. Monack, *Mater. Prot. Perform.* **12** (1973) 11.
[2] G. Engelhardt and D.D. Macdonald, *Corrosion* **54** (1998) 469.
[3] G. Engelhardt and D.D. Macdonald, *Corros. Sci.* **46** (2004) 2755.
[4] D.D. Macdonald, C. Liu, M. Urquidi-Macdonald, G.H. Stickford, B. Hindin, A.K. Agrawal and K. Krist, *Corrosion* **50** (1994) 761.
[5] A. Turnbull, L.N. McCartney and S. Zhou, *Corros. Sci.* **48** (2006) 2084.
[6] A. Anderko, N. Sridhar and D.S. Dunn, *Corros. Sci.* **46** (2004) 1583.
[7] D.S. Dunn, G.A. Cragnolino and N. Sridhar, *Corrosion* **56** (2000) 90.
[8] N. Sridhar and G.A. Cragnolino, *Corrosion* **49** (1993) 885.
[9] T.T. Lunt, J.R. Scully, V. Brusamarello, A.S. Mikhailov and J.L. Hudson, *J. Electrochem. Soc.* **149** (2002) B163.
[10] C. Punckt, M. Bolscher, H.H. Rotermund, A.S. Mikhailov, L. Organ, N. Budiansky, J.R. Scully and J.L. Hudson, *Science* **305** (2004) 1133.
[11] J. Newman and K.E. Thomas-Alyea, *Electrochemical Systems*, Third ed., Wiley, Hoboken, NJ, 2004.
[12] S.M. Sharland, *Corros. Sci.* **27** (1987) 289.
[13] A. Turnbull, Br. *Corros. J.* **28** (1993) 297.
[14] S.M. Sharland, C.P. Jackson and A.J. Diver, *Corros. Sci.* **29** (1989) 1149.
[15] S.M. Sharland and P.W. Tasker, *Corros. Sci.* **28** (1988) 603.
[16] A. Turnbull and M.K. Gardner, *Corros. Sci.* **22** (1982) 661.
[17] J.N. Harb and R.C. Alkire, *J. Electrochem. Soc.* **138** (1991) 3568.
[18] J.N. Harb and R.C. Alkire, *Corros. Sci.* **29** (1989) 31.
[19] M.L. Kronenberg, J.C. Banter, E. Yeager and F. Hovorka, *J. Electrochem. Soc.* **110** (1963) 1007.
[20] J.N. Harb and R.C. Alkire, *J. Electrochem. Soc.* **138** (1991) 2594.
[21] M.W. Verbrugge, D.R. Baker and J. Newman, *J. Electrochem. Soc.* **140** (1993) 2530.
[22] G. Engelhardt and H.H. Strehblow, *Corros. Sci.* **36** (1994) 1711.
[23] H.K. Kuiken, J.J. Kelly and P.H.L. Notten, *J. Electrochem. Soc.* **133** (1986) 1217.
[24] G. Engelhardt, M. Urquidi-Macdonald and D.D. Macdonald, *Corros. Sci.* **39** (1997) 419.
[25] G. Engelhardt and D.D. Macdonald, *Corros. Sci.* **46** (2004) 1159.
[26] M. Verhoff and R. Alkire, *J. Electrochem. Soc.* **147** (2000) 1349.
[27] M. Verhoff and R. Alkire, *J. Electrochem. Soc.* **147** (2000) 1359.
[28] E.G. Webb and R.C. Alkire, *J. Electrochem. Soc.* **149** (2002) B286.
[29] M. Kamrunnahar, R.D. Braatz and R.C. Alkire, *J. Electrochem. Soc.* **151** (2004) B90.
[30] J.R. Gray, C. Homescu, L.R. Petzold and R.C. Alkire, *J. Electrochem. Soc.* **152** (2005) B277.
[31] N.J. Laycock and S.P. White, *J. Electrochem. Soc.* **148** (2001) B264.
[32] N.J. Laycock, S.P. White, J.S. Noh, P.T. Wilson and R.C. Newman, *J. Electrochem. Soc.* **145** (1998) 1101.

[33] T. Hakkarainen, *Mater. Sci. Forum* **8** (1986) 81.
[34] G.T. Gaudet, W.T. Mo, T.A. Hatton, J.W. Tester, J. Tilly, H.S. Isaacs and R.C. Newman, *AIChE J.* **32** (1986) 949.
[35] T. Hakkarainen, in: A. Turnbull (Ed.), *Corrosion Chemistry within Pits, Crevices and Cracks*, Her Majesty's Stationery Office, London, 1987, p. 17.
[36] P.C. Pistorius and G.T. Burstein, *Philos. Trans. R. Soc. Lond., Ser. A* **341** (1992) 531.
[37] U. Steinsmo and H.S. Isaacs, *J. Electrochem. Soc.* **140** (1993) 643.
[38] R.S. Alwitt, H. Uchi, T.R. Beck and R.C. Alkire, *J. Electrochem. Soc.* **131** (1984) 13.
[39] D. Goad, *J. Electrochem. Soc.* **144** (1997) 1965.
[40] K.R. Hebert, *J. Electrochem. Soc.* **148** (2001) B236.
[41] Y.S. Tak and K.R. Hebert, *J. Electrochem. Soc.* **141** (1994) 1453.
[42] Y.S. Tak, E.R. Henderson and K.R. Hebert, *J. Electrochem. Soc.* **141** (1994) 1446.
[43] N. Sinha and K.R. Hebert, *J. Electrochem. Soc.* **147** (2000) 4111.
[44] Y. Tak, N. Sinha and K.R. Hebert, *J. Electrochem. Soc.* **147** (2000) 4103.
[45] K. Hebert and R. Alkire, *J. Electrochem. Soc.* **135** (1988) 2146.
[46] Y. Zhou and K.R. Hebert, *J. Electrochem. Soc.* **145** (1998) 3100.
[47] J.R. Galvele, *J. Electrochem. Soc.* **123** (1976) 464.
[48] K. Hebert and R. Alkire, *J. Electrochem. Soc.* **135** (1988) 2447.
[49] J.O.M. Bockris and A.K.N. Reddy, *Modern Electrochemistry*, Plenum, New York, 1977.
[50] K.R. Hebert, *Proc. – Electrochem. Soc.* **99–14** (1999) 54.
[51] R.A. Robinson and R.H. Stokes, *Electrolyte Solutions; The Measurement and Interpretation of Conductance, Chemical Potential, and Diffusion in Solutions of Simple Electrolytes*, 2nd ed., Butterworths, London, 1959.
[52] R.H. Perry and C.H. Chilton (Eds.), *Perry's Chemical Engineers' Handbook*, 5th ed., McGraw-Hill, New York, 1973.
[53] J. Newman, *J. Electrochem. Soc.* **113** (1966) 1235.
[54] J. Newman, *J. Electrochem. Soc.* **113** (1966) 501.
[55] J.B. Jorcin, C. Blanc, N. Pebere, B. Tribollet and V. Vivier, *J. Electrochem. Soc.* **155** (2008) C46.
[56] N. Dimitrov, J.A. Mann and K. Sieradzki, *J. Electrochem. Soc.* **146** (1999) 98.
[57] M.B. Vukmirovic, N. Dimitrov and K. Sieradzki, *J. Electrochem. Soc.* **149** (2002) B428.
[58] C.R. Christensen and F.C. Anson, *Anal. Chem.* **35** (1963) 205.
[59] A.T. Hubbard and F.C. Anson, *Anal. Chem.* **36** (1964) 723.
[60] A.T. Hubbard and F.C. Anson, *Anal. Chem.* **38** (1966) 58.
[61] A.T. Hubbard and F.C. Anson, in: A.J. Bard (Ed.), *Electroanalytical Chemistry*, Marcel Dekker, New York, 1970, pp. 129.
[62] C. Fiaud, M. Keddam, A. Kadri and H. Takenouti, *Electrochim. Acta* **32** (1987) 445.
[63] E. Remita, E. Sutter, B. Tribollet, F. Ropital, X. Longaygue, C. Taravel-Condat and N. Desamais, *Electrochim. Acta* **52** (2007) 7715.
[64] K. Micka, K. Kratochvilova and J. Klima, *Electrochim. Acta* **42** (1997) 1005.
[65] T. Jacobsen and K. West, *Electrochim. Acta* **40** (1995) 255.
[66] R. de Levie, in: P. Delahay (Ed.), *Advances in Electrochemistry and Electrochemical Engineering*, New York, Interscience, 1967, pp. 329.
[67] C. Gabrielli, M. Keddam, N. Portail, P. Rousseau, H. Takenouti and V. Vivier, *J. Phys. Chem. B* **110** (2006) 20478.

8

Density-Functional Theory in External Electric and Magnetic Fields

Ednilsom Orestes[1,2], Henrique J. P. Freire[2], and Klaus Capelle[2]

[1] *Departamento de Química e Física Molecular, Instituto de Química de São Carlos, Universidade de São Paulo, Caixa Postal 780, 13560-970 São Carlos, São Paulo, Brazil, eorestes@ifsc.usp.br*
[2] *Departamento de Física e Informática, Instituto de Física de São Carlos, Universidade de São Paulo, Caixa Postal 369, 13560-970 São Carlos, São Paulo, Brazil, freire@ifsc.usp.br; capelle@ifsc.usp.br*

I. SCOPE OF THIS CHAPTER

Density-functional theory (DFT) is one of the most widely used quantum mechanical approaches for calculating the structure and properties of matter on an atomic scale. It is nowadays routinely applied for calculating physical and chemical properties of molecules that are too large to be treatable by wave-function-based methods. The problem of determining the many-body wave function of a real system rapidly becomes prohibitively complex.[1] Methods such as configuration interaction (CI) expansions, coupled cluster (CC) techniques or Møller–Plesset (MP) perturbation theory thus become harder and harder to apply. Computational complexity here is related to questions such as how many atoms there are in the molecule, how many electrons each atom contributes, how many basis functions are required to adequately describe these electrons, how many competing minima there are in the potential-energy surface determining the molecular geometry, and whether any additional external fields are present. The description of the many-body wave function in CI, CC and MP techniques depends sensitively on these questions, and becomes very difficult for systems with more than a few electrons.

DFT shifts attention from the many-electron wave function to the charge density, which is much simpler to obtain and to interpret. In practice, this charge density is expressed in terms of single-particle orbitals, similar to those employed in the Hartree–Fock approximation, but the way these orbitals are obtained, used and interpreted is quite different. It is one aim of this chapter to describe these differences. Simple density functionals, such as the local-density approximation (LDA), are computationally cheaper even than the Hartree–Fock approximation, but typically provide more accurate total energies and related quantities. If the demands on accuracy are higher than can be satisfied by the LDA – as they normally are in quantum chemistry – beyond-LDA functionals such as gradient-dependent approximations or hybrid functionals are available. Today, generalized-gradient approximations (GGAs) and hybrids are standard tools for applications of DFT to molecular properties, such as geometries, vibration frequencies, binding energies and ionization energies. The dependence of such properties on external electric and magnetic fields is also accessible. The inclusion of such fields in the formalism of DFT is also described in this chapter.

As demands on accuracy increase, density functionals become more complex. Exact exchange, self-interaction corrections (SICs) or kinetic-energy densities are examples of ingredients of such more complex functionals. These, and some related approaches, are also briefly described here.

In spite of all its successes, DFT also suffers from severe limitations. As the systems become larger and larger, DFT, too, encounters computational limitations, and must yield to semiempirical modelling. At the other end of the spectrum, as demands on accuracy grow beyond that attained by today's best functionals (quantified below), wave-function-based methods such as CI, CC or MP reign supreme. Simultaneous progress in computing technology and construction of density functionals, however, has pushed both frontiers of DFT further and further out. As a consequence, many problems of interest in physics and chemistry, including in electrochemistry,[2–6] are now accessible via DFT.

Typical applications of DFT to electrochemical problems have been thoroughly reviewed in a previous volume of this series,[2] and a wealth of useful information on the performance of DFT in the calculation of electric polarizabilities and hyperpolarizabilities of atoms, molecules and clusters is contained in Ref. [3]. These sources, however, give only very brief descriptions of the formalism of DFT, and the nature of the different approximations used in a typical DFT calculation. The scope of the present chapter is complementary: we are very short on applications, limiting us to giving a few representative numbers for simple systems, but describe the conceptual structure of DFT, the theorems it is based on, and how these theorems can be turned into practical tools for solving chemical and physical problems.

The literature on DFT is large, and rich in excellent reviews and overviews. Some representative examples of full reviews and systematic collections of research papers are Refs. [7–21]. This chapter is much less detailed and advanced than those treatments. Introductions to DFT that are more similar in spirit to the present one (but differ in emphasis and selection of topics) are the contribution of M. Levy in Ref. [11, page 3], the one of S. Kurth and J. P. Perdew in Refs. [17 page 8, 18 page 1] that by Argaman and Makov [22] and that by one of the present authors.[23]

II. ELEMENTS OF THE QUANTUM MECHANICS OF MANY-ELECTRON SYSTEMS

In this section we recapitulate a few selected elements of the quantum mechanics of many-electron systems that are useful for understanding and using DFT.

1. Hamiltonians and Wave Functions

To get a first idea of what DFT is about, it is useful to take a step back and recall some elementary quantum mechanics. In quantum mechanics we learn that all information we can possibly obtain about a given system in a pure state is contained in the system's wave function, Ψ. Here we will exclusively be concerned with the electronic structure of atoms, molecules and solids. Within the Born–Oppenheimer approximation, the nuclear degrees of freedom appear only in the form of a potential $v(\mathbf{r})$ acting on the electrons, so the wave function depends only on the electronic coordinates. (Strictly speaking, $v(\mathbf{r})$ is a potential energy, not a potential as the word is used in electrostatics, but the terminology 'potential' is very common.)

Nonrelativistically, this wave function is calculated from Schrödinger's equation, which for a single electron moving in a potential $v(\mathbf{r})$ reads

$$\left[-\frac{\hbar^2 \nabla^2}{2m} + v(\mathbf{r})\right] \Psi(\mathbf{r}) = \epsilon \Psi(\mathbf{r}). \tag{1}$$

Both the Born–Oppenheimer approximation and the nonrelativistic approximation can be avoided by formulating the problem from the outset without them. In the former case, this leads to a combined DFT for electrons and nuclei;[24] in the latter case it results in relativistic DFT (RDFT).[20,25–27] Alternatively, relativistic or nonadiabatic corrections can be added to the standard formalism via perturbation theory.

If there is more than one electron (i.e. one has a many-body problem), Schrödinger's equation becomes

$$\left[\sum_i^N \left(-\frac{\hbar^2 \nabla_i^2}{2m} + v(\mathbf{r}_i)\right) + \sum_{i<j} U(\mathbf{r}_i, \mathbf{r}_j)\right] \Psi(\mathbf{r}_1, \mathbf{r}_2, \ldots, \mathbf{r}_N)$$
$$= E \Psi(\mathbf{r}_1, \mathbf{r}_2, \ldots, \mathbf{r}_N), \tag{2}$$

where N is the number of electrons and $U(\mathbf{r}_i, \mathbf{r}_j)$ is the electron–electron interaction. For a Coulomb system one has

$$\hat{U} = \sum_{i<j} U(\mathbf{r}_i, \mathbf{r}_j) = \sum_{i<j} \frac{q^2}{|\mathbf{r}_i - \mathbf{r}_j|}. \tag{3}$$

Note that this is the same operator for any system of particles interacting via the Coulomb interaction, just as the kinetic energy operator

$$\hat{T} = -\frac{\hbar^2}{2m} \sum_i \nabla_i^2 \tag{4}$$

is the same for any nonrelativistic system.

Whether the system under study is an atom, a molecule or a solid thus depends only on the potential $v(\mathbf{r}_i)$. For an atom, e.g.,

$$\hat{V} = \sum_i v(\mathbf{r}_i) = \sum_i \frac{Qq}{|\mathbf{r}_i - \mathbf{R}|}, \quad (5)$$

where $\mathbf{R}$ denotes the nuclear position, relative to some fixed coordinate system, and Q is nuclear charge. (Here and below we adopt the convention that in terms of the elementary charge $e > 0$ and the atomic number Z, the nuclear charge is $Q = Ze$ and the charge on the electron is $q = -e$.) For a molecule or a solid one has

$$\hat{V} = \sum_i v(\mathbf{r}_i) = \sum_{ik} \frac{Q_k q}{|\mathbf{r}_i - \mathbf{R}_k|}, \quad (6)$$

where the sum on k extends over all nuclei in the system, each with charge $Q_k = Z_k e$ and position $\mathbf{R}_k$. It is only the spatial arrangement of the $\mathbf{R}_k$ (together with the corresponding boundary conditions) that distinguishes, fundamentally, a molecule from a solid. For this reason one sometimes says that $\hat{T}$ and $\hat{U}$ are 'universal', while $\hat{V}$ is system-dependent, or 'nonuniversal'.

Similarly, it is only through the term $\hat{U}$ that the (essentially simple) single-body quantum mechanics of (1) differs from the extremely complex many-body problem posed by (2). These properties are built into DFT in a very fundamental way.

The usual quantum-mechanical approach to Schrödinger's equation (SE) can be summarized by the following sequence,

$$v(\mathbf{r}) \stackrel{SE}{\Longrightarrow} \Psi(\mathbf{r}_1, \mathbf{r}_2, \ldots, \mathbf{r}_N) \stackrel{\langle\Psi|\cdots|\Psi\rangle}{\Longrightarrow} \text{observables}, \quad (7)$$

i.e. one specifies the system by choosing $v(\mathbf{r})$, plugs it into Schrödinger's equation, solves that equation for the wave function Ψ and then calculates observables by taking expectation values of operators with this wave function. One among the observables that are calculated in this way is the particle density:

$$n(\mathbf{r}) = N \int d^3 r_2 \int d^3 r_3 \cdots \int d^3 r_N \, \Psi^*(\mathbf{r}, \mathbf{r}_2, \ldots, \mathbf{r}_N) \Psi(\mathbf{r}, \mathbf{r}_2, \ldots, \mathbf{r}_N). \quad (8)$$

Many powerful methods for solving Schrödinger's equation have been developed during decades of struggling with the many-body problem. In physics, e.g., one has diagrammatic perturbation theory (based on Feynman diagrams and Green's functions), while in chemistry one often uses CI methods, which are based on systematic expansion in Slater determinants. A host of more special techniques also exists. The problem with these methods is the great demand they place on one's computational resources: it is simply impossible to apply them efficiently to large and complex systems. Nobody has ever calculated the chemical properties of a 100-atom molecule with full CI, or the electronic structure of a real semiconductor using nothing but Green's functions.

A simple estimate of the computational complexity of this task is to imagine a real-space representation of Ψ on a mesh, in which each coordinate is discretized by using 20 mesh points (which is not very much). For N electrons, Ψ becomes a function of $3N$ coordinates (ignoring spin, and taking Ψ to be real), and 20^{3N} values are required to describe Ψ on the mesh. For $N = 10$ electrons, approximately 10^{39} data values are required. A CD-ROM disc holds approximately 700 MB. If each value is represented by 1 byte, one would thus require approximately 1.5×10^{30} discs, which, at an average thickness of 1.2 mm per CD, would result in a pile of CDs about 10 times the diameter of the known universe (20 billion lightyears).

The density $n(\mathbf{r})$ is a function of three coordinates, and requires 20^3 values on the same mesh, which conveniently fits on any standard storage medium. The same applies to single-particle orbitals, which by definition also depend on three coordinates (plus spin) only. N such orbitals, used to build the density, require $20^3 N$ values. (A CI calculation also employs unoccupied orbitals, and requires more values.) For $N = 10$ electrons, the many-body wave function thus requires $20^{30}/20^3 \approx 10^{35}$ times more storage space than the density, and $20^{30}/(10 \times 20^3) \approx 10^{34}$ times more than sets of single-particle orbitals. Clever use of symmetries can reduce these ratios, but the full many-body wave function remains, in practice, unaccessible for real systems with more than a few electrons.

It is here where DFT provides a viable alternative. DFT explicitly recognizes that nonrelativistic Coulomb systems differ only by their potential $v(\mathbf{r})$, and supplies a prescription for dealing with the universal operators $\hat{T}$ and $\hat{U}$ once and for all. We will see that in practice this prescription can be implemented only approximately.

Still, these approximations retain a high degree of universality in the sense that they often work well for more than one type of system. For this reason, the density-functional approach forms the basis of the large majority of electronic-structure calculations in physics and chemistry. Much of what we know about the electric, magnetic and structural properties of materials has been calculated using DFT, and the extent to which DFT has contributed to the molecular sciences is reflected by the 1998 Nobel Prize in Chemistry, which was awarded to Walter Kohn,[1] the founding father of DFT, and John Pople,[28] who was instrumental in implementing DFT in computational chemistry.

The density-functional approach can be summarized by the sequence[7,29]

$$n(\mathbf{r}) \Longrightarrow \Psi(\mathbf{r}_1, \ldots, \mathbf{r}_N) \Longrightarrow v(\mathbf{r}), \tag{9}$$

i.e. knowledge of $n(\mathbf{r})$ implies knowledge of the wave function and the potential, and hence of all other observables. We stress that, contrary to what is sometimes claimed in the literature, the sequence of mappings in DFT is not $n(\mathbf{r}) \Longrightarrow v(\mathbf{r}) \Longrightarrow \Psi(\mathbf{r}_1, \ldots, \mathbf{r}_N)$. All proofs of the Hohenberg–Kohn theorem prove that the wave function is a density functional. The properties of the potential follow in a second step. This issue is addressed in more detail in Sect. III.1.

Although sequence (9) describes the logical structure of DFT, it does not really represent what is done in actual applications of it, which typically proceed along rather different lines, and do not make explicit use of many-body wave functions. The rest of this chapter attempts to explain both the conceptual structure and some of the many possible shapes and disguises under which this structure appears in applications.

2. Density Matrices and Density Functionals

It is a fundamental postulate of quantum mechanics that the wave function contains all possible information about a system in a pure state at zero temperature, whereas at nonzero temperature, or in a general mixed state, this information is contained in the density matrix of quantum statistical mechanics. Normally, this is much more information that one can handle: for a system with $N = 100$ particles the many-body wave function is an extremely complicated function of 300 spatial and 100 spin variables that would be impossible to manipulate algebraically or to extract any information from, even if it were possible to calculate it in the first place. For this reason one searches for less complicated objects to formulate the theory. Such objects should contain the experimentally relevant information, such as energies and densities, but do not need to contain explicit information about the coordinates of every single particle. One class of such objects are *reduced density matrices*.

For a general quantum system at temperature T, the density operator in a canonical ensemble is defined as

$$\hat{\gamma} = \frac{\exp(-\beta \hat{H})}{\text{Tr}[\exp(-\beta \hat{H})]}, \quad (10)$$

where $\text{Tr}[\cdot]$ is the trace and $\beta = 1/(k_B T)$. Standard textbooks on statistical physics show how this operator is obtained in other ensembles, and how it is used to calculate thermal and quantum expectation values. Here we focus on the relation to DFT. To this end we write $\hat{\gamma}$ in the energy representation as

$$\hat{\gamma} = \frac{\sum_i \exp(-\beta E_i) |\Psi_i\rangle \langle \Psi_i|}{\sum_i \exp(-\beta E_i)}, \quad (11)$$

where $|\Psi_i\rangle$ is eigenfunction of $\hat{H}$, and the sum is over the entire spectrum of the system, each state being weighted by its Boltzmann weight $\exp^{-\beta E_i}$. At zero temperature only the ground-state contributes to the sums, so that

$$\hat{\gamma} = |\Psi_i\rangle \langle \Psi_i|. \quad (12)$$

The coordinate-space matrix element of this operator for an N-particle system is

$$\langle x_1, x_2, \ldots, x_N | \hat{\gamma} | x'_1, x'_2, \ldots, x'_N \rangle$$
$$= \Psi(x_1, x_2, \ldots, x_N)^* \Psi(x'_1, x'_2, \ldots, x'_N)$$
$$=: \gamma(x_1, x_2, \ldots, x_N; x'_1, x'_2, \ldots, x'_N), \quad (13)$$

which shows the connection between the density matrix and the wave function. (We use the usual abbreviation $x = \mathbf{r}s$ for space and spin coordinates.) The expectation value of a general N-particle operator $\hat{O}$ is obtained from $O =$

Density-Functional Theory in External Electric and Magnetic Fields 349

$$\langle \hat{O} \rangle = \int dx_1 \int dx_2 \cdots \int dx_N \Psi(x_1, x_2, \ldots, x_N)^* \hat{O} \, \Psi(x_1, x_2, \ldots, x_N),$$

which for multiplicative operators becomes

$$\langle \hat{O} \rangle = \int dx_1 \int dx_2 \cdots \int dx_N \, \hat{O} \, \gamma(x_1, x_2, \ldots, x_N; x_1, x_2, \ldots, x_N) \quad (14)$$

and involves only the function $\gamma(x_1, x_2, \ldots, x_N; x_1, x_2, \ldots, x_N)$, which is the diagonal element of the matrix γ. Most operators we encounter in quantum mechanics are one- or two-particle operators and can be calculated from reduced density matrices that depend on fewer than $2N$ variables. Note that in this context, expressions such as 'two-particle operator' and 'two-particle density matrix' refer to the number of particles involved in the definition of the operator (two in the case of an interaction, one for a potential energy, etc.), not to the total number of particles present in the system.

The reduced two-particle density matrix is defined as

$$\gamma_2(x_1, x_2; x_1', x_2')$$
$$= \frac{N(N-1)}{2} \int dx_3 \int dx_4 \cdots \int dx_N \gamma(x_1, x_2, x_3, x_4, \ldots, x_N;$$
$$x_1', x_2', x_3, x_4, \ldots, x_N), \quad (15)$$

where $N(N-1)/2$ is a convenient normalization factor. This density matrix determines the expectation value of the particle–particle interaction, of static correlation and response functions, of the exchange and correlation holes, and related quantities. The pair-correlation function $g(x, x')$, e.g., is obtained from the diagonal element of $\gamma_2(x_1, x_2; x_1', x_2')$ according to $\gamma_2(x_1, x_2; x_1, x_2) =: n(x_1)n(x_2)g(x, x')$.

Similarly, the single-particle density matrix is defined as

$$\gamma(x_1, x_1')$$
$$= N \int dx_2 \int dx_3 \int dx_4 \cdots \int dx_N \, \gamma(x_1, x_2, x_3, x_4, \ldots, x_N;$$
$$x_1', x_2, x_3, x_4, \ldots, x_N)$$
$$= N \int dx_2 \int dx_3 \int dx_4 \cdots \int dx_N \, \Psi^*(x_1, x_2, x_3, \ldots, x_N)$$
$$\times \Psi(x_1', x_2, x_3, \ldots, x_N). \quad (16)$$

The structure of reduced density matrices is quite simple: all coordinates that γ does not depend upon are set equal in Ψ and Ψ^*, and integrated over.

In the special case that the wave function Ψ is a Slater determinant, i.e. the wave function of N noninteracting fermions, the single-particle density matrix can be written in terms of the orbitals comprising the determinant, as

$$\gamma(x, x') = \sum_j \phi_j^*(x)\phi_j(x'), \tag{17}$$

which is known as the Dirac (or Dirac–Fock) density matrix.

The usefulness of the single-particle density matrix becomes apparent when we consider how one would calculate the expectation value of a multiplicative single-particle operator $\hat{A} = \sum_i^N a(\mathbf{r}_i)$, such as the potential $\hat{V} = \sum_i^N v(\mathbf{r}_i)$:

$$\langle \hat{A} \rangle = \int dx_1 \cdots \int dx_N \, \Psi^*(x_1, x_2, \ldots, x_N) \left[\sum_i^N a(x_i) \right] \Psi(x_1, x_2, \ldots, x_N)$$

$$= N \int dx_1 \cdots \int dx_N \, \Psi^*(x_1, x_2, \ldots, x_N) \, a(x_1) \, \Psi(x_1, x_2, \ldots, x_N)$$

$$= \int dx \, a(x) \, \gamma(x, x), \tag{18}$$

which is a special case of (14). The second line follows from the first by exploiting the property that the fermionic wave function Ψ changes sign upon interchange of two of its arguments. The last equation implies that if one knows $\gamma(x, x)$, one can calculate the expectation value of any multiplicative single-particle operator in terms of it, regardless of the number of particles present in the system. For nonmultiplicative single-particle operators (such as the kinetic energy, which contains a derivative) one requires the full single-particle matrix $\gamma(x, x')$ and not only $\gamma(x, x)$. The simplification is enormous, and reduced density matrices are very popular in, e.g., computational chemistry for precisely this reason. More details are given in, e.g., Ref. [8]. The full density operator, (10), on the other hand, is the central quantity of quantum statistical mechanics.

It is not possible to express expectation values of two-particle operators, such as the interaction itself, or the full Hamiltonian (i.e. the total energy), explicitly in terms of the single-particle density matrix $\gamma(\mathbf{r}, \mathbf{r}')$. For this purpose one requires the two-particle density matrix.

Apparently even less information is contained in the particle density $n(\mathbf{r})$, which is obtained by summing the diagonal element of $\gamma(x, x')$ over the spin variable,

$$n(\mathbf{r}) = \sum_s \gamma(\mathbf{r}s, \mathbf{r}s). \tag{19}$$

This equation follows immediately from comparing (8) with (16). We can define an alternative density operator, $\hat{n}$, by requiring that the same equation must also be obtained by substituting $\hat{n}(\mathbf{r})$ into (18), which holds for any single-particle operator. This requirement implies that $\hat{n}(\mathbf{r}) = \sum_i^N \delta(\mathbf{r} - \mathbf{r}_i)$. The expectation value of $\hat{n}$ is the particle density, and therefore $\hat{n}$ is often also called the 'density operator'. This concept must not be confused with any of the various density matrices or the density operator of statistical physics, (10).

The particle density is an even simpler function than $\gamma(x, x')$: it depends on one set of coordinates x only, it can easily be visualized as a three-dimensional charge distribution and it is directly accessible in experiments. These advantages, however, seem to be more than compensated for by the fact that one has integrated out an enormous amount of specific information about the system in going from wave functions to density.

The great surprise of DFT is that in fact no information has been lost at all, at least as long as one considers the system only in its ground state: according to the Hohenberg–Kohn theorem,[29] which is discussed in detail in Sect. III, the ground-state density $n(x)$ completely determines the ground-state wave function $\Psi(x_1, x_2, \ldots, x_N)$. The Runge–Gross theorem, which forms the basis of time-dependent DFT (TD-DFT),[30] similarly guarantees that the time-dependent density contains the same information as the time-dependent wave function.

Hence, in the ground state, the function of one variable $n(\mathbf{r})$ is equivalent to the function of N variables $\Psi(x_1, x_2, \ldots, x_N)$. This property shows that we have only integrated out *explicit* information on our way from wave functions via density matrices to densities. *Implicitly* all the information that was contained in the ground-state wave function is still contained in the ground-state density. Part of the art of practical DFT is how to get this implicit information out, once one has obtained the density!

3. Functionals and Their Derivatives

Since according to (9) the wave function is determined by the density, we can write it as $\Psi = \Psi[n](\mathbf{r}_1, \mathbf{r}_2, \ldots, \mathbf{r}_N)$, which indicates that Ψ is a function of its N spatial variables, but a *functional* of the function $n(\mathbf{r})$. In principle, therefore, observables in DFT are calculated, explicitly or implicitly, as functionals of the ground-state density.

Functionals

Functionals are a most useful mathematical tool. In general, a functional $F[n]$ can be defined as a rule for going from a function to a number, just as a function $y = f(x)$ is a rule (f) for going from a number (x) to a number (y). A simple example of a functional is the particle number,

$$N = \int d^3r \, n(\mathbf{r}) = N[n], \tag{20}$$

which is a rule for obtaining the number N, given the function $n(\mathbf{r})$. Note that the name given to the argument of n is completely irrelevant, since the functional depends on the *function* itself, not on its variable. Hence, we do not need to distinguish $F[n(\mathbf{r})]$ from, e.g., $F[n(\mathbf{r}')]$. Another important case is that in which the functional depends on a parameter, such as in

$$v_H[n](\mathbf{r}) = q^2 \int d^3r' \, \frac{n(\mathbf{r}')}{|\mathbf{r} - \mathbf{r}'|}, \tag{21}$$

which is a rule that for any value of the parameter $\mathbf{r}$ associates a value $v_H[n](\mathbf{r})$ with the function $n(\mathbf{r}')$. Equation (21) is the so-called Hartree potential, which we will repeatedly encounter below.

Functional Derivative

Given a function of one variable, $y = f(x)$, one can think of two types of variations of y, one associated with x, the other with f. For a fixed functional dependence $f(x)$, the ordinary differential dy measures how y changes as a result of a variation $x \to x + dx$ of the variable x. This is the variation studied in ordinary calculus.

Similarly, for a fixed point x, the functional variation δy measures how the value y at this point changes as a result of a variation in the functional form $f(x)$. This is the variation studied in variational calculus.

The derivative formed in terms of the ordinary differential, df/dx, measures the first-order change of $y = f(x)$ upon changes of x, i.e. the slope of the function $f(x)$ at x:

$$f(x + dx) = f(x) + \frac{df}{dx}dx + O(dx^2). \qquad (22)$$

The functional derivative measures, similarly, the first-order change in a functional upon a functional variation of its argument:

$$F[f(x)+\delta f(x)] = F[f(x)] + \int \frac{\delta F[f]}{\delta f(x)} \delta f(x)\, dx + O(\delta f^2), \qquad (23)$$

where the integral arises because the variation in the functional F is determined by variations in the function at all points in space. The first-order coefficient (or 'functional slope') is defined to be the functional derivative $\delta F[f]/\delta f(x)$.

The functional derivative allows us to study how a functional changes upon changes in the form of the function it depends on. A general expression for obtaining functional derivatives of a functional $F[n] = \int f(n, n', n'', n''', \ldots; x) dx$ with respect to $n(x)$ is[8]

$$\frac{\delta F[n]}{\delta n(x)} = \frac{\partial f}{\partial n} - \frac{d}{dx}\frac{\partial f}{\partial n'} + \frac{d^2}{dx^2}\frac{\partial f}{\partial n''} - \frac{d^3}{dx^3}\frac{\partial f}{\partial n'''} + \cdots. \qquad (24)$$

This expression is frequently used in DFT to obtain exchange–correlation potentials from exchange–correlation energies.

III. THE HOHENBERG–KOHN THEOREM

1. Enunciation and Discussion of the Hohenberg–Kohn Theorem

At the heart of DFT is the Hohenberg–Kohn theorem,[29] which states that for ground states (8) can be inverted: given a *ground-state* density $n_0(\mathbf{r})$ it is possible, in principle, to calculate the corresponding *ground-state* wave function $\Psi_0(\mathbf{r}_1, \mathbf{r}_2, \ldots, \mathbf{r}_N)$. This means that Ψ_0 is a functional of n_0. Consequently, all ground-state observables are functionals of n_0 too. If Ψ_0 can be calculated from n_0 and vice versa, both functions are equivalent and contain exactly the same information. At first sight this seems impossible: how can a function of one (vectorial) variable $\mathbf{r}$ be equivalent to a function of N (vectorial) variables $\mathbf{r}_1 \ldots \mathbf{r}_N$? How can one arbitrary variable contain the same information as N arbitrary variables?

The crucial fact which makes this possible is that knowledge of $n_0(\mathbf{r})$ implies implicit knowledge of much more than that of an arbitrary function $f(\mathbf{r})$. The ground-state wave function Ψ_0 must not only reproduce the ground-state density, but it must also minimize the energy. For a given ground-state density $n_0(\mathbf{r})$, we can write this requirement as

$$E_{v,0} = \min_{\Psi \to n_0} \langle \Psi | \hat{T} + \hat{U} + \hat{V} | \Psi \rangle, \qquad (25)$$

where $E_{v,0}$ denotes the ground-state energy in potential $v(\mathbf{r})$. The preceding equation tells us that, for a given density $n_0(\mathbf{r})$, the ground-state wave function Ψ_0 is that which reproduces this $n_0(\mathbf{r})$ *and* minimizes the energy.

For an arbitrary density $n(\mathbf{r})$, we define the functional

$$E_v[n] = \min_{\Psi \to n} \langle \Psi | \hat{T} + \hat{U} + \hat{V} | \Psi \rangle. \qquad (26)$$

If n is a density different from the ground-state density n_0 in potential $v(\mathbf{r})$, then the Ψ that produce this n are different from the ground-state wave function Ψ_0, and according to the variational principle the minimum obtained from $E_v[n]$ is higher than (or equal to) the ground-state energy $E_{v,0} = E_v[n_0]$. Thus, the functional $E_v[n]$ is minimized by the ground-state density n_0, and its value at the minimum is $E_{v,0}$.

The total-energy functional can be written as

$$E_v[n] = \min_{\Psi \to n} \langle \Psi | \hat{T} + \hat{U} | \Psi \rangle + \int d^3r\, n(\mathbf{r}) v(\mathbf{r}) =: F[n] + V[n], \qquad (27)$$

where the internal-energy functional $F[n] = \min_{\Psi \to n} \langle \Psi | \hat{T} + \hat{U} | \Psi \rangle$ is independent of the potential $v(\mathbf{r})$, and is thus determined only by the structure of the operators $\hat{U}$ and $\hat{T}$. This universality of the internal-energy functional allows us to define the ground-state wave function Ψ_0 as that antisymmetric N-particle function that delivers the minimum of $F[n]$ *and* reproduces n_0. If the ground state is nondegenerate, this double requirement uniquely determines Ψ_0 in terms of $n_0(\mathbf{r})$, without our having to specify $v(\mathbf{r})$ explicitly. Note that this is exactly the opposite of the conventional prescription of specifying the Hamiltonian via $v(\mathbf{r})$, and obtaining Ψ_0 from solving Schrödinger's equation, without our having to specify $n(\mathbf{r})$ explicitly.

Equations (25)–(27) constitute the constrained-search proof of the Hohenberg–Kohn theorem, given independently by Levy[31] and Lieb.[32] The original proof by Hohenberg and Kohn[29] proceeded by assuming that Ψ_0 was not determined uniquely by n_0 and showed that this produced a contradiction to the variational principle. Both proofs, by constrained search and by contradiction, are elegant and simple. In fact, it is a bit surprising that it took 38 years from Schrödinger's first papers on quantum mechanics[33] to Hohenberg and Kohn's 1964 paper containing their famous theorem.[29]

For future reference we now provide a commented summary of the content of the Hohenberg–Kohn theorem. This summary consists of four statements:

(1) The nondegenerate ground-state wave function is a unique functional of the ground-state density:

$$\Psi_0(\mathbf{r}_1, \mathbf{r}_2, \ldots, \mathbf{r}_N) = \Psi[n_0(\mathbf{r})]. \qquad (28)$$

This is the essence of the Hohenberg–Kohn theorem. As a consequence, the ground-state expectation value of any observable $\hat{O}$ is a functional of $n_0(\mathbf{r})$ too:

$$O_0 = O[n_0] = \langle \Psi[n_0] | \hat{O} | \Psi[n_0] \rangle. \qquad (29)$$

This equation is sometimes called the '*first Hohenberg–Kohn theorem*'.

(2) Perhaps the most important observable is the ground-state energy. This energy

$$E_{v,0} = E_v[n_0] = \langle \Psi[n_0] | \hat{H} | \Psi[n_0] \rangle, \qquad (30)$$

where $\hat{H} = \hat{T} + \hat{U} + \hat{V}$, has the variational property

$$E_v[n_0] \leq E_v[n'], \tag{31}$$

where n_0 is the ground-state density in potential $\hat{V}$ and n' is some other density. Equation (31) is so important for practical applications of DFT that it is sometimes called the '*second Hohenberg–Kohn theorem*'.

If the ground state is degenerate, several of the degenerate ground-state wave functions may produce the same density, so a unique functional $\Psi[n]$ does not exist, but by definition these wave functions all yield the same energy, so the functional $E_v[n]$ continues to exist and to be minimized by n_0. A universal functional $F[n]$ can also still be defined.[7,34]

In performing the minimization of $E_v[n]$, one takes into account the constraint that the total particle number N is an integer by means of a Lagrange multiplier, replacing the constrained minimization of $E_v[n]$ by an unconstrained one of $E_v[n] - \mu N$. Since $N = \int d^3 r\, n(\mathbf{r})$, this leads to

$$\frac{\delta E_v[n]}{\delta n(\mathbf{r})} = \mu = \frac{\partial E}{\partial N}, \tag{32}$$

where μ is the chemical potential.

(3) Recalling that the kinetic and interaction energies of a nonrelativistic Coulomb system are described by universal operators, we can also write E_v as

$$E_v[n] = T[n] + U[n] + V[n] = F[n] + V[n], \tag{33}$$

where $T[n]$ and $U[n]$ are *universal functionals* (defined as expectation values of the type (29) of $\hat{T}$ and $\hat{U}$), independent of $v(\mathbf{r})$.

On the other hand, the potential energy in a given potential $v(\mathbf{r})$ is the expectation value of (6),

$$V[n] = \int d^3 r\, n(\mathbf{r}) v(\mathbf{r}), \tag{34}$$

and is obviously nonuniversal (it depends on $v(\mathbf{r})$, i.e. on the system under study), but very simple: once the system has been specified, i.e. $v(\mathbf{r})$ is known, the functional $V[n]$ is known explicitly.

(4) There is a fourth substatement to the Hohenberg–Kohn theorem, which shows that if $v(\mathbf{r})$ is not held fixed, the functional $V[n]$ becomes universal: the ground-state density determines not only the ground-state wave function Ψ_0, but, up to an additive constant, also the potential $V = V[n_0]$. This is simply proven by writing Schrödinger's equation as

$$\hat{V} = \sum_i v(\mathbf{r}_i) = E_k - \frac{(\hat{T} + \hat{U})\Psi_k}{\Psi_k}, \qquad (35)$$

which shows that any eigenstate Ψ_k (and thus, in particular, the ground state $\Psi_0 = \Psi[n_0]$) determines the potential operator $\hat{V}$ up to an additive constant, the corresponding eigenenergy. As a consequence, the explicit reference to the potential v in the energy functional $E_v[n]$ is not necessary, and one can rewrite the ground-state energy as

$$E_0 = E[n_0] = \langle \Psi[n_0]|\hat{T} + \hat{U} + \hat{V}[n_0]|\Psi[n_0]\rangle. \qquad (36)$$

Another consequence is that n_0 now determines not only the ground-state wave function but also the complete Hamiltonian (the operators $\hat{T}$ and $\hat{U}$ are fixed), and thus all excited states too:

$$\Psi_k(\mathbf{r}_1, \mathbf{r}_2, \ldots, \mathbf{r}_N) = \Psi_k[n_0], \qquad (37)$$

where k labels the entire spectrum of the many-body Hamiltonian $\hat{H}$.

Originally the fourth statement was considered to be as sound as the other three. However, it has become clear very recently, as a consequence of work of Eschrig and Pickett[35] and of Capelle and Vignale,[34,36,37] that there are significant exceptions to it. In fact, the fourth substatement holds only when one formulates DFT exclusively in terms of the charge density, as we have done up to this point. It does not hold when one works with spin densities (spin-density-functional theory, SDFT) or current densities (current-density-functional theory, CDFT). (In Sect. VII we will discuss these formulations of DFT.) In these (and some other) cases the densities still determine the wave function, but they do not uniquely determine the corresponding potentials. This so-called *nonuniqueness problem* has been discovered only recently, and its consequences are now beginning to be explored.[34–45] It is clear, however, that the fourth substatement is, from a practical point of view, the least important of the four, and most applications of DFT do not have to be reconsidered as a consequence of its eventual failure. (But some do: see Refs. [34, 36, 37, 45] for examples.)

Another conceptual problem with the Hohenberg–Kohn theorem, much better known and more studied than nonuniqueness, is representability. To understand what representability is about, consider the following two questions: (1) How does one know, given an arbitrary function $n(\mathbf{r})$, that this function can be represented in the form (8), i.e. that it is a density arising from an antisymmetric N-body wave function $\Psi(\mathbf{r}_1, \mathbf{r}_2, \ldots, \mathbf{r}_N)$? (2) How does one know, given a function that can be written in the form (8), that this density is a ground-state density of a local potential $v(\mathbf{r})$? The first of these questions is known as the N-representability problem; the second is known as the v-representability problem. Note that these are quite important questions: if one should find, e.g., in a numerical calculation, a minimum of $E_v[n]$ that is not N-representable, then this minimum is not the physically acceptable solution to the problem at hand. Luckily, the N-representability problem of the single-particle density has been solved: any nonnegative function can be written in terms of some antisymmetric $\Psi(\mathbf{r}_1, \mathbf{r}_2, \ldots, \mathbf{r}_N)$ in the form (8).[46, 47]

No similarly general solution is known for the v-representability problem: The Hohenberg–Kohn theorem only guarantees that there cannot be *more than one* potential for each density, but does not exclude the possibility that there is *less than one*, i.e. zero, potentials capable of producing that density. However, it is known that in *discretized* systems every density is *ensemble* v-representable, which means that a local potential with a degenerate ground state can always be found, such that the density $n(\mathbf{r})$ can be written as a linear combination of the densities arising from each of the degenerate ground states.[48–50]

It is not clear if one of the two restrictions ('discretized systems' and 'ensemble') can be relaxed in general (yielding 'in continuum systems' and 'pure state', respectively), but it is known that one may not relax both: there are densities in continuum systems that are not representable by a single nondegenerate antisymmetric function that is the ground state of a local potential $v(\mathbf{r})$.[7, 48–50] In any case, the constrained search algorithm of Levy and Lieb shows that v-representability in an interacting system is not required for the proof of the Hohenberg–Kohn theorem. The related question of simultaneous v-representability in a noninteracting system, which appears in the context of the Kohn–Sham formulation of DFT, is discussed below (61).

2. A Simple Example: Thomas–Fermi Theory

After these abstract considerations, let us now consider one way in which one can make practical use of DFT. Assume we have specified our system (i.e. $v(\mathbf{r})$ is known). Assume further that we have reliable approximations for $U[n]$ and $T[n]$. In principle, all one then has to do is to minimize the sum of kinetic, interaction and potential energies

$$E_v[n] = T[n]+U[n]+V[n] = T[n]+U[n]+\int d^3r\, n(\mathbf{r})v(\mathbf{r}) \quad (38)$$

with respect to $n(\mathbf{r})$. Wave functions are not required in the process. The minimizing function $n_0(\mathbf{r})$ is the system's ground-state charge density and the value $E_{v,0} = E_v[n_0]$ is the ground-state energy. Assume now that $v(\mathbf{r})$ depends on a parameter a. This can be, e.g., the lattice constant in a solid or the angle between two atoms in a molecule. Calculation of $E_{v,0}$ for many values of a allows one to plot the curve $E_{v,0}(a)$ and to find the value of a that minimizes it. This value, a_0, is the ground-state lattice constant or angle. In this way one can calculate quantities such as molecular geometries and sizes, lattice constants, unit cell volumes, charge distributions and total energies. By looking at the change of $E_{v,0}(a)$ with a, one can, moreover, calculate compressibilities, phonon spectra and bulk moduli (in solids) and vibrational frequencies (in molecules). By comparing the total energy of a composite system (e.g. a molecule) with that of its constituent systems (e.g. individual atoms), one obtains dissociation energies. By calculating the total energy for systems with one electron more or less, one obtains electron affinities and ionization energies. By appealing to the Hellman–Feynman theorem, one can calculate forces on atoms from the derivative of the total energy with respect to the nuclear coordinates. All this follows from DFT without our having to solve the many-body Schrödinger equation and without our having to make a single-particle approximation. For comments on the most useful additional possibility to also calculate single-particle band structures of solids, see Sect. V.2.

In theory it should be possible to calculate *all* observables, since the Hohenberg–Kohn theorem guarantees that they are all functionals of $n_0(\mathbf{r})$. In practice, one does not know how to do this explicitly. Another problem is that the minimization of $E_v[n]$ is, in general, a tough numerical problem on its own. And, moreover, one needs reliable approximations for $T[n]$ and $U[n]$ to begin with. In the next section, on the Kohn–Sham equations, we will see one widely used method for addressing these problems. Before looking at it, however, it is worthwhile recalling an older, but still occasionally useful, alternative – the Thomas–Fermi approximation, which, historically, predates modern DFT and the Hohenberg–Kohn theorem.

In this approximation one writes

$$U[n] \approx E_\mathrm{H}[n] = \frac{q^2}{2} \int \mathrm{d}^3 r \int \mathrm{d}^3 r' \frac{n(\mathbf{r})n(\mathbf{r}')}{|\mathbf{r}-\mathbf{r}'|}, \qquad (39)$$

i.e. approximates the full interaction energy by the Hartree energy, the electrostatic interaction energy of the fixed charge distribution $n(\mathbf{r})$. The kinetic energy can be approximated as

$$T[n] \approx T^\mathrm{LDA}[n] = \int \mathrm{d}^3 r\, t^\mathrm{hom}(n(\mathbf{r})), \qquad (40)$$

where $t^\mathrm{hom}(n)$ is the kinetic-energy density of a homogeneous interacting system with (constant) density n. Since it refers to interacting electrons $t^\mathrm{hom}(n)$ is not known explicitly, and further approximations are called for. As it stands, however, this formula is already a first example of a LDA. In this type of approximation one imagines the real inhomogeneous system (with density $n(\mathbf{r})$ in potential $v(\mathbf{r})$) to be decomposed into small cells in each of which $n(\mathbf{r})$ and $v(\mathbf{r})$ are approximately constant. In each cell (i.e. locally) one can then use the per-volume energy of a homogeneous system to approximate the contribution of the cell to the real inhomogeneous one. Making the cells infinitesimally small and summing over all of them yields (40).

For a noninteracting system (specified by subscript s, for 'single particle'), the function $t_s^\mathrm{hom}(n)$ is known explicitly,

$$t_s^\mathrm{hom}(n) = 3\hbar^2(3\pi^2)^{2/3} n^{5/3}/(10m) \qquad (41)$$

(see also Sect. VI.1). This is exploited to further approximate

$$T[n] \approx T^{\text{LDA}}[n] \approx T_s^{\text{LDA}}[n] = \int d^3r \, t_s^{\text{hom}}(n(\mathbf{r})), \qquad (42)$$

where $T_s^{\text{LDA}}[n]$ is the LDA to $T_s[n]$, the kinetic energy of noninteracting electrons of density n. Equivalently, it may be considered the noninteracting version of $T^{\text{LDA}}[n]$. (The quantity $T_s[n]$ will reappear below, when we discuss the Kohn–Sham equations.)

The Thomas–Fermi approximation consists in combining (39) with (42) and minimizing the resulting energy functional

$$E[n] = T[n] + U[n] + V[n] \approx E^{\text{TF}}[n] = T_s^{\text{LDA}}[n] + E_H[n] + V[n]. \qquad (43)$$

The Thomas–Fermi approximation to the dielectric constant, which provides a simple description of static screening, is obtained by minimizing $E^{\text{TF}}[n]$ with respect to n and linearizing the resulting relation between $v(\mathbf{r})$ and $n(\mathbf{r})$. It thus involves one more approximation (the linearization) compared with what is called the 'Thomas–Fermi approximation' in DFT.[51] In two dimensions no linearization is required and both uses of the name Thomas-Fermi become equivalent.[51]

A major defect of the Thomas–Fermi approximation is that within it molecules are unstable: the energy of a set of isolated atoms is lower than that of the bound molecule. This fundamental deficiency and the lack of accuracy resulting from neglect of correlations in (39) and from using the local approximation (42) for the kinetic energy limit the practical use of the Thomas–Fermi approximation in its own right. However, it is found to be a most useful starting point for a large body of work on improved approximations in chemistry and physics.[14,52] More recent approximations for $T[n]$ can be found in Refs. [53–55], in the context of orbital-free DFT. The extension of the local-density concept to the exchange–correlation energy is at the heart of many modern density functionals (see Sect. VI.1).

IV. THE EXCHANGE–CORRELATION ENERGY

1. Definition of the Exchange–Correlation Energy

The Thomas–Fermi approximation (42) for $T[n]$ is not very good. A more accurate scheme for treating the kinetic-energy functional of interacting electrons, $T[n]$, is based on decomposing it into one part that represents the kinetic energy of noninteracting particles of density n, i.e., the quantity above called $T_s[n]$, and one that represents the remainder, denoted $T_c[n]$ (the subscripts s and c stand for 'single particle' and 'correlation', respectively). T_s is defined by minimizing the expectation value of the kinetic-energy operator $\hat{T}$ over all wave functions yielding the density $n(\mathbf{r})$, i.e.

$$T_s[n] = \min_{\Phi \to n} \langle \Phi | \hat{T} | \Phi \rangle = \langle \Phi[n] | \hat{T} | \Phi[n] \rangle. \tag{44}$$

Similarly, the full kinetic energy is defined as $T[n] = \langle \Psi[n] | \hat{T} | \Psi[n] \rangle$, where $\Psi[n]$ minimizes $\hat{T} + \hat{U}$ and yields $n(\mathbf{r})$. All consequences of antisymmetrization (i.e. exchange) are described by employing single-determinantal wave functions in defining T_s. The difference $T_c := T - T_s$ is then a pure correlation effect.

$T_s[n]$ is not known exactly as a functional of n [and using the LDA to approximate it leads one back to the Thomas–Fermi approximation (42)], but it is easily expressed in terms of the single-particle orbitals $\phi_i(\mathbf{r})$ of a noninteracting system with density n, as

$$T_s[n] = -\frac{\hbar^2}{2m} \sum_i^N \int d^3r \, \phi_i^*(\mathbf{r}) \nabla^2 \phi_i(\mathbf{r}), \tag{45}$$

because for noninteracting particles the total kinetic energy is just the sum of the individual kinetic energies. Since all $\phi_i(\mathbf{r})$ are functionals of n, this expression for T_s is an explicit orbital functional but an implicit density functional, $T_s[n] = T_s[\{\phi_i[n]\}]$, where the notation indicates that T_s depends on the full set of occupied orbitals ϕ_i, each of which is a functional of n. Other such orbital functionals will be discussed in Sect. VI. We now rewrite the exact energy functional as

$$E[n] = T[n] + U[n] + V[n] = T_s[\{\phi_i[n]\}] + E_H[n] + E_{xc}[n] + V[n], \tag{46}$$

where by definition the exchange–correlation functional $E_{xc} \leq 0$ contains the differences $T_c = T - T_s$ and $W_{xc} := U - E_H$. Unlike (43), (46) is formally exact, but of course E_{xc} is unknown – although the Hohenberg–Kohn theorem guarantees that it is a density functional. E_{xc} is defined as $T_c + W_{xc}$, but it is often decomposed as

$E_\text{xc} = E_\text{x} + E_\text{c}$, where E_x is due to the Pauli principle (exchange energy) and $E_\text{c} := T_\text{c} + U - E_\text{H} - E_\text{x} =: T_\text{c} + W_\text{c}$ is due to correlations. This definition of E_c shows that part of the correlation energy E_c is due to the difference T_c between the noninteracting and the interacting kinetic energies, which is occasionally called the 'correlation kinetic energy' or the 'kinetic energy of correlation'. Similarly, W_c may be called the 'potential energy of correlation'. In Sect. IV.2 we give a simple physical interpretation of the various contributions to E_c.

The exchange energy can be written explicitly in terms of the single-particle orbitals as

$$E_\text{x}[\{\phi_i[n]\}] = -\frac{q^2}{2} \sum_{jk} \int d^3r \int d^3r' \frac{\phi_j^*(\mathbf{r})\phi_k^*(\mathbf{r}')\phi_j(\mathbf{r}')\phi_k(\mathbf{r})}{|\mathbf{r} - \mathbf{r}'|}, \tag{47}$$

which is known as the Fock term, but no general exact expression in terms of the density is known. Note that (47) is the same functional of the orbitals used also in Hartree–Fock theory, but the orbitals themselves are different, of course: the Hartree–Fock orbitals are solutions of the Hartree–Fock integro-differential equation, with an integral operator representing exchange, whereas the Kohn–Sham orbitals are solutions of the Kohn–Sham equation, with a local multiplicative potential representing exchange and correlation effects. Moreover, the Hartree–Fock energy functional is minimized with respect to the orbitals, while the DFT energy functional is minimized with respect to the density, of which the orbitals are (highly nonlocal) functionals.

For the correlation energy no general explicit expression is known, neither in terms of orbitals nor in terms of densities. A simple way to understand the origin of correlation is to recall that the Hartree energy is obtained in a variational calculation in which the many-body wave function is approximated as a product of single-particle orbitals. Use of an antisymmetrized product (a Slater determinant) produces the Hartree energy and the exchange energy.[8,56,57] The correlation energy is defined as the difference between the full ground-state energy (obtained with the correct many-body wave function) and the one obtained from the (Hartree–Fock or Kohn–Sham) Slater determinant. As a consequence of the variational principle, $E_\text{c} \leq 0$. The Hartree–Fock and the Kohn–Sham

Slater determinants are not identical, since they are composed of different single-particle orbitals, and thus the definition of exchange and correlation energy in DFT and in conventional quantum chemistry is slightly different. Quantitatively, however, the difference between both definitions of E_c is rather small.[58]

2. Interpretation of the Exchange–Correlation Energy

Recalling the quantum mechanical interpretation of the wave function as a probability amplitude, we see that a product form of the many-body wave function corresponds to treating the probability amplitude of the many-electron system as a product of the probability amplitudes (orbitals) of individual electrons. Mathematically, the probability of a composed event is the product of the probabilities of the individual events, provided the individual events are independent of each other. If the probability of a composed event is not equal to the probability of the individual events, these individual events are said to be correlated. Correlation is thus a general mathematical concept describing the fact that certain events are not independent. It can also be defined in classical physics, and in applications of statistics to problems outside science. Exchange, on the other hand, is due to the indistinguishability of particles, and is a true quantum phenomenon, without any analogue in classical physics.

The Hartree energy E_H is obtained from a product wave function, and thus describes the interaction energy of a fixed charge distribution $n(\mathbf{r})$. In quantum mechanics, charges never have an absolutely fixed position in space, but undergo quantum fluctuations, which ultimately are due to the uncertainty principle. In the presence of particle–particle interactions, the particles coordinate their fluctuations such as to further minimize their total energy, relative to the value obtained from a (hypothetical) fixed distribution. The individual probability amplitudes are now not independent anymore, and the particles become correlated in the sense described above. The correlation energy E_c is the additional energy lowering obtained from these synchronized quantum fluctuations. The kinetic energy $T_c = T - T_s$ associated with these fluctuations is positive, but this is compensated for by the gain in interaction energy $W_c = U - (E_H + E_x) \leq 0$, so that $E_c = T_c + W_c \leq 0$, in accordance with the variational principle.

Clearly E_c is an enormously complex object, and DFT would be of little use if one had to know it exactly to make calculations. The practical advantage of writing $E[n]$ in the form (46) is that the unknown functional $E_{xc}[n]$ is typically much smaller than the known terms T_s, E_H and V. One can thus hope that reasonably simple approximations for $E_{xc}[n]$ will provide useful results for $E[n]$. Some successful approximations are discussed in Sect. VI. Clearly, in the construction of approximations it is useful to have at hand a list of exact properties of E_{xc}.

3. Selected Exact Properties of the Exchange–Correlation Energy

As we have already seen, the variational principle immediately provides the inequalities

$$E_{xc} = (T - T_s) + (U - E_H) = T_c + W_{xc} = E_x + E_c \leq E_x \leq 0. \quad (48)$$

Among the known properties of these functionals are the coordinate-scaling conditions first obtained by Levy and Perdew[59]

$$E_x[n_\lambda] = \lambda E_x[n], \quad (49)$$
$$E_c[n_\lambda] > \lambda E_c[n] \quad \text{for } \lambda > 1, \quad (50)$$
$$E_c[n_\lambda] < \lambda E_c[n] \quad \text{for } \lambda < 1, \quad (51)$$

where $n_\lambda(\mathbf{r}) = \lambda^3 n(\lambda \mathbf{r})$ is a scaled density integrating to total particle number N.

Another important property of the exact functional is the one-electron limit

$$E_c[n^{(1)}] \equiv 0, \quad (52)$$
$$E_x[n^{(1)}] \equiv -E_H[n^{(1)}], \quad (53)$$

where $n^{(1)}$ is a one-electron density. These latter two conditions, which are satisfied within the Hartree–Fock approximation, but not by standard local-density and gradient-dependent functionals, ensure that there is no artificial self-interaction of one electron with itself.

The Lieb–Oxford bound,[60–62]

$$E_x[n] \geq E_{xc}[n] \geq -1.68 e^2 \int d^3 r\, n(\mathbf{r})^{4/3}, \quad (54)$$

establishes a lower bound on the exchange–correlation energy, which is satisfied,e.g, by the LDA,[63,64] the GGAs PW91[65] and PBE,[66] and the Tao, Perdew, Staroverov, Scuseria meta-GGA.[67] On the other hand, earlier GGAs[68] and meta-GGAs[69] and other functionals containing fitting parameters[70–73] are not guaranteed to satisfy the bound for all possible densities. The success of some of these latter functionals shows that, while the bound is doubtlessly obeyed by the exact functional, satisfaction is not a necessary condition for good performance in practice.

The fact that both exchange and correlation tend to keep electrons apart has given rise to the concept of an exchange–correlation hole, $n_{xc}(\mathbf{r}, \mathbf{r}')$, describing the reduction of probability for encountering an electron at $\mathbf{r}'$, given one at $\mathbf{r}$. The exchange–correlation energy can be written as a Hartree-like interaction between the charge distribution $n(\mathbf{r})$ and the exchange–correlation hole $n_{xc}(\mathbf{r}, \mathbf{r}') = n_x(\mathbf{r}, \mathbf{r}') + n_c(\mathbf{r}, \mathbf{r}')$,

$$E_{xc}[n] = \frac{q^2}{2} \int d^3r \int d^3r' \frac{n(\mathbf{r}) n_{xc}(\mathbf{r}, \mathbf{r}')}{|\mathbf{r} - \mathbf{r}'|}, \quad (55)$$

which defines n_{xc}. The exchange component $E_x[n]$ of the exact exchange–correlation functional describes the energy lowering due to antisymmetrization (i.e. the tendency of like-spin electrons to avoid each other). It gives rise to the exchange hole $n_x(\mathbf{r}, \mathbf{r}')$, which obeys the sum rule $\int d^3r' n_x(\mathbf{r}, \mathbf{r}') = -1$. The correlation component $E_c[n]$ accounts for the additional energy lowering arising because electrons with opposite spins also avoid each other. The resulting correlation hole integrates to zero, so the total exchange–correlation hole satisfies $\int d^3r' n_{xc}(\mathbf{r}, \mathbf{r}') = -1$.

One of the most intriguing properties of the exact functional, which has resisted all attempts at describing it in local or semilocal approximations, is the derivative discontinuity of the exchange–correlation functional with respect to the total particle number,[74–76]

$$\left. \frac{\delta E_{xc}[n]}{\delta n(\mathbf{r})} \right|_{N+\delta} - \left. \frac{\delta E_{xc}[n]}{\delta n(\mathbf{r})} \right|_{N-\delta} = v_{xc}^+(\mathbf{r}) - v_{xc}^-(\mathbf{r}) = \Delta_{xc}, \quad (56)$$

where δ is an infinitesimal shift of the electron number N, and Δ_{xc} is a system-dependent, but $\mathbf{r}$-independent shift of the exchange–correlation potential $v_{xc}(\mathbf{r})$ as it passes from the electron-poor to the electron-rich side of integer N. The noninteracting kinetic-energy functional has a similar discontinuity, given by

$$\left.\frac{\delta T_s[n]}{\delta n(\mathbf{r})}\right|_{N+\delta} - \left.\frac{\delta T_s[n]}{\delta n(\mathbf{r})}\right|_{N-\delta} = \epsilon_{N+1} - \epsilon_N = \Delta_{KS}, \qquad (57)$$

where ϵ_N and ϵ_{N+1} are the Kohn–Sham single-particle energies of the highest occupied and lowest unoccupied eigenstate, corresponding to the Kohn–Sham highest occupied molecular orbital (HOMO) and lowest unoccupied molecular orbital (LUMO), respectively. These eigenvalues are discussed in more detail in Sect. V.2.

The kinetic-energy discontinuity is thus simply the Kohn–Sham single-particle gap Δ_{KS}, or HOMO–LUMO gap, whereas the exchange–correlation discontinuity Δ_{xc} is a many-body effect. The true fundamental gap $\Delta = E(N+1) + E(N-1) - 2E(N)$ is the discontinuity of the total ground-state energy functional,[7,74–76]

$$\Delta = \left.\frac{\delta E[n]}{\delta n(\mathbf{r})}\right|_{N+\delta} - \left.\frac{\delta E[n]}{\delta n(\mathbf{r})}\right|_{N-\delta} = \Delta_{KS} + \Delta_{xc}. \qquad (58)$$

Since all terms in E other than E_{xc} and T_s are continuous functionals of $n(\mathbf{r})$, the fundamental gap is the sum of the Kohn–Sham gap and the exchange–correlation discontinuity. Standard density functionals (LDA and GGA) predict $\Delta_{xc} = 0$, and thus often underestimate the fundamental gap. The fundamental and Kohn–Sham gaps are also illustrated in Fig. 1.

All these properties serve as constraints or guides in the construction of approximations for the functionals $E_x[n]$ and $E_c[n]$. Many other similar properties are known. A useful overview of scaling properties is the contribution of M. Levy in Ref. [21 page 11].

V. THE KOHN–SHAM EQUATIONS

DFT can be implemented in many ways. The minimization of an explicit energy functional, discussed up to this point, is not normally the most efficient among them. Much more widely used is the Kohn–Sham approach. Interestingly, this approach owes its success and popularity partly to the fact that it does not exclusively work in terms of the particle (or charge) density, but brings a special kind of wave function (single-particle orbitals) back into the game. As a consequence, DFT then formally looks like a single-particle theory, although many-body effects are still included via the so-called exchange–correlation functional. We will now see in some detail how this is done.

1. Self-Consistent Single-Particle Equations and Ground-State Energies

Since T_s is now written as an orbital functional one cannot directly minimize (46) with respect to n. Instead, one commonly employs a scheme suggested by Kohn and Sham[77] for performing the minimization indirectly. This scheme starts by writing the minimization as

$$0 = \frac{\delta E[n]}{\delta n(\mathbf{r})} = \frac{\delta T_s[n]}{\delta n(\mathbf{r})} + \frac{\delta V[n]}{\delta n(\mathbf{r})} + \frac{\delta E_H[n]}{\delta n(\mathbf{r})} + \frac{\delta E_{xc}[n]}{\delta n(\mathbf{r})}$$
$$= \frac{\delta T_s[n]}{\delta n(\mathbf{r})} + v(\mathbf{r}) + v_H(\mathbf{r}) + v_{xc}(\mathbf{r}) \,. \tag{59}$$

As a consequence of (34), $\delta V/\delta n = v(\mathbf{r})$, the 'external' potential the electrons move in. This potential is called 'external' because it is external to the electron system and is not generated self-consistently from the electron–electron interaction, as v_H and v_{xc}. It comprises the lattice potential and any additional truly external field applied to the system as a whole. The term $\delta E_H/\delta n$ simply yields the Hartree potential, introduced in (21). For the term $\delta E_{xc}/\delta n$, which can only be calculated explicitly once an approximation for E_{xc} has been chosen, one commonly writes v_{xc}.

Consider now a system of noninteracting particles moving in the potential $v_s(\mathbf{r})$. For this system the minimization condition is simply

$$0 = \frac{\delta E_s[n]}{\delta n(\mathbf{r})} = \frac{\delta T_s[n]}{\delta n(\mathbf{r})} + \frac{\delta V_s[n]}{\delta n(\mathbf{r})} = \frac{\delta T_s[n]}{\delta n(\mathbf{r})} + v_s(\mathbf{r}), \tag{60}$$

since there are no Hartree and exchange–correlation terms in the absence of interactions. The density solving this Euler equation is $n_s(\mathbf{r})$. Comparing this with (59), we find that both minimizations have the same solution $n_s(\mathbf{r}) \equiv n(\mathbf{r})$ if v_s is chosen to be

$$v_s(\mathbf{r}) = v(\mathbf{r}) + v_H(\mathbf{r}) + v_{xc}(\mathbf{r}). \tag{61}$$

Consequently, one can calculate the density of the interacting (many-body) system in potential $v(\mathbf{r})$, described by a many-body Schrödinger equation of the form (2), by solving the equations of a noninteracting (single-body) system in potential $v_s(\mathbf{r})$. The question whether such a potential $v_s(\mathbf{r})$ always exists in the mathematical sense is called the 'noninteracting v-representability problem'. It is known that every interacting ensemble v-representable density is also noninteracting ensemble v-representable, but, as mentioned in Sect. III.1, only in discretized systems has it been proven that all densities are interacting ensemble v-representable. It is not known if interacting ensemble-representable densities may be noninteracting pure state representable (i.e by a single determinant), which would be convenient (but is not necessary) for Kohn–Sham calculations.

The Schrödinger equation of this auxiliary system,

$$\left[-\frac{\hbar^2 \nabla^2}{2m} + v_s(\mathbf{r})\right] \phi_i(\mathbf{r}) = \epsilon_i \, \phi_i(\mathbf{r}), \tag{62}$$

yields orbitals that reproduce the density $n(\mathbf{r})$ of the original system (these are the same orbitals employed in (45)),

$$n(\mathbf{r}) \equiv n_s(\mathbf{r}) = \sum_i^N f_i \, |\phi_i(\mathbf{r})|^2, \tag{63}$$

where f_i is the occupation of the ith orbital.

Equations (61)–(63) are the celebrated Kohn–Sham equations. They replace the problem of minimizing $E[n]$ by that of solving a single-body Schrödinger equation. (Recall that the minimization of $E[n]$ originally replaced the problem of solving the many-body Schrödinger equation!)

The techniques for solving the Kohn–Sham equation are similar to those used for the Hartree–Fock equation. Commonly, the orbitals are expanded in basis functions, and a secular equation is solved to find the energetically optimal coefficients of the expansion. The literature on suitable basis sets is enormous and cannot be reviewed here. It is common practice to use in DFT calculations the same basis sets developed for Hartree–Fock and CI calculations, although it is not always clear (in particular in the case of correlation-consistent bases) if this is always an optimal choice.

Since both v_H and v_{xc} depend on n, which depends on the ϕ_i, which in turn depend on v_s, the problem of solving the Kohn–Sham equations is a nonlinear one. The usual way of solving such problems is to start with an initial guess for $n(\mathbf{r})$, calculate the corresponding $v_s(\mathbf{r})$ and then solve the differential equation (62) for the ϕ_i. From these one calculates a new density, using (63), and starts again. The process is repeated until it converges. The technical name for this procedure is 'self-consistency cycle'. Different convergence criteria (such as convergence in the energy, the density or some observable calculated from these) and various convergence-accelerating algorithms (such as mixing of old and new effective potentials) are in common use. Only rarely does it requires more than a few dozen iterations to achieve convergence, and even rarer are cases where convergence seems unattainable, i.e. a self-consistent solution of the Kohn–Sham equation cannot be found.

Once one has a converged solution n_0, one can calculate the total energy from (46) or, equivalently and more conveniently, from

$$E_0 = \sum_i^N \epsilon_i - \frac{q^2}{2} \int d^3r \int d^3r' \frac{n_0(\mathbf{r})n_0(\mathbf{r}')}{|\mathbf{r}-\mathbf{r}'|}$$
$$- \int d^3r\, v_{xc}(\mathbf{r})n_0(\mathbf{r}) + E_{xc}[n_0]. \qquad (64)$$

All terms on the right-hand side of (64) except for the first, involving the sum of the single-particle energies, are sometimes known as double-counting corrections, in analogy to a similar equation valid within Hartree–Fock theory. These corrections give mathematical meaning to the common statement that *the whole is more than the sum of its parts*.

Equation (64) follows from writing $V[n]$ in (46) by means of (61) as

$$V[n] = \int d^3r\, v(\mathbf{r})n(\mathbf{r}) = \int d^3r\, [v_s(\mathbf{r}) - v_H(\mathbf{r}) - v_{xc}(\mathbf{r})]n(\mathbf{r})$$
$$= V_s[n] - \int d^3r\, [v_H(\mathbf{r}) + v_{xc}(\mathbf{r})]n(\mathbf{r}), \quad (65)$$

and identifying the energy of the noninteracting (Kohn–Sham) system as $E_s = \sum_i^N \epsilon_i = T_s + V_s$.

2. Single-Particle Eigenvalues and Excited-State Energies

Equation (64) shows that E_0 is not simply the sum of all ϵ_i. If E_0 can be written approximately as $\sum_i^N \tilde{\epsilon}_i$ (where the $\tilde{\epsilon}_i$ are not the same as the Kohn–Sham eigenvalues ϵ_i), the system can be described in terms of N weakly interacting quasiparticles, each with energy $\tilde{\epsilon}_i$. Fermi-liquid theory in metals and effective-mass theory in semiconductors are examples of this type of approach. The $\tilde{\epsilon}_i$, however, are not the same as the Kohn–Sham eigenvalues ϵ_i, nor are they necessarily approximated by them.

In fact, it should be clear from our derivation of (62) that the ϵ_i are introduced as completely artificial objects: they are the eigenvalues of an auxiliary single-body equation whose eigenfunctions (orbitals) yield the correct density. It is only this density that has strict physical meaning in the Kohn–Sham scheme. The Kohn–Sham eigenvalues, on the other hand, in general bear only a semiquantitative resemblance to the true energy spectrum,[78] but are not to be trusted quantitatively.

The main exception to this rule is the highest occupied Kohn–Sham eigenvalue. Denoting by $\epsilon_N(M)$ the Nth eigenvalue of a system with M electrons, one can show rigorously that $\epsilon_N(N) = -I$, the negative of the first ionization energy of the N-body system, and $\epsilon_{N+1}(N+1) = -A$, the negative of the electron affinity of the same N-body system.[75,79,80] These relations hold for the exact functional only. When calculated with an approximate functional of the LDA or GGA type, the highest eigenvalues usually do not provide good approximations to the experimental I and A. Better results for these observables are obtained by calculating them as total-energy differences, according to $I = E_0(N-1) - E_0(N)$ and $A = E_0(N) - E_0(N+1)$, where $E_0(N)$ is the ground-state energy of the N-body system. Alternatively, SICs can be used to obtain improved ionization energies and electron affinities from Kohn–Sham eigenvalues.[81]

Figure 1 illustrates the role played by the highest occupied and lowest unoccupied Kohn–Sham eigenvalues, and their relation to observables. For molecules, HOMO(N) is the HOMO of the N-electron system, HOMO(N+1) is that of the $N+1$-electron system and LUMO(N) is the LUMO of the N-electron system. In solids with a gap, the HOMO and LUMO become the top of the valence band and the bottom of the conduction band, respectively, whereas in metals they are both identical to the Fermi level. The vertical lines

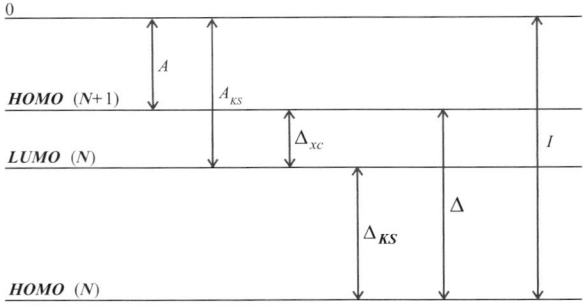

Figure 1. Schematic representation of some important Kohn–Sham eigenvalues relative to the vacuum level, denoted by 0, and their relation to observables. I is the ionization energy of the many-body system, which is equal to that of the Kohn–Sham system. A is the many-body electron affinity, A_{KS} is the electron affinity of the Kohn–Sham system, Δ is the fundamental gap, Δ_{KS} is the single-particle highest occupied molecular orbital–lowest unoccupied molecular orbital gap and Δ_{xc} is the derivative discontinuity. (Reprinted from Ref. [23]. Copyright © 2006 with permission from the *Brazilian Journal of Physics*.)

indicate the Kohn–Sham (single-particle) gap Δ_{KS}, the fundamental (many-body) gap Δ, the derivative discontinuity of the exchange–correlation functional, Δ_{xc}, the ionization energy of the interacting N-electron system $I(N) = -\epsilon_N(N)$ (which is also the ionization energy of the Kohn–Sham system $I_{KS}(N)$), the electron affinity of the interacting N-electron system $A(N) = -\epsilon_{N+1}(N+1)$ and the Kohn–Sham electron affinity $A_{KS}(N) = -\epsilon_{N+1}(N)$.

Given the auxiliary nature of the other Kohn–Sham eigenvalues, it comes as a welcome surprise that in many situations the Kohn–Sham eigenvalues ϵ_i do, empirically, provide a reasonable first approximation to the actual energy levels of extended systems. This approximation is behind most band-structure calculations in solid-state physics, and often gives results that agree well with experimental photoemission and inverse photoemission data,[82] but much research remains to be done before it is clear to what extent such conclusions can be generalized, and how situations in which the Kohn–Sham eigenvalues are good starting points for approximating the true excitation spectrum are to be characterized microscopically.[83,84] Several more rigorous approaches to excited states in DFT, which do not require the Kohn–Sham eigenvalues to have physical meaning, have been developed:

1. The early suggestion of Gunnarsson and Lundqvist[85] to use a symmetry-dependent exchange–correlation functional to calculate the lowest-energy excited state of each symmetry class has been implemented approximately by von Barth,[86] but suffers from lack of knowledge of the symmetry dependence of the functional. More recent work on this dependence is provided in Ref. [87].
2. Extrema of the ground-state functional $E_v[n]$ above the minimum energy correspond to densities of excited states, but the excited states obtained in this way do not cover the entire spectrum of the many-body Hamiltonian.[88] Moreover, if the minimization of the functional $E_v[n]$ is done (as usual) by solving a Kohn–Sham equation, there may occur spurious self-consistent solutions of these equations with $E > E_0$ that do not extremize $E_v[n]$ at all, and it is not always possible to tell if a given high-energy Kohn–Sham solution is of this spurious nature, or really a higher extremum of $E_v[n]$. For these reasons the path via higher-lying extrema is rarely used. However, an interesting application within semiconductor physics was recently reported.[89]
3. Excitation energies can also be calculated from excited-state densities. A mapping between excited-state densities and the external potential, and a Kohn–Sham-like scheme for calculating the density and energy of a specific excited state of interest have been constructed,[90,91] but the definition of a proper Kohn–Sham procedure is not trivial.[92]
4. A systematic approach to excited states in DFT is ensemble DFT, developed by Theophilou[93] and further elaborated by Oliveira et al.[94] In this formalism the functional depends on the particular choice for the ensemble, and a simple approximation for this dependence is available.[94] Some applications of this method have been worked out by Nagy[95] and a recent analysis of this method is presented in Ref. [96].
5. The most widely used DFT approach to excited states is TD-DFT. The time-dependent generalization of the Hohenberg–Kohn theorem, the Runge–Gross theorem, cannot be proven along the lines of the original Hohenberg–Kohn theorem, but requires a different approach.[30,97] For recent reviews of TD-DFT, see Ref. [98]. Excited states were first extracted from TD-DFT in Refs. [99, 100]. This approach is now implemented in standard quantum-chemical DFT program

packages[101,102] and is increasingly applied also in solid-state physics.[103]
6. Recently, excitation energies have also been extracted from ground-state DFT by means of generator coordinates.[232]

Most band-structure calculations in solid-state physics do not employ any of these techniques (which are computationally very expensive for solids), but are actually calculations of the Kohn–Sham eigenvalues ϵ_i. This simplification has proved enormously successful, but one must be aware of the fact that it takes the auxiliary single-body equation (62) literally as an approximation to the many-body Schrödinger equation. DFT, practised in this mode, is not a rigorous many-body theory anymore, but a mean-field theory (albeit one with a very sophisticated mean field $v_s(\mathbf{r})$). A more reliable alternative is provided by the GW method, which in principle is independent of DFT, but in practice normally takes DFT results as a starting point.[103–105]

The energy gap obtained in Kohn–Sham band-structure calculations is the one called the 'HOMO–LUMO' gap in molecular calculations, i.e. the difference between the energies of the highest occupied and the lowest unoccupied single-particle states. Neglect of the derivative discontinuity Δ_{xc}, defined in (56), by standard local and semilocal exchange–correlation functionals leads to an underestimate of the gap (the so-called band-gap problem), which is most severe in transition-metal oxides and other strongly correlated systems. SICs provide a partial remedy for this problem.[106–108]

A partial justification for the interpretation of the Kohn–Sham eigenvalues as the starting point for approximations to quasi-particle energies, common in band-structure calculations can be given by comparing the Kohn–Sham equation with other self-consistent equations of many-body physics. Among the simplest such equations are the Hartree equation

$$\left[-\frac{\hbar^2 \nabla^2}{2m} + v(\mathbf{r}) + v_H(\mathbf{r}) \right] \phi_i^H(\mathbf{r}) = \epsilon_i^H \phi_i^H(\mathbf{r}), \qquad (66)$$

and the Hartree–Fock equation

$$\left[-\frac{\hbar^2 \nabla^2}{2m} + v(\mathbf{r}) + v_H(\mathbf{r}) \right] \phi_i^{HF}(\mathbf{r}) - q^2 \int d^3 r' \frac{\gamma(\mathbf{r}, \mathbf{r}')}{|\mathbf{r} - \mathbf{r}'|} \phi_i^{HF}(\mathbf{r}')$$
$$= \epsilon_i^{HF} \phi_i^{HF}(\mathbf{r}), \qquad (67)$$

where $\gamma(\mathbf{r}, \mathbf{r}')$ is the density matrix of (16). It is a fact known as Koopman's theorem[57] that the Hartree–Fock eigenvalues ϵ_i^{HF} can be interpreted as unrelaxed electron-removal energies (i.e. ionization energies of the ith electron, neglecting reorganization of the remaining electrons after removal). As mentioned above, in DFT only the highest occupied eigenvalue corresponds to an ionization energy, but unlike in the Hartree–Fock approximation this energy includes relaxation effects.

The Kohn–Sham equation (62) includes both exchange and correlation via the multiplicative operator v_{xc}. Both exchange and correlation are normally approximated in DFT, whereas the Hartree–Fock approximation accounts for exchange exactly, through the integral operator containing $\gamma(\mathbf{r}, \mathbf{r}')$, but neglects correlation completely. A possibility to treat exchange exactly in DFT is offered by the optimized effective potential (OEP) method discussed in Sect. VI.3. In practice, DFT results are typically at least as good as Hartree–Fock ones and are often comparable to much more complex wave-function-based methods.

The interpretation of the Kohn–Sham eigenvalues as single-particle energies is suggestive and useful, but certainly not necessary for DFT to work: if the Kohn–Sham equations are only used to obtain the density, and all other observables, such as total energies, are calculated from this density, then the Kohn–Sham equations in themselves are not an approximation at all, but simply a very useful mathematical tool.

VI. AN OVERVIEW OF APPROXIMATE EXCHANGE–CORRELATION FUNCTIONALS

As described at the beginning, the standard formulation of DFT is based on the Born–Oppenheimer approximation and the nonrelativistic approximation. Although both of these approximations can be relaxed within DFT, neither is specific to DFT. In addition to these, there are basically three more specific types of approximations involved in a DFT calculation.

One is conceptual, and concerns the interpretation of Kohn–Sham eigenvalues and orbitals as physical energies and wave functions. This approximation is optional – if one does not want to make it, one simply does not attach meaning to the eigenvalues of (62). The pros and cons of this procedure were discussed in Sect. V.

The second type of approximation is numerical, and concerns methods for actually solving the differential equation (62). A main aspect here is the selection of suitable basis functions, which we do not address in the present text, because the selection follows the same criteria that are well known in the context of Hartree–Fock or CI calculations.

The third type of approximation involves constructing an expression for the unknown exchange–correlation functional $E_{xc}[n]$, which contains all many-body aspects of the problem (see (46)). The present section is intended to give the reader an idea of what types of functionals exist, and to describe what their main features are, separately for local functionals (Thomas–Fermi, LDA and $X\alpha$; Sect. VI.1), semilocal, or gradient-dependent, functionals [gradient-expansion approximation (GEA) and GGA; Sect. VI.2), and nonlocal functionals [hybrids, orbital functionals such as meta-GGAs, exact exchange (EXX) and SIC, and integral-dependent functionals such as average-density approximation (ADA); Sect. VI.3]. Section VI.4 contains a brief description of selected aspects of the performance of some of the functionals described in Sects. VI.1–3. For more details on functional construction, testing and extensive comparisons of a wide variety of functionals, the reader is referred to the reviews[7–21] or to the original papers cited below.

1. Local Functionals: LDA

Historically (and in many applications also practically) the most important type of approximation is the LDA. To understand the concept of a LDA recall first how the noninteracting kinetic energy $T_s[n]$ is treated in the Thomas–Fermi approximation. In a homogeneous system one knows that, per volume,

$$t_s^{\text{hom}}(n) = \frac{3\hbar^2}{10m}(3\pi^2)^{2/3}n^{5/3}, \qquad (68)$$

where $n = const$ and the change from a capital T to a lower-case t is used to indicate quantities per volume. In an inhomogeneous system, with $n = n(\mathbf{r})$, one approximates locally

$$t_s(\mathbf{r}) \approx t_s^{\text{hom}}(n(\mathbf{r})) = \frac{3\hbar^2}{10m}(3\pi^2)^{2/3}n(\mathbf{r})^{5/3} \qquad (69)$$

and obtains the full kinetic energy by integration over all space:

$$T_s^{\text{LDA}}[n] = \int d^3r \, t_s^{\text{hom}}(n(\mathbf{r})) = \frac{3\hbar^2}{10m}(3\pi^2)^{2/3} \int d^3r \, n(\mathbf{r})^{5/3}. \tag{70}$$

For the kinetic energy the approximation $T_s[n] \approx T_s^{\text{LDA}}[n]$ is much inferior to the exact treatment of T_s in terms of orbitals, offered by the Kohn–Sham equations, but the LDA concept turned out to be highly useful for another component of the total energy (46), the exchange–correlation energy $E_{\text{xc}}[n]$. For the exchange energy $E_{\text{x}}[n]$ the procedure is very simple, since the per-volume exchange energy of the homogeneous electron liquid is known exactly,[7,8]

$$e_{\text{x}}^{\text{hom}}(n) = -\frac{3q^2}{4}\left(\frac{3}{\pi}\right)^{1/3} n^{4/3}, \tag{71}$$

so that

$$E_{\text{x}}^{\text{LDA}}[n] = -\frac{3q^2}{4}\left(\frac{3}{\pi}\right)^{1/3} \int d^3r \, n(\mathbf{r})^{4/3}. \tag{72}$$

This is the LDA for E_{x}. If one adds this term to the Thomas–Fermi expression (43), one obtains the so-called Thomas–Fermi–Dirac approximation to $E[n]$. If one multiplies it with an adjustable parameter α, one obtains the so-called $X\alpha$ approximation to $E_{\text{xc}}[n]$. These approximations are not much used today in DFT.

The LDA for the correlation energy $E_{\text{c}}[n]$ formally consists in

$$E_{\text{c}}[n] \approx E_{\text{c}}^{\text{LDA}}[n] = \int d^3r \, e_{\text{c}}^{\text{hom}}(n)|_{n\to n(\mathbf{r})} = \int d^3r \, e_{\text{c}}^{\text{hom}}(n(\mathbf{r})), \tag{73}$$

so that

$$E_{\text{xc}}[n] \approx E_{\text{xc}}^{\text{LDA}}[n] = \int d^3r \, e_{\text{xc}}^{\text{hom}}(n)|_{n\to n(\mathbf{r})} = \int d^3r \, e_{\text{xc}}^{\text{hom}}(n(\mathbf{r})), \tag{74}$$

where $e_{\text{xc}}^{\text{hom}} = e_{\text{x}}^{\text{hom}} + e_{\text{c}}^{\text{hom}}$ is the per-volume exchange–correlation energy of the uniform electron liquid. The corresponding exchange–correlation potential is simply

$$v_{\text{xc}}^{\text{LDA}}[n](\mathbf{r}) = \left.\frac{\partial e_{\text{xc}}^{\text{hom}}(n)}{\partial n}\right|_{n\to n(\mathbf{r})}. \tag{75}$$

This approximation for $E_{xc}[n]$ has proved amazingly successful, even when applied to systems that are quite different from the electron liquid that forms the reference system for the LDA. A partial explanation for this success of the LDA is systematic error cancellation: typically, the LDA underestimates E_c but overestimates E_x, resulting in unexpectedly good values of E_{xc}. This error cancellation is not accidental, but systematic, and is caused by the fact that for any density the LDA exchange–correlation hole satisfies the correct sum rule on the exchange–correlation hole $n_{xc}(\mathbf{r}, \mathbf{r}')$, $\int d^3 r' n_{xc}^{LDA}(\mathbf{r}, \mathbf{r}') = -1$, which is only possible if integrated errors in n_x^{LDA} cancel with those of n_c^{LDA}.

Unlike for the exchange energy, the correlation energy $e_c^{hom}[n]$ of the uniform electron liquid is not known exactly: the determination of the correlation energy of a homogeneous interacting electron system is already a difficult many-body problem on its own! Early approximate expressions for $e_c^{hom}(n)$ were based on applying perturbation theory (e.g. the random-phase approximation) to this problem.[85,109] These approximations became outdated with the advent of highly precise quantum Monte Carlo calculations for the electron liquid, by Ceperley and Alder.[110] Modern expressions for $e_c^{hom}(n)$[63,64,111] are parameterizations of these data. These expressions are implemented in most standard DFT program packages and in typical applications give almost identical results. On the other hand, the earlier parameterizations of the LDA, based on perturbation theory,[85,109] can occasionally deviate substantially from the quantum Monte Carlo ones, and are better avoided. In the chemistry literature, the combination of the LDA for exchange, (72), with the LDA for correlation, (73), in the parameterization in Ref. [64] is known as the Slater exchange plus Vosko–Wilk–Nusair correlation version 5 (SVWN5) functional.

For many decades the LDA has been applied in, e.g., calculations of band structures and total energies in solid-state physics. In quantum chemistry it is much less popular, because it fails to provide results that are accurate enough to permit a quantitative discussion of the chemical bond in molecules (so-called chemical accuracy requires calculations with an error of not more than about 1 kcal/mol = 0.0434 eV/particle). For the same reason, electrochemical calculations rarely use the LDA, but employ more refined functionals, as described below.

At this stage it may be worthwhile to recapitulate what practical DFT does, and where the LDA enters its conceptual structure: what real systems, such as atoms, molecules, clusters and solids, have in common is that they are simultaneously inhomogeneous (the electrons are exposed to spatially varying electric fields produced by the nuclei) and interacting (the electrons interact via the Coulomb interaction). The way DFT, in the LDA, deals with this inhomogeneous many-body problem is by decomposing it into two simpler (but still highly nontrivial) problems: the solution of a spatially homogeneous interacting problem (the homogeneous electron liquid) yields the exchange–correlation energy $e_{xc}^{hom}(n)$, and the solution of a spatially inhomogeneous noninteracting problem (the inhomogeneous electron gas described by the Kohn–Sham equations) yields the particle density. Both steps are connected by the local-density potential (75), which shows how the exchange–correlation energy of the uniform interacting system enters the equations for the inhomogeneous noninteracting system.

The particular way in which the inhomogeneous many-body problem is decomposed and the various possible improvements on the LDA are behind the success of DFT in practical applications of quantum mechanics to real materials. Some such improvements on the LDA are discussed in the next two sections.

2. Semilocal Functionals: GEA, GGA and Beyond

In the LDA one exploits knowledge of the density at point $\mathbf{r}$. Any real system is spatially inhomogeneous, i.e. it has a spatially varying density $n(\mathbf{r})$, and it would clearly be useful to also include information on the rate of this variation in the functional. A first attempt at doing this was the so-called GEAs. In this class of approximation one tries to systematically calculate gradient corrections of the form $|\nabla n(\mathbf{r})|$, $|\nabla n(\mathbf{r})|^2$, $\nabla^2 n(\mathbf{r})$, etc. to the LDA. A famous example is the lowest-order gradient correction to the Thomas–Fermi approximation for $T_s[n]$,

$$T_s[n] \approx T_s^W[n] = T_s^{LDA}[n] + \frac{\hbar^2}{8m} \int d^3r \, \frac{|\nabla n(\mathbf{r})|^2}{n(\mathbf{r})}. \qquad (76)$$

The second term on the right-hand side is called the 'Weizsäcker term'. If one adds this term to the Thomas–Fermi expression (43), one obtains the so-called Thomas–Fermi–Weizsäcker approximation to $E[n]$. In a systematic gradient expansion, the 8 in the denominator is replaced by 72.[7,8] Similarly, in

$$E_x[n] \approx E_x^{\text{GEA}(2)}[n] = E_x^{\text{LDA}}[n] - \frac{10q^2}{432\pi(3\pi^2)^{1/3}} \int d^3r \frac{|\nabla n(\mathbf{r})|^2}{n(\mathbf{r})^{4/3}} \tag{77}$$

the second term on the right-hand side is the lowest-order gradient correction to $E_x^{\text{LDA}}[n]$.

Remarkably, the form of this term is fully determined already by dimensional analysis. In $E_x^{\text{GEA}(2)} = q^2 \int d^3r \, f(n, |\nabla n|^2)$ the function f must have dimensions (length)$^{-4}$. Since the dimensions of n and $|\nabla n|^2$ are (length)$^{-3}$ and (length)$^{-8}$, respectively, and to second order no higher powers or higher derivatives of n are allowed, the only possible combination is $f \propto |\nabla n(\mathbf{r})|^2/n^{4/3}$.

In practice, the inclusion of low-order gradient corrections almost never improves on the LDA, and often even worsens it. Higher-order corrections (e.g. $\propto |\nabla n(\mathbf{r})|^\alpha$ or $\propto \nabla^\beta n(\mathbf{r})$ with $\alpha, \beta > 2$), on the other hand, are exceedingly difficult to calculate, and little is known about them. In this situation it was a major breakthrough when it was realized, in the early 1980s, that instead of power-series-like systematic gradient expansions one could experiment with more general functions of $n(\mathbf{r})$ and $\nabla n(\mathbf{r})$, which need not proceed order by order. Such functionals, of the general form

$$E_{\text{xc}}^{\text{GGA}}[n] = \int d^3r \, f(n(\mathbf{r}), \nabla n(\mathbf{r})), \tag{78}$$

have become known as GGAs.

Different GGAs differ in the choice of the function $f(n, \nabla n)$. Note that this makes different GGAs much more different from each other than the different parameterizations of the LDA: essentially there is only one correct expression for $e_{\text{xc}}^{\text{hom}}(n)$, and the various parameterizations of the LDA[63,64,85,109,111] are merely different ways of writing it. On the other hand, depending on the method of construction employed for obtaining $f(n, \nabla n)$ one can obtain very different GGAs.

Nowadays, the most popular (and most reliable) GGAs are PBE (denoting the parameter-free functional proposed in 1996 by Perdew, Burke and Ernzerhof[66]) and B88LYP (denoting the combination of Becke's 1988 one-parameter exchange functional B88[70] with the four-parameter correlation functional of Lee, Yang and Parr[71]). Many other GGA-type functionals are also available, and new ones continue to appear. A useful collection of explicit expressions for some GGAs can be found in the appendix of Ref., [112] and more detailed discussion of some selected GGAs and their performance is given in Ref. [113] and in the chapter by S. Kurth and J. P. Perdew in Refs. [17 page 8, 18 page 1].

3. Orbital Functionals and Other Nonlocal Approximations: Hybrids, Meta-GGA, SIC, OEP, etc.

In spite of these advances, the quest for more accurate functionals is ongoing, and both in chemistry and in physics various beyond-GGA functionals have appeared. One important line of research has led to hybrid functionals, mixing a fraction of Hartree–Fock exchange into the DFT exchange functional (other mixtures are also possible). The most popular functional in quantum chemistry, B3LYP,[114] is such a hybrid, combining exact exchange with LDA exchange, the B88 exchange GGA[70] and the LYP correlation GGA.[71] The original construction employed the PW91 parameter-free correlation GGA,[65] but nowadays it is mostly used together with the four-parameter LYP GGA for correlation, *without* recalculating the mixing parameters. In spite of the label '3', the B3 functional actually contains eight parameters: three are mixing parameters, one parameter is contained in the B88 GGA for exchange[70] and four are contained in the LYP GGA for correlation.[71]

The construction of a hybrid functional involves a certain amount of empiricism in the choice of functionals that are mixed and in the optimization of the weight factors given to the mixed functionals. Formally, this might be considered a drawback, but in practice B3LYP has proven to be the most successful approximate functional for chemical applications.

More extreme examples of this semiempirical mode of construction of functionals are Becke's 1997 hybrid functional,[115] which contains ten adjustable parameters, and the functionals in Refs. [116, 117], each of which contains more than 20 parameters. A more recent example is DF07,[73] which promises a unified treatment of long-range (static) and short-range (dynamical) correlations, with a special correction added to account for dispersion (van der Waals) interactions. This functional contains five mixing parameters, in addition to three parameters contained in the input functionals.

Another recent beyond-GGA development is the emergence of so-called meta-GGAs, which depend, in addition to the density and its derivatives, also on the Kohn–Sham kinetic-energy density $\tau(\mathbf{r})$[67,69,118–120]

$$\tau(\mathbf{r}) = \frac{\hbar^2}{2m} \sum_i |\nabla \phi_i(\mathbf{r})|^2, \tag{79}$$

so E_{xc} can be written as $E_{xc}[n(\mathbf{r}), \nabla n(\mathbf{r}), \tau(\mathbf{r})]$. The additional degree of freedom provided by τ is used to satisfy additional constraints on E_{xc}, such as a self-interaction-corrected correlation functional, recovery of the fourth-order gradient expansion for exchange in the limit of slowly varying densities and a finite exchange potential at the nucleus.[67,118,119] In several recent tests,[67,118,119,121–124] meta-GGAs have given favourable results, even when compared with the best GGAs, but the full potential of this type of approximation is only beginning to be explored systematically. Of course, meta-GGAs can also be used as components of hybrids. A recent successful example is given in Ref. [119].

As we have seen in the case of T_s, it can be much easier to represent a functional in terms of single-particle orbitals than directly in terms of the density. Such functionals are known as orbital functionals, and (45) constitutes a simple example. Another important orbital-dependent functional is the exchange energy (Fock term) of (47). The meta-GGAs and hybrid functionals mentioned above are also orbital functionals, because they depend on the kinetic energy density (79), and on a combination of the orbital functional (47) with ordinary GGAs, respectively.

Still another type of orbital functional is the SIC proposed in Ref. [111] (PZ-SIC),

$$E_{xc}^{\text{approx,SIC}}[n_\uparrow, n_\downarrow] = E_{xc}^{\text{approx}}[n_\uparrow, n_\downarrow] \\ - \sum_{i,\sigma} \left(E_H[n_{i\sigma}] - E_{xc}^{\text{approx}}[n_{i\sigma}, 0] \right), \tag{80}$$

which subtracts, orbital by orbital, the contribution the Hartree and the exchange–correlation functionals would make if there was only one electron in the system. This correction can be applied on top of any approximate density functional, and ensures that the resulting corrected functional satisfies $E_{xc}^{\text{approx,SIC}}[n^{(1)}, 0] = -E_H[n^{(1)}]$ for a one-electron system, as demanded by (52) and (53). The LDA is exact for a completely uniform system, and thus is self-interaction-free in this limit, but neither it nor common GGAs or hybrids satisfy the requirement of freedom from self-interaction in general. This self-interaction is particularly critical for localized states, such as the d states in transition-metal oxides. For such systems PZ-SIC has been shown to greatly improve the uncorrected LDA,[106,107] but for thermochemistry PZ-SIC does not seem to be significant.[125]

Unfortunately the PZ-SIC approach, which minimizes the corrected energy functional with respect to the orbitals, does not lead to Kohn–Sham equations of the usual form, because the resulting effective potential is different for each orbital. As a consequence, various specialized algorithms for minimizing the PZ-SIC energy functional have been developed. For more details on these algorithms and some interesting applications in solid-state physics see Refs. [106–108] and for applications to thermochemistry see Refs. [126–130]. For finite systems, PZ-SIC has also been implemented by means of the OEP,[81] which produces a common local potential for all orbitals, and is discussed in the next paragraph. A detailed review of implementations and applications of PZ-SIC can be found in the contribution of Temmerman et al.[19] Alternatives to the PZ-SIC formulation in Ref. [111] have recently been analysed in Refs. [131, 132], with a view to either improving the results obtained with PZ-SIC or simplifying the implementation of the correction.

Since hybrid functionals, meta-GGAs, SIC, the Fock term and all other orbital functionals depend on the density only implicitly, via the orbitals $\phi_i[n]$, it is not possible to directly calculate the functional derivative $v_{xc} = \delta E_{xc}/\delta n$. Instead one must use indirect approaches to minimize $E[n]$ and obtain v_{xc}. In the case of the kinetic-energy functional $T_s[\{\phi_i[n]\}]$, this indirect approach is simply the Kohn–Sham scheme, described in Sect. V. In the case of orbital expressions for E_{xc}, the corresponding indirect scheme is known as the OEP[133] or, equivalently, the optimized-potential model.[134] The minimization of the orbital functional with respect to the density is achieved by repeated application of the chain rule for functional derivatives,

$$\begin{aligned}
v_{xc}[n](\mathbf{r}) &= \frac{\delta E_{xc}^{orb}[\{\phi_i\}]}{\delta n(\mathbf{r})} \\
&= \int d^3 r' \int d^3 r'' \sum_i \left[\frac{\delta E_{xc}^{orb}[\{\phi_i\}]}{\delta \phi_i(\mathbf{r}')} \frac{\delta \phi_i(\mathbf{r}')}{\delta v_s(\mathbf{r}'')} \frac{\delta v_s(\mathbf{r}'')}{\delta n(\mathbf{r})} + \text{c.c.} \right],
\end{aligned} \quad (81)$$

where E_{xc}^{orb} is the orbital functional (e.g. the Fock term) and v_s is the Kohn–Sham effective potential. Further evaluation of (81) gives rise to an integral equation that determines the $v_{xc}[n]$ belonging to the chosen orbital functional $E_{xc}[\{\phi_i[n]\}]$.[133,135] As an alternative to solving the full OEP integral equation, Krieger, Li and Iafrate (KLI) have proposed a simple but surprisingly accurate approximation that greatly facilitates implementation of the OEP.[133]

The application of the OEP methodology to the Fock term (47), either with or without the KLI approximation, is also known as the EXX method. The OEP-EXX equations have been solved for atoms[133,134,136] and solids,[137,138] with very encouraging results. Other orbital-dependent functionals that have been treated within the OEP scheme are the PZ-SIC[81] and the Colle–Salvetti functional.[136] A detailed review of the OEP and its KLI approximation is given in Ref. [135].

The high accuracy attained by complex orbital functionals implemented via the OEP, and the fact that it is easier to devise orbital functionals than explicit density functionals, makes the OEP concept attractive, but the computational cost of solving the OEP integral equation is a major drawback. However, this computational cost is significantly reduced by the KLI approximation[133] and other recently proposed simplifications.[139–141] In the context of the EXX method (i.e. using the Fock exchange term as an orbital functional) the OEP is a viable way to proceed. For more complex orbital functionals, additional simplifications may be necessary.[133,139–141]

A further reduction of computational complexity is achieved by not evaluating the orbital functional self-consistently, via (81), but only once, using the orbitals and densities of a converged self-consistent LDA or GGA calculation. This 'post-GGA' or 'post-LDA' strategy completely avoids the OEP and has been used both for hybrid functionals and for meta-GGAs.[69,115,120,142] A drawback of post methods is that they provide only approximations to the self-consistent total energies, not to eigenvalues, effective potentials, orbitals or densities. An attempt to maintain the formal simplicity of post methods, while perserving at least some corrections to the eigenvalues, is the scaled-self-consistency approach.[143,144]

In the case of hybrid functionals, still another mode of implementation has become popular. This alternative, which also avoids solution of (81), is to calculate the derivative of the hybrid functional with respect to the single-particle orbitals, and not with respect to the density as in (81). The resulting single-particle equation is of Hartree–Fock form, with a nonlocal potential, and with a weight factor in front of the Fock term. Strictly speaking, the orbital derivative is not what the Hohenberg–Kohn theorem demands, but rather a Hartree–Fock-like procedure, but in practice it is a convenient and successful approach. This scheme, in which self-consistency is obtained with respect to the single-particle orbitals, can be considered an evolution of the Hartree–Fock Kohn–Sham method,[8] and is how hybrids are commonly implemented. Recently, it has also been used for meta-GGAs.[67,118,119] For occupied orbitals, results obtained from orbital self-consistency differ little from those obtained from density self-consistency, implemented via the OEP.

Apart from orbital functionals, which are implicit nonlocal density functionals because the orbitals depend on the density in a nonlocal way, there is also a class of explicit nonlocal density functionals. Such nonlocal density functionals take into account, at any point $\mathbf{r}$, not only the density at that point, $n(\mathbf{r})$, and its derivatives, $\nabla n(\mathbf{r})$, etc., but also the behaviour of the density at different points $\mathbf{r}' \neq \mathbf{r}$, by means of integration over physically relevant regions of space. A typical example is

$$E_{\text{xc}}^{\text{ADA}}[n] = \int d^3r \, n(\mathbf{r}) \epsilon_{\text{xc}}^{\text{hom}}(\bar{n}(\mathbf{r})), \tag{82}$$

where $\epsilon_{\text{xc}}^{\text{hom}}$ is the per-particle exchange–correlation energy of the homogeneous electron liquid. In the LDA one would have $\bar{n}(\mathbf{r}) \equiv n(\mathbf{r})$, but in the ADA one takes[145]

$$\bar{n}(\mathbf{r}) = \int d^3r' \, n(\mathbf{r}') w[n](|\mathbf{r} - \mathbf{r}'|), \tag{83}$$

where $w[n](|\mathbf{r} - \mathbf{r}'|)$ is a weight function that samples the density not only semilocally, as do the GGAs, but over a volume determined by the range of w.

The dependence of the ADA and the closely related weighted-density approximation (WDA)[145] on $\bar{n}(\mathbf{r})$, the integral over $n(\mathbf{r})$, instead of on derivatives, as in the GGAs, is the reason why such functionals are called 'nonlocal'. In practice, this integral makes the functionals computationally expensive, and in spite of their great promise they are used much less than GGAs. However, recent comparisons of ADA and WDA with LDA and GGAs for low-dimensional systems[122,146] and for bulk silicon[147] have shown that nonlocal integral-dependent density functionals can outperform local and semilocal approximations.

4. Performance of Approximate Functionals: A Few Examples

No systematic attempt at comparing explicit functionals can be made here. Extensive comparisons of a wide variety of functionals can be found in Refs. [7–21]. For pure illustrative purposes only, Tables 1–5 show ground-state energies of the argon atom and of the water

Table 1.
Ground-state energy in atomic units of the Ar atom ($Z = 18$), obtained with some representative density functionals and related methods. The Hartree–Fock and exact exchange values are from Krieger et al. (third of Ref. [133]), average-density approximation (*ADA*) and weighted-density approximation (*WDA*) values are from Gunnarsson et al.,[145] as reported in Ref. [7], and the local density approximation (*LDA*)–self-interaction correction (*SIC*)(PZ) value is from Perdew and Zunger.[111] The experimental value is based on,[231] and is given to fewer significant digits than the calculated values, because of relativistic and quantum electrodynamical effects (Lamb shift) that are automatically included in the experimental result but not in the calculated values. (Reprinted from Ref. [23]. Copyright © 2006 with permission from the *Brazilian Journal of Physics*.)

Method	$-E$/au
Thomas–Fermi	625.7
Hartree–Fock	526.818
Exact exchange	526.812
LDA (exchange only)	524.517
LDA (VWN)	525.946
LDA (PW92)	525.940
LDA-SIC(PZ)	528.393
ADA	527.322
WDA	528.957
GGA (B88LYP)	527.551
Experiment	527.6

Table 2.
Ground-state energy, in atomic units, of the H_2O molecule, obtained with some representative density functionals and basis sets. The calculation was performed with the Gaussian03[160] electronic structure package. For comparison, the total energy for H_2O obtained with full configuration interaction (FCI/cc-pVDZ) is -76.242 hartree.[161]

Basis set	LDA (SVWN5)	GGA (PBE)	Hybrid (B3LYP)
STO-3G	−74.730	−75.223	−75.310
6-311G	−75.851	−76.328	−76.416
cc-pVDZ	−75.853	−76.332	−76.419
aug-cc-pVQZ	−75.911	−76.386	−76.472

Table 3.
Same as Table 2, for the H_2O^+ ion. FCI/cc-pVDZ total energy is -75.733 hartree.[161]

Basis set	LDA (SVWN5)	GGA (PBE)	Hybrid (B3LYP)
STO-3G	−74.376	−74.871	−74.954
6-311G	−75.351	−75.871	−75.956
cc-pVDZ	−75.391	−75.883	−75.968
aug-cc-pVQZ	−75.428	−75.916	−76.002

Table 4.
Same as Table 2, for the H_3O^+ ion. Calculated Multi-Reference Double-Excitation configuration interaction/cc-pVTZ total energy is -76.567 hartree at a ground state equilibrium geometry C_{3v} with H-O-H angle 120° and O–H bond length 0.096 nm.[162]

Basis set	LDA (SVWN5)	GGA (PBE)	Hybrid (B3LYP)
STO-3G	−75.103	−75.590	−75.676
6-311G	−76.140	−76.619	−76.706
cc-pVDZ	−76.136	−76.614	−76.701
aug-cc-pVQZ	−76.176	−76.655	−76.742

Table 5.
Same as Table 2, for the OH$^-$ ion. FCI/cc-pVDZ total energy is -75.623 hartree.[163]

Basis set	LDA (SVWN5)	GGA (PBE)	Hybrid (B3LYP)
STO-3G	−73.806	−74.302	−74.395
6-311G	−75.189	−75.658	−75.746
cc-pVDZ	−75.159	−75.633	−75.721
aug-cc-pVQZ	−75.293	−75.760	−75.842

molecule. Owing to the importance of the paradigmatic dissociation reaction $2H_2O \rightarrow H_3O^+ + OH^-$ for electrochemistry, we also include data on the ionic dissociation products. Although such very limited comparisons cannot substitute a complete analysis, the data in Tables 1–5 do illustrate the quality, or lack thereof, of common density functionals, combined with common basis sets.

Accurate ground-state energies are the key to obtaining many other properties, such as geometries, ionization energies and vibrational frequencies, and functionals are usually constructed (or, in the case of semiempirical functionals, fitted) to give as good as possible energies. Modern high-level functionals predict molecular energies to within less than approximately 4 kcal/mol,[67, 73, 118, 119] which is close to the desired chemical accuracy of 1 kcal/mol = 0.0434 eV/particle. On the other hand, energy gaps in solids can be wrong by 100%.

Ionization energies can, in principle, be obtained from the highest occupied Kohn–Sham eigenvalue (see Sect. V.2), but for local and semilocal approximations, this does not yield reliable results because the asymptotic effective potential obtained from these approximations decays exponentially, and not as the correct $1/r$. As a consequence, the outermost electron is too weakly bound, and ionization energies are underestimated. SICs or other fully nonlocal functionals are needed to improve this behaviour and the resulting eigenvalues.[81, 111] Ionization energies obtained as total-energy differences, on the other hand, are much more reliable.

Electron affinities are harder to obtain than ionization energies, because within local and semilocal approximations the $N+1$'st electron often is not bound at all, so wrong affinities result both from eigenvalues and from total energies. By exploiting error cancellation between basis-set effects (which artificially bind the extra electron) and the wrong asymptotic tail of LDA and GGA potentials (which leads to underbinding of the extra electron), one can nevertheless obtain reliable affinities from standard functionals.[148]

Another consequence of the wrong asymptotic behaviour of local and semilocal functionals is that static polarizabilities are frequently overestimated by standard functionals. Much useful information on the performance of approximate density functionals in calculations of electric polarizabilities and hyperpolarizabilities of atoms, molecules and clusters can be found in Schwerdtfeger,[149] Fuentealba[150] and Pouchen et al.[151]

Quite generally, current functionals give acceptable results for lengths and strengths of all main types of chemical bonds (covalent, ionic, metallic and hydrogen bridge). The LDA tends to slightly (approximately 1%) underestimate bond lengths, while the GGA overestimates them by a similar margin. Hybrid functionals and other orbital-dependent functionals further improve this situation: bond lengths of molecules can be predicted by high-level density functionals with an average error of less than 0.001 nm.

For van der Waals (dispersion) interactions, however, many common GGAs and the LDA fail. The PBE GGA[66] is a partial exception[121,152] because it works reasonably well near the equilibrium distance of the van der Waals bond, but PBE recovers only the short-range behaviour and does not describe correctly the long-range asymptotic regime of the van der Waals interaction. To describe these very weak interactions, a variety of more specialized approaches has been developed within DFT,[73,153–159] but it is too early to say whether one of them will evolve into a reliable standard method, comparable to the LDA or the GGA.

As the systems become larger and larger, DFT encounters computational limitations, and must yield to semiempirical modelling. At the other end of the spectrum, as demands on accuracy grow beyond that attained by today's best functionals, wave-function-based methods such as CI, CC or MP reign supreme. Simultaneous progress in computing technology and construction of density functionals, however, has pushed both frontiers of DFT further and further out.

VII. EXTERNAL ELECTRIC AND MAGNETIC FIELDS

In static fields, one can separately deal with electric and magnetic phenomena. Electric fields couple to the charge density. Magnetic fields can be further classified according to whether they couple to the spin degrees of freedom (spin-only magnetic fields, as in pure Zeeman coupling), to the orbital degrees of freedom (current-only magnetic fields, as in classical physics) or to both. Although only the latter case corresponds to physical magnetic fields, in the absence of spin–orbit coupling (which is a relativistic effect) spins and currents represent independent degrees of freedom, and the coupling of external fields to them can be described separately. Moreover, quite frequently the coupling to currents (mediated by a vector potential in the Hamiltonian) is much weaker than that to the spins. In this situation one can in a first approximation neglect currents and vector potentials altogether, and deal with charges and spins only. This is the situation for which SDFT was designed.[85, 109]

Below we first offer some brief remarks on SDFT, and then turn to the more complex problems posed by orbital currents, described by CDFT. Both, SDFT and CDFT also have a relativistic version, on which we make some brief comments in Sect. VII.2. Next, we briefly consider the simpler (but still not trivial) case of external electric fields. Finally, we describe the peculiar problems arising in the calculation of the electric polarization and the orbital magnetization of extended systems. We do not address, in this chapter, time-dependent electromagnetic fields, for which we refer the reader to reviews of the large and flourishing field of TD-DFT.[164–166]

1. Magnetic Fields Coupling to the Spins: SDFT

Up to this point we have discussed DFT in terms of the charge (or particle) density $n(\mathbf{r})$ as a fundamental variable. To reproduce the correct charge density of the interacting system in the noninteracting (Kohn–Sham) system, one must apply to the latter the effective Kohn–Sham potential $v_s = v + v_H + v_{xc}$, in which the last two terms simulate the effect of the electron–electron interaction on the charge density. This form of DFT, which is the one proposed originally,[29] could also be called 'charge-only' DFT. It is not the most widely used DFT in practical applications. Much more common is a formulation that employs one density for each spin, $n_\uparrow(\mathbf{r})$ and $n_\downarrow(\mathbf{r})$, i.e. works with two fundamental variables. To reproduce both of these in the noninteracting system, one must now apply two effective potentials, $v_{s,\uparrow}(\mathbf{r})$ and $v_{s,\downarrow}(\mathbf{r})$. More generally, one requires one effective potential for each density-like quantity to be reproduced in the Kohn–Sham system. Such potentials and corresponding densities are called 'conjugate variables'.

The formulation of DFT in terms of $n_\uparrow(\mathbf{r})$ and $n_\downarrow(\mathbf{r})$ is known as collinear SDFT.[85,109] Its fundamental variables $n_\uparrow(\mathbf{r})$ and $n_\downarrow(\mathbf{r})$ can be used to calculate the charge density $n(\mathbf{r})$ and the spin-magnetization density $m(\mathbf{r})$ from

$$n(\mathbf{r}) = n_\uparrow(\mathbf{r}) + n_\downarrow(\mathbf{r}), \tag{84}$$
$$m(\mathbf{r}) = \mu_0(n_\uparrow(\mathbf{r}) - n_\downarrow(\mathbf{r})), \tag{85}$$

where $\mu_0 = q\hbar/2mc$ is the Bohr magneton for particles of charge q. More generally, the Hohenberg–Kohn theorem of SDFT states that in the presence of a magnetic field $\boldsymbol{B}(\mathbf{r})$ that couples only to the electron spin (via the familiar Zeeman term $\int d^3r\, m(\mathbf{r})\boldsymbol{B}(\mathbf{r})$) the ground-state wave function and all ground-state observables are unique functionals of n and m or, equivalently, of $n_\uparrow$ and $n_\downarrow$. Almost the entire further development of the Hohenberg–Kohn theorem and the Kohn–Sham equations can be immediately rephrased for SDFT, just by attaching a suitable spin index to the densities. For this reason we could afford the luxury of exclusively discussing 'charge-only' DFT in the preceding sections, without missing any essential aspects of SDFT.

There are, however, some exceptions to this simple rule. One is the fourth statement of the Hohenberg–Kohn theorem, as discussed in Sect. III.1. Another is the construction of functionals. For the exchange energy it is known, e.g., that[167]

$$E_\text{x}^\text{SDFT}[n_\uparrow, n_\downarrow] = \frac{1}{2}\left(E_\text{x}^\text{DFT}[2n_\uparrow] + E_\text{x}^\text{DFT}[2n_\downarrow]\right). \tag{86}$$

In analogy to the coordinate scaling of (49)–(51), this property is often called 'spin-scaling', and it can be used to construct a SDFT exchange functional from a given DFT exchange functional. In the context of the local-spin-density approximation, von Barth and Hedin[109] wrote the exchange functional in terms of an interpolation between the unpolarized and fully polarized electron gas which by construction satisfies (86). Alternative interpolation procedures can be found in Ref. [64]. GGA exchange functionals also satisfy (86) by construction. For the correlation energy no scaling relation of the type (86) holds, so in practice correlation functionals are either directly constructed in terms of the spin densities or written using, without formal justification, the same interpolation already used in the exchange functional. In the case of the local-spin-density approximation, this latter procedure was introduced in Ref., [109] and further analysed and improved in Ref. [64].

The Kohn–Sham equations of SDFT are

$$\left[-\frac{\hbar^2 \nabla^2}{2m} + v_{s\sigma}(\mathbf{r}) \right] \phi_{i\sigma}(\mathbf{r}) = \epsilon_{i\sigma} \phi_{i\sigma}(\mathbf{r}), \qquad (87)$$

where $v_{s\sigma}(\mathbf{r}) = v_\sigma(\mathbf{r}) + v_H(\mathbf{r}) + v_{xc,\sigma}(\mathbf{r})$. In a nonrelativistic calculation the Hartree term does not depend on the spin label [spin–spin dipolar interactions are a relativistic effect of order $(1/c)^2$, as are current–current interactions], but in the presence of an externally applied magnetic field $v_\sigma(\mathbf{r}) = v(\mathbf{r}) - \sigma \mu_0 B$ (where $\sigma = \pm 1$). Finally,

$$v_{xc,\sigma}(\mathbf{r}) = \frac{\delta E_{xc}^{SDFT}[n_\uparrow, n_\downarrow]}{\delta n_\sigma(\mathbf{r})}. \qquad (88)$$

The exchange–correlation magnetic field $B_{xc} = \mu_0(v_{xc,\downarrow} - v_{xc,\uparrow})$ is the origin of, e.g., ferromagnetism in transition metals.

References to recent work *with* SDFT include almost all practical DFT calculation: SDFT is by far the most widely used form of DFT. In fact, SDFT has become synonymous with DFT to such an extent that often no distinction is made between the two, i.e. a calculation referred to as a DFT one is most often really a SDFT one Some recent work *on* SDFT is described in Ref. [168]. A more detailed discussion of SDFT can be found in Refs., [7, 8, 85] and a particularly clear exposition of the construction of exchange–correlation functionals for SDFT is the contribution of Kurth and Perdew in Refs. [17, 18].

If the direction of the spins is not uniform in space, we are dealing with *noncollinear magnetism*. Noncollinear spin structures appear, e.g., as canted or helical spin configurations in rare-earth compounds, as helical spin-density waves, or as domain walls in ferromagnets. To describe these, one requires a formulation of SDFT in which the spin magnetization is not a scalar, as above, but a three-component vector $\mathbf{m}(\mathbf{r})$. Different proposals for extending SDFT to this situation are available.[169–171]

2. Brief Remarks on Relativistic DFT

Frequently, noncollinearity is due to spin–orbit coupling. Although spin–orbit terms can be added as a perturbation to the equations of SDFT, a complete description requires a relativistic formulation. A generalization of DFT that does account for spin–orbit coupling and other relativistic effects is RDFT.[20,25,26] Here the fundamental variable is the relativistic four-component current j^μ and the Kohn–Sham equation is now of the form of the single-particle Dirac equation, instead of the Schrödinger equation.

In practice, one typically employs so-called relativistic SDFT (R-SDFT), in which a Gordon decomposition of the four-current is performed to separate orbital and spin degrees of freedom, and only the spin magnetization is maintained as a fundamental variable. This reduced RDFT is not Lorentz-invariant (which is not essential in solid-state physics and quantum chemistry, where a preferred frame is provided by the laboratory) and does not account for orbital magnetism (except, possibly, induced by spin–orbit coupling, which can be treated as a perturbation). Relativistic CDFT (R-CDFT) can overcome both limitations, but is less frequently used since it is more complicated to implement, and less is known about approximate four-current functionals.

In both R-SDFT and R-CDFT there are also many subtle questions involving renormalizability and the use of the variational principle in the presence of negative-energy states. For details on these problems and their eventual solution the reader is referred to the chapters by E. Engel et al., [12 page 1, 21 page 65] and to the book by Eschrig.[20] A didactical exposition of RDFT, together with representative applications in atomic and condensed-matter physics, can be found in the book by Strange,[27] and a recent numerical implementation of R-SDFT is presented in Ref. [172].

3. Magnetic Fields Coupling to Spins and Currents: CDFT

There are at least four conceptually distinct ways in which orbital magnetism can appear in a physical system. One is the presence of external magnetic fields $\mathbf{B}(\mathbf{r})$, whose vector potential $\mathbf{A}(\mathbf{r})$ enters the Hamiltonian via the usual minimal substitution in the kinetic energy

$$\frac{\hat{p}^2}{2m} \longrightarrow \frac{1}{2m}\left(\hat{p} - \frac{q}{c}\mathbf{A}(\mathbf{r})\right)^2. \tag{89}$$

This substitution is, formally, easy to perform in the many-body and the Kohn–Sham Hamiltonians of SDFT, but the presence of the vector potential complicates the task of solving these equations. Maintaining gauge invariance is not trivial in approximate calculations. Moreover, in extended systems the vector potential breaks translational invariance, so Bloch's theorem cannot be used anymore.[173,174] Still, couplings of external vector potentials become relevant in many situations, and ways to deal with them have been developed, e.g. for the calculation of nuclear magnetic shielding tensors and spin–spin coupling constants.[175–179]

A second way in which orbital magnetism can appear is due to current–current interactions, which are part of the Breit interaction[27,180] and therefore a relativistic effect. The nonretarded part of this interaction is

$$-\frac{q^2}{c^2} \int d^3r \int d^3r' \frac{\mathbf{j}_p(\mathbf{r}) \cdot \mathbf{j}_p(\mathbf{r}')}{|\mathbf{r} - \mathbf{r}'|} = -\frac{q}{c} \int d^3r \, \mathbf{j}_p(\mathbf{r}) \cdot \mathbf{A}_H(\mathbf{r}), \quad (90)$$

which describes the Hartree-like coupling of currents to the self-induced vector potential, corresponding to the Amperian currents of classical electrodynamics.

Third, in the presence of spin–orbit coupling a nonzero spin magnetization can induce an orbital magnetization. Orbital magnetic moments induced by spin–orbit coupling can be treated even within SDFT or R-SDFT, if spin–orbit coupling is added to the Hamiltonian. Such magnetic moments become important, e.g., in magnetic solids, where spin–orbit coupling produces phenomena such as magnetocrystalline anisotropy[181,182] or magnetooptical effects such as dichroism.[183,184]

A consistent treatment of terms of second order in v/c requires a relativistic formulation, and therefore both the Hartree vector potential and spin–orbit coupling are often neglected in nonrelativistic CDFT.

Finally, and most intriguingly, orbital magnetism can also occur spontaneously in a system with pure Coulomb interactions if the system minimizes its energy in a current-carrying state.[185,186] The resulting currents are, in principle, functionals of the charge and spin densities, because the original formulation of the SDFT Hohenberg–Kohn theorem applies in the absence of magnetic fields, but SDFT provides no explicit prescription for how to calculate the spontaneous orbital currents and their effect on observables.

This situation has changed with the advent of nonrelativistic CDFT, developed by Vignale and Rasolt,[185, 187] which describes spontaneous currents by introducing in the Kohn–Sham equations a self-consistent exchange–correlation vector potential $\mathbf{A}_{xc}$, which can be nonzero also in the absence of external magnetic fields and of relativistic effects.

In addition to possibly appearing spontaneously, $\mathbf{A}_{xc}(\mathbf{r})$ also becomes nonzero as soon as currents are induced by one of the other three mechanisms described previously. In this case, it constitutes a correction to the external or Hartree vector potentials, or spin–orbit terms, already present in the Hamiltonian. CDFT, with some approximation for $\mathbf{A}_{xc}$, has been applied to the calculation of the effects of orbital magnetism in atoms,[188–190] quantum dots,[191–193] molecules[194, 195] and solids.[186, 196–198]

To briefly describe the formalism of CDFT, we first recall the form of the traditional Kohn–Sham equation,

$$\left[-\frac{\hbar^2}{2m}\nabla^2 + v_{\rm s}^{\rm d}(\mathbf{r}) \right] \phi_k(\mathbf{r}) = \epsilon_k^{\rm d}\phi_k(\mathbf{r}). \tag{91}$$

Here an upper index d denotes DFT, and the effective single-particle potential, $v_{\rm s}^{\rm d}(\mathbf{r})$, is defined as

$$v_{\rm s}^{\rm d}(\mathbf{r}) = v(\mathbf{r}) + v_{\rm H}(\mathbf{r}) + v_{\rm xc}^{\rm d}(\mathbf{r}), \tag{92}$$

where $v(\mathbf{r})$ is the external potential, $v_{\rm H}$ is the Hartree-type electrostatic potential and $v_{\rm xc}^{\rm d}(\mathbf{r})$ is the exchange–correlation potential, in which the entire complexity of the many-body problem is hidden.[7, 8] By construction, the single-particle orbitals ϕ_k, solving the eigenvalue problem (91), reproduce the density of the interacting system via

$$n(\mathbf{r}) = \sum_k \phi_k^*(\mathbf{r})\phi_k(\mathbf{r}). \tag{93}$$

On the other hand, the paramagnetic current density,

$$\mathbf{j}_{\rm p}^{\rm KS}(\mathbf{r}) = \frac{\hbar}{2mi} \sum_k \left[\phi_k^*(\mathbf{r})\nabla\phi_k(\mathbf{r}) - \phi_k(\mathbf{r})\nabla\phi_k^*(\mathbf{r}) \right], \tag{94}$$

following from these orbitals, is, a priori, not guaranteed to have any relation with the true current density of the interacting system.

The corresponding equations of CDFT have a slightly more complicated form:

$$\left[\frac{1}{2m}\left(\frac{\hbar}{i}\nabla - \frac{q}{c}\mathbf{A}_s(\mathbf{r})\right)^2 + V_s^c(\mathbf{r})\right]\psi_k(\mathbf{r}) = \epsilon_k^c \psi_k(\mathbf{r}), \quad (95)$$

where an upper index c denotes CDFT,

$$V_s^c(\mathbf{r}) = v_s^c(\mathbf{r}) + \frac{q^2}{2mc^2}\left(\mathbf{A}(\mathbf{r})^2 - \mathbf{A}_s(\mathbf{r})^2\right), \quad (96)$$

$$v_s^c(\mathbf{r}) = v(\mathbf{r}) + v_H(\mathbf{r}) + v_{xc}^c(\mathbf{r}) \quad (97)$$

and

$$\mathbf{A}_s(\mathbf{r}) = \mathbf{A}(\mathbf{r}) + \mathbf{A}_{xc}(\mathbf{r}). \quad (98)$$

Here $\mathbf{A}_{xc}$ and v_{xc}^c are the exchange–correlation scalar and vector potentials of CDFT, respectively.[185,187] $\mathbf{A}_{xc}$, in particular, is a gauge-invariant functional of the densities $n(\mathbf{r})$ and $\mathbf{j}_p(\mathbf{r})$, written $\mathbf{A}_{xc}[n, \mathbf{j}_p](\mathbf{r})$. By setting $\mathbf{A}_{xc} \equiv 0$, one recovers from CDFT the equations of (S)DFT in an external vector potential $\mathbf{A}$. If observables for $\mathbf{A}_{xc} \equiv 0$, but $\mathbf{B} = \nabla \times \mathbf{A} \neq 0$, are considered functionals of the external field $\mathbf{B}$, one arrives at a formulation known as BDFT.[199,200] The novel feature of CDFT is that $\mathbf{A}_{xc}$ accounts for the orbital degrees of freedom in the single-particle equations, even in the absence of external fields, and allows calculation of orbital currents directly from a set of Kohn–Sham-type equations.

The single-particle orbitals ψ_k, solving the more complicated eigenvalue problem (95), reproduce by construction both the density

$$n(\mathbf{r}) = \sum_k \psi_k^*(\mathbf{r})\psi_k(\mathbf{r}), \quad (99)$$

and the paramagnetic current density

$$\mathbf{j}_p(\mathbf{r}) = \frac{\hbar}{2mi}\sum_k \left[\psi_k^*(\mathbf{r})\nabla\psi_k(\mathbf{r}) - (\nabla\psi_k^*(\mathbf{r}))\psi_k(\mathbf{r})\right] \quad (100)$$

of the interacting many-body system. The paramagnetic current alone is not gauge-invariant, but the gauge-invariant orbital current, $\mathbf{j}_{orb}(\mathbf{r})$, is simply obtained from

$$\mathbf{j}_{\mathrm{orb}}(\mathbf{r}) = \mathbf{j}_{\mathrm{p}}(\mathbf{r}) - \frac{q}{mc} n(\mathbf{r}) \mathbf{A}(\mathbf{r}). \tag{101}$$

An important property of the effective potential of DFT, $v_{\mathrm{s}}^{\mathrm{d}}(\mathbf{r})$, [which is *not* shared by the CDFT potentials $V_{\mathrm{s}}^{\mathrm{c}}(\mathbf{r})$ and $\mathbf{A}_{\mathrm{s}}(\mathbf{r})$] is its *uniqueness*: for any given ground-state density $n(\mathbf{r})$ there is, up to an irrelevant additive constant, at most *one* such local multiplicative potential.[7]

For $\mathbf{B} = 0$ the natural choice of gauge is $\mathbf{A} = 0$ (the same gauge universally chosen in solving (91)). The CDFT Kohn–Sham equation (95) then takes the form

$$\left[-\frac{\hbar^2}{2m} \nabla^2 + v_{\mathrm{s}}^{\mathrm{c}}(\mathbf{r}) + \hat{Y}(\mathbf{r}) \right] \psi_k(\mathbf{r}) = \epsilon_k^{\mathrm{c}} \psi_k(\mathbf{r}), \tag{102}$$

where the operator $\hat{Y}(\mathbf{r})$ is defined by

$$\hat{Y}(\mathbf{r}) = -\frac{\hbar q}{2mci} \left(\nabla \mathbf{A}_{\mathrm{xc}}(\mathbf{r}) + \mathbf{A}_{\mathrm{xc}}(\mathbf{r}) \nabla \right). \tag{103}$$

A perturbative treatment of $\hat{Y}$ has been suggested in Ref. [201].

Any CDFT calculation requires an approximation for the current dependence of the E_{xc} functional. For the homogeneous three-dimensional electron liquid in strong uniform magnetic fields, the exchange energy is known exactly,[202] and the correlation energy has been calculated within the random-phase approximation[203] and the self-consistent local-field-corrected Singwi–Tosi–Land–Sjolander scheme.[204] In weak fields, where linear-response theory applies, the exchange–correlation energy can be expressed in terms of the magnetic susceptibility, for which many-body calculations are available from.[205] The exchange–correlation energy of two-dimensional electron liquids in uniform magnetic fields has been much studied in the context of the fractional quantum Hall effect in quasi-two-dimensional semiconductor heterostructures, but as such systems are less relevant for chemical applications, we refrain from describing this work here.

All these results for homogeneous electron liquids in uniform magnetic fields can be used to construct LDAs to CDFT. The effect of nonuniform magnetic fields in inhomogeneous many-electron systems is then obtained by solving the CDFT Kohn–Sham equations with this LDA. Alternatively, the exchange–correlation energy in special nonuniform magnetic fields has also been calculated directly, by Harris and Grayce,[199] but such calculations are very hard to generalize to magnetic fields of arbitrary form.

Finally, we note that the nonrelativistic time-dependent CDFT, recently formulated by Vignale et al.[206] does not include the coupling to external magnetic fields. Time-dependent CDFT is thus very different from static CDFT, which does include this coupling.

4. Electric Fields

Static external electric fields are much simpler to include in the basic formalism of DFT than magnetic fields and vector potentials, since they couple to the charge density only. The effective single-particle potential in the Kohn–Sham equations is a sum of several terms,

$$v_s(\mathbf{r}) = v_{ext}(\mathbf{r}) + v_H(\mathbf{r}) + v_x(\mathbf{r}) + v_c(\mathbf{r}), \qquad (104)$$

where the Hartree, exchange and correlation terms arise self-consistently from the electron–electron interaction. The potential normally referred to as 'external' potential $v_{ext}(\mathbf{r})$ in DFT is external in the sense that it does not arise from the electrons it is acting on, but, within the Born–Oppenheimer approximation, from the protons in the nuclei.

Static electric fields applied in the laboratory, e.g. by placing the system between plates of a capacitor or contacting it with electrodes, are external to both the electronic and the nuclear subsystems, and can simply be added to $v_{ext}(\mathbf{r})$. No generalizations of the Hohenberg–Kohn theorem are required to perform this step, and the behaviour of observables of *finite systems* as a function of external electric fields can be studied by means of the conventional Kohn–Sham equations of SDFT. In fact, SDFT is now commonly used in a wide variety of electrochemical calculations. Representative examples are found in Refs., [3–6] as well as in a previous volume of this series.[2]

An interesting complication, requiring special care, occurs in generalizing DFT in external electric or magnetic fields to the case of extended systems, which we discuss in the next section.

5. Polarization and Magnetization

The proper quantum mechanical treatment of the electric polarization of a dielectric material and of the orbital magnetization of a magnetic material is rather subtle.[207] To illustrate the main complication arising in the context of DFT, let us consider a homogeneous electric field $\mathbf{E} = -\nabla \Phi$, entering the Hamiltonian via the corresponding electric potential

$$\hat{H}_{\text{el}} = \sum_i q_i \Phi(\mathbf{r}_i) = -\sum_i q_i \mathbf{r}_i \cdot \mathbf{E}, \qquad (105)$$

where the sum is over all particles in the system, with charges q_i and masses m_i, and $\mathbf{E}$ is independent of $\mathbf{r_i}$. This potential diverges for $|\mathbf{r}_i| \to \infty$. As a consequence, there is no stable quantum mechanical ground state in homogeneous electric fields.[208–210] The variational principle then does not apply, and the Hohenberg–Kohn theorem does not hold. The potential of the homogeneous electric field is therefore not determined by the charge density. Equivalently, the homogeneous electric field itself is not determined by its conjugate variable, the macroscopic electric polarization, related to n via $\nabla \cdot \mathbf{P} = n$.

Since perfectly homogeneous electric fields do not exist in nature, it seems that this lack of a Hohenberg–Kohn theorem would be of little consequence for real materials and systems. However, the electronic structure of real dielectrics is very frequently calculated under the assumption that the system is infinitely extended, and for such macroscopic dielectrics it thus appears that DFT in its usual formulation cannot be applied.

One obvious solution is to treat the system as finite, in which case the boundary conditions to the electric field (e.g. capacitor plates) and the surface of the sample must be taken into account explicitly during the calculation. As an alternative, an additional variable can be included in the density-functional formalism to account for the information not contained in the charge density. Perhaps the simplest choice is the homogeneous external electric field itself.[211,212] Alternatively, it has been proposed to include the macroscopic polarization among the basic variables of DFT, leading to a density-and-polarization-functional theory,[210,213–217] but it has been argued that no combination of the purely local variables $n(\mathbf{r})$, $\mathbf{P}(\mathbf{r})$ and $\mathbf{E}(\mathbf{r})$ fully determines the corresponding Kohn–Sham potential.[212]

An alternative approach that avoids many of the problems associated with electric polarization in dielectrics is TD-DFT. In the time-dependent case, the time change of the polarization induces a current, which may be considered an ultra-nonlocal functional of the charge density, and has been successfully used as an alternative additional variable for the description of dielectric properties of both solids and molecular systems.[216–218]

In homogeneous magnetic fields, a very similar, though somewhat less explored, set of questions arises. The magnetization is a quantity analogous to the electric polarization. In a spin-only magnetic field, such as used in SDFT, the homogeneous magnetic field couples to the spin magnetization via the Zeeman term

$$\hat{H}_Z = -\frac{\hbar}{2c} \sum_i \frac{q_i}{m_i} \vec{\sigma}_i \cdot \mathbf{B}. \qquad (106)$$

For a system of electrons, $\hbar q_i/2m_i c = -\hbar e/2mc$. The contribution of this term to the total energy does not diverge in infinite homogeneous systems, unlike the coupling to electric fields, and therefore the calculation of spin magnetizations is more straightforward than that of electric polarizations. The orbital magnetization, however, arises from the coupling to vector potentials, which in linear order is given by

$$\hat{H}_{\text{orb}} = -\frac{1}{2c} \sum_i \frac{q_i}{m_i} \left[\mathbf{A}(\mathbf{r}_i) \cdot \mathbf{p}_i + \mathbf{p}_i \cdot \mathbf{A}(\mathbf{r}_i) \right], \qquad (107)$$

$$= -\frac{1}{2c} \sum_i \frac{q_i}{m_i} \mathbf{B} \times \mathbf{r}_i \cdot \mathbf{p}_i, \qquad (108)$$

where we used the gauge $\mathbf{A} = \frac{1}{2} \mathbf{B} \times \mathbf{r}$. This coupling again features a linear term in the position operator, and diverges for $|\mathbf{r}| \to \infty$. The calculation of orbital magnetizations in DFT is thus subject to the same questioning as that of electric polarizations, but that of spin magnetizations is not. Highly nonlocal, e.g. orbital-dependent, functionals of the current density are required to calculate orbital magnetizations of extended systems.[219–222]

For finite systems, the complications arising in infinite periodic systems disappear, but now the surface prohibits application of Bloch's theorem, and calculations on finite clusters must be carried out for the specific geometry of the system under study. In either case, the orbitals used to calculate orbital currents and orbital magnetizations should, rigorously, be those of CDFT. SDFT orbitals calculated in the presence of external vector potentials (but no exchange–correlation vector potentials) are not guaranteed to reproduce the correct many-body orbital currents and magnetizations, regardless of whether the system is finite or infinite. How large the influence of $\mathbf{A}_{xc}$ is in a given system is a separate question. The empirical success of SDFT with external vector potentials in calculating nuclear-magnetic shielding tensors[175–178] and the smallness of $\mathbf{A}_{xc}$ in such calculations,[194, 195, 223] as well as in calculations of magnetic moments of solids[196, 198] and of ionization energies of current-carrying states,[189, 189] suggest that CDFT corrections are small in most cases, and SDFT provides at least a useful starting point.

VIII. OUTLOOK

Extensions of DFT to time-dependent, magnetic, relativistic and a multitude of other situations involve more complicated Hamiltonians than the basic ab initio many-electron Hamiltonian defined in (2)–(6). Frequently, the inverse strategy is also useful: instead of attempting to achieve a more complete description of the many-body system under study by adding additional terms to the Hamiltonian, it can be advantageous to reduce the complexity of the ab initio Hamiltonian by replacing it by simpler models, which focus on specific aspects of the full many-body problem. DFT can be applied to such model Hamiltonians too, once a suitable density-like quantity has been identified as a basic variable. Following pioneering work by Gunnarsson and Schönhammer,[224] LDA-type approximations have, e.g., recently been formulated and exploited for the Hubbard,[225] the delta-interaction[226] and the Heisenberg[227] models. Common aspects and potential uses of DFT for model Hamiltonians are described in Ref. [228].

Still another way of using DFT which does not depend directly on approximate solution of Kohn–Sham equations is the quantification and clarification of traditional chemical concepts, such as electronegativity,[8] hardness, softness, Fukui functions and other reactivity indices[8,229] or aromaticity.[230] The true potential of DFT for this kind of investigation is only beginning to be explored, but holds much promise.

All extensions of DFT face the same formal questions (e.g. interacting and noninteracting v-representability of the densities, meaning of the Kohn–Sham eigenvalues, nonuniqueness of the Kohn–Sham potentials) and practical problems (e.g. how to efficiently solve the Kohn–Sham equations, how to construct accurate approximations to E_{xc}, how to treat systems with very strong correlations or with a very large number of electrons) as does standard (S)DFT. These questions and problems, however, have never stopped DFT from advancing,[7–21] and at present DFT emerges as the method of choice for solving a wide variety of quantum mechanical problems in chemistry and physics, and in many situations, such as large and inhomogeneous systems, it is the only applicable first-principles method at all.

ACKNOWLEDGEMENTS

This work was supported financially by FAPESP and CNPq. We thank Daniel Vieira for providing the original version of Fig. 1 and the *Brazilian Journal of Physics* for permission to use Fig. 1 and Table 1, which were originally published in Ref. [23].

REFERENCES

[1] W. Kohn, Rev. Mod. Phys. **71**, 1253 (1999).
[2] M. T. M. Koper, in Modern aspects of electrochemistry, edited by C. G. Vayenas, B. E. Conway, and R. E. White, vol. 36, chap. 2 (Kluwer/Plenum, New York, 2003).
[3] G. Maroulis, ed., *Atoms, Molecules and Clusters in Electric Fields: Theoretical Approaches to the Calculation of Electric Polarizability* (Imperial College Press, London, 2006).
[4] W. Schmickler, Annu. Rep. Prog. Chem., Sect. C **95**, 117 (1999).
[5] M. Jacoby, Chem. Eng. News **82**, 25 (2004).
[6] Z. Shia, J. Zhanga, Z.-S. Liua, H. Wanga, and D. P. Wilkinson, Electrochim. Acta **51**, 1905 (2006).
[7] R. M. Dreizler and E. K. U. Gross, *Density Functional Theory* (Springer, Berlin, 1990).
[8] R. G. Parr and W. Yang, *Density-Functional Theory of Atoms and Molecules* (Oxford University Press, Oxford, 1989).
[9] W. Koch and M. C. Holthausen, *A Chemist's Guide to Density Functional Theory* (Wiley, New York, 2001).

[10] R. O. Jones and O. Gunnarsson, Rev. Mod. Phys. **61**, 689 (1989).
[11] J. M. Seminario, ed., *Recent Developments and Applications of Modern DFT* (Elsevier, Amsterdam, 1996).
[12] R. F. Nalewajski, ed., Density functional theory I–IV, *Topics in Current Chemistry*, vols. 180–183 (Springer, Berlin, 1996).
[13] V. I. Anisimov, ed., *Strong Coulomb Correlations in Electronic Structure Calculations: Beyond the Local Density Approximation* (Gordon & Breach, Amsterdam, 1999).
[14] N. H. March, *Electron Density Theory of Atoms and Molecules* (Academic, London, 1992).
[15] B. B. Laird, R. B. Ross, and T. Ziegler, eds., *Chemical Applications of Density Functional Theory* (American Chemical Society, Washington, 1996).
[16] D. P. Chong, ed., *Recent Advances in Density Functional Methods* (World Scientific, Singapore, 1995).
[17] D. Joulbert, ed., Density functionals: theory and applications, *Lecture Notes in Physics*, vol. 500 (Springer, Berlin, 1998).
[18] C. Fiolhais, F. Nogueira, and M. Marques, eds., A primer in density functional theory, *Lecture Notes in Physics*, vol. 620 (Springer, Berlin, 2003).
[19] J. F. Dobson, G. Vignale, and M. P. Das, eds., *Density Functional Theory: Recent Progress and New Directions* (Plenum, New York, 1998).
[20] H. Eschrig, *The Fundamentals of Density Functional Theory* (Teubner, Leipzig, 1996).
[21] E. K. U. Gross and R. M. Dreizler, eds., *Density Functional Theory* (Plenum, New York, 1995).
[22] N. Argaman and G. Makov, Am. J. Phys. **68**, 69 (2000).
[23] K. Capelle, Braz. J. Phys. **36**, 1318 (2006), also available from arXiv:cond-mat/0211443.
[24] T. Kreibich and E. K. U. Gross, Phys. Rev. Lett. **86**, 2984 (2001).
[25] A. K. Rajagopal and J. Callaway, Phys. Rev. B **7**, 1912 (1973).
[26] A. H. MacDonald and S. H. Vosko, J. Phys. C **12**, 2977 (1979).
[27] P. Strange, *Relativistic Quantum Mechanics with Applications in Condensed Matter and Atomic Physics* (Cambridge University Press, Cambridge, 1998).
[28] J. A. Pople, Rev. Mod. Phys. **71**, 1267 (1999).
[29] P. Hohenberg and W. Kohn, Phys. Rev. **136**, B864 (1964).
[30] E. Runge and E. K. U. Gross, Phys. Rev. Lett. **52**, 997 (1984).
[31] M. Levy, Phys. Rev. A **26**, 1200 (1982).
[32] E. H. Lieb, in Density functional methods in physics, edited by R. M. Dreizler and J. da Providencia (Plenum, New York, 1985).
[33] E. Schrödinger, Ann. Phys. **79**, 361, 489, 734 (1926); **80**, 437 (1926); **81**, 109 (1926).
[34] K. Capelle, C. A. Ullrich, and G. Vignale, Phys. Rev. A **76**, 012508 (2007).
[35] H. Eschrig and W. E. Pickett, Solid State Commun. **118**, 123 (2001).
[36] K. Capelle and G. Vignale, Phys. Rev. Lett. **86**, 5546 (2001).
[37] K. Capelle and G. Vignale, Phys. Rev. B **65**, 113106 (2002).
[38] O. Gritsenko and E. J. Baerends, J. Chem. Phys. **120**, 8364 (2004).
[39] N. Argaman and G. Makov, Phys. Rev. B **66**, 052413 (2002).
[40] N. Gidopolous, in The fundamentals of density matrix and density functional theory in atoms, molecules and solids, edited by N. Gidopoulos and S. Wilson (Kluwer, Boston, 2003), Progress in Theoretical Chemistry and Physics.
[41] W. Kohn, A. Savin, and C. A. Ullrich, Int. J. Quantum Chem. **100**, 20 (2004).
[42] C. A. Ullrich, Phys. Rev. B **72**, 073102 (2005).

43. T. Gál, Phys. Rev. B **75**, 235119 (2007).
44. N. I. Gidopoulos, Phys. Rev. B **75**, 134408 (2007).
45. W. E. Pickett and H. Eschrig, J. Phys. Condens. Matter **19**, 315203 (2007).
46. T. L. Gilbert, Phys. Rev. B **12**, 2111 (1975).
47. J. E. Harriman, Phys. Rev. A **24**, 680 (1981).
48. J. T. Chayes, L. Chayes, and M. B. Ruskai, J. Stat. Phys. **38**, 497 (1985).
49. C. A. Ullrich and W. Kohn, Phys. Rev. Lett. **89**, 156401 (2002); **87**, 093001 (2001).
50. P. E. Lammert, J. Chem. Phys. **125**, 074114 (2006).
51. A. P. Favaro, J. V. B. Ferreira, and K. Capelle, Phys. Rev. B **73**, 045133 (2006).
52. N. H. March, *Self-consistent Fields in Atoms* (Pergamon, Oxford, 1975).
53. L. W. Wang and M. P. Teter, Phys. Rev. B **45**, 13196 (1992).
54. M. Foley and P. A. Madden, Phys. Rev. B **53**, 10589 (1996).
55. B. J. Zhou, V. L. Ligneres, and E. A. Carter, J. Chem. Phys. **122**, 044103 (2005).
56. E. K. U. Gross, E. Runge, and O. Heinonen, *Many Particle Theory* (Adam Hilger, Bristol, 1991).
57. A. Szabo and N. S. Ostlund, *Modern Quantum Chemistry* (McGraw-Hill, New York, 1989).
58. E. K. U. Gross, M. Petersilka, and T. Grabo, in Chemical applications of density-functional theory, edited by B. B. Laird, R. B. Ross, and T. Ziegler (American Chemical Society, Washington, 1996a), ACS Symposium Series.
59. M. Levy and J. P. Perdew, Phys. Rev. A **32**, 2010 (1985).
60. E. H. Lieb and S. Oxford, Int. J. Quantum Chem. **19**, 427 (1981).
61. G. K.-L. Chan and N. C. Handy, Phys. Rev. A **59**, 3075 (1999).
62. M. M. Odashima, K. Capelle, and S. B. Trickey, J. Chem. Theory Comput. **5**, 798 (2009).
63. J. P. Perdew and Y. Wang, Phys. Rev. B **45**, 13244 (1992).
64. S. H. Vosko, L. Wilk, and M. Nusair, Can. J. Phys. **58**, 1200 (1980).
65. J. P. Perdew, J. A. Chevary, S. H. Vosko, K. A. Jackson, M. R. Pederson, D. J. Singh, and C. Fiolhais, Phys. Rev. B **46**, 6671 (1992).
66. J. P. Perdew, K. Burke, and M. Ernzerhof, Phys. Rev. Lett. **77**, 3865 (1996); **78**, 1396(E) (1997).
67. J. Tao, J. P. Perdew, V. N. Staroverov, and G. E. Scuseria, Phys. Rev. Lett. **91**, 146401 (2003).
68. J. P. Perdew and Y. Wang, Phys. Rev. B **33**, 8800 (1986).
69. J. P. Perdew, S. Kurth, A. Zupan, and P. Blaha, Phys. Rev. Lett. **82**, 2544 (1999).
70. A. D. Becke, Phys. Rev. A **38**, 3098 (1988).
71. C. Lee, W. Yang, and R. G. Parr, Phys. Rev. B **37**, 785 (1988).
72. Y. Zhang and W. Yang, Phys. Rev. Lett. **80**, 890 (1998).
73. A. D. Becke and E. R. Johnson, J. Chem. Phys. **127**, 124108 (2007).
74. L. J. Sham and M. Schlüter, Phys. Rev. Lett. **51**, 1888 (1983).
75. J. P. Perdew, R. G. Parr, M. Levy, and J. L. Balduz, Phys. Rev. Lett. **49**, 1691 (1982).
76. J. P. Perdew and M. Levy, Phys. Rev. Lett. **51**, 1884 (1983).
77. W. Kohn and L. J. Sham, Phys. Rev. **140**, A1133 (1965).
78. W. Kohn, A. D. Becke, and R. G. Parr, J. Phys. Chem. **100**, 12974 (1996).
79. C. O. Almbladh and U. von Barth, Phys. Rev. B **31**, 3231 (1985).
80. M. Levy, J. P. Perdew, and V. Sahni, Phys. Rev. A **30**, 2745 (1984).
81. J. Chen, J. B. Krieger, Y. Li, and G. J. Iafrate, Phys. Rev. A **54**, 3939 (1996).
82. M. Lüders, A. Ernst, W. M. Temmerman, Z. Szotek, and P. J. Durham, J. Phys. Condens. Matter **13**, 8587 (2001).
83. J. Muskat, A. Wander, and N. M. Harrison, Chem. Phys. Lett. **342**, 397 (2001).
84. A. Savin, C. J. Umrigar, and X. Gonze, Chem. Phys. Lett. **288**, 391 (1998).

[85] O. Gunnarsson and B. Lundqvist, Phys. Rev. B **13**, 4274 (1976).
[86] U. von Barth, Phys. Rev. A **20**, 1693 (1979).
[87] A. Görling, Phys. Rev. Lett. **85**, 4229 (2000).
[88] J. P. Perdew and M. Levy, Phys. Rev. B **31**, 6264 (1985).
[89] H. J. P. Freire and J. C. Egues, Phys. Rev. Lett. **99**, 026801 (2007).
[90] A. Görling, Phys. Rev. A **54**, 3912 (1996); **59**, 3359 (1999).
[91] M. Levy and A. Nagy, Phys. Rev. Lett. **83**, 4361 (1999).
[92] P. Samal and M. K. Harbola, J. Phys. B **39**, 4065 (2006).
[93] A. K. Theophilou, J. Phys. C **12**, 5419 (1979).
[94] E. K. U. Gross, L. N. Oliveira, and W. Kohn, Phys. Rev. A **37**, 2805 (1988); **37**, 2809 (1988); **37**, 2821 (1988).
[95] Á. Nagy, Phys. Rev. A **49**, 3074 (1994); **42**, 4388 (1990).
[96] N. I. Gidopoulos, P. G. Papaconstantinou, and E. K. U. Gross, Phys. Rev. Lett. **88**, 033003 (2002).
[97] R. van Leeuwen, Phys. Rev. Lett. **82**, 3863 (1999).
[98] E. K. U. Gross, J. F. Dobson, and M. Petersilka in Ref. [12]; K. Burke and E. K. U. Gross in Ref. [17]
[99] M. Petersilka, U. J. Gossmann, and E. K. U. Gross, Phys. Rev. Lett. **76**, 1212 (1996). See also T. Grabo, M. Petersilka, and E. K. U. Gross, J. Mol. Struct. (Theochem) **501**, 353 (2000).
[100] M. E. Casida, in Ref. [16]; J. Jamorski, M. E. Casida, and D. R. Salahub, J. Chem. Phys. **104**, 5134 (1996).
[101] S. J. A. van Gisbergen, J. G. Snijders, and E. J. Baerends, Comput. Phys. Commun. **118**, 119 (1999).
[102] R. E. Stratmann, G. E. Scuseria, and M. J. Frisch, J. Chem. Phys. **109**, 8218 (1998).
[103] G. Onida, L. Reining, and A. Rubio, Rev. Mod. Phys. **74**, 601 (2002).
[104] W. G. Aulbur, L. Jönsson, and J. W. Wilkins, Solid State Phys. **54**, 1 (1999).
[105] F. Aryasetiawan and O. Gunnarsson, Rep. Prog. Phys. **61**, 237 (1998).
[106] A. Svane and O. Gunnarsson, Phys. Rev. Lett. **65**, 1148 (1990); **72**, 1248 (1994).
[107] Z. Szotek, W. M. Temmermann, and H. Winter, Phys. Rev. Lett. **72**, 1244 (1994).
[108] P. Strange, A. Svane, W. M. Temmermann, Z. Szotek, and H. Winter, Nature **399**, 756 (1999).
[109] U. von Barth and L. Hedin, J. Phys. C **5**, 1629 (1972).
[110] D. M. Ceperley and B. J. Alder, Phys. Rev. Lett. **45**, 566 (1980).
[111] J. P. Perdew and A. Zunger, Phys. Rev. B **23**, 5048 (1981).
[112] C. Filippi, C. J. Umrigar, and M. Taut, J. Chem. Phys. **100**, 1290 (1994).
[113] P. Ziesche, S. Kurth, and J. P. Perdew, Comp. Mat. Sci. **11**, 122 (1998).
[114] P. J. Stephens, F. Devlin, C. Chabalowski, and M. Frisch, J. Phys. Chem **98**, 11623 (1994).
[115] A. D. Becke, J. Chem. Phys. **107**, 8554 (1997). See also A. D. Becke, J. Comp. Chem. **20**, 63 (1999).
[116] D. J. Tozer and N. C. Handy, J. Chem. Phys. **108**, 2545 (1998).
[117] T. van Voorhis and G. E. Scuseria, J. Chem. Phys. **109**, 400 (1998).
[118] J. Tao, J. P. Perdew, V. N. Staroverov, and G. E. Scuseria, Phys. Rev. Lett. **120**, 6898 (2004); Phys. Rev. B **69**, 075102 (2004).
[119] V. N. Staroverov, G. E. Scuseria, J. Tao, and J. P. Perdew, J. Chem. Phys. **119**, 12129 (2003).
[120] A. D. Becke, J. Chem. Phys. **104**, 1040 (1996).
[121] J. Tao and J. P. Perdew, J. Chem. Phys. **122**, 114102 (2005).
[122] Y.-H. Kim, I.-H. Lee, S. Nagaraja, J.-P. Leburton, R. Q. Hood, and R. M. Martin, Phys. Rev. B **61**, 5202 (2000).

[123] S. Kurth, J. P. Perdew, and P. Blaha, Int. J. Quantum Chem. **75**, 889 (1999).
[124] C. Adamo, M. Ernzerhof, and G. E. Scuseria, J. Chem. Phys. **112**, 2643 (2000).
[125] O. A. Vydrow and G. E. Scuseria, J. Chem. Phys. **121**, 8187 (2004).
[126] J. Gräfenstein, E. Kraka, and D. Cremer, J. Chem. Phys. **120**, 524 (2004).
[127] S. Patchkovskii and T. Ziegler, J. Chem. Phys. **116**, 7806 (2002).
[128] S. Goedecker and C. J. Umrigar, Phys. Rev. A **55**, 1765 (1997).
[129] B. G. Johnson, C. A. Gonzales, P. M. W. Gill, and J. A. Pople, Chem. Phys. Lett. **221**, 100 (1994).
[130] G. I. Csonka and B. G. Johnson, Theor. Chem. Acc. **99**, 158 (1998).
[131] U. Lundin and O. Eriksson, Int. J. Quantum Chem. **81**, 247 (2001).
[132] C. Legrand, E. Suraud, and P.-G. Reinhard, J. Phys. B **35**, 1115 (2002).
[133] J. B. Krieger, Y. Li, and G. J. Iafrate, Phys. Rev. A **45**, 101 (1992); **46**, 5453 (1992); **47**, 165 (1993).
[134] E. Engel and S. H. Vosko, Phys. Rev. A **47**, 2800 (1993).
[135] T. Grabo, T. Kreibich, S. Kurth, and E. K. U. Gross, in Ref. [13]
[136] T. Grabo and E. K. U. Gross, Int. J. Quantum Chem. **64**, 95 (1997); Chem. Phys. Lett. **240**, 141 (1995).
[137] T. Kotani, Phys. Rev. Lett. **74**, 2989 (1995).
[138] M. Stadele, J. A. Majewski, P. Vogl, and A. Görling, Phys. Rev. Lett. **79**, 2089 (1997); M. Stadele, M. Moukara, J. A. Majewski, P. Vogl, and A. Görling, Phys. Rev. B **59**, 10031 (1999); Y.-H. Kim, M. Stadele, and R. M. Martin, Phys. Rev. A **60**, 3633 (1999).
[139] S. Kümmel and J. P. Perdew, Phys. Rev. B **68**, 035103 (2003).
[140] W. Yang and Q. Qu, Phys. Rev. Lett. **89**, 143002 (2002).
[141] V. N. Staroverov, G. E. Scuseria, and E. R. Davidson, J. Chem. Phys. **125**, 081104 (2006).
[142] A. D. Becke, J. Chem. Phys. **98**, 5648 (1993).
[143] M. Cafiero and C. Gonzalez, Phys. Rev. A **71**, 042505 (2005).
[144] M. P. Lima, L. S. Pedroza, A. J. R. da Silva, A. Fazzio, D. Vieira, H. J. P. Freire, and K. Capelle, J. Chem. Phys. **126**, 144107 (2007).
[145] O. Gunnarsson, M. Jonson, and B. I. Lundqvist, Phys. Rev. B **20**, 3136 (1979).
[146] A. Cancio, M. Y. Chou, and R. O. Hood, Phys. Rev. B **64**, 115112 (2002).
[147] P. García-González, Phys. Rev. B **62**, 2321 (2000).
[148] N. Rösch and S. B. Trickey, J. Chem. Phys. **106**, 8940 (1997).
[149] P. Schwerdtfeger, in Ref. [3]
[150] P. Fuentealba, in Ref. [3]
[151] C. Pouchana, D. Y. Zhang, and D. Begue, in Ref. [3]
[152] D. C. Patton and M. R. Pederson, Phys. Rev. A **56**, 2495 (1997).
[153] W. Kohn, Y. Meir, and D. E. Makarov, Phys. Rev. Lett. **80**, 4153 (1998).
[154] M. Lein, J. F. Dobson, and E. K. U. Gross, J. Comp. Chem. **20**, 12 (1999).
[155] J. F. Dobson and B. P. Dinte, Phys. Rev. Lett. **76**, 1780 (1996).
[156] Y. Andersson, D. C. Langreth, and B. I. Lundqvist, Phys. Rev. Lett. **76**, 102 (1996).
[157] E. R. Johnson and A. D. Becke, J. Chem. Phys. **123**, 024101 (2005).
[158] A. D. Becke and E. R. Johnson, J. Chem. Phys. **123**, 154101 (2005).
[159] A. D. Becke and E. R. Johnson, J. Chem. Phys. **127**, 154108 (2007).
[160] M. J. Frisch et al., *Gaussian 03, Revision C.02*, Gaussian, Inc., Wallingford, CT, 2004.
[161] J. Olsen, P. Jørgensen, H. Koch, A. Balkova, and R. J. Bartlett, J. Chem. Phys. **104**, 8007 (1996).
[162] F. D. Giacomo, F. A. Gianturco, F. Raganelli, and F. Schneider, J. Chem. Phys. **101**, 3952 (1994).

[163] J. M. L. Martin, Spectrochim. Acta A **57**, 875 (2001).
[164] E. K. U. Gross, J. F. Dobson, and M. Petersilka, in Density functional theory I–IV, edited by R. F. Nalewajski, *Topics in Current Chemistry*, vols. 180–183 (Springer, Berlin, 1996).
[165] K. Burke and E. K. U. Gross, in Density functionals: theory and applications, edited by D. Joulbert, *Lecture Notes in Physics*, vol. 500 (Springer, Berlin, 1998).
[166] M. A. L. Marques, C. A. Ullrich, F. Nogueira, A. Rubio, K. Burke, and E. K. U. Gross, eds., Time-dependent density functional theory, *Lecture Notes in Physics*, vol. 706 (Springer, Berlin, 2006).
[167] G. L. Oliver and J. P. Perdew, Phys. Rev. A **20**, 397 (1979).
[168] K. Capelle and V. L. Libero, Int. J. Quantum Chem. **105**, 679 (2005).
[169] L. M. Sandratskii, Adv. Phys. **47**, 91 (1998).
[170] L. Nordström and D. Singh, Phys. Rev. Lett. **76**, 4420 (1996).
[171] K. Capelle and L. N. Oliveira, Phys. Rev. B **61**, 15228 (2000).
[172] E. Engel, T. Auth, and R. M. Dreizler, Phys. Rev. B **64**, 235126 (2001).
[173] A. Trellakis, Phys. Rev. Lett. **91**, 056405 (2003).
[174] W. Cai and G. Galli, Phys. Rev. Lett. **92**, 186402 (2004).
[175] R. M. Dickson and T. Ziegler, J. Phys. Chem. **100**, 5286 (1996).
[176] G. Schreckenbach and T. Ziegler, J. Phys. Chem. **99**, 606 (1995).
[177] G. Magyarfalvi and P. Pulay, J. Chem. Phys. **119**, 1350 (2003).
[178] S. N. Maximoff and G. E. Scuseria, Chem. Phys. Lett. **390**, 408 (2004).
[179] V. G. Malkin, O. L. Malkina, and D. R. Salahub, Chem. Phys. Lett. **204**, 80 (1993).
[180] P. Pyykkö, Adv. Quantum Chem. **11**, 353 (1978).
[181] O. Eriksson and J. Wills, in The augmented spherical wave method: a comprehensive treatment, edited by H. Dreyssé, *Lecture Notes in Physics*, vol. 535, pp. 247–285 (Springer, Berlin/Heidelberg, 1999).
[182] M. Eisenbach, B. L. Györffy, G. M. Stocks, and B. Újfalussy, Phys. Rev. B **65**, 144424 (2002).
[183] H. Ebert, Rep. Prog. Phys. **59**, 1665 (1996).
[184] K. Capelle, E. K. U. Gross, and B. L. Györffy, Phys. Rev. Lett. **78**, 3753 (1997).
[185] G. Vignale and M. Rasolt, Phys. Rev. B **37**, 10685 (1988).
[186] M. Rasolt and F. Perrot, Phys. Rev. Lett. **69**, 2563 (1992).
[187] G. Vignale and M. Rasolt, Phys. Rev. Lett. **59**, 2360 (1987).
[188] E. Orestes, T. Marcasso, and K. Capelle, Phys. Rev. A **68**, 022105 (2003).
[189] E. Orestes, A. B. F. da Silva, and K. Capelle, Int. J. Quantum Chem. **103**, 516 (2005).
[190] J. Tao and J. P. Perdew, Phys. Rev. Lett. **95**, 196403 (2005).
[191] M. Ferconi and G. Vignale, Phys. Rev. B **50**, 14722 (1994).
[192] O. Steffens, U. Rössler, and M. Suhrke, Europhys. Lett. **42**, 529 (1998).
[193] M. Pi, M. Barranco, A. Emperador, E. Lipparini, and L. Serra, Phys. Rev. B **57**, 14783 (1998).
[194] S. M. Colwell and N. C. Handy, Chem. Phys. Lett. **217**, 271 (1994).
[195] A. M. Lee, S. M. Colwell, and N. C. Handy, Chem. Phys. Lett. **229**, 225 (1994).
[196] H. Ebert, M. Battocletti, and E. K. U. Gross, Europhys. Lett. **40**, 545 (1997).
[197] G. Vignale, Phys. Rev. B **47**, 10105 (1993).
[198] S. Sharma, S. Pittalis, S. Kurth, S. Shallcross, J. K. Dewhurst, and E. K. U. Gross, Phys. Rev. B **76**, 100401 (2007).
[199] C. J. Grayce and R. A. Harris, Phys. Rev. A **50**, 3089 (1994).
[200] F. R. Salsbury, Jr and R. A. Harris, J. Chem. Phys. **108**, 6102 (1998).
[201] K. Capelle, Phys. Rev. A **60**, R733 (1999).
[202] R. W. Danz and M. L. Glasser, Phys. Rev. B **4**, 94 (1971).

[203] P. Skudlarski and G. Vignale, Phys. Rev. B **48**, 8547 (1993).
[204] Y. Takada and H. Goto, J. Phys. Condens. Matter **10**, 11315 (1998).
[205] G. Vignale, M. Rasolt, and D. J. W. Geldart, Phys. Rev. B **37**, 2502 (1988).
[206] G. Vignale, C. A. Ullrich, and S. Conti, Phys. Rev. Lett. **79**, 4878 (1997).
[207] R. Resta, Rev. Mod. Phys. **66**, 899 (1994).
[208] G. Nenciu, Rev. Mod. Phys. **63**, 91 (1991).
[209] R. W. Nunes and D. Vanderbilt, Phys. Rev. Lett. **73**, 712 (1994).
[210] X. Gonze, P. Ghosez, and R. W. Godby, Phys. Rev. Lett. **74**, 4035 (1995).
[211] P. Umari and A. Pasquarello, Int. J. Quantum Chem. **101**, 666 (2005).
[212] D. Vanderbilt, Phys. Rev. Lett. **79**, 3966 (1997).
[213] R. Resta, Phys. Rev. Lett. **77**, 2265 (1996).
[214] X. Gonze, P. Ghosez, and R. W. Godby, Phys. Rev. Lett. **78**, 294 (1997).
[215] R. M. Martin and G. Ortiz, Phys. Rev. B **56**, 1124 (1997).
[216] F. Kootstra, P. L. de Boeij, and J. G. Snijders, J. Chem. Phys. **112**, 6517 (2000).
[217] G. F. Bertsch, J.-I. Iwata, A. Rubio, and K. Yabana, Phys. Rev. B **62**, 7998 (2000).
[218] M. van Faassen, P. L. de Boeij, R. van Leeuwen, J. A. Berger, and J. G. Snijders, J. Chem. Phys. **118**, 1044 (2003).
[219] D. Xiao, J. Shi, and Q. Niu, Phys. Rev. Lett. **95**, 137204 (2005).
[220] J. Shi, G. Vignale, D. Xiao, and Q. Niu, Phys. Rev. Lett. **99**, 197202 (2007).
[221] T. Thonhauser, D. Ceresoli, D. Vanderbilt, and R. Resta, Phys. Rev. Lett. **95**, 137205 (2005).
[222] D. Ceresoli, T. Thonhauser, D. Vanderbilt, and R. Resta, Phys. Rev. B **74**, 024408 (2006).
[223] A. M. Lee, N. C. Handy, and S. M. Colwell, J. Chem. Phys. **103**, 10095 (1995).
[224] O. Gunnarsson and K. Schönhammer, Phys. Rev. Lett. **56**, 1968 (1986).
[225] N. A. Lima, M. F. Silva, L. N. Oliveira, and K. Capelle, Phys. Rev. Lett. **90**, 146402 (2003). N. A. Lima, L. N. Oliveira, and K. Capelle, Europhys. Lett. **60**, 601 (2002). M. F. Silva, N. A. Lima, A. L. Malvezzi, and K. Capelle, Phys. Rev. B **71**, 125130 (2005).
[226] R. J. Magyar and K. Burke, Phys. Rev. A **70**, 032508 (2004).
[227] V. L. Líbero and K. Capelle, Phys. Rev. B **68**, 024423 (2003). P. E. G. Assis, V. L. Libero, and K. Capelle, Phys. Rev. B **71**, 052402 (2005). See also Ref. [168]
[228] V. L. Libero and K. Capelle, cond-mat/0506206.
[229] H. Chermette, J. Comp. Chem. **20**, 129 (1999).
[230] F. D. Proft and P. Geerlings, Chem. Rev. **101**, 1451 (2001).
[231] Veillard and Clementi, J. Chem. Phys. **49**, 2415 (1968).
[232] E. Orestes, A. B. F. da Silva, and K. Capelle, Physical Chemistry – Chemical Physics, *accepted* (2009). Available in electronic form as arXiv:0712.1586.

9

Acoustic Microscopy Applied to Nanostructured Thin Film Systems

Chiaki Miyasaka

University of Windsor, Windsor, ON, Canada
The Pennsylvania State University, University Park, PA, N9B 3P4, 16802, USA,
cmiyasaka1@yahoo.com

I. INTRODUCTION

The present volume is devoted to the issues of modeling and numerical simulations in electrochemistry. With the continuing development of more and more powerful computer hardware and software systems, the nature of modeling keeps evolving and expanding. Workers in industry and academia keep developing, testing, and understanding and producing new products. Those in most cases require new materials which benefit from modeling as it obviates the need to actually try every possible new material. Indeed, owing to the growth in the development of material science and technology, the requirements for high-quality, reliable materials have become more stringent. That is so especially in the nano 3D space industries. It is more often than not difficult for conventional materials to completely meet those new more stringent requirements. In this case, closer study of nanostructured materials is often called for.

However, such study/search is not possible if no data for securing reliability and safety of the nanostructured materials can be assured.

Generally speaking, when it comes to evaluating "conventional" materials, test samples for service life time estimation, quality, and safety analyses are collected. Data showing material characteristics are typically secured by means of optical and electron microscopes through destructive tests such as pull, compression, bending, torsion, creep, fatigue, impact, and corrosion. The same methods may be applied to a nanoscaled material.

Currently, scanning probe microscopy (e.g., atomic force microscopy) is often used for testing instead of optical and electron microscopes. A few advanced destructive techniques have been reported for the evaluation of nanostructured material.[1–3] However, they are still not good enough for the evaluation of a nanoscaled thin film system (such as electrochemically/electrolessly deposited metal films) owing to critical defects such as voids, delaminations, and debonding sites of the system which often exist in the interior. Therefore, microscopy designed for surface analysis may not be suitable for application to the system directly.

Attempts at exposing the defects located in the interior by cutting or eliminating the system in layers may cause the following problems:

1. The shape of the defects might be changed when cutting or eliminating layers of the system.
2. There is no guarantee that all defects are exposed.
3. Certain specimens are difficult to cut or it is difficult to eliminate layers.
4. When the sizes/dimensions of the defects are less than a 1 μm, it is difficult to see their 3D distributions by eliminating layers.

At least two nondestructive evaluation methods have been suggested for the visualization of the internal structure of an opaque nanoscaled thin film system and to characterize the same by instruments utilizing basic features of ultrasound (e.g., reflection, transmission, refraction, and diffraction). One is a laser-based ultrasonic technique which is known as "picosecond acoustics."[4–8] Acoustic waves with frequencies ranging from approximately 10 GHz to 1.0 THz, corresponding to acoustic wavelengths in the range from approximately 5 to 500 nm may be generated and detected with a single microscope objective. The detection of acoustic echoes on reflection from buried interfaces reveals details of bonding, sound velocities, film thicknesses, ultrasonic attenuation and ultrafast acoustic generation mechanisms. However, this method cannot form a highly resolved image showing internal structure. Second is ultrasonic atomic force microscopy,[9,10] which is a dynamic operation mode of the atomic force microscope[11] that permits the measurement of elastic properties and the visualization of the surface and/or the subsurface of the system with high spatial resolution.[12–14] Although this method is promising for characterizing the system, development of the apparatus itself is still in progress at the time of writing this chapter.

The mechanical scanning acoustic reflection microscope[15] (SAM) utilizes ultrasound to produce magnified images of the microscopic structures of materials. In other words, the SAM is an instrument that subjects an object to ultrasound, and detects the variations of the elastic properties of the object. The elastic properties are determined by the molecular arrangement of the material, molecule size, intermolecular force, and the like, with each material having its unique characteristics. Accordingly, each object reacts in its own characteristic way when subjected to ultrasonic waves. The waves that return from the object may be converted into an image. Since the ultrasonic waves penetrate into the object interior, not only the surface, but also the interior can be visualized. Since the beam is focused (up to submicrometer size) and is of high frequency (100 MHz to 3 GHz), the resolution of the system is on the order of that of optical microscopes. In addition to the image formation, the SAM can also detect the amplitude and the phase of the reflected wave. By analyzing those, one can quantitatively determined the elastic properties of the specimen.

The SAM has been used to characterize thin or thick film systems,[16–34] but not nanoscaled ones.

This article is presented with the view to clarify whether scanning acoustic microscopy, which is one of the more advanced ultrasonic imaging technologies, can be applied to nanoscaled electrochemically deposited thin film systems (e.g., electroless deposition of ultrathin metal film systems).

II. PRINCIPLE OF THE SCANNING ACOUSTIC MICROSCOPE

1. Imaging Mechanism

In this application, instead of a pulse wave, a tone-burst wave is used (see Fig. 1), and the frequency domain is such that the wavelengths of the ultrasound (i.e., in water range from 15.0 to 1.5 μm; see Table 1). The penetration depth of the waves is limited by attenuation. The SAM is used for penetrations substantially up to 300 μm.

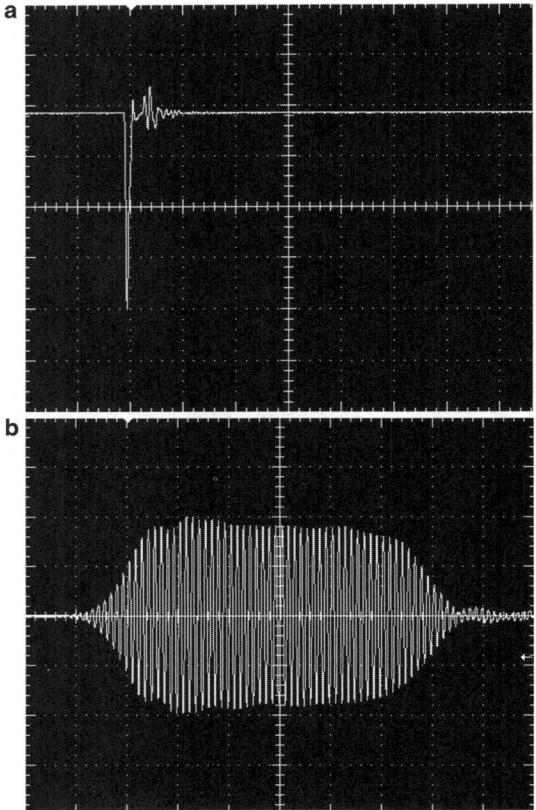

Figure 1. Waveform. (**a**) Pulse wave; (**b**) tone-burst wave.

Table 1.

Wavelength in water at various frequencies.

100 MHz	200 MHz	400 MHz	600 MHz	800 MHz	1.0 GHz
15.0 μm	7.5 μm	3.7 μm	2.5 μm	1.8 μm	1.5 μm

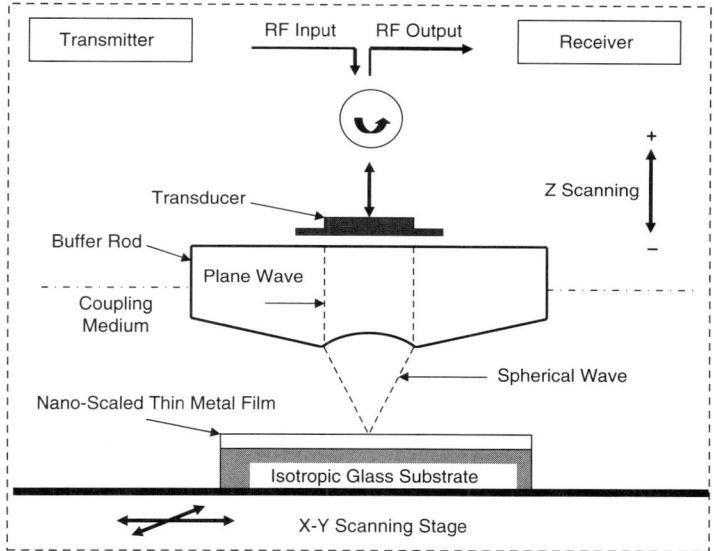

Figure 2. A mechanical scanning acoustic reflection microscope.

Figure 2 shows the schematic diagram of the SAM. Referring to Fig. 2, we describe the imaging mechanism of the SAM below.

An electrical signal is generated by an RF tone-burst source. In the beginning, a Colpitts oscillator, or the like, was used as the RF tone-burst source. Nowadays, however, burst waves are gated out from continuous waves by a single pole double throw switch for frequency stability. The output (i.e., voltage) of the source is approximately 10 V.

The electrical signal is transmitted to a piezoelectric transducer located on the top of a buffer rod through a circulator (or the single pole double throw switch). The electrical signal is converted to an acoustic signal (i.e., ultrasonic plane wave) at the transducer. The ultrasonic plane wave travels through the buffer rod to a spherical recess (hereinafter called simply the "lens") located at the bottom of the buffer rod, wherein the lens is coated by the acoustic impedance matching layer which is the so-called acoustic antireflection coating (AARC). The lens converts the ultrasonic plane wave to an ultrasonic spherical wave (i.e., ultrasonic beam). The ultrasonic beam is focused within the specimen, and reflected from the specimen. The reflected ultrasonic beam, which carries acoustic information of the specimen, is again converted to an ultrasonic plane wave by the lens. The ultrasonic plane wave returns to the transducer through the buffer rod. The ultrasonic plane wave is again converted to an electric signal at the transducer. The voltage of the electric signal ranges from 300 mV to 1 V. When the operating frequencies range from 100 MHz to 1 GHz, the corresponding insertion loss is approximately 30 and 80 dB. Therefore, the electric signal must be amplified by 30–80 dB at a receiver. Furthermore, the electric signal comprises transmission leaks, internal reflections from the interface between the lens and the AARC, and reflections from the specimen. Therefore, the reflections must be selected by a rectangular wave from a double balanced mixer, the so-called the first gate. Then, the peak of the amplitude of the electric signal is detected by a circuit, which includes a diode and a capacitor (i.e., the peak detection technique). The gate noise is removed by using the second gate existing within the first gate (the blanking technique). The peak-detected signal is stored in a memory through an analog-to-digital signal converter. The stored signal is again converted into an analog signal by a digital-to-analog signal converter. This flow of processes allows the information that is collected at a single spot on a specimen to be displayed as intensity at a certain point on the TV monitor.

To form a 2D acoustic image, an acoustic lens and/or an X–Y stage is mechanically scanned across a given area of the specimen.

The acoustic lens is able to translate axially along the z direction by variation of the distance between the specimen and the lens for subsurface visualization. That is, when the surface of the specimen is visualized, the acoustic lens is focused on the specimen (we denote $z = 0$ μm), and when a subsurface of the specimen is visualized, the acoustic lens is mechanically defocused toward the specimen (we denote $z = -x$ μm, where x is the defocused distance).

2. Description of Acoustic Lens

The acoustic lens is made of a piezoelectric transducer and a buffer rod. The transducer is deposited on the top of the buffer rod, and a lens covered by an AARC is located on the bottom of the buffer rod.

(i) Piezoelectric Transducer

The role of the piezoelectric transducer is to convert electric signals to acoustic signals, and back again. Zinc oxide is used typically for the transducer at frequencies of 100 MHz or more. The transducer is sputtered onto the electrode (lower portion) deposited on the polished surface of the buffer rod. Electrode is made from either chromium–gold or titanium–gold. The electrode (upper portion) is deposited on the zinc oxide using a mask plate for matching the center of the electrode (upper portion) and that of the lens.

When an electric signal is incident upon the transducer, the transducer is excited at its resonance frequency f_r. The condition for resonance is expressed as

$$\frac{2n-1}{f_r} = \frac{2d}{c}, \tag{1}$$

where d is the thickness of the transducer, n is an integer positive number, and c is the velocity of the longitudinal wave set up in the transducer.

Therefore, by modification of (1), the thickness of the transducer is determined as

$$d = \left(\frac{2n-1}{2}\right)\left(\frac{c}{f_r}\right) = \left(\frac{2n-1}{2}\right)\lambda, \tag{2}$$

where λ is the wavelength of the acoustic wave.

However, it ought to be remembered that the relationship between the acoustic impedance of the buffer rod and that of the transducer is also considered to determine the thickness. Let us denote the acoustic impedances of the buffer rod and the transducer as Z_1, and Z_2, respectively. Then, the relationship of these impedances with each other can be either $Z_1 > Z_2$ or $Z_1 < Z_2$. The thickness (d) is determined as $\lambda/4$ when the relation is $Z_1 > Z_2$, and as $\lambda/2$ when $Z_1 < Z_2$.

(ii) Buffer Rod

The diameter of the buffer rod is set to be larger than the diameter of the transducer. The length of the buffer rod is generally chosen to be longer than the near-field region (i.e., Fresnel zone), which is in turn calculated through the following equation:

$$N = \frac{D^2}{4\lambda}, \tag{3}$$

where N is the near-field region, D is the diameter of the transducer, and λ is the wavelength in the material used for the buffer rod.

Considering the presence of acoustic energy loss, a material having a high velocity compared with that of a coupling medium and low attenuation (e.g., fused quartz or sapphire) is generally selected. The selection of the material directly relates to the design of the lens in terms of spherical aberration that needs to be minimized.

(iii) Lens

The paraxial focal distance of the acoustic lens (F_0) is approximately expressed as follows:

$$F_0 = \frac{R}{(1 - C')}, \tag{4}$$

$$C' = \frac{C_2}{C_1}, \tag{5}$$

where R is the radius of curvature of the surface of the lens, C_1 is the longitudinal wave velocity in the buffer rod, and C_2 is the longitudinal wave velocity of the coupling medium.

The radius of curvature of the surface of the lens is determined by considering the operating frequency and attenuation coefficient. Table 2 shows the range of the radii corresponding to various frequencies. The aperture angle of the lens (denoted as θ_α) is determined by the focal distance. When the focal distance and the aperture angle are known the radius of the aperture (r) is determined by the following equation:

$$r = F_0 \sin\left(\frac{\theta_\alpha}{2}\right). \tag{6}$$

Referring to Fig. 3, we describe the spherical aberration of the acoustic lens below.

Table 2.
Range of radius due to frequency.

Frequency	Radius
100 MHz	1–2 mm
200 MHz	500 μm to 1 mm
400 MHz	500 μm
800–1,000 MHz	125 μm
1.5–3 GHz	50 μm

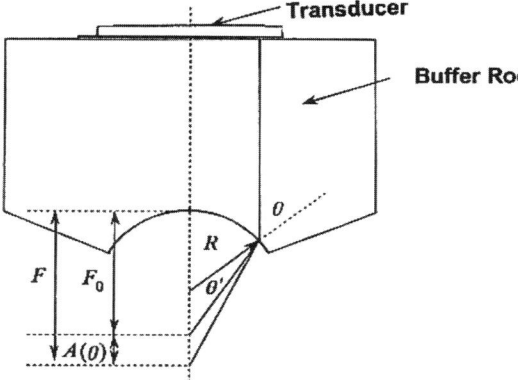

Figure 3. Calculation of the spherical aberration of the acoustic lens. $A(\theta)$ is the spherical aberration, F_0 is the paraxial focal distance of the acoustic lens, F is the zonal focal distance, R is the radius of curvature of the surface of the lens, θ is the incident angle of the acoustic wave, and θ' is the refracted angle of the acoustic wave.

When the acoustic wave from the lens is emitted into the specimen, the following equation holds by Snell's law:

$$C_1 \sin \theta' = C_2 \sin \theta, \qquad (7)$$

where θ is the incident angle of the acoustic wave and θ' is the refracted angle of the acoustic wave.

Then, the zonal focal distance (F) is expressed as follows:

$$F = R \left((1 - \cos \theta) + \frac{\sin \theta}{\tan (\theta - \theta')} \right). \qquad (8)$$

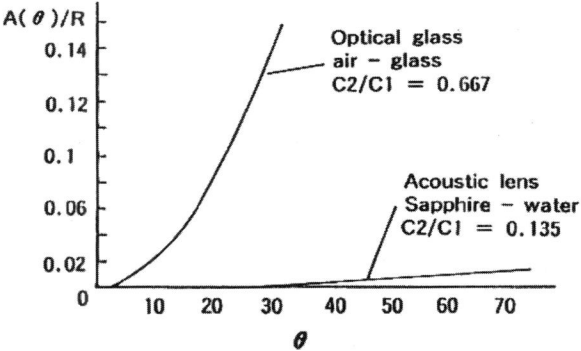

Figure 4. Spherical aberration of optical and acoustic lenses. $A(\theta)$ is the spherical aberration, R is the radius of curvature of the surface of the lens, C_1 is the longitudinal wave velocity of the buffer rod, and C_2 is the longitudinal wave velocity of the coupling medium.

Therefore, the spherical aberration, $A(\theta)$, is expressed as follows:

$$A(\theta) = F_0 - F. \tag{9}$$

For example, when sapphire and water are used for a buffer rod and a coupling medium, respectively, $A(\theta)$ is calculated to be $0.003 R$. For an acoustic lens operating at a frequency of 1.0 GHz, R is approximately 100 μm; therefore, $A(\theta)$ is calculated as 0.3 μm. Considering the wavelength of the ultrasonic wave having a frequency of 1.0 GHz in water (i.e., 1.5 μm), this value is small enough for a single spherical lens to form a well-focused image. Therefore, sapphire (z-cut) is considered as the best material although it is difficult to make the spherical recess by mechanical polish.

For reference, a comparison of the ratio of the spherical aberration and the radius, $A(\theta)/R$, between the acoustic lens and the optical lens is shown in Fig. 4. The spherical aberration of the acoustic lens is much smaller than that of the optical lens. This is an important advantage.

(iv) Acoustic Antireflection Coating

When acoustic waves are emitted from sapphire into water, used as a coupling medium directly, 88% of the acoustic waves will be reflected back from the interface between the lens and water. This is due to their impedance mismatch. Thus, it becomes necessary to coat the lens with an AARC. The thickness of the AARC should be $\lambda/4$, where λ is the wavelength of the acoustic wave to have an optimized result, and the acoustic impedance Z of the AARC is

$$Z = \sqrt{Z_1 Z_2}, \quad (10)$$

where Z_1 is the impedance of the lens rod and Z_2 is the impedance of the coupling medium.

Evaporated silicon oxide is typically used as the material for the AARC although silicon oxide does not completely satisfy (10).

III. RESOLUTION

Two types of resolutions must be considered for the SAM. One is a lateral resolution (Δr), and the other is a vertical resolution ($\Delta \rho$). These may be expressed as follows:

$$\Delta r = F\lambda = F\left(\frac{v_w}{f}\right), \quad (11)$$

$$\Delta \rho = 2F^2 \lambda = \left(2F^2\right)\left(\frac{v_w}{f}\right) = \left[2\left(\frac{f_o}{D}\right)^2\right]\left(\frac{v_w}{f}\right)$$
$$= \left[2\left(\frac{1}{2\tan\theta}\right)^2\right]\left(\frac{v_w}{f}\right) = \left(\frac{1}{2(\tan\theta)^2}\right)\left(\frac{v_w}{f}\right), \quad (12)$$

where F is a constant related to the lens geometry, λ is the wavelength in the coupling medium (i.e., water), f is the frequency of the wave generated by the transducer, v_w is the longitudinal wave velocity in the coupling medium, f_o is the focal distance of the lens, D is the diameter of the lens aperture, and θ is half of the aperture angle of the lens.

Figure 5b shows the surface and demonstrates the resolution at a frequency of 1.0 GHz. The resolution is seen to be substantially the same as that obtained by a conventional optical microscope (Fig. 5a). However, it is not of a quality sufficient for evaluating the nanoscaled thin film system. This is the main reason for not using the SAM for the evaluation of nanoscaled thin film systems. It is very important

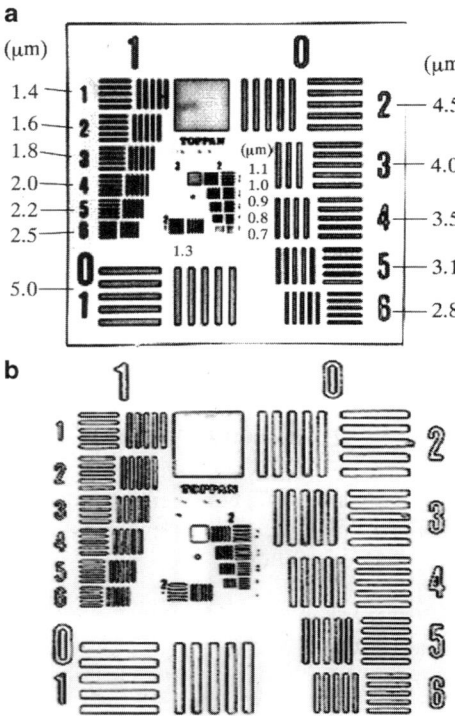

Figure 5. Resolution. (**a**) Optical image of a standard specimen having patterns for measuring resolution (i.e., resolution chart) for the scanning acoustic microscope in C-scan mode using a tone-burst wave; (**b**) acoustic image. The acoustic lens is focused onto the surface of the resolution chart. The acoustic lens is operated at a frequency of 1.0 GHz.

to note that the actual resolutions in the images, however, tend to be higher than the calculated resolutions. That is significant and it means that contrast is as important as the calculated resolution in determining the actual resolution in the image. We discuss the contrast factor in a separate section later.

Resolution of an image formed by the SAM is determined by the frequency of the wave, the velocity in the coupling medium, and the lens geometry. Therefore, three approaches were pursued by many workers to increase resolution.

The first approach was to raise the acoustic frequency. By means of this approach, Hadimiooglu et al achieved a resolution of 0.2 μm by operating at about 4.4 GHz with the specimen in boiling water to minimize the attenuation of the coupling medium.[35]

The second approach was to use lowvelocity coupling media such as liquid nitrogen[36] and ethanol[37] to increase resolution. Foster et al were able to obtain a resolution better than 500 Å using superfluid liquid helium as the coupling medium at a temperature of 0.2°K at 4.2 GHz.[38] Muha et al. were able to obtain a resolution of 150 Å using an acoustic microscope operating at 15.3 GHz, wherein the coupling medium was pressurized superfluid ^{4}He, and wherein the temperature of the medium was less than 0.9 K.[39] These approaches are excellent in increasing resolution on the surface of a specimen, but not necessarily in its interior. In the first approach, the ultrasonic wave is attenuated in proportion to the square of its frequency; therefore, when the highfrequency ultrasonic wave is used as a probe, the wave will not penetrate the interior of the specimen, and the internal information may not be obtained. In the second approach, the acoustic impedances between the coupling medium and the specimen are very different, so most of the ultrasonic waves are reflected at the interface between the coupler and the specimen, and it is unlikely that an image from any significant depth will be obtained. Thus, acoustic microscopy's most important and unique feature, namely obtaining subsurface information may not be realized by these two approaches. However, since the ultrasonic penetration depth is not so crucial for the case of a nanoscaled thin film system, those techniques could be suitable for application to the system.

The third approach, which was to design a new acoustic lens, has been considered to be the best approach for maintaining the most important feature of the SAM. By means of the third approach, the following new acoustic lenses were proposed. Chubachi et al developed the concave transducer to increase the resolution by removing a spherical aberration.[40] Davids et al., Atalar et al., Yakub et al., and Miyasaka et al. proposed new acoustic lenses having restricted apertures using Rayleigh waves[41] a nonspherical aperture using Lamb waves,[42] a pinhole aperture for near-field imaging,[43] and a centersealed aperture substantially using shear waves,[44] respectively. A new lens may have to be designed for the nanoscaled thin film system to optimize the quality of the acoustic image.

IV. PRINCIPLE OF QUANTITATIVE DATA ACQUISITION

In 1979, Weglein et al. found that a change in the voltage of the transducer deposited on the top of acoustic lens, which is known as the $V(z)$ curve, when defocusing the acoustic lens toward a specimen, is uniquely related to the elastic properties of the specimen.[45] This phenomenon was modeled using ray tracing by Parmon et al.,[46] and using Fourier optics by Atalar.[47] These workers also found that the velocity of surface acoustic waves traveling within a spot of an ultrasonic beam could be obtained by measuring the distance between the periods of the $V(z)$ curves. By the establishment of these fundamentally significant theories, the SAM became the universally accepted apparatus not only for imaging but also for quantitative data acquisition.

1. $V(z)$ Curve

The transducer output voltage may be periodic with the axial motion as the acoustic lens advances from the focal plane toward the specimen. The period of this variation is characteristic of the specimen's elastic material properties involved and results from interaction between two ray components, shown in Fig. 6, that radiate into the liquid from the solid–liquid interface. Specifically, Fig. 7 shows an example of a $V(z)$ curve for fused quartz. An acoustic lens having an aperture angle of $120°$ and a working distance of $310\,\mu m$ was used at an operating frequency at $400\,MHz$. The specimen was located in the water tank. The temperature was set at $22.3°C$ in the coupling medium (i.e., distilled water) and it was measured by a thermocouple. The temperature was substantially stabilized (changes less than $\pm 0.1°C$). The movement of the acoustic lens along the Z-axis was monitored by a laser-based measuring instrument. The transducer output voltage was periodic with axial motion as the acoustic lens advanced from the focal plane toward the specimen. The contrast changed in accordance with the same period.

The $V(z)$ curve is expressed as follows:

$$V(z) = C^{-1} \int_0^\infty u^2(r) P^2(r) R\left(\frac{r}{f}\right) \exp\left[i2kz\sqrt{1-\left(\frac{r}{f}\right)^2}\right] r\,dr, \tag{13}$$

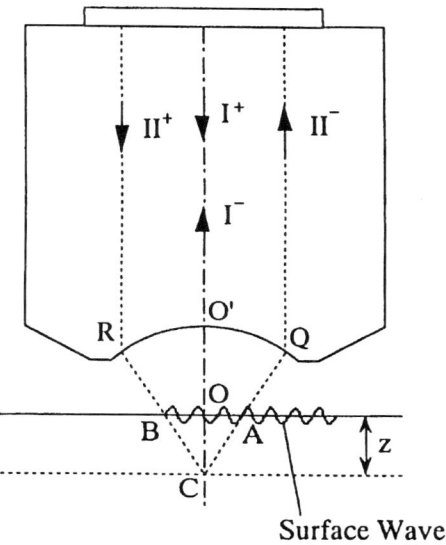

Figure 6. Cross-sectional geometry of the spherical acoustic lens, explaining the mechanism of the $V(z)$ curves.

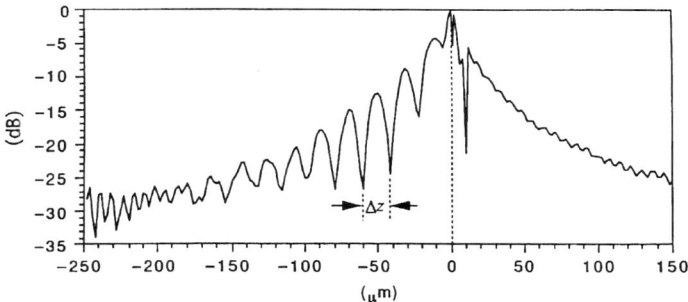

Figure 7. $V(z)$ curve for fused quartz; specimen fused quartz, coupling medium distilled water, temperature of the coupling medium 22.3°C (change less than ±0.1°C). The parameters of the acoustic lens are as follows: frequency 400 MHz, aperture angle 120°, and working distance 310 μm.

where
$$C = \int_0^\infty u^2(r) P^2(r) r \, dr, \tag{14}$$

where u is the acoustic field, P is the pupil function defined as the lens geometry, R is the reflectance function, k is the wave number in the coupling medium, and f is the focal length.

Equations (13) and (14) are expressed using $r = f \sin \theta$ as follows:

$$V(z) = C^{-1} \int_0^\infty u^2(\theta) P^2(\theta) R(\theta) \exp(i2kz \cos \theta) \sin \theta \cos \theta d\theta, \tag{15}$$

$$C = \int_0^{\theta_0} u^2(\theta) P^2(\theta) \sin \theta \cos \theta d\theta, \tag{16}$$

where θ is half the aperture angle of the lens.

When $k_z = k \cos \theta$ is used, (15) and (16) are expressed as follows:

$$V(z) = C^{-1} \int_k^{k \cos \theta_0} Q^2(k_z) R(k_z) \exp(i2k_z z) \, dk_z, \tag{17}$$

$$C = \int_k^{k \cos \theta_0} Q^2(k_z) \, dk_z, \tag{18}$$

$$Q^2(k_2) = u^2(k_z) P^2(k_z) k_z. \tag{19}$$

From (17), the following equation is obtained:

$$\mathcal{F}^{-1}\{V(z)\} = C^{-1} Q^2(k_z) R(k_z), \tag{20}$$

where $\mathcal{F}^{-1}\{\}$ is the inverse Fourier transform.

The material characterization is implemented by using (20) to monitor amplitude and phase changes of $R(k_z)$.

2. Phase Change

Figure 8 shows amplitude and phase changes of the reflectance function due to the incident angles of the acoustic waves from water to fused quartz.

Figure 8a shows the relation between the incident angle and the amplitude of the ultrasonic beam emitted from the acoustic lens

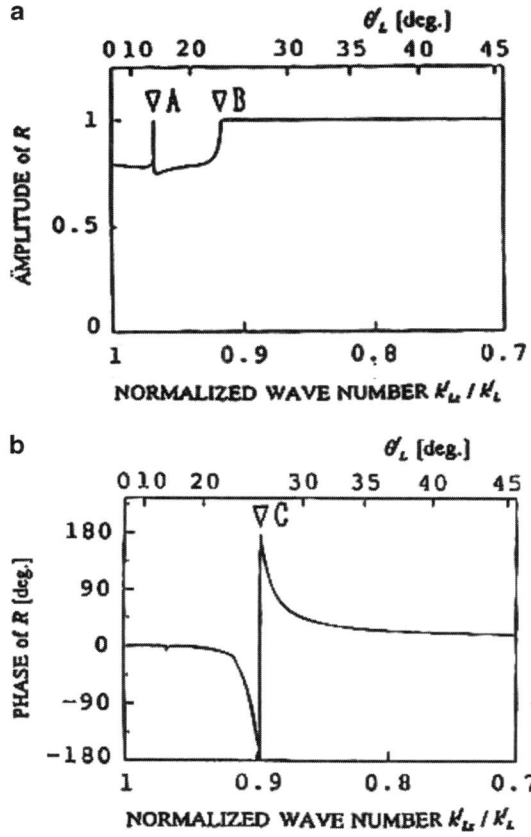

Figure 8. Reflectance function. (**a**) Amplitude; (**b**) phase specimen fused quartz, coupling medium distilled water, temperature of the coupling medium 22.3°C (change less than ±0.1°C). The parameters of the acoustic lens are as follows: frequency 400 MHz, aperture angle 120°, and working distance 310 μm.

operating at a frequency of 400 MHz onto the specimen (i.e., fused quartz) via the coupling medium (i.e., water). When the incident angle is close to the critical angle of the longitudinal wave (i.e., 14.58°), the amplitude of the reflectance function becomes abruptly strong, and becomes the maximum value at the critical angle. After passing the critical angle of the longitudinal wave, when the incident angle increases, the amplitude decreases to about 0.8. Then, the incident angle is close to the critical angle of the shear wave (i.e., 23.49°), the amplitude is again abruptly strong, and assumes the maximum value at the critical angle. After passing the critical angle of the shear wave, the amplitude remains constant.

Figure 8b shows the relation between the incident angle and the phase of the ultrasonic beam emitted from the acoustic lens operating at the frequency of 400 MHz to the specimen (i.e., fused quartz) via the coupling medium (i.e., water). The phase changes marginally at the critical angle of the longitudinal wave, but changes significantly in the neighborhood of the Rayleigh critical angle. This significant phase change is the key factor of the contrast change.

Using a ray-tracing technique, one may understand this mechanism understood as follows. The period of this variation results from interference between the two components. Figure 6 shows one component which is spectrally reflected at normal incidence, while the second one undergoes a lateral shift on incidence and reradiates at the critical phase-matching angle for the surface acoustic wave (also referred to as "leaky Rayleigh waves").

When the acoustic wave is focused onto the surface of the specimen, the phases of the acoustic waves traveling path I and path II are identical. Let this phase be denoted as Φ_f. When the acoustic lens is defocused toward the specimen by a distance z, the phase changes of the waves traveling paths I and II are expressed, respectively, as

$$\Phi_I = \Phi_f - \left(\frac{2\overline{OC}}{\lambda_W}\right) 2\pi = \Phi_f - \frac{4\pi z}{\lambda_W}, \tag{21}$$

$$\Phi_{II} = \Phi_f - \left(\frac{2\overline{AC}}{\lambda_W}\right) 2\pi + \left(\frac{2\overline{AB}}{\lambda_R}\right) 2\pi + \pi$$

$$= \Phi_f - \frac{4\pi z}{\lambda_W \cos \theta_R} + \frac{4\pi \tan \theta_R}{\lambda_R} + \pi, \tag{22}$$

where λ_W is a wavelength of the coupling medium (i.e., water), λ_R is the wavelength, and θ_R is the Rayleigh critical angle.

The phase difference is calculated by the following:

$$\Delta\Phi = \Phi_{II} - \Phi_I = 4\pi z \left(\frac{1 - \frac{1}{\cos \theta_R}}{\lambda_W + \frac{\tan \theta_R}{\lambda_R}}\right) + \pi. \tag{23}$$

Applying Snell's law, we obtain the following equation:

$$\frac{V_W}{V_R} = \frac{\lambda_W}{\lambda_R} = \sin\theta_R, \quad (24)$$

where V_R is the surface acoustic wave (e.g., Rayleigh wave) velocity on the surface of the specimen with a penetration depth of about one wavelength:

$$\lambda_R = \frac{\lambda_W}{\sin\theta_R}. \quad (25)$$

Then, inserting (24) into (23), we obtain the quantitative contrast factor as follows:

$$\Delta\Phi = 4\pi z \left(\frac{1-\cos\theta_R}{\lambda_W}\right) + \pi. \quad (26)$$

3. Theory of the Surface Acoustic Wave Velocity Measurement

When $(\Phi_{II} - \Phi_I) = (2n-1)\pi$, where n is a positive integer number, the $V(z)$ curve is at a local minimum. Hence, the period of $V(z)$ is obtained as follows:

$$\Delta z = \frac{\lambda_W}{2(1-\cos\theta_R)}. \quad (27)$$

By rewriting (25), we obtain

$$V_R = \frac{V_W}{\sin\theta_R}. \quad (28)$$

λ_W is expressed as

$$\lambda_W = \frac{V_W}{f}, \quad (29)$$

where λ is a frequency used for an acoustic lens.

Therefore, from (27) to (29), we finally obtain the following equation for calculating the surface acoustic wave velocity:

$$V_R = \frac{V_W}{\sqrt{1-\left(1-\frac{1}{2}\frac{V_W}{\Delta z f}\right)^2}}. \quad (30)$$

4. Optimizing Measurement Precision

Assume $1 \gg \frac{V_W}{2f\Delta z}$, then (30) is approximately expressed as follows:

$$V_R \cong \sqrt{V_W f \cdot \Delta z} \qquad (31)$$

To enhance the precision of the surface acoustic wave velocity measurement by the $V(z)$ curve technique, the two sides of (31) are differentiated and after taking logarithms of both sides, we obtain the following equation:

$$\frac{dV_R}{V_R} = \frac{1}{2}\frac{dV_W}{V_W} + \frac{1}{2}\frac{df}{f} + \frac{1}{2}\frac{d\Delta z}{\Delta z}. \qquad (32)$$

Equation (32) shows that the errors in the measurement of the surface acoustic wave velocity are the sum of the errors in the values of the velocity of the coupling medium, the frequency of the acoustic wave, and the distance of the period. Therefore, to minimize the measurement error, it is necessary to maintain constant temperature for the coupling medium to stabilize the frequency of the acoustic wave, and measure accurately the movement of the acoustic lens along the Z-axis.

Many of refinements in the techniques for precision measurement have been reported. Liang et al developed a SAM which can obtain complex $V(z)$ curves based on a nonparaxial formulation of the $V(z)$ integral, and a reflectance function of a liquid–solid interface by inverting the $V(z)$ curves formed at frequencies of 10 MHz or less.[48] Endo et al. improved the above-mentioned SAM in terms of a mechanical movement and an increased frequency range (up to 3.00 GHz) to obtain higherprecision measurement.[49]

Quantitative data (i.e., velocities of surface acoustic waves) obtained by a spherical lens, which is commonly used for acoustic imaging, may not be expected to provide information on the elastic anisotropy of materials. A cylindrical lens (i.e., line-focus lens) and a measuring method were developed to overcome this issue.[50–56] This technique is especially useful when the substrate is anisotropic in the nanoscaled thin film system.

V. CONTRAST

Acoustic properties (i.e., reflection coefficient, attenuation, and velocity of acoustic wave), and surface condition (i.e., surface roughness and discontinuities) of the specimen are factors in forming acoustic images. For a nanoscaled thin film system, (1) deference in the velocity of the surface acoustic wave propagating through the portion of the system and (2) increase of the amplitude of the acoustic wave caused by returning of the acoustic wave from the discontinuity located within the system are important for contrast factor.

1. Reflectance Function

Contrast in an image formed by a SAM for the simplest structure of the nanoscaled film system is mathematically expressed as $V(z)$ as follows (see Fig. 9):

$$v(z) = e^{i2k_0[z+f(1+\bar{c}^2)]} V(z). \tag{33}$$

Omitting $e^{i2k_0[z+f(1+\bar{c}^2)]}$, we can express the $V(z)$ curve as follows:[47]

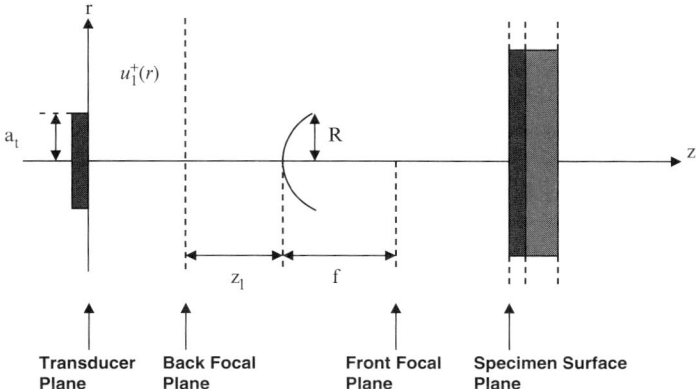

Figure 9. An acoustic lens for expressing the $V(z)$ curve with an angular-spectrum approach to a nanoscaled thin film system. a_t is the radius of the transducer, z_1 is the distance from the transducer to the back focal plane, R is the radius of the aperture of the lens, f is the focal distance of the lens, and u_1^+ is the acoustic field at the back focal plane.

$$V(z) = \int_0^\infty \left[u_1^+(r)\right]^2 [P(r)]^2 R\left(\frac{r}{f}\right) e^{-ik_0 z \left(\frac{r}{f}\right)^2} r \, dr, \quad (34)$$

where $u_1^+(r)$ is the an acoustic field at the back focal plane, $R\left(\frac{r}{f}\right)$ is the reflectance function, $e^{-ik_0 z \left(\frac{r}{f}\right)^2}$ is the lens defocusing factor, and $P(r)$ is the pupil function and is expressed as follows:

$$P(r) = \begin{cases} 1 & r \leq R \\ 0 & r > R \end{cases} \quad (35)$$

Note that $u_1^+(r)$ may be used as a Gaussian function determined by the size of the transducer for the calculation.[49]

From (34), the contrast depends on the reflection reflectance function of a specimen. The reflectance function is determined by the structure of the specimen. Now, we need to assume two cases, i.e., the system has a good adhesive condition at the interface between the film and the substrate (case I), and the system has a bad adhesive condition (i.e., delamination) at the interface (case II).

2. Reflectance Function for Layered Media

Case I

Referring to Fig. 10a, the relations among a particle velocity, a particle displacement, and a stress are as follows:

$$v^x = \frac{\partial \Phi}{\partial x} - \frac{\partial \Psi}{\partial z} \quad (36)$$

$$v^z = \frac{\partial \Phi}{\partial z} - \frac{\partial \Psi}{\partial x} \quad (37)$$

$$u^x = \frac{1}{-i\omega} \cdot v^x \quad (38)$$

$$u^z = \frac{1}{-i\omega} \cdot v^z \quad (39)$$

$$z^x = \lambda \cdot \left(\frac{\partial u^x}{\partial x} + \frac{\partial u^z}{\partial z}\right) + 2\mu \cdot \frac{\partial u^z}{\partial z} \quad (40)$$

$$z^z = \mu \cdot \left(\frac{\partial u^x}{\partial z} + \frac{\partial u^z}{\partial x}\right) \quad (41)$$

Acoustic Microscopy Applied to Nanostructured Thin Film Systems

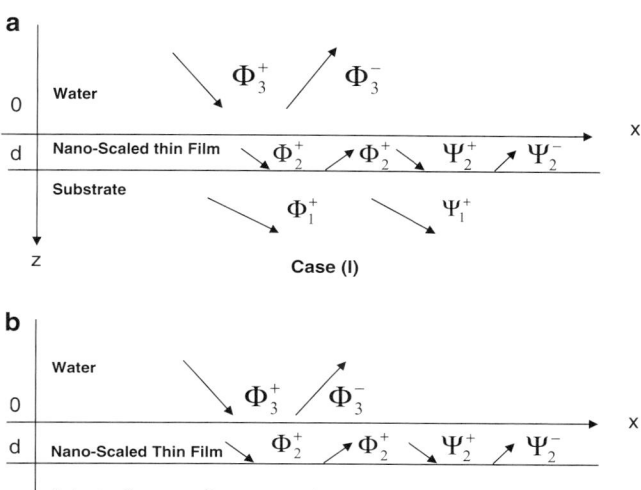

Figure 10. Acoustic wave propagating within the nanoscaled thin film system. (**a**) The system has good adhesion at the interface. (**b**) The system has poor adhesion, such as delamination.

where ω is an angular frequency, d is a thickness of the nanoscaled thin film, Φ is a potential of a longitudinal wave, Ψ is a potential of a shear wave, u^x is a particle displacement (i.e., X-axis component), u^z is a particle displacement (i.e., Z-axis component), v^x is a particle velocity (i.e., X-axis component), v^z is a particle velocity (i.e., Z-axis component), z^x is a stress (i.e., X-axis component), z^z is a stress (i.e., Z-axis component), and λ and μ are Lamé's constant.

The potential of a longitudinal wave in water is expressed as follows:

$$\Phi_3 = \left(\Phi_3^+ e^{i\alpha_3 z} + \Phi_3 - e^{-i\alpha_3 z} \right) e^{i(\sigma x - \omega t)}. \tag{42}$$

The potentials of longitudinal and shear waves in the thin film are expressed as follows:

$$\Phi_2 = \left(\Phi_2^+ e^{i\alpha_2(z-d)} + \Phi_2^- e^{-i\alpha_2(z-d)} \right) e^{i(\sigma x - \omega t)}, \tag{43}$$

$$\Psi_2 = \left(\Psi_2^+ e^{i\beta_2(z-d)} + \Psi_2^- e^{-i\beta_2(z-d)} \right) e^{i(\sigma x - \omega t)}. \tag{44}$$

The potentials of longitudinal and shear waves in the substrate are expressed as follows:

$$\Phi_1 = \Phi_1^+ e^{i\alpha_1(z-d)} e^{i(\sigma x - \omega t)}, \tag{45}$$

$$\Psi_1 = \Psi_1^+ e^{i\beta_1(z-d)} e^{i(\sigma x - \omega t)}. \tag{46}$$

where k is the wave number of the longitudinal wave, κ is the wave number of the shear wave, α is the longitudinal wave propagation vector (i.e., Z-axis component), β is the shear wave propagation vector (i.e., Z-axis component).

In this case, we can have the following equations:

$$\alpha^2 + \sigma^2 = k^2 \tag{47}$$

$$\beta^2 + \sigma^2 = \kappa^2 \tag{48}$$

The boundary conditions are (1) when $z = 0$, Z-axis components of the particle velocities are equal, (2) Z-axis components of stresses applied to the Z plane are equal, and (3) X-axis components of stresses applied to the Z plane are equal, and their values are zero. Then we have the following equations:

$$v_3^z \big|_{z=0} = v_2^z \big|_{z=0} \tag{49}$$

$$z_3^z \big|_{z=0} = z_2^z \big|_{z=0} \tag{50}$$

$$z_3^x \big|_{z=0} = z_2^x \big|_{z=0} = 0 \tag{51}$$

$$v_2^z \big|_{z=d} = v_1^z \big|_{z=d} \tag{52}$$

$$v_2^x \big|_{z=d} = v_1^x \big|_{z=d} \tag{53}$$

$$z_2^z \big|_{z=d} = z_1^z \big|_{z=d} \tag{54}$$

$$z_2^x \big|_{z=d} = z_1^x \big|_{z=d} \tag{55}$$

Solving the above equations, we obtain the reflectance function (i.e., $R_1 = \Phi_3^- / \Phi_3^+$) as follows:

$$R_1 = \frac{(d_{31}d_{43} - d_{33}d_{41}) + \frac{\lambda_3 k_3^2}{\alpha_3 \omega}(d_{21}d_{43} - d_{23}d_{41})}{(d_{31}d_{43} - d_{33}d_{41}) - \frac{\lambda_3 k_3^2}{\alpha_3 \omega}(d_{21}d_{43} - d_{23}d_{41})}, \tag{56}$$

where

$$d_{21} = C_{21} + C_{22} \tag{57}$$

$$d_{23} = C_{23} + C_{24} \tag{58}$$

$$d_{31} = C_{31} + C_{32} \tag{59}$$

$$d_{33} = C_{33} + C_{34} \tag{60}$$

$$d_{41} = C_{41} + C_{42} \tag{61}$$

$$d_{43} = C_{43} + C_{44} \tag{62}$$

Note that

$$C = A_2(-d) \begin{bmatrix} 1 & 0 & 0 & 0 \\ 0 & 1 & 0 & 0 \\ 0 & 0 & 1 & 0 \\ 0 & 0 & 0 & \frac{\mu_1}{\mu_2} \end{bmatrix} P_1(0) = \begin{bmatrix} C_{11} & C_{12} & C_{13} & C_{14} \\ C_{21} & C_{22} & C_{23} & C_{24} \\ C_{31} & C_{32} & C_{33} & C_{34} \\ C_{41} & C_{42} & C_{43} & C_{44} \end{bmatrix}, \tag{63}$$

where

$$A_2(-d) = P_2(-d) P_2^{-1}(0), \tag{64}$$

$$P_{n:n=1,2,3}(x) = \begin{bmatrix} i\sigma \cos(\alpha x) & -\sigma \sin(\alpha x) & -i\beta \cos(\beta x) & \beta \sin(\beta x) \\ -\sigma \sin(\alpha x) & i\sigma \cos(\alpha x) & -\sigma \sin(\beta x) & i\sigma \cos(\beta x) \\ \frac{-i(\lambda k^2 + 2\mu\alpha^2)}{\omega} \cos(\alpha x) & \frac{\lambda k^2 + 2\mu\alpha^2}{\omega} \sin(\alpha x) & \frac{-i2\mu\alpha\beta}{\omega} \cos(\beta x) & \frac{2\mu\alpha\beta}{\omega} \sin(\beta x) \\ \frac{\alpha\omega}{\omega} \sin(\alpha x) & \frac{-i\alpha\sigma}{\omega} \cos(\alpha x) & \frac{\sigma^2 + \beta^2}{2\omega} \sin(\beta x) & \frac{-i(\sigma^2 + \beta^2)}{2\omega} \cos(\beta x) \end{bmatrix}. \tag{65}$$

Case II

In case II, only the boundary conditions are different (see Fig. 10b):

$$v_3^z \big|_{z=0} = v_2^z \big|_{z=0} \tag{66}$$

$$z_3^z \big|_{z=0} = z_2^z \big|_{z=0} \tag{67}$$

$$z_3^x \big|_{z=0} = z_2^x \big|_{z=0} = 0 \tag{68}$$

$$v_2^z \big|_{z=d} = v_1^z \big|_{z=d} \tag{69}$$

$$z_2^z \big|_{z=d} = z_1^z \big|_{z=d} \qquad (70)$$

$$z_2^x \big|_{z=d} = z_1^x \big|_{z=d} = 0 \qquad (71)$$

Therefore, similarly, we obtain the reflectance function as follows:

$$R_2 = \frac{m_{32} - \frac{\lambda_1 k_1^2}{\alpha_1 \omega} m_{33} + \frac{\lambda_3 k_3^2}{\alpha_3 \omega} \left(m_{22} - \frac{\lambda_1 k_1^2}{\alpha_1 \omega} m_{23} \right)}{m_{32} - \frac{\lambda_1 k_1^2}{\alpha_1 \omega} m_{33} - \frac{\lambda_3 k_3^2}{\alpha_3 \omega} \left(m_{22} - \frac{\lambda_1 k_1^2}{\alpha_1 \omega} m_{23} \right)}, \qquad (72)$$

where

$$m_{22} = a_{22} - \frac{a_{21} a_{42}}{a_{41}} \qquad (73)$$

$$m_{23} = a_{23} - \frac{a_{21} a_{43}}{a_{41}} \qquad (74)$$

$$m_{32} = a_{32} - \frac{a_{31} a_{41}}{a_{41}} \qquad (75)$$

$$m_{33} = a_{32} - \frac{a_{31} a_{43}}{a_{41}} \qquad (76)$$

Note that a_{ij} are components of the matrix $A_2(-d)$.

When a specimen is a multilayered nanoscaled thin film system (see Fig. 11), this theory is easily extended as follows:

$$\Phi(t) = \left(\Phi_{(t)}^+ e^{i\alpha(t)[z-z(t)]} + \Phi_{(t)}^- e^{-i\alpha(t)[z-z(t)]} \right) e^{i(\sigma x - \omega t)}, \qquad (77)$$

$$\psi(t) = \left(\Psi_{(t)}^+ e^{i\beta(t)[z-z(t)]} + \Psi_{(t)}^- e^{-i\beta(t)[z-z(t)]} \right) e^{i(\sigma x - \omega t)}, \qquad (78)$$

$$\begin{bmatrix} v_{(t)}^x \\ v_{(t)}^z \\ z_{(t)}^z \\ \frac{1}{2\mu_{(t)}} z_{(t)}^x \end{bmatrix}_{z=z(t-1)} = A_{(t)}[-d(t)] \begin{bmatrix} v_{(t)}^x \\ v_{(t)}^z \\ z_{(t)}^z \\ \frac{1}{2\mu_{(t)}} z_{(t)}^x \end{bmatrix}_{z=z(t)}, \qquad (79)$$

$$A_{(t)}[-d(t)] = P_{(t)}[-d(t)] P_{(t)}^{-1}(0), \qquad (80)$$

Acoustic Microscopy Applied to Nanostructured Thin Film Systems

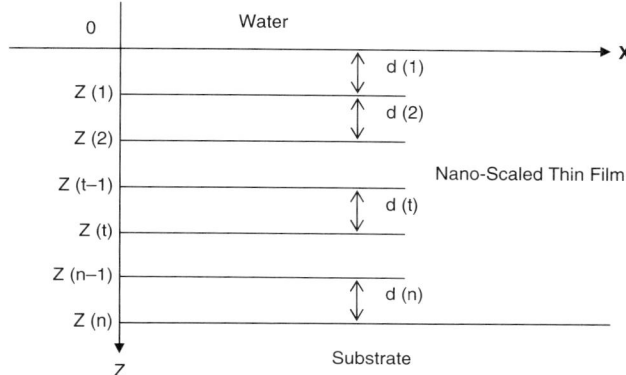

Figure 11. A multilayered and nanoscaled thin film system.

$$\begin{bmatrix} v^x_{(t)} \\ v^z_{(t)} \\ z^z_{(t)} \\ \frac{1}{2\mu_{(t)}} z^x_{(t)} \end{bmatrix}_{z=z(t-1)} = A_{(t)}\left[-d(t)\right] \begin{bmatrix} 1 & 0 & 0 & 0 \\ 0 & 1 & 0 & 0 \\ 0 & 0 & 1 & 0 \\ 0 & 0 & 0 & \frac{\mu_{(2)}}{\mu_{(1)}} \end{bmatrix} A_{(2)}\left[-d(2)\right]$$

$$\cdots \begin{bmatrix} 1 & 0 & 0 & 0 \\ 0 & 1 & 0 & 0 \\ 0 & 0 & 1 & 0 \\ 0 & 0 & 0 & \frac{\mu_{(n)}}{\mu_{(n-1)}} \end{bmatrix} A_{(n)}\left[-d(n)\right] \begin{bmatrix} v^x_{(n)} \\ v^z_{(n)} \\ z^z_{(n)} \\ \frac{1}{2\mu_{(n)}} z^x_{(n)} \end{bmatrix}_{z=z(n)}. \quad (81)$$

Then, we need to use the following matrix A_{total} for the calculation:

$$A_{\text{total}} = A_{(1)}\left[-d(1)\right] \begin{bmatrix} 1 & 0 & 0 & 0 \\ 0 & 1 & 0 & 0 \\ 0 & 0 & 1 & 0 \\ 0 & 0 & 0 & \frac{\mu_{(2)}}{\mu_{(1)}} \end{bmatrix} A_{(2)}\left[-d(2)\right] \times$$

$$\cdots \times \begin{bmatrix} 1 & 0 & 0 & 0 \\ 0 & 1 & 0 & 0 \\ 0 & 0 & 1 & 0 \\ 0 & 0 & 0 & \frac{\mu_{(n)}}{\mu_{(n-1)}} \end{bmatrix} A_{(n)}\left[-d(n)\right]. \quad (82)$$

3. Contrast Enhancement Caused by Discontinuities

When a specimen includes an elastic discontinuity, such as an edge, a step, a crack, or a joint interface, an acoustic image of the elastically discontinuous and peripheral portions visualized by the SAM shows unique contrast such as fringes or black stripes when the lens is defocused toward the specimen. As a model shown in Fig. 12, this type of contrast appears as an interference effect of surface acoustic waves incident on and reflected from elastic discontinuities.[57–59] Since the thickness of the film is about 100 nm, the same effect may not be remarkable at the operating frequency ranging from 0.6 to 1 GHz for the discontinuities. However, this technique is useful for enhancing existence of surface microcracks that conventional microscopes may not be able to visualize

With use of a ray tracing technique, the directions of acoustic waves emitted from the lens onto the specimen having the discontinuity (i.e., vertical crack as ξ) through a coupling medium are shown in Fig. 13.

The acoustic fields are expressed by $u_i^{\pm}$ or $U_i^{\pm}$ ($I = 0, 1, 2,$ and 3), where $u_i^{\pm}$ is the spatial distribution and $U_i^{\pm}$ is the frequency distribution, where numbers 0, 1, 2, and 3 represent the transducer plane, the back focal plane, the front focal plane, and the surface of the specimen, respectively. The $\pm$ superscripts indicate that the acoustic field travels in the direction from the acoustic lens to the specimen or from the specimen to the acoustic lens,

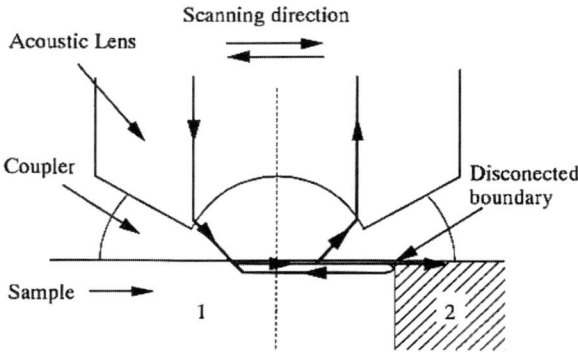

Figure 12. Mechanism of an extra surface acoustic wave generation from a discontinuity such as a crack.

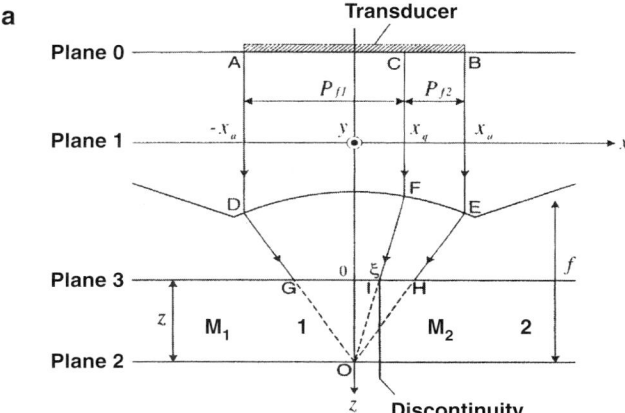

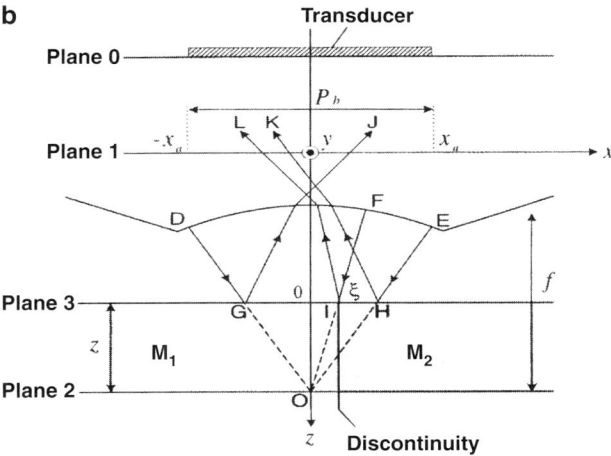

Figure 13. Acoustic waves reflected from a discontinuity. (**a**) Acoustic waves incident onto a material having a discontinuity; (**b**) acoustic waves reflected from a material having a discontinuity.

respectively. Using the Fourier transformation, we can express the relation between $u_i^{\pm}$ and $U_i^{\pm}$ as follows:

$$U_i^{\pm}(k_x, k_y) = \mathcal{F}\left\{u_i^{\pm}(x, y)\right\}, \quad (83)$$

$$u_i^{\pm}(x, y) = \mathcal{F}^{-1}\left\{U_i^{\pm}(k_x, k_y)\right\}. \quad (84)$$

The pupil function of the lens is again defined as follows:

$$P_f(x, y) = \text{circ}(r) = \begin{cases} 1 & r < 1 \\ 0 & r \geq 1 \end{cases}, \tag{85}$$

where r is the radius of the lens.

In this case, the pupil function may be divided into the following two parts for analysis of its acoustic fields:[60]

$$P_{f1}(x, y; z, \xi) = \begin{cases} 1 & (-x_a \leq x \leq x_q) \\ 0 & (x_q \leq x \leq x_a) \end{cases}, \tag{86}$$

$$P_{f2}(x, y; z, \xi) = \begin{cases} 1 & (-x_q \leq x \leq x_a) \\ 0 & (x_q \leq x \leq x_a) \end{cases}, \tag{87}$$

where $|x_a|$ is the radius of the whole pupil function, and x_q is expressed as follows:

$$x_q = -\frac{\xi f}{z}, \tag{88}$$

where f is the focal length of the lens.

When the acoustic waves are reflected back from the specimen, the pupil function must be considered within the region expressed as $-x_a \leq x \leq x_a$ in accordance with the directions of wave propagation shown in Fig. 13b. Using the angular spectrum approach,[47] we can express the intensity of the wave at the transducer of the acoustic lens as follows:

$$V(\xi, z) = V_1(\xi, z) + V_2(\xi, z), \tag{89}$$

where $z < 0$, $V_1(\xi, z)$ is the intensity relating to the acoustic wave incident on medium 1 passing through P_{f1}, and $V_2(\xi, z)$ is the intensity relating to the acoustic wave incident on medium 2 passing through P_{f2}. It is needless to say that medium 1 and medium 2 can be the same materials.

$V_1(\xi, z)$ and $V_2(\xi, z)$ are expressed as follows:

$$V_1(\xi, z) = \iint_{-\infty}^{\infty} u_1^+(-x, -y) u_1^+(x, y) P_{f1}(-x, -y; z, \xi) P_b(x, y)$$

$$\times R_1\left(\frac{x}{f}, \frac{y}{f}\right) \exp\left(i2kz\sqrt{\frac{1-(x^2+y^2)}{f^2}}\right) dx\, dy, \tag{90}$$

$$V_2(\xi, z) = \iint_{-\infty}^{\infty} u_1^+(-x, -y) u_1^+(x, y) P_{f1}(-x, -y; z, \xi) P_b(x, y)$$
$$\times \mathcal{R}_2\left(\frac{x}{f}, \frac{y}{f}\right) \times \exp\left(i2kz\sqrt{\frac{1-(x^2+y^2)}{f^2}}\right) dx\, dy, \tag{91}$$

where k is the wave number in the coupling medium, and $\mathcal{R}_1$ and $\mathcal{R}_2$ are the reflectance functions of medium 1 and medium 2, respectively. P_b is the pupil function of the acoustic lens for the acoustic wave reflected from the specimen and is expressed as follows:

$$P_b(x, y; z, \xi) = \begin{cases} 1 & (x^2 + y^2 \leq x_a^2) \\ 0 & \text{otherwise} \end{cases}. \tag{92}$$

Considering the excited surface acoustic wave at the jointed interface, we can express a new acoustic field of the plane as follows:

$$\hat{u}_3^-(x, y) = u_3^-(x, y) + \rho u_{3R}^-(2\xi - x, y), \tag{93}$$

where u_{3R}^- is an acoustic field of the surface acoustic wave excited at the jointed interface and ρ is the reflectivity of surface acoustic wave at the jointed interface. ρ may present the characteristic of the discontinuity. Then, (93) is rewritten by (84) as follows:

$$\hat{U}_3^-(k_x, k_y) = U_3^-(k_x, k_y) + \rho \exp(-i2k_x\xi) U_{3R}^-(-k_x, k_y). \tag{94}$$

The relation between U_3^+ and U_3^- is expressed as follows:

$$U_3^-(k_x, k_y) = \mathcal{R}(k_x, k_y) U_3^+(k_x, k_y). \tag{95}$$

The relation between u_{3R}^+ and u_{3R}^- is expressed as follows:

$$U_{3R}^-(k_x, k_y) = \mathcal{R}_R(k_x, k_y) U_3^+(k_x, k_y), \tag{96}$$

where $\mathcal{R}_R$ is a reflectance function for the surface acoustic wave excited on the discontinuity.[61,62]

Using (14) and (15), (13) is rewritten as follows:

$$\hat{U}_3^- (k_x, k_y) = \mathcal{R}(k_x, k_y) U_3^+ (k_x, k_y)$$
$$+ \rho \exp(-i2k_x\xi) \mathcal{R}_R (-k_x, k_y) U_3^+ (-k_x, k_y) \quad (97)$$

The field propagations from 0 to 3 and 3 to 0 have been formulated in the $V(z)$ analysis by Atalar.[47] Using (95) instead of the condition of surface reflection $U_3^- = \mathcal{R} U_3^+$, we can express (89) as follows:

$$V(\xi, z)$$
$$= \iint_{-\infty}^{\infty} \left[\begin{array}{c} P_{f1}(-x, -y; z, \xi) \mathcal{R}_1 \left(\frac{x}{f}, \frac{y}{f}\right) \\ + \rho \exp\left[-i2k\xi\left(\frac{x}{f}\right)\right] P_{f1}(-x, -y) \\ \times \mathcal{R}_{R1}\left(-\frac{x}{f}, \frac{y}{f}\right) \end{array} \right] u_1^+(-x, -y)$$
$$\times u_1^+ (x, y) P_b (x, y) \exp\left(\frac{-ikz(x^2 + y^2)}{f^2}\right) dx\,dy$$

$$+ \iint_{-\infty}^{\infty} \left[\begin{array}{c} P_{f2}(-x, -y; z, \xi) \mathcal{R}_2 \left(\frac{x}{f}, \frac{y}{f}\right) \\ + \rho \exp\left[-i2k\xi\left(\frac{x}{f}\right)\right] P_{f2}(-x, -y) \\ \times \mathcal{R}_{R2}\left(-\frac{x}{f}, \frac{y}{f}\right) \end{array} \right] u_1^+(-x, -y)$$
$$\times u_1^+ (x, y) P_b (x, y) \exp\left(\frac{-ikz(x^2 + y^2)}{f^2}\right) dx\,dy. \quad (98)$$

4. Computer Simulation

Based on (56) and (72), we have completed the computer simulation (see Figs. 14, 15). The simulation is implemented as the following conditions such as parameters of the acoustic lens and the specimen (see Tables 3,4).

Figure 14 shows the sensitivities of the $V(z)$ curves in accordance with the acoustic lenses, operating frequencies at 1, 0.8, and 0.6 GHz, respectively, for the nanoscaled thin film system, wherein the film material is copper, the thickness of the film is 120 nm, and

a 1000 MHz

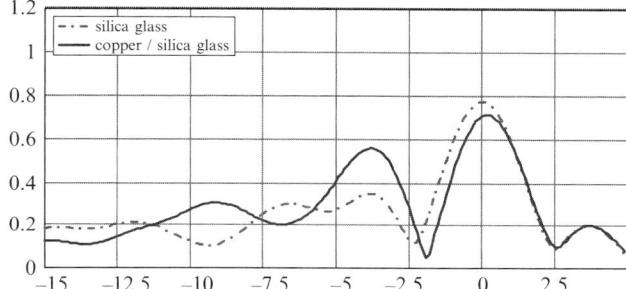

b 800 MHz

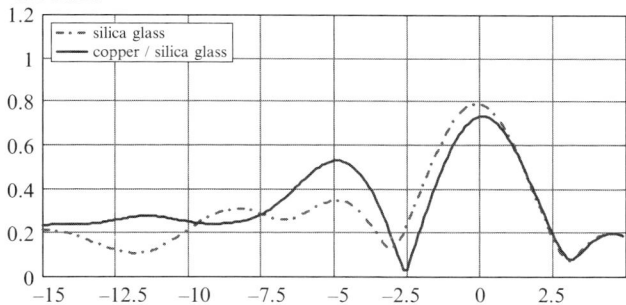

c 600 MHz

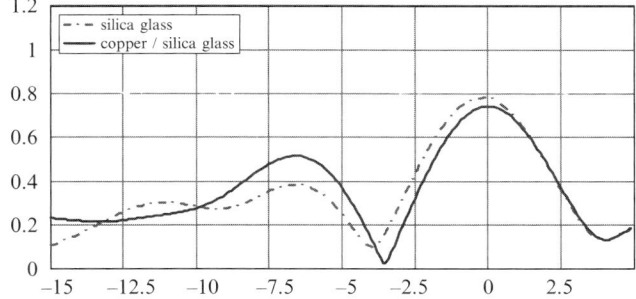

Figure 14. $V(z)$ curves calculated in accordance with the acoustic lenses, operating at frequencies of 1, 0.8, and 0.6 GHz, respectively, for the nanoscaled thin film system, wherein the film material is copper, the thickness of the film is 120 nm, and the substrate is silica glass. The $V(z)$ curves illustrated with the *solid lines* are for the system and those illustrated with the *broken lines* are for the silica glass.

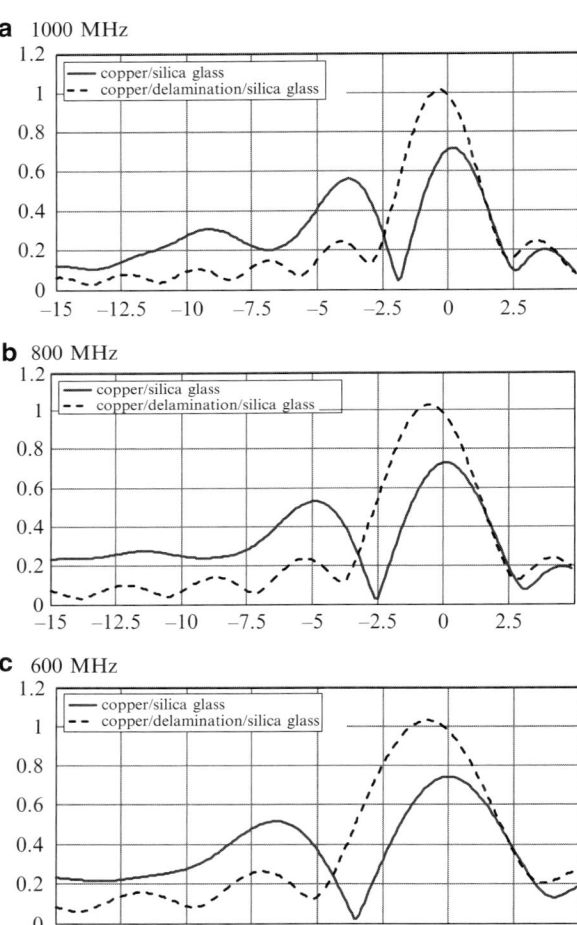

Figure 15. The $V(z)$ curves in accordance with acoustic lenses operating at frequencies of 1, 0.8, and 0.6 GHz, respectively, for the nanoscaled thin film system having the delamination (i.e., defective system) at the interface between the film and the silica glass substrate, wherein the film material is copper, and the thickness of the film is 120 nm. The $V(z)$ curves illustrated with *solid lines* are for the system having no delamination at the interface (i.e., normal system) and those illustrated with *broken lines* are for the defective system.

Table 3.
Parameters of acoustic lenses used for the computer simulation.

Frequency	1.0 GHz
Buffer rod	Sapphire
Radius of the transducer	216.5 μm
Distance from the transducer to the back focal plane	5,033 μm
Focal distance	144.38 μm
Aperture angle	120°
Frequency	0.8 GHz
Buffer rod	Sapphire
Radius of the transducer	216.5 μm
Distance from the transducer to the back focal plane	5,033 μm
Focal distance	144.38 μm
Aperture angle	120°
Frequency	0.6 GHz
Buffer rod	Sapphire
Radius of the transducer	361.00 μm
Distance from the transducer to the back focal plane	7,690.00 μm
Focal distance	288.76 μm
Aperture angle	120°

Table 4.
Parameters of the specimen (film) used for the computer simulation.

Material, copper	
Thickness	120 nm
Substrate, silica glass	
Type of defect	Delamination
Thickness	1.0 nm

the substrate is silica glass. The $V(z)$ curves illustrated with the solid lines are for the system and the broken lines are for the silica glass. As can be appreciated, those acoustic lenses can visualize a flaw when the lenses are defocused within a few micrometers from the surface.

Figure 15 shows the sensitivities of the $V(z)$ curves in accordance with acoustic lenses operating frequencies at 1.0, 0.8, and 0.6 GHz, respectively, for the system having the delamination (i.e., defective system) at the interface between the film and the substrate. The $V(z)$ curves illustrated with solid lines are for the system having no delamination at the interface (i.e., normal system) and the broken lines are for the defective system. The results of the simulation inform us that the delaminated portion of the system forms the acoustic image with high intensity compared with a portion having good adhesion.

5. Experimental Result

Figure 16 shows the optical image of the nanoscaled thin film system, wherein the film material is copper, the thickness of the film is 120 nm, and the substrate is silica glass, and wherein the image is formed with an optical microscope (Olympus, model BH2) with × 50 objective lens. Portions of the system where the films have come off (i.e., dark areas) are observed. Cracks are observed at the upper-left corner.

Figure 17 shows the acoustic images of the system. Since the film is so thin, the details of the subsurface of the system, such as delaminated portions and poor adhesive portions, in addition to those

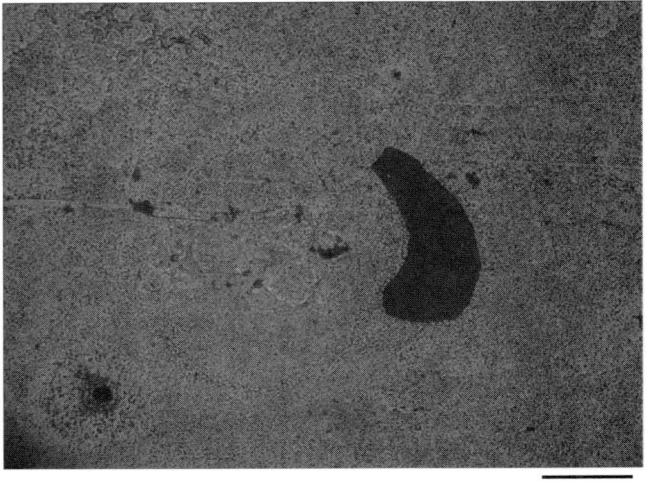

50μm

Figure 16. Optical image. A × 50 objective lens is used to form the image.

a

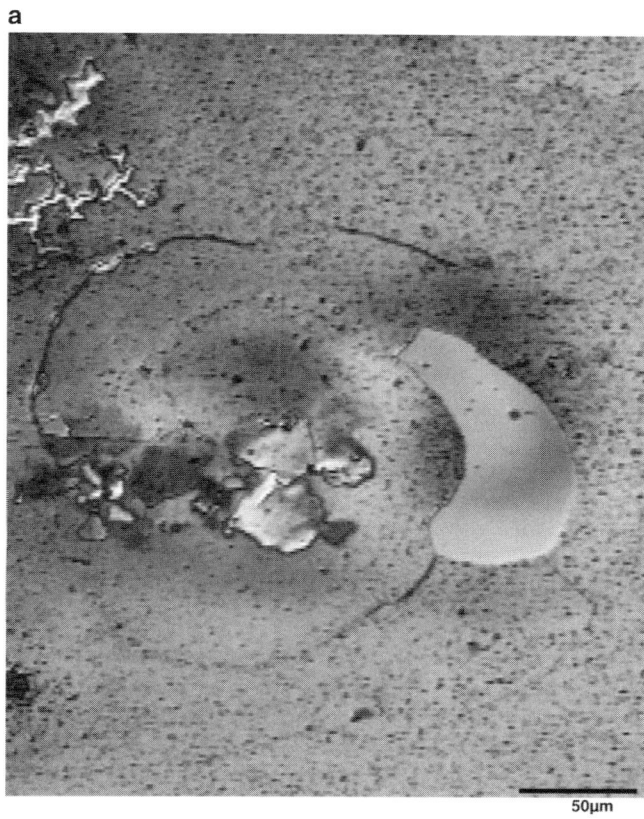

Figure 17. Acoustical image. (**a**) the acoustic lens is focused at the surface (denoted as $Z = 0\,\mu m$); (**b**) the acoustic lens is mechanically defocused by $3\,\mu m$ toward the specimen (denoted as $Z = -3\,\mu m$). The acoustic lens is operated at a frequency of 1 GHz.

of the surface, such as cracks and the elevated portion of the film, are visualized when the acoustic lens is focused at the surface (see Fig. 17a). The cracks are clearly observed when the acoustic lens is defocused toward the specimen. This means that extra surface acoustic waves are generated from the cracks and enhanced the existence of the cracks in the acoustic image (see Fig. 17b).

b

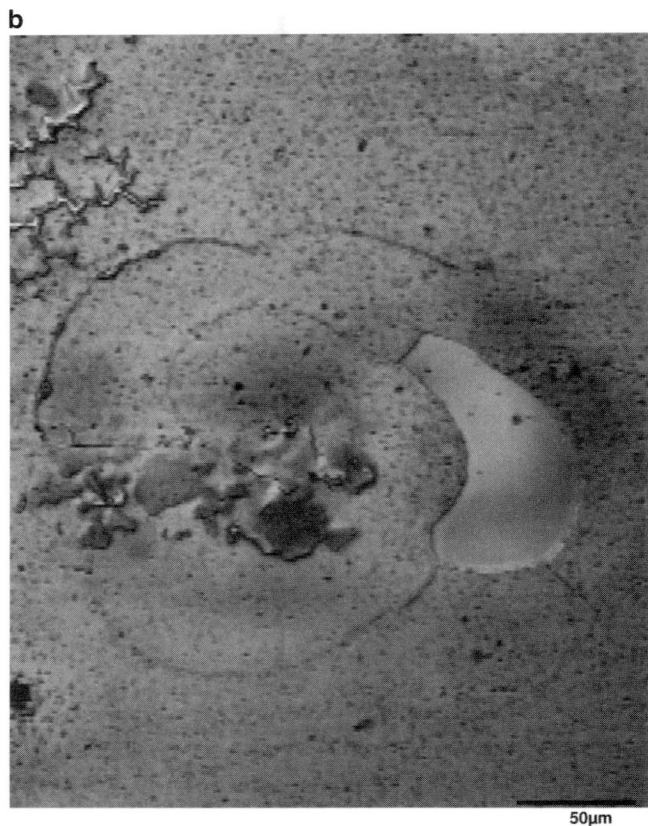

50μm

Figure 17. (continued)

Figures 18, 19, and 20 show flaws of the systems, wherein the film materials are nickel, cobalt, and nickel–cobalt, and wherein the thickness of the film is 120 nm, and the substrate is silica glass. An acoustic lens operating at 600 MHz is used to visualize the flaws.

The experimental results show that the SAM with an operating frequency ranging from 0.6 to 1.0 GHz may be able to detect the flaws of the system.

Acoustic Microscopy Applied to Nanostructured Thin Film Systems 447

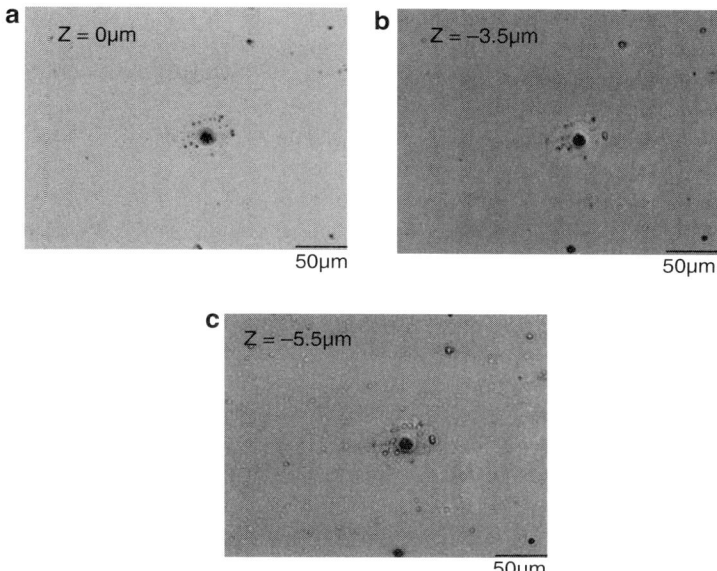

Figure 18. Nickel film; frequency 600 MHz; scanning width $X = 0.25$ mm.

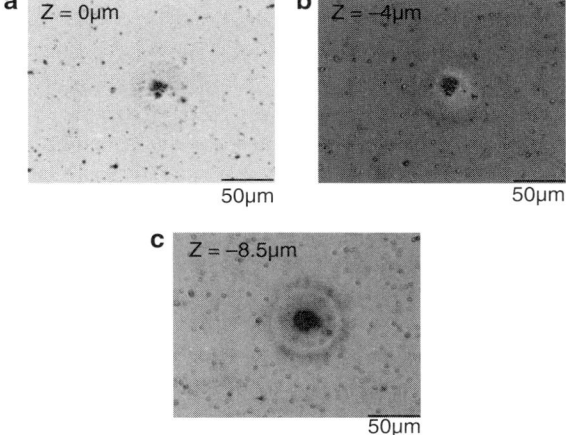

Figure 19. Cobalt specimen; frequency 600 MHz; scanning width $X = 0.25$ mm.

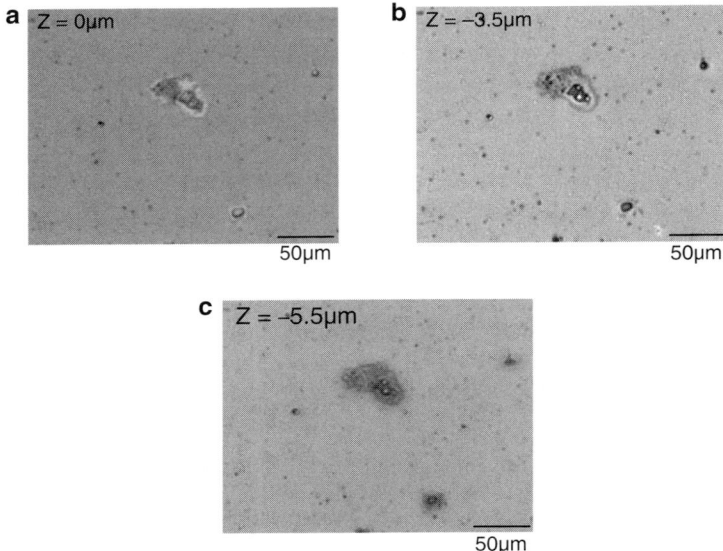

Figure 20. Nickel–cobalt specimen; frequency 600 MHz; scanning width $X = 0.25$ mm.

VI. CONCLUSION

In this article, (1) the imaging principles, including the description of the acoustic lens, (2) resolution, and (3) the contrast mechanism through $V(z)$ curve analyses of layered media for the SAM were described. Through the analyses, contrast is a key issue of the imaging factor, not resolution. Computer simulations were implemented to estimate the sensitivity of detection for the flaws (i.e., delamination) of the nanoscaled thin film system. The simulation results show that the SAM is applicable to the system. Finally, experiments were implemented to confirm that the SAM operating frequencies ranging from 0.6 to 1.0 GHz can be applicable to visualize the flaws of the system.

ACKNOWLEDGMENTS

C.M. thanks M. Schlesinger for his useful advice, and E. Buakulin for developing the simulation software.

REFERENCES

[1] J. R. Greer, W. C. Oliver, and W. D. Nix, *Acta Mater* **53** (2005), 1821.
[2] Y. Isono, T. Namazu, and N. Terayama, *J Microelectromech Syst* **15**(1) (2006), 169.
[3] J.-W. Lee, J.-C. Huang, and J.-G. Duh, *Tamkang J Sci Eng* **7**(4) (2004), 237.
[4] C. Thomsen, H. T. Grahn, H. J. Maris, and J. Tauc., *Phys Rev B* **34** (1986), 4129.
[5] H. T. Grahn, H. J. Maris, J. Tauc, and B. Abeles, *Phys Rev B* **38** (1988), 6066.
[6] O. B. Wright and K. Kawashima, *Phys Rev Lett* **69** (1992), 1668.
[7] H.-N. Lin, H. J. Maris, and L. B. Freund, *J Appl Phys* **73**(1) (1992), 37.
[8] B. Bonello, B. Perrin, and C. Rossignol, *J Appl Phys* **83**(6) (1998), 3081.
[9] K. Yamanaka, H. Ogiso, and O. Kolosov, *Appl Phys Lett* **64**(2) (1994), 178.
[10] K. Yamanaka and S. Nakano, *Jpn J Appl Phys* **35**(5) (1996), 3787.
[11] G. Binnig, C. F. Quate, and Ch. Gerber, *Phys Rev Lett* **56**(1986), 930.
[12] K. Yamanaka, S. Nakano, and H. Ogiso, *International Symposium on Measurement Technology and Intelligent Instruments* (1996), 451.
[13] S. Amerio, A. V. Goldade, U. Rabe, V. Scherer, B. Bhusan, and W. Arnold, *Thin Solid Films* **392**(2001), 75.
[14] B. Gauitier and V. *Bornand, Thin Solid Films* **515** (2006), 1592.
[15] A. Atalar, C. F. Quate, and H. K. Wickramasinge, *Appl Phys Lett* **31** (1977), 791.
[16] R. D. Weglein, *IEEE Trans Sonics* **SU27**(2) (1980), 82.
[17] R. C. Bray, C. F. Quate, J. Calhoun, and R. Kock, *Thin Solid Films* **74** (1980), 295.
[18] R. D. Weglein, *Electron Lett* **18**(22) (1982), 1003.
[19] C. C. Lee, C. S. Tsai, and X. Cheng, *IEEE Trans Sonics Ultrasonics* **SU-32**(2)(1985), 248.
[20] J. Kushibiki and N. Chubachi, *Electron Lett* **23**(12)(1987), 652.
[21] R. C. Addison, M. C. Somekh, J. M. Rowe, and G. A. D. Briggs, *SPIE 768 Pattern recognition and acoustical imaging* (ed. L. A. Ferrari) (1987), 275.
[22] J. Kushibiki, T. Ishikawa, and N. Chubachi, *Appl Phys Lett* **57**(19) (1990), 1967.
[23] R. D. Weglein and A. K. Mal, *Surf Coat Technol* **47** (1991), 667.
[24] A. Okada, C. Miyasaka, and T. Nomura, *JIM* **33**(1) (1992), 73.
[25] T. Kundu, *J Appl Mech* **59** (1992), 54.
[26] Y. Sasaki, T. Endo, T. Yamagishi, and M. Sakai, *IEEE Trans* **UFFC-39**(5) (1992), 638.
[27] T. Endo, C. Abe, M. Sakai, and M. Ohono, *Proceedings Ultrasonic International* 93 *Conference* (1993), 45.
[28] D. Achenbach, J. D. Kim, and Y. C. Lee, *Advances in Acoustic Microscopy*, vol. **1**, Plenum, New York, (1995), 153.
[29] S. Parthasarathi, B. R. Tittmann, and R. J. Ianno, *Thin Solid Films* **300** (1997), 42.
[30] A. Doghmane, Z. Hadjoub, and F. Hadjoub, *Thin Solid Films* **310** (1997), 203.
[31] D. Rats, J. von Stebut, and F. Augereau, *Thin Solid Films* **355–356** (1999), 347.
[32] Z. Guo, J. D. Achenbach, A. Madan, K. Martin, and M. E. Graham, *Thin Solid Films* **394** (2001), 189.

[33] M. J. Bamber, K. E. Cooke, A. B. Mann, and B. Derby, *Thin Solid Films* **398–399** (2001), 299.
[34] F. Zhang, S. Krishnaswarmy, D. Fei, D. A. Rebinsky, and B. Feng, *Thin Solid Films* **503** (2006), 250.
[35] B. Hadimioglu and C. F. Quate, *Appl Phys Lett* **43** (1983), 1006.
[36] K. Karaki and M. Sakai, *Ultrasonic Technology 1987, Toyohashi International Conference on Ultrasonic Technology*, Toyohashi, Japan, MYU Research, Tokyo (1987), 25.
[37] K. Yamanaka, Y. Nagata, and T. Koda, *Ultrasonics Int* **89** (1989), 744.
[38] Foster and D. Rugar, *Appl Phys Lett* **42** (1983), 869.
[39] M. S. Muha, A. A. Moulthrop, G. C. Kozlowski, and B. Hadimioglu, *Appl Phys Lett* **56** (1990), 1019.
[40] N. Chubachi, J. Kushibiki, T. Sannomia, and Y. Iyama, *Proc IEEE Ultrasonics Symp* (1979), 415.
[41] D. A. Davids, P. Y. We, and D. Chizhik, *Appl Phys Lett* **54**(17) (1989), 1639.
[42] A. Atalar, H. Koymen, and L.Degertekin, *Proc IEEE Ultrasonics Symp* (1990), 359.
[43] B. T. Khuri-Yakub, C. Cinbis, and P. A. Reinholdtsen, *Proc IEEE Ultrasonics Symp* (1989), 805.
[44] C. Miyasaka, B. R. Tittmann, and M. Ohno, *Res Nondestr Eval* **11**(1999), 97.
[45] R. D. Weglein, *Appl Phys Lett* **34** (1979), 179.
[46] W. Parmon and H. L. Bertoni, *Electron Lett* **15** (1979), 684.
[47] A. Atalar, *J Appl Phys* **49** (1978), 5130.
[48] K. Liang, G. S. Kino, and B. T. Khuri-Yakub, *IEEE Trans* **SU-32** (1985), 213.
[49] T. Endo, Y. Sasaki, T. Yamagishi, and M. Sakai, *Jpn Appl Phys* **31** (1992), 160.
[50] J. Kushibiki, K. Horii, and N. Chubachi, *Electron Lett* **19** (1983), 404.
[51] J. Kushibiki, T. Kobayashi, H. Ishiji, and N. Chubachi, *Appl Phys Lett* **61**(18) (1992), 2164.
[52] J. Kushibiki, M. Miyashita, and N. Chubachi, *IEEE Photonics Technol Lett* **8**(11) (1996), 1516.
[53] J. Kushibiki and M. Miyashita, *Jpn J Appl Phys* **36**(7B) (1997), 959.
[54] J. Kushibiki and N. Chubachi, *IEEE Trans Sonics Ultrasonics* **SU-32**(2) (1985), 189.
[55] Z. L. Li, *IEEE Trans* **UFCC-40**(6) (1993), 680.
[56] J. Kushibiki and M. Arakawa, *IEEE Trans* **UFCC-45**(2) (1998), 421.
[57] K. Yamanaka and Y. Enomoto, *J Appl Phys* **53**(1982), 846.
[58] C. Iett, M. G. Somekh, and G. A. D. Briggs, *Proc R Soc Lond* **A393** (1984), 171.
[59] G. A. D. Briggs, P. J. Jenkins, and M. Hoppe, "How fine a surface crack can you see in a scanning acoustic microscope?," *J Microsc* **159** (1990), 15–32.
[60] M. Ohono, C. Miyasaka, and B. R. Tittmann, "Pupli function splitting method in calculating acoustic microscopic signals for elastic discontinuities," *J Wave Motion Sound* **33** (2001), 309–320.
[61] C. Miyasaka, B. R. Tittmann, and S. Tanaka, "Characterization of stress at a ceramic/metal joint interface by the V(z) technique of scanning acoustic microscopy," *J Pressure Vessel Technol* **124**(3) (2002), 336.
[62] C. Miyasaka, B. R. Tittmann, and S. Tanaka, *Nondestr Test Eval* **18** (2002), 131.

10

Current Distribution in Electrochemical Cells: Analytical and Numerical Modeling

Uziel Landau

Department of Chemical Engineering, Case Western Reserve University, Cleveland, OH 44106, USA, uziel.landau@case.edu

I. INTRODUCTION AND OVERVIEW

The topic of current distribution modeling is central to the analysis of electrochemical systems and has been addressed in textbooks,[1] reviews (e.g., Refs. [2–4]) and numerous journal publications. Newman's textbook[1] provides a meticulous and comprehensive treatment of the subject. Prentice and Tobias[2] present a review of the early (up to about 1980) publications in the area. Dukovic's more recent review[3] is very comprehensive, providing critical analysis of both the electrochemical and the numerical aspects of the topic. A recent review by Schlesinger[4] focuses primarily on the numerical techniques. The present monograph introduces the fundamental processes and equations underlying the modeling of the current distribution, and critically analyzes common assumptions and approximations. Focus is placed on discussing scaling parameters for the characterization of the current distribution. Commonly used algorithms for numerical determination of the current distribution are compared and a few numerical implementations are discussed. Lastly, the modeling of the current distribution in some special configurations and applications is introduced, emphasizing recent publications.

II. SIGNIFICANCE OF MODELING THE CURRENT DISTRIBUTION

The current distribution is among the most significant parameters characterizing the operation of the electrochemical cell. The current density on the electrodes is directly proportional to the reaction rate and its distribution critically affects the electrochemical process. In electroplating, the deposit thickness distribution, and properties such as the deposit surface texture and its morphology are directly linked to the current distribution. When multiple simultaneous electrode reactions are present, such as in alloy deposition or in hydrogen co-evolution, the alloy composition in the former case and the current efficiency in the latter are controlled by the overpotential distribution, which, as discussed below, is directly related to the current distribution. Electrolytic processes which do not involve deposition are also strongly affected by the current distribution. Examples include optimized utilization of catalytic electrodes and the need to prevent the current density from surging on electrode sections, on separators, and on membranes. The power required for operating an electrochemical cell, and particularly the ohmic loss are also dependent on the current distribution. Lastly, the correct interpretation of experimental data hinges on understanding the range of current densities to which the tested electrode has been subjected.

The current distribution can be analyzed on different scales. The *macroscopic* current *distribution*, where the distribution is resolved on length scale on the order of centimeters, is important in characterizing the deposit thickness uniformity on a plated part, or in selective plating, where a nonuniform current distribution is sought. The *microscale* distribution, on the other hand, where the current density is resolved on submillimeter length scales, affects primarily parameters such as the deposit texture and roughness, nucleation, and deposition within micron-scale and nanoscale features.

For many applications, numerical simulation capability which provides the current distribution in a given configuration and plating conditions, or for a given set of such parameters, is sufficient. However, for predictive process design and for scale-up of cells and processes (and scale-down of industrial processes for laboratory testing) analytical models that elucidate the dependence of the current distribution on the process parameters are more beneficial.

III. EXPERIMENTAL DETERMINATION OF THE CURRENT DISTRIBUTION

Electroplating processes, where a solid deposit is formed and its thickness can be directly measured, provide a relatively convenient means for determination of the current distribution. The deposit thickness can be measured by a number of commercially available devices, based on, e.g., X-ray fluorescence, beta backscatter, magnetic properties, or controlled dissolution. A direct probe based on the induced field associated with the current flow has recently been introduced. Optical and electron microscopy of cross-sectioned deposits provide a common means for measuring the deposit thickness. Once the deposit thickness, d, has been measured, it can be related to the current density, i, through Faraday's law:

$$d = \frac{Mt}{Fn\rho}\varepsilon_F i. \quad (1)$$

Here, t is the plating time, F is Faraday's constant, M, ρ, and n are the plated metal atomic weight, its density, and the number of electrons transferred in the deposition reaction, respectively, and ε_F is the faradaic efficiency, accounting for side ("parasitic") reactions. Metals noble to hydrogen are typically plated from aqueous solutions at $\varepsilon_F \sim 1$ corresponding to close to 100% faradaic efficiency (unless driven to the limiting current). When the faradaic efficiency is less than 100%, (1) can be used to determine the fradaic current efficiency once the current density has been evaluated.

For the case of redox or gas-evolving reactions, where no solid deposit is formed, the current distribution on the electrodes can be determined using segmented electrodes, or insulated probe electrodes. Here, the electrode on which the current distribution is sought is sectioned into multiple, electrically isolated segments, to which the current may be individually fed and measured. If multiple, narrow, segments are provided, the average segmental current densities, obtained by dividing the segmental currents by the corresponding segmental areas, provide an approximation to the current distribution. For the segmented electrode to resemble a continuous electrode, all segments must be coplanar and essentially equipotential. To ensure that the potential of all segments is within a few millivolts, multichannel potentiostats must be used. A less costly approach is to connect each segment to the common bus via a very low (typically milliohm) shunt resistor, which enables the measurement of the current, yet introduces insignificant voltage variation.

IV. ANALYTICAL DERIVATION OF THE CURRENT DISTRIBUTION

This topic is covered in significant detail in Newman's textbook.[1] A summary, relevant to the ensuing discussion, is provided here.

1. The Current Density

The current density is directly related to the ionic flux, N_j, in the electrochemical cell. The flux is typically described in terms of three major components: diffusion of ions across a concentration gradient, migration of charged ions down the electric field, and transport of ions due to bulk electrolyte convection. Consequently, the flux of an ionic species j is given by

$$N_j = -D_j \nabla c_j - u_j z_j F c_j \nabla \Phi + c_j v. \quad (2)$$

The current density is determined by assigning the charge Fz_j to the flux of each species j and summing over all ionic species:

$$i = F \sum_j z_j N_j \quad (3)$$

Substituting (2) into (3),

$$i = -F \sum_j z_j D_j \nabla c_j - F^2 \left(\sum_j u_j z_j^2 c_j \right) \nabla \Phi + F \left(\sum_j z_j c_J \right) v. \quad (4)$$

Electroneutrality, expressed as

$$\sum_j z_j c_j = 0 \quad (5)$$

is present throughout the cell (except for the vanishingly thin double layer) and renders the last term on the right of (4) zero, providing for the total current density

$$i = -F \sum_j z_j D_j \nabla c_j - F^2 \left(\sum_j u_j z_j^2 c_j \right) \nabla \Phi. \qquad (6)$$

We also recognize[1] that the electrolyte conductivity, κ, is given by

$$\kappa \equiv F^2 \left(\sum_j u_j z_j^2 c_j \right) \qquad (7)$$

Hence, we can rewrite (6) as

$$i = -F \sum_j z_j D_j \nabla c_j - \kappa \nabla \Phi. \qquad (8)$$

Equation (8) indicates that the current density is determined by both the potential and the concentration gradients. The explicit velocity term is absent from (8) (owing to electroneutrality); however, convection still affects the current density by controlling the concentration field. It should also be noted that the electrode kinetics which do not appear explicitly in (8) establish the boundary conditions required for its solution. As subsequently shown, the electrode kinetics may influence quite significantly the current distribution.

While representing the current density as a function of the potential and concentration distributions, (8) does *not* provide the necessary relationships required for solving the distribution. This is derived from the constitutive equations described below.

2. Material Balance

The governing equations for a cell with diffusion, migration, and convection are derived by performing a material balance on a volume element, for each of the ionic species:

$$\frac{\partial c_j}{\partial t} = -\nabla \cdot N_j + R_j. \qquad (9)$$

R_j is the rate of species j generation due to a homogeneous reaction within the volume element. Such reactions are uncommon in electrochemical cells (but may be encountered, e.g., when a complex dissociates, releasing the ionic species j), and therefore we set identically $R_j = 0$.

When the flux expression (2) is substituted into (9), the general equation ("Nernst–Planck") for the concentration and potential fields is obtained:

$$\frac{\partial c_j}{\partial t} + v \cdot \nabla c_j = F \nabla \cdot (z_j u_j c_j \nabla \Phi) + \nabla \cdot (D_j \nabla c_j). \qquad (10)$$

For a multicomponent electrolyte with j ionic species, (10) represents a system of j equations, one for each ionic species. Since the potential is present in each equation, there are a total of $j + 2$ unknowns (j species concentrations, c_j, the electrostatic potential, Φ, and the fluid velocity, v). The extra equations required for solving the system are the electroneutrality condition (5) and the momentum equation, which describes the fluid velocity at all locations within the cell. The momentum equation is typically represented in terms of the "Navier–Stokes" approximation:[5]

$$V_j \frac{\partial v_i}{\partial x_j} = -\frac{1}{\rho}\frac{\partial P}{\rho \partial x_i} + (\nu + \nu_T)\frac{\partial^2 v_i}{\partial x_j^2} + \frac{\partial \nu_T}{\partial x_j}\left(\frac{\partial v_i}{\partial x_j} + \frac{\partial v_j}{\partial x_i}\right). \qquad (11)$$

The Navier–Stokes equation is written here for a Cartesian two-dimensional coordinate system where i and j represent the two axes. Accordingly, v_i and v_j are the velocity components in the directions i and j. P is the hydrostatic pressure, and ν and ν_T are the molecular and the turbulent kinematic viscosity, respectively.[5] For systems involving forced convection, the fluid flow equations are typically decoupled from the electrochemical process, and can be solved separately.

The set of $j + 2$ equations (j equations (10) plus electroneutrality equation (5) plus the momentum equation (11)) fully describe the current, potential, and concentration distributions in the cell as a function of time. This set of equations must be solved subject to the electrochemical boundary conditions.

3. Boundary Conditions

Two types of boundaries are present in electrochemical cells: (i) insulating boundaries and (ii) electrodes:

(i) *Insulator*: Here, no current may flow into the boundary and, accordingly, the current density normal to the insulating boundary is specified as zero:

$$i_n = 0. \tag{12}$$

(ii) *Electrodes*: On an electrode, an expression for the reaction kinetics that relates the potential to the normal current must be provided:

$$i_n|_e = f(c_j, \Phi)|_e. \tag{13}$$

Electrode Kinetics and Overpotentials

Typically, the current density is related to the overpotential, η, which is the driving force for the electrochemical reaction. The overall overpotential at the electrode is given by

$$\eta = V - E - \Phi, \tag{14}$$

where V is the electrode potential. E is the thermodynamic equilibrium potential corresponding to the condition of no current flow, and Φ is the electrostatic potential within the solution next the electrode, measured at the outer edge of the mass transport (or concentration) boundary layer.

The total overpotential at the electrode can be further resolved into two overpotential components, η_s and η_c, The first, η_s, is the surface (or "activation") overpotential, which relates directly to the kinetics of the electrode processes. The second overpotential component, η_c, is the concentration overpotential, accounting for the voltage dissipation associated with transport limitations.

The surface overpotential, η_s, is typically related to the current density through the Butler–Volmer equation:[6]

$$i_n = i_{0,e} \left[\exp\left(\frac{\alpha_A F}{RT}\eta_s\right) - \exp\left(-\frac{\alpha_C F}{RT}\eta_s\right) \right]. \tag{15}$$

Equation (15) incorporates three empirically measured parameters: the exchange current density, i_0 (given here in terms of its value on the electrode, $i_{0,e}$), and the anodic and cathodic transfer coefficients, α_A and α_C, respectively. These parameters are obtained from polarization measurements.[7] Often, but not always, $\alpha_A + \alpha_C = n$. It should be noted that while the Butler–Volmer equation correlates well many electrode reactions, there are numerous others, particularly when carried in the presence of plating additives, which do not follow it.

The Butler–Volmer equation in the form presented by (15) relates to pure electrode kinetics and does not consider transport limitations, which cause the concentration at the electrode, c_e, to vary from its bulk value, c_b. This concentration variation affects mostly two parameters: the exchange current density, i_0, which is a function of the concentration, and the overpotential, η, which now also includes the component associated with transport limitations, η_c. We can write

$$\eta = \eta_s + \eta_c, \tag{16}$$

where the concentration overpotential, η_c, is given by

$$\eta_c = \frac{RT}{nF} \ln \frac{c_e}{c_b}. \tag{17}$$

It should be noted that the division of the total overpotential into a "pure" kinetics component (η_s) and a mass transport component (η_c) as presented by (16) is somewhat arbitrary and is used mainly to characterize the two types of dissipative processes. Both terms are strongly coupled and it is very difficult to directly measure either component separately. While chemical engineers often discuss the two overpotential components separately,[1] chemists (e.g., Refs. [6, 7]) tend to combine both terms together and characterize the electrochemical system in terms of the total overpotential, η.

To account for the concentration variations which are always present at electrodes in current-carrying cells, a correction must be introduced into (15). Either of two approaches is typically taken. The first more characteristic to engineering publications,[1] presents (15) as

$$i = i_{0,c_b} \left(\frac{C_e^o}{C_b^o}\right)^\gamma \left(\frac{C_e^R}{C_b^R}\right)^\delta \left(e^{\frac{\alpha_A F}{RT}\eta_s} - e^{-\frac{\alpha_c F}{RT}\eta_s}\right). \tag{18}$$

Written here for the general first-order reaction,

$$O + n\bar{e} \Leftrightarrow R. \tag{19}$$

γ and δ are parameters adjusting the value of i_0 from its bulk value to its value at the electrode and are generally determined empirically. If the reduced species does not dissolve within the electrolyte (as in most plating systems), the concentration ratio for the reduced species involving δ is identically 1. Furthermore,[1] for many divalent ions, $\gamma \sim 1/2$.

Chemists typically tend to account for the concentration variation at the electrode through a modified Butler–Volmer equation of the form[6,7]

$$i = i_{0,c_b} \left[\left(\frac{C_e^R}{C_b^R} \right) e^{\frac{\alpha_A F}{RT} \eta} - \left(\frac{C_e^o}{C_b^o} \right) e^{-\frac{\alpha_C F}{RT} \eta} \right]. \tag{20}$$

It can be shown that the Tafel approximation of the cathodic branches of the two forms of the modified Butler–Volmer equation, ((18) and (20)) are identical when

$$\gamma = 1 - \frac{\alpha_C}{n}. \tag{21}$$

4. General Solution Procedure

Once the fluid-flow equations (11) have been solved and the velocity components have been specified within the cell, the system of j transport equations (10), in conjunction with the electroneutrality condition (5), is solved. The boundary condition at insulating boundaries is specified by substituting (8), representing the current approaching the boundary from the electrolyte, into (12), stating that the current on the insulating boundary is zero:

$$i_n|_{\text{insulator}} = -\kappa \nabla \Phi - F \sum z_j D_j \nabla c_j = 0. \tag{22}$$

On electrodes, we equate the current approaching the electrode from the solution side (8) to the current entering the electrode, subject to the reaction kinetics equation:

$$i_n|_{\text{electrode}} = -\kappa \nabla \Phi - F \sum z_j D_j \nabla c_j = f(c_j, \Phi)|_e. \tag{23}$$

The function $f(c_j, \Phi)|_e$ appearing on the right-hand side of (23) corresponds to the kinetics expressions given by either (18) or (20). We recall that when (18) is applied, the overpotential is given by

$$\eta_s = V - E - \Phi - \eta_c. \qquad (24)$$

When applying the kinetics expression (20), we use the total overpotential as given by (14).

Clearly, the procedure outlined above is complex. It requires solution of the flow field, in conjunction with the determination of the distribution of the electrostatic potential and of all species concentrations within the cell. In addition to the mathematical complexity, the transport properties (diffusivities, mobility) for all species must be given. This is further complicated by the fact that most practical electrolytes are concentrated and hence transport interactions between the species must be accounted for, requiring the application of the more complex concentrated electrolyte theory.[1] Additionally, the electrode kinetics parameters must be known. However, as discussed below, simplifications are often possible, since most operating cells are typically controlled by either the electric potential distribution *or* by the concentration distribution (in conjunction with the electrode kinetics), and only a few systems are influenced about equally by both.

5. Thin Boundary Layer Approximation

A common assumption in engineering modeling of electrochemical cells is the thin boundary layer approximation.[8] Accordingly, concentration variations are assumed to be limited to a boundary layer along the electrodes, which is considered to be much thinner than the well-mixed bulk electrolyte region. This decoupling, depicted schematically in Fig. 1, eliminates all terms involving concentration gradients in (10) when the latter is applied to the bulk region. The concentration effects are now all lumped within the thin boundary layer and incorporated in the boundary conditions.

Following this procedure, and discarding all terms involving concentration gradients, the Nernst–Planck equation (10) reduces in the bulk to the Laplace equation for the potential:

$$\nabla^2 \Phi = 0. \qquad (25)$$

Current Distribution in Electrochemical Cells

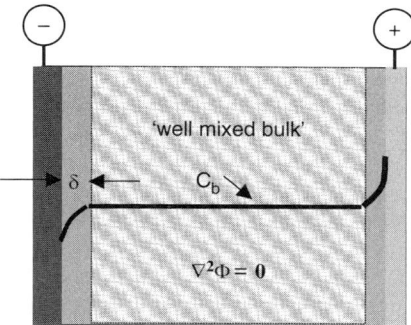

Figure 1. An electrochemical cell, depicting the thin boundary layer approximation. The bulk of the cell is well mixed and all concentration variations are assigned to a thin boundary layer next to the electrodes. Typically, the boundary layer thickness, δ, is far thinner with respect to the bulk than illustrated here. In the region of uniform concentration, the Laplace equation for the potential holds.

In deriving (25), we also set the transient concentration term on the left in (10) to zero, thus considering only steady-state (or pseudo-steady-state) processes. One may still apply the approximation of the thin boundary layer, as stated by (25), to transient problems, allowing the concentration within the thin boundary layer to vary with time. Detailed discussion of this class of problems is, however, outside the scope of this review and can be found in publications focusing on transients (e.g., Refs. [9–12]).

Equation (25) is solved for the bulk region by applying, in the absence of concentration gradient in the bulk, simplified boundary conditions.

On insulators we have a simplified form of (22):

$$i_n|_{\text{insulator}} = -\kappa \nabla \Phi|_{\text{ins}} = 0 \rightarrow \nabla \Phi|_{\text{ins}} = 0. \qquad (26)$$

On electrodes we apply (23), recognizing that now, in the absence of concentration gradients in the bulk, the current on the solution side is driven only by electric migration:

$$i_n|_{\text{electorde}} = -\kappa \nabla \Phi = f(c_j, \Phi)|_{\text{e}}. \qquad (27)$$

As before, the right-hand side on (27) is given by either (18) or (20). Alternatively, we can specify a potential balance at the electrode by applying (lumped across the thin boundary layer) either (14) or (24), rearranged in the form of (28):

$$\Phi = V - E - \eta \quad \text{or} \quad \Phi = V - E - \eta_s - \eta_c. \qquad (28)$$

Since the solution on the electrolyte side involves only the potential Φ, as given by the Laplace equation (25), matching the potential Φ in the solution, at the outer edge of the diffusion layer with the corresponding potential satisfying the electrochemical kinetics on the electrode (28), is occasionally easier.

An important additional advantage of invoking the thin boundary layer approximation is that the velocity field needs no longer to be fully computed. The concentration variation next to the electrode, required for the determination of the concentration overpotential and for correcting the kinetics expression, can often be accounted for with acceptable accuracy by determining just the thickness of the concentration boundary layer, typically reported in terms of the equivalent stagnant Nernst diffusion layer, δ_N, along the electrode. The latter is often available from correlations,[13] textbooks,[14] experimental measurements,[15] or interpretation of computational fluid dynamics software output. Obviously, δ_N may vary with position along the electrode.

V. COMMON APPROXIMATIONS FOR THE CURRENT DISTRIBUTION

Even when applying the thin boundary layer approximation, the equations required for solving the current and potential distributions in the electrochemical cell yield a nonlinear system requiring iterative solution. The reason is that the boundary conditions incorporate the unknown term (the electrostatic potential or the current density). While this presents no serious hurdle for computer-implemented numerical solutions, analytical solutions of nonlinear systems are difficult and generally require a linearization procedure. To analytically characterize features of the current distribution, some simplifying approximations are frequently applied. These are summarized in Table 1, and are discussed below.

Current Distribution in Electrochemical Cells

Table 1.
Common approximations for modeling the current distribution.

Approximation	Prevailing overpotential	Controlling equation	Boundary conditions
Primary	$\eta_\Omega \gg \eta_s + \eta_c$	$\nabla^2 \Phi = 0$	$\Phi = V - E$
Secondary	$\eta_\Omega \sim \eta_s >> \eta_c$	$\nabla^2 \Phi = 0$	$\Phi = V - E - \eta_s$
Mass transport	$\eta_c \gg \eta_\Omega + \eta_s$	$\frac{\partial C}{\partial t} + v\nabla C = D\nabla^2 C$	$c_e \ll c_b$ or $c_b \sim 0$
Tertiary[a] (no approximation)	$\eta_\Omega \sim \eta_s \sim \eta_c$	$\nabla^2 \Phi = 0$	$\Phi = V - E - \eta_s - \eta_c$

[a] The tertiary distribution represents the formal thin boundary layer solution with no further approximation.

1. Primary Distribution: $\eta_\Omega \gg \eta_s + \eta_c$

Here we assume that the prevailing overpotential is associated with the ohmic drop within the electrolyte,

$$\eta_\Omega \equiv \Delta\Phi = (\Phi_A - \Phi_c) \gg \eta_s + \eta_c. \qquad (29)$$

This assumption is tantamount to stating that the electrode reactions are perfectly reversible and kinetics and mass transport limitations are both negligible. Since the ohmic overpotential is given by[1]

$$\eta_\Omega = i\frac{l}{\kappa} \qquad (30)$$

we may conclude that the primary distribution is likely to prevail in cells with large interelectrode gap, l, and low conductivity. Since the surface and mass transport overpotentials must be small (in comparison with the ohmic overpotential), it is further expected that a distribution close to primary will prevail on highly catalytic electrodes or kinetically reversible reactions, at high temperatures (e.g., in molten salts, where the kinetics are very fast) and in systems operating far from the limiting current (i.e., with negligible mass transport limitations).

Consequently, the Laplace equation (25) is solved subject to the following boundary conditions:
Insulator:
$$\nabla\Phi = 0 \qquad (31)$$

Electrode:

$$V - E - \Phi = 0 \rightarrow \Phi = V - E = \text{constant.} \qquad (32)$$

Equation (32) is derived from (28) by setting the surface and concentration overpotentials identically to zero ($\eta_s + \eta_c = 0$), since their magnitude is insignificant compared with that of Φ. As a consequence of this simplification, the potential in solution next to the electrode, Φ differs from the electrode potential, V, only by a constant (the equilibrium potential, E). This provides for a simpler and more direct solution of the Laplace equation. The primary distribution has some unique characteristics. While the magnitudes of the average and the local current densities depend linearly on the applied potential across the cell and on the conductivity, the *current distribution depends only on the geometry*. Perhaps the most notable characteristic of the primary distribution is that it is typically nonuniform, and at a certain cell location it may exhibit singularities. These are points within the cell where the potential gradient, which, according to (27) is proportional to the current density, will approach infinity. Hence, while the potential within the cell must be bound between the anode and cathode potentials (after subtraction of the equilibrium potential), the primary current density can approach infinitely large values at singularities. The singularities are located at intersections of cell boundaries: (1) an electrode and an insulator intersecting at an angle larger than 90° and (2) two electrodes intersecting at an angle larger than 180°. The physical interpretation of this behavior is related to the fact that while the potential driving force is finite, the local resistance becomes vanishingly small at the singularities, driving the current density to infinitely large values, irrespective of the counter electrode position, or the applied voltage.

Other "special" geometric configurations of interest are (1) intersection of an electrode and an insulator forming an angle smaller than 90° and (2) two intersecting electrodes forming an angle smaller than 180°. In these two latter cases, the local current density tends to zero at the intersection point, indicating an infinitely large local resistance. In all those cases, the magnitude of the intersection angle affects only how sharply the current density approaches its limiting values, but not the nature of the limit itself. Practical cell boundaries do not intersect at a mathematically sharp point. Nonetheless, if the radius of curvature formed by the intersecting boundaries is small, and the intersection involves the geometrical features described above, the current density will still become either very large or vanishingly small at these points.

While clearly being an approximation, the primary current distribution represents the worst (least uniform) current distribution a system may exhibit. The presence of kinetics limitations, as discussed below, will typically lead to more uniform distributions. It should also be stated that when the system is subject to mass transport control under stagnant diffusion (typical for microscale features), the controlling equation, as discussed below, is the Laplace equation for the concentration:

$$\nabla^2 c = 0 \qquad (33)$$

Equation (33), in complete analogy to (25) (the Laplace equation for the potential), also follows the "primary distribution," and exhibits the same characteristics as described above.

Practical Implications Associated with the Characteristics of the Primary Current Distribution

- Small nodules or roughness elements on electrodes typically form a sharp tip with small radius of curvature, causing the (primary) current density at the tip to become very high. As a consequence, these sharp tips tend to propagate very rapidly,[39–42] restricted only by the kinetics limitations at the tip. Consequently, electrodes with very reversible kinetics, e.g., lithium, silver, lead, and zinc, tend to evolve needles and dendritic growth quite readily, while less reversible metals, e.g., nickel and iron, do not.
- Electrodes bounded by coplanar insulators (forming 180° intersection) tend to exhibit high current densities close to the edge (limited only by the kinetics); therefore, such cell features should be avoided where possible. Slight embedment of the electrode will provide a finite current at the edge, while presenting only minimal resistance to flow (typically minimized by secondary flow eddies).
- Nonuniformities in the primary distribution usually originate with geometric "perturbations" and typically do not propagate very far. In cylindrical or spherical fields, generated by, e.g., an insulating bubble, or spherical or cylindrical electrodes, the current distribution perturbation diminishes to a

small fraction (few percent) of its maximal value once the distance from the curved surface exceeds about three diameters. As a consequence, to effectively shield a curved electrode or an edge region, the shield must be placed very close to the surface. By the same token, to selectively plate a small feature using a small-diameter wire or a sharp tip electrode, one must place the tip very close to the surface.

The primary distribution is not unique to electrochemical systems and other physical systems exhibit the very same distribution. Textbooks available in these areas provide information that can be directly applied to electrochemical systems operating under conditions approaching the primary distribution. Examples include heat transfer by conduction,[16] diffusion in solids,[17] electrostatics,[18] potential (ideal) flow,[19] and mathematical texts on the theory of complex variables and conformal mapping. A comprehensive discussion of the primary current distribution in electrochemical systems is provided by Newman.[1]

2. Secondary Distribution: $\eta_\Omega + \eta_s \gg \eta_c$

Here, both ohmic and kinetics irreversibilities are considered; however, mass transport limitations are assumed to be negligible. Since significant mass transport effects are present only when operating close to the limiting current, e.g., in very dilute solutions, the secondary distribution presents a valid approximation for most electrochemical systems. Here, the Laplace equation, (25), is solved subject to the following boundary conditions:

Insulator:
$$\nabla \Phi = 0 \qquad (31)$$

Electrode:
$$V - E - \Phi = \eta_s. \qquad (34)$$

η_s and i are related by

$$i = i_0 \left[\exp\left(\frac{\alpha_A F}{RT} \eta_s \right) - \exp\left(-\frac{\alpha_C F}{RT} \eta_s \right) \right]. \qquad (15)$$

It is significantly more complicated to analytically solve the secondary distribution (than the primary distribution), since the presence of the kinetics overpotential renders the system nonlinear.

Current Distribution in Electrochemical Cells

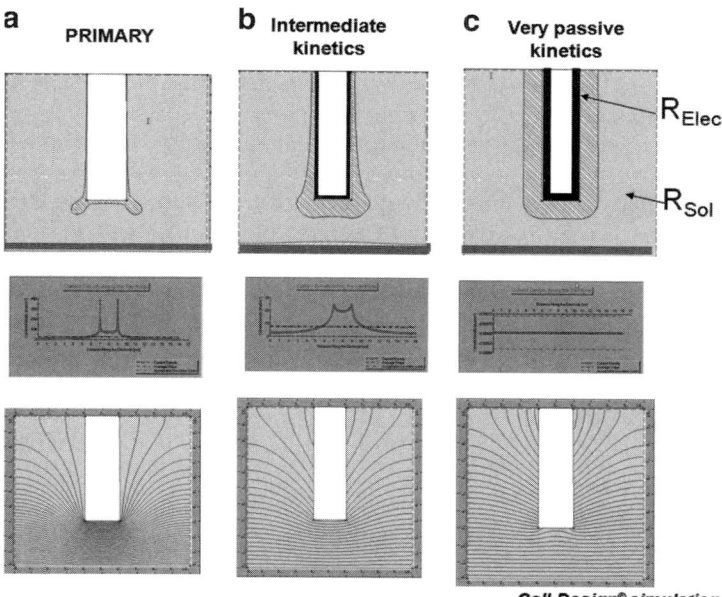

Figure 2. Examples of numerical solutions for the cathodic current distribution on a plate electrode immersed in a cell with the counter electrode at the *bottom*. Three cases are compared: (**a**) (*left column*) completely reversible kinetics (primary distribution); (**b**) (*center*) intermediate kinetics ($Wa \sim 0.2$); (**c**) (*right column*) irreversible kinetics ($Wa \sim 10$). The *top row* provides a comparison of the current distribution or the deposit profile on the cathode (*cross-hatched region*). The *center row* provides the current distribution along the electrode ("stretched"). The *bottom row* provides the corresponding potential distributions. It is evident that the current distribution uniformity increases as the electrode kinetics become more passivated (Cell-Design software simulations[68]).

Available analytical solutions include among many others the disk electrode,[20] Wagner's solution of the current distribution at a corner,[21] current distribution within a through hole[22] and a blind via, and within a thin electrolyte layer.[23] Most of the analytical solutions employ linearization about some average current density. Computer-implemented numerical solutions of the secondary distribution (example provided in Fig. 2) are readily obtained and do not require linearization. A few examples include modeling the current distribution in the wedge[24] and Hull cells,[25–27] the rotating Hull cell,[28] inside a channel,[29] in pattern plating of printed circuits,[30,31] inside a via,[32] and in the characterization of anodization applications.[33]

The secondary distribution incorporates the effects of the ohmic resistance, which gives the primary distribution its nonuniform characteristics, in combination with the surface resistance associated with the limited reversibility of the electrode kinetics. The latter (with the exception of special cases, involving, e.g., the use of unique additives) leads to a uniform distribution. While it is difficult to derive accurate analytical solutions for the secondary distribution, we can characterize the degree of nonuniformity by evaluating the relative magnitude of the resistances associated with the surface and ohmic dissipative processes.

Accordingly, the degree of uniformity of the secondary distribution can be characterized in terms of a dimensionless number, named after Carl Wagner, representing the ratio of the surface to the ohmic resistance:

$$W_a = \frac{R_s^*}{R_\Omega^*}. \tag{35}$$

The resistances in (35) are specific resistances (per unit area, $\Omega\,\text{cm}^2$), corresponding to the local slope of the polarization curve, $\partial \eta / \partial i$.

The ohmic resistance, which is a constant, independent of the current density, is given by

$$\eta_\Omega = IR = i\frac{l}{\kappa} \rightarrow R_\Omega^* = \left|\frac{\partial \eta_\Omega}{\partial i}\right| = \frac{l}{\kappa}. \tag{36}$$

The surface (activation) resistance depends on the current density in a relatively complex manner, and therefore it is convenient to consider the system in terms of two separate regimes: the Tafel ("high-field") and linear (low current density) approximations.[6]

In the Tafel regime, we consider the Butler–Volmer equation in terms of two subregions, one for high anodic polarization ($\eta_S \gg \alpha_A F/RT$), the second for high cathodic polarization ($-\eta_S \gg \alpha_C F/RT$). It can be shown that these approximations are valid when $|i|_{\text{avg}} \gg i_0$. The mathematical representations for the anodic and cathodic Tafel regions are similar:

$$\eta_{s,T} \approx \frac{RT}{\alpha F} \ln \frac{i}{i_0} \equiv a + b \ln i, \tag{37}$$

where $b = RT/\alpha F$ (2.3b is the "Tafel slope"), a is a constant, and $\alpha = \alpha_A$ for anodic polarization and $\alpha = \alpha_C$ for the cathodic process. We now can derive the surface resistance for the Tafel regime:

$$R_{S,T}^* = \left|\frac{\partial \eta_s}{\partial i}\right| = \frac{RT}{\alpha F |i|}. \tag{38}$$

For the linear polarization regime (equivalent to low current density or micropolarization, $|i|_{avg} \ll i_0$), recognizing that here $|\eta_S| \ll (\alpha_A + \alpha_C)F/RT$, we can linearize the Butler–Volmer equation:

$$\eta_{s,L} \approx \frac{RT}{(\alpha_A + \alpha_C) F} \frac{i}{i_0} \equiv b'i, \tag{39}$$

where b' is the linear polarization slope. The linear regime surface resistance is

$$R_{S,L}^* = \left|\frac{\partial \eta_s}{\partial i}\right| = \frac{RT}{(\alpha_A + \alpha_C)Fi_0} = \frac{RT}{nFi_0} = b'. \tag{40}$$

Unlike the constant ohmic resistance, the surface (kinetics) resistance decreases with increasing current density in the Tafel range, but is a constant in the linear regime.

Substituting the resistances (36)–(40) into the *Wa* number expression (35), we get the following:

Tafel:
$$Wa = \frac{\kappa b}{li} \quad \text{and} \quad b \equiv \frac{RT}{\alpha F}. \tag{41}$$

Linear:
$$Wa = \frac{\kappa b'}{l} \quad \text{and} \quad b' \equiv \frac{RT}{nFi_0}. \tag{42}$$

b (the Tafel polarization slope) and b' (linear polarization slope) incorporate the kinetics parameters.

A large *Wa* number designates a large surface resistance and a small ohmic resistance, leading to a uniform current distribution. By the same token, a small *Wa* number is indicative of prevalent ohmic overpotential, leading to a nonuniform distribution, approaching in the limit a primary distribution. Since the *Wa* number characterizes the current distribution uniformity, a relevant question is how large must the Wa number be to ensure a high degree of uniformity, and, conversely, how low should it be to indicate nonuniformity. Inspecting numerically derived distributions, we find that when $Wa > 5$, the distribution is quite uniform, and when $Wa < 0.2$, most distributions exhibit significant nonuniformity. This also becomes quite evident by inspecting Newman's classical analytical derivation[20] of the current distribution on a disk electrode under mixed surface and ohmic control ("secondary distribution"). The current distribution is given here in terms of a parameter J which is equivalent to $1/Wa$.

An important issue is the clear identification of the characteristic length, l, in the Wa number. In some configurations the correct selection of l is intuitively evident, but in many others it is not. For example, the selection of the characteristic length in the plating of blind vias has been controversial, with some authors selecting the via depth, L, as the characteristic length (e.g., Ref. [34]), while others (e.g., Ref. [22]) indicate that the proper characteristic length for this configuration is L^2/r, where r is the via radius. Akolkar and Landau,[35] pointed out that the characteristic length can be unambiguously identified only through an analytical solution.

3. Mass Transport Controlled Distribution: $\eta_c \gg \eta_\Omega + \eta_a$

This approximation pertains to systems where the ohmic losses within the electrolyte and the kinetic limitations on the electrode are considered to be negligible as compared with mass transport limitations. Instead of solving the Laplace equation (25) for the potential, which is a common approximation to the more general Nernst–Planck equation (10), we need to solve the latter (10) for the case when the potential gradients are negligible as compared with concentration gradients, i.e.,

$$\nabla c \gg \nabla \Phi. \tag{43}$$

Subject to (43), (10) can be approximated by

$$\frac{\partial c_j}{\partial t} + v \cdot \nabla c_j = \nabla \cdot (D_j \nabla c_j). \tag{44}$$

The boundary conditions are reformulated, recognizing that since the electric field driven current becomes negligible, the current density given by (8) now takes the form

$$i|_n = -F \sum_j z_j D_j \nabla c_j. \tag{45}$$

On the cell boundaries we have the following:
Insulator:
$$i = 0 \to \nabla C_R = 0. \tag{46}$$

Electrode:

$$V - E = \eta_c = \frac{RT}{nF} \ln \frac{RT}{nF} \ln \left(1 - \frac{i}{i_L}\right). \tag{47}$$

In deriving (47), we combine (14), (16), and (17), recognizing that η_s and Φ are negligible compared with η_c. We furthermore recognize that the concentration overpotential (in boundary condition (47)) cannot be significant unless i approaches i_L. For this to happen we must have

$$c_e \ll c_b \quad \text{or} \quad c_e \sim 0. \tag{48}$$

Inspecting (44) and its boundary conditions (46) and (48), we recognize that it is identical to the convective diffusion equation common in representing transport problems in nonelectrochemical systems.[5] Accordingly, invoking the mass transport control approximation causes the problem to lose all its electrochemical characteristics, transforming it to a transient diffusion problem. Obtaining general analytical solutions to the transient convective diffusion problem is complex since it requires solving the transient concentration distribution in the cell in conjunction with the fluid flow.

A common simplification, which often does not detract from the usefulness of the solution, is to consider only the steady-state form of (44):

$$v \cdot \nabla c_j = \nabla \cdot (D_j \nabla c_j). \tag{49}$$

Analytical solutions of (49) have been presented for special configurations where the velocity profile is well established, e.g., the rotating disk electrode or laminar flow in a channel.[1] An important simplification has been proposed by Levich,[8] who recognized that electrochemical systems are typically characterized by a concentration boundary layer that is much thinner than the corresponding velocity boundary layer. Levich recognized that in such systems it is not necessary to solve the entire velocity field, and knowing the velocity gradient at the electrode is sufficient for obtaining analytical solutions. Newman presents a number of such cases.[1]

A different approximation that applies to systems undergoing transient polarization can be invoked. Assuming a stagnant solution (no velocity) within the region of varying concentration where the approximation is applied (usually the boundary layer), (44) simplifies to

$$\frac{\partial c_R}{\partial t} = D_R \nabla^2 c_R. \tag{50}$$

The subscript R in (50) indicates that it is applied to the reactant ion. Also, it is assumed that the reactant diffusivity is constant, independent of the concentration. Equation (50) is known as Fick's second law. Its solution is particularly relevant to problems of transient (periodic) current and potential applications such as pulse and periodic reverse waveforms when the time constant is in the range of the concentration profile relaxation time, i.e., about 0.01 s or longer.[9–12]

Lastly, we can consider systems under steady-state (or pseudo-steady-state) mass transport control with no flow (stagnant). Here, within the region of varying concentrations, where the approximation is applied (usually the boundary layer), (44) simplifies to

$$\nabla^2 c_R = 0. \tag{51}$$

Equation (51) is the Laplace equation applied to the reactant concentration. It is identical in form to the primary distribution and the characteristics of the primary distribution discussed above apply here as well. The approximation stated by (51) pertains particularly to small features such as plating within narrow trenches or vias. Often on these scales, convective transport is not effective, and the ionic transport progresses mainly through diffusion.[36] Furthermore, the ohmic drop on these scales is negligible, rendering (51) the relevant controlling equation.

4. Tertiary Distribution: $\eta_c \sim \eta_\Omega \sim \eta_s$ ("Mixed Control")

This is the most general case. Here, none of the dissipative processes are controlling, and therefore no mechanism is assumed to be negligible. The Nernst–Planck equation, (10), or its thin boundary layer approximation (the Laplace equation (25)), is solved subject to the boundary conditions (22) and (23). Just about all solutions are numerical. Methods of solutions and some examples are discussed in subsequent sections.

VI. SCALING ANALYSIS OF ELECTROCHEMICAL CELLS

It is evident from the previous discussion that significantly simplified modeling of the current distribution can often be achieved once scaling analysis has identified the controlling dissipative mechanisms in the cell. It has been shown that the current distribution in most common systems can be characterized in terms of three major dissipative processes: ohmic (within the electrolyte across the cell), mass transport (across the concentration boundary layer), and surface activation (on the electrode). These are designated in terms of the corresponding resistances: R_Ω^*, R_C^*, and R_S^*. The *Wa* number characterizes the current distribution in terms of the relative importance of two of the three resistances: the surface (R_S^*) and the ohmic (R_Ω^*) resistances. Clearly, more complete characterization of the system requires the comparison of two additional resistance ratios and the formulation of two additional dimensionless parameters.[36]

The relative significance of the ohmic (R_Ω^*) and the mass transport (R_C^*) resistances is important in terms of characterizing the prevailing transport mechanism in the cell. Equation (8) indicates that the current flow is due to two mechanisms: mass transport down a concentration gradient and migration due to the electric field. The relative resistances associated with these will indicate which of the two processes affects the current density (at the given conditions) more significantly, or, to which of the two transport processes the current density is more sensitive. Accordingly, a dimensionless parameter representing the ratio of the mass transport to the ohmic resistance has been formulated[36] and designated as the Tobias number (after Charles W. Tobias):

$$To = \frac{R_C^*}{R_\Omega^*} = \frac{\kappa}{l} \frac{RT}{nFi_L \left(1 - \frac{i}{i_L}\right)}. \qquad (52)$$

When $To \gg 1$, mass transport prevails over ohmic migration, validating the approximations indicated by (43)–(51). On the other hand, when $To \ll 1$, ohmic migration prevails, and the use of the Laplace equation for the potential alone (primary distribution) or in combination with the electrode kinetics (secondary distribution), disregarding mass transport, is justified. Further inspection of the Tobias number indicates that it becomes much larger than 1, irrespective of the fraction of the limiting current (i/i_L), for length scales significantly smaller than a critical length scale given by[36]

$$l_{\text{crit}} = \frac{\kappa RT}{nFi_L \left(1 - \frac{i}{i_L}\right)}, \qquad (53)$$

i.e., on very small length scales, $l \ll l_{\text{crit}}$, mass transport will always prevail over ohmic migration. This conclusion is consistent with observations that cells with large interelectrode gaps are typically more sensitive to the ohmic resistance than to mass transport (providing that some circulation is present), and that the ohmic resistance is insignificant in comparison with diffusion in very small (e.g., submillimeter or smaller) features. We find that for electrochemical systems with $\kappa \sim 0.1 - 1\,\text{S/cm}$, $i_L \sim 0.1 - 1\,\text{A/cm}^2$, and $n = 1$ or 2, the critical length below which mass transfer prevails over the ohmic effects is on the order of $0.01 - 2.5\,\text{mm}$ (depending on the magnitude of κ, i_L, and n). As the current density approaches the limiting current, the critical scale, l_{crit}, below which mass transport dominates over the ohmic resistance becomes larger.

It should be noted that the analysis associated with the Tobias number compares only the ohmic with the mass transport processes, while the relative importance of the electrode kinetics is not considered. Typically, the electrode kinetics effect is accounted for through the *Wa* number, which compares the resistances associated with the kinetics and the ohmic processes. However, in systems with large Tobias number (e.g., small scale features and systems operating close to their limiting current), the ohmic resistance is insignificant as compared with the mass transport effects. Accordingly, the scaling analysis based on the *Wa* number is irrelevant for such systems, and one should consider instead a dimensionless parameter that compares the two relevant resistances: kinetics and mass transport. The leveling parameter, L, represents this ratio:[36]

$$L = \frac{R_s^*}{R_c^*} = \frac{i_L \left(1 - \frac{i}{i_L}\right)}{i_o} \quad \text{(linear polarization)}, \tag{54}$$

$$L = \frac{n i_L \left(1 - \frac{i}{i_L}\right)}{\alpha i} \quad \text{(Tafel polarization)}. \tag{55}$$

For the convenience of obtaining an explicit expression, the leveling parameter has been formulated separately for linear polarization (54) and for the Tafel regime (55).

The leveling parameter is analogous to the *Wa* number since it compares the resistance associated with the surface reactions with the resistance associated with the transport process, except that here the conventional, but irrelevant, ohmic resistance has been replaced by the relevant mass transport resistance. It should be pointed out that both the latter resistances (ohmic and mass transport) relate to the transport process within the electrolyte; hence, they are sensitive to the cell configuration and typically lead to nonuniformities in the current distribution. The leveling parameter, being applicable to systems where mass transport rather than the ohmic resistance prevails, typically pertains to small-scale systems. Here, $L \gg 1$ indicates that the kinetics resistance, which typically leads to uniformity, is significantly larger than the mass transport resistance, which leads to nonuniformity. The leveling parameter can be used for the determination of the degree of uniformity of the current distribution on the microscale (where ohmic effects are negligible) and hence it characterizes the "smoothness" of the deposit texture. To ensure level deposits, one must have $L \gg 1$. This corresponds in the linear polarization regime to[36]

$$\frac{i}{i_L} \ll 1 - \frac{i_0}{i_L} \quad \text{(for smooth deposition under linear polarization)}, \tag{56}$$

or

$$i \ll i_L - i_0. \tag{57}$$

Linear polarization is encountered when $i < i_0$, i.e., mostly in systems with relatively large exchange current density. Such systems exhibit according to (57) a very limited operation range, $i_L - i_0$, where smooth deposition is expected. This is consistent with observations indicating that highly reversible metals such as lithium, silver, tin, and lead, or the deposition from high-temperature molten salts (all characterized by high exchange current densities), tend to produce rough, nodular, or dendritic deposits.[37,38]

In the Tafel regime we obtain for smooth deposition ($L \gg 1$)

$$\left(\frac{i}{i_L}\right) < \frac{1}{1 + \frac{\alpha}{n}} \quad \text{(smooth deposition under Tafel polarization)}. \tag{58}$$

This expression explains the common observation that rough deposits are produced as the limiting current is approached. Equation (58) indicates that for a relatively reversible electrode reaction with two-electron transfer, i.e., $n = 2$, and $\alpha = 1$, $(i/i_L) < 0.66$. Accordingly, for smooth deposition the current density may not exceed about 66% of the limiting current. On the other hand, for a highly irreversible electrode reaction with $\alpha = 0.1$ (which can be realized through the addition of strongly inhibiting additives) and $n = 2$, $(i/i_L) = 0.95$; i.e. the current density can reach 95% of the limiting current, while still producing smooth deposits.

Implications of the Leveling Parameter for the Prevention of Rough, Nodular, and Dendritic Deposit Texture

Low exchange current density, i_0, in the linear regime and a low transfer coefficient, α, in the Tafel regime promote smooth electrodeposition. These parameters may be controlled through the use of inhibiting additives. Equations (57) and (58) indicate that for smooth deposition it is beneficial to operate at a low fraction of the limiting current, i/i_L, i.e., low current density and a high limiting current. The latter is promoted by high reactant concentration and vigorous agitation.

The radius of curvature at the tip of a surface roughness element or a propagating dendrite is typically very small, on the order of 10^{-5} cm.[39–42] It can be shown (36) that such systems will not support smooth deposition unless the exchange current density, i_0, is kept below $20 \, \text{mA/cm}^2$. This explains the difficulty of obtaining smooth deposition of lithium, silver, tin, zinc, and lead since all have significantly larger exchange current densities. It can be further shown[36] that such elements form their own characteristic diffusion layer, which is inversely proportional to the radius of curvature, leading to extremely high limiting currents and low mass transport resistance. This causes such features to grow very fast, at a rate limited only by their kinetics resistance.

VII. TRANSPORT EFFECTS ON KINETICALLY CONTROLLED SYSTEMS

According to the analysis presented above, systems under ohmic and kinetics control (secondary distribution) should not be sensitive to transport effects. Yet numerous systems operating at high *Wa* number, where the resistance to the surface kinetics is dominant, exhibit noticeable effects of flow and agitation. This is particularly evident in plating applications, where flow markings can be observed in deposits near surface protrusions, even far from the limiting current, i.e., when the concentration overpotential and the transport resistance are negligible. The explanation for this puzzling behavior hinges on more careful examination of the Butler–Volmer equation, which is often applied to such systems. Although the flow does not appear explicitly in the Butler–Volmer equation, the exchange current density is a function of the surface concentration, which, in turn, depends on transport. This becomes evident upon inspection of the "chemists" form of the Butler–Volmer equation (20), which can be approximated for deposition in the Tafel regime by

$$i = i_{0,c_b}\left(1 - \frac{i}{i_L}\right) e^{-\frac{\alpha_C F}{RT}\eta}, \tag{59}$$

i.e., the current density is *linearly* proportional to the fraction of the limiting current (or to the surface concentration). This is in contrast with the logarithmic dependence of the concentration overpotential on the fraction of the limiting current. Unlike the logarithmic dependence, which does not become appreciable until very close to the limiting value, the linear dependence shown by (59) indicates dependence on flow at all current densities.

Accordingly, systems which may be far from the limiting current, i.e., exhibiting negligible mass transport overpotential, and which are under kinetics control may still exhibit dependence on the flow, through the mediated effect of the flow on the kinetics as discussed above. An example has been provided by Galasco et al.,[43] who studied the pattern plating by copper of an isolated line, which typically exhibits faster growth, often leading to its extension above the photolithographic mask, and occasionally resulting in a "mushrooming" effect. Under typical conditions, such a system operates under a very high *Wa* number (or a high leveling parameter), expected to lead to uniform deposition. Yet, when considering

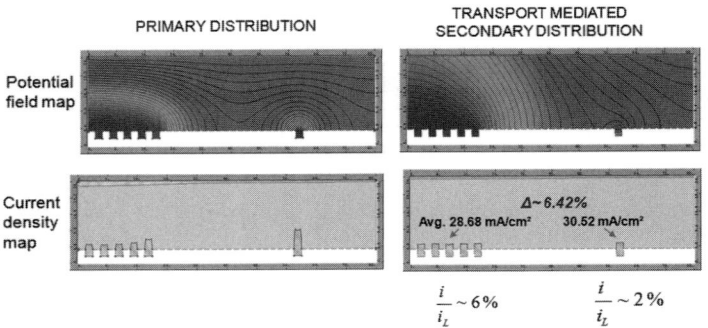

Figure 3. Numerical simulation of the potential (*top*) and the current density (*bottom*) in pattern plating of line clusters (*left*) and an isolated line (*right*). Patterned lines (shown here in a cross section) are 2 mm wide × 2 mm high. Flow at a linear velocity of 10 cm/s is approaching from the left. Copper kinetics are assumed with $i_0 = 10^{-3}$ A/cm^2, $\alpha_C = 0.5$. Figures on the *left* correspond to the primary distribution. The figures on the *right* correspond to the secondary distribution, accounting for the mediated effect of transport on the kinetics. The height of the *cross-hatched bars* represent the current density. While the primary distribution indicates a much higher plating rate of the isolated line, analysis[43] indicates that the process is carried out under prevailing kinetics control ($\eta_S \sim 180$ mV, $\eta_\Omega \sim 2$ mV). The reason the isolated line is plated about 6.5% faster than the cluster is the mediated transport effect on the kinetics (Cell-Design simulations).

the mediated effect of transport on the kinetics, one finds that the isolated feature exhibits faster growth kinetics owing to the higher surface concentration near the isolated feature. This is illustrated in the numerical simulations shown in Fig. 3.

VIII. COMPARISON OF ANALYTICAL AND NUMERICAL SOLUTIONS

Closed-form analytical solutions offer the advantage of providing the functionality dependence of the distribution on the system parameters, elucidating also the significant dimensionless parameters. These are important for predictive design, scaling, optimization, and for determining the system's sensitivity to variations in the different parameters. However, most practical systems involve complex configurations that are intractable to analytical solutions. By contrast, numerical solutions that have been implemented in software packages can simulate a complex system fast, accurately, and with minimal user expertise. The major downside of numerically implemented solutions is that these are simulation tools, lacking predictive capability. Therefore, a number of cell configurations and operating conditions must be explored to discern a trend or to identify acceptable or optimized conditions. Furthermore, since no *a-priori* simplification is usually provided, a large set of process parameters (many of which may not be critical) must be typically provided. In the early years when desktop or minicomputers had just become accessible, great emphasis was placed on computational speed and on memory requirements.[24] Today, with the advent of powerful desktop computers, many of these considerations have become irrelevant and significantly more complex systems can now be readily solved.

IX. A SIMPLIFIED SOLUTION ALGORITHM

The procedure outlined below describes an algorithm for numerical solution of the current distribution. It is typically not the one implemented in practical software since is not efficient. However, for a conceptual description of the procedure it is the simplest to discuss, as it is not hampered by extraneous mathematical considerations. The thin boundary layer approximation is invoked and rather than a complete solution of the flow field, it is assumed that the equivalent stagnant concentration boundary layer, δ_N, along the electrode is available (may vary with position). The computational routine consists of the following steps applied to numerous points along the electrodes:

1. Guess:

   i along electrodes (an acceptable initial guess is $i = 0$). (60)

2. Calculate:

$$\eta_s^{\text{cathode}} = \frac{RT}{\alpha_C F} \ln \frac{i_0}{i}; \quad \eta_s^{\text{anode}} = \frac{RT}{\alpha_A F} \ln \frac{i}{i_0}. \quad (61)$$

3. Calculate:

$$\eta_c = \frac{RT}{nF} \ln \left(1 - \frac{i \delta_N}{nFDC_b}\right). \quad (62)$$

4. Calculate the potential in solution along the electrodes:

$$\Phi(x, y, z) = V - \eta_s - \eta_c - E \quad (63)$$

5. Solve:

$$\nabla^2 \Phi = \frac{\partial^2 \Phi}{\partial x^2} + \frac{\partial^2 \Phi}{\partial y^2} + \frac{\partial^2 \Phi}{\partial z^2} = 0 \quad \text{(Cartesian coordinate system)}. \tag{64}$$

Boundary conditions:

$$\text{Insulator}: \quad \frac{\partial \Phi}{\partial n}\bigg|_{\text{ins}} = 0. \tag{65}$$

$$\text{Electrode}: \quad \Phi|_e = V - \eta_s - \eta_c - E. \tag{66}$$

$$\text{Get } \Phi(x, y, z). \tag{67}$$

6. Calculate:

$$i = -\kappa \frac{\partial \Phi}{\partial n}\bigg|_{\text{electrode}}. \tag{68}$$

7. Check and update step 1 (60) as necessary. Updating may involve direct replacement of the previous guess with the newly computed values. More advanced updating algorithms, e.g., Newton–Raphson, may be applied for faster convergence.

X. NUMERICAL PROCEDURES FOR SOLVING THE LAPLACE EQUATION

The modeling of the current distribution in a general-geometry cell nearly always requires a numerical solution. The following discussion focuses on the thin boundary layer approximation, with the overpotential components lumped within a thin boundary layer which may be of a varying thickness. The Laplace equation for the potential with nonlinear boundary conditions must be solved. Similar considerations typically apply to the more comprehensive solution of the Nernst–Planck equation (10); however, the need to account for the convective fluid flow in the latter case makes the application of the boundary methods more complex. We focus our brief discussion on the most common methods: the finite-difference method, the finite-element method, and the boundary-element method, schematically depicted in Fig. 4. Since the finite-difference method is the simplest to implement and the best known technique, it is discussed in somewhat more detail.

Common Methods for Solving Laplace's Equation

- Analytical
- Finite difference
- Finite element
- Boundary elements methods
- Orthogonal collocations
- Conformal mapping

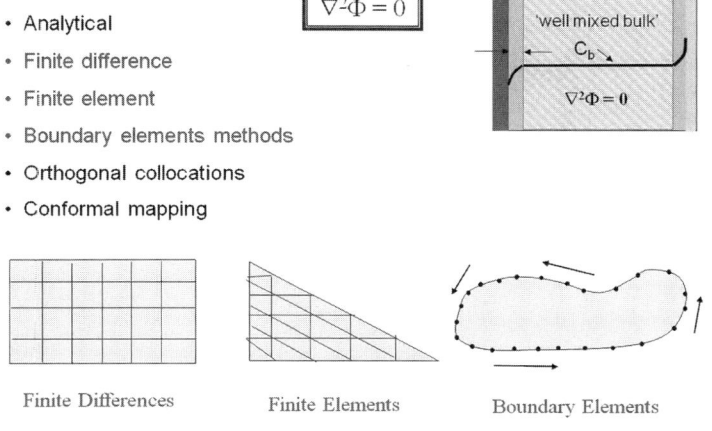

Finite Differences   Finite Elements   Boundary Elements

Figure 4. Numerical techniques for solving the Laplace equation.

A significantly more comprehensive discussion of the numerical algorithms has been provided recently by Schlesinger.[4]

1. The Finite-Difference Method

The finite-difference technique is based on dividing the cell region into a grid of square or rectangular elements (Fig. 5) and solving the finite-difference equation for the potential at each grid point. Various methods are available for solving the system of equations. The technique is relatively simple to implement and its accuracy depends on the mesh size and order of the difference formula. In general, a finer grid leads to more accurate results since the error is proportional to h^2 (from the Taylor expansion), where h is the linear grid dimension. However, the computational effort and the storage requirements rapidly increase with a finer grid.

The grid may be rectangular such that h_x need not be equal to h_y. Also, the grid size may vary with position in the cell to accommodate necessary resolution in different regions. An extension is an adaptive grid, where the grid is adjusted to accommodate certain geometrical features (e.g., a corner) or variations in the current density.

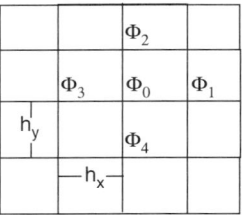

Figure 5. A meshed region. The potentials assigned to an arbitrary central point and its four closest neighbors are depicted.

The Laplace equation is first written in terms of a finite-difference approximation. For simplicity, a two-dimensional Cartesian coordinate system is assumed:

$$\nabla^2 \varphi = \frac{\partial^2 \varphi}{\partial x^2} + \frac{\partial^2 \varphi}{\partial y^2} = 0 \cong \frac{\Delta^2 \varphi}{\Delta x^2} + \frac{\Delta^2 \varphi}{\Delta y^2}. \quad (69)$$

The second derivatives are approximated in terms of the forward (70) and backward (71), difference equations:

$$\frac{\Delta_f \phi}{\Delta x} = \frac{\phi_1 - \phi_0}{h_x}, \quad (70)$$

$$\frac{\Delta_b \phi}{\Delta x} = \frac{\phi_0 - \phi_3}{h_x}. \quad (71)$$

The second derivatives at the central point are given by

$$\frac{\Delta_c \phi^2}{(\Delta x)^2} = \frac{\Delta_f \phi / \Delta x - \Delta_b \phi / \Delta x}{h_x} = \frac{\phi_1 - 2\phi_0 + \phi_3}{h_x^2}, \quad (72)$$

$$\frac{\Delta_c \phi^2}{(\Delta y)^2} = \frac{\phi_2 - 2\phi_0 + \phi_4}{h_y^2}. \quad (73)$$

The finite-difference form of the Laplace equation at any grid point (i, j) is accordingly

$$\frac{\varphi_{i+1,j} + \varphi_{i-1,j} - 2\varphi_{i,j}}{\Delta x^2} + \frac{\varphi_{i,j+1} + \varphi_{i,j-1} - 2\varphi_{i,j}}{\Delta y^2} = 0, \quad (74)$$

Current Distribution in Electrochemical Cells

where $\Delta x = h_x$ is the spacing between grid points in the x direction, $\Delta y = h_y$ is the spacing between grid points in the y direction, i is the grid point index in the x direction, and j is the grid point index in the y direction.

Interestingly, if we select $\Delta x = \Delta y$ we find from (74) that

$$\frac{\varphi_{i+1,j} + \varphi_{i-1,j} + \varphi_{i,j+1} + \varphi_{i,j-1}}{4} = \varphi_{i,j}, \tag{75}$$

or

$$\phi_{i+1,j} + \phi_{i-1,j} + \phi_{i,j+1} + \phi_{i,j-1} - 4\phi_{i,j} = 0, \tag{76}$$

i.e., the potential ϕ at any point is the arithmetic average of its four closest neighbors.

Applying the finite-difference form of the Laplace equation (75) to each grid point in the cell yields a large set of equations to be solved. For example, on a 30 × 30 grid there are 900 equations to be solved. However, each equation has at most five unknowns. The unknown potentials are solved by a numerical iterative procedure. When calculating the secondary or tertiary distribution, exponential terms will appear at the boundaries. These terms are linearized about the potential of the previous iteration. The iterations are repeated until the potentials converge to within the specified error, typically 10^{-4} or smaller.

Once the potential field has been solved, the current is determined from

$$x \text{ component: } i_x = -\kappa \nabla \Phi = -\kappa \left(\frac{\Phi_{i,j+1} - \Phi_{i,j-1}}{2\Delta x} \right), \tag{77}$$

$$y \text{ component: } i_y = -\kappa \Delta \Phi = -\kappa \left(\frac{\Phi_{i+1,j} - \Phi_{i-1,j}}{2\Delta y} \right). \tag{78}$$

(i) Methods of Solving the Finite-Difference Equation

A number of numerical techniques are available for solving the large set of algebraic equations of the form of (75) or (76). Selection depends to a large extent on the convenience of the programmer, the type of computer, and the required speed.

(1) Relaxation method
An initial potential is guessed and assigned to each grid point. Typically, the same potential, e.g., $\phi = 0$ or $\phi = (V_A + V_C)/2$, is assigned to all points. Since the guess is not perfect, the right-hand side of (76) will not be zero, but a finite residual R. The residuals at all points are then calculated and the largest residual is identified. This residual is then set to zero by reassigning the potential ϕ_{ij} for this point per (76). All the other potentials are left intact. The residuals are then calculated again and the procedure listed above is repeated until all residuals are within the specified error tolerance.
  (a) Advantages
    This method uses a minimum of memory space. For N points only N real numbers are needed for storing the potentials.
  (b) Disadvantages
    The method is iterative and requires many repeat steps, exceeding by a large factor the number of points.
(2) Overrelaxation or underrelaxation
The same procedure as in the relaxation method is followed except that the potentials at each point are overcorrected or undercorrected according to certain rules. This often provides a faster convergence.
(3) Matrix method
The Laplace equation is written for each point and the resulting matrix equations for all points are solved using Gauss elimination. If we set

$$\lambda = \Delta x / \Delta y, \tag{79}$$

we can rewrite (76) as

$$\left(\Phi_{i-1,j} + \Phi_{i+1,j} - 2\Phi_{ij}\right)\lambda^2 + \left(\Phi_{i,j-1} + \Phi_{i,j+1} - 2\Phi_{ij}\right) = 0. \tag{80}$$

Each of the five points has a (different) coefficient. This system of equation is put into an $N \times N$ matrix which is solved.
  (a) Advantages
    Using an efficient Gauss elimination routine, one can solve the matrix equations quickly. Since all points are coupled and only five points at a time are used, the matrix

should be pentadiagonal. If a recursion formula for such a set of equations is employed, the system can be solved very quickly.
(b) Disadvantages
There are large memory storage requirements: for N points, at least $N^2 + N$ real numbers must be stored.

2. The Finite-Element Method

The finite-element technique is based on dividing the cell domain into polygonal sections. The potential within each of the elements is assumed to be a linear combination of the value at the vertices. However, unlike the finite-difference method, which solves the finite-difference approximation of the Laplace equation, the finite-elements method seeks a solution for the potential distribution within the cell, which best fits the Laplace equation and the boundary conditions. The degree of accuracy is similar to that of the finite-difference method; however, curved boundaries and narrow corners can be described with more precision and ease. On the other hand, the presence of electrochemical nonlinear boundary conditions leads to ill-conditioned matrix equations which are more difficult to solve than the finite-difference system.

3. Boundary-Element Methods

The boundary-element method is based on applying Green's theorem to convert the Laplace equation in the bulk to an integral equation along the cell boundaries. Efficient solutions can be obtained, particularly for complex and moving boundaries; however, separated regions in the cell are more difficult to model than with the grid methods.

The technique is similar to the eigenfunction method in that the solution is written as an integral on the boundary using Green's theorem, combined with a Green function solution of the Laplace equation. The Green function in this case is the fundamental solution,

$$G(x; y; X_f Y_f) = \frac{1}{2} \ln\left[(x - X_F)^2 + (y - Y_F)^2\right], \quad (81)$$

written at any fixed point in the cell $F = (X_F, Y_F)$. The solution is expressed as a well-conditioned integral equation for either the potential or the current specified on a segment of the boundary. A spline function approximation to the current and potential on the boundary is used and substituted into the integral equations. This results in a matrix system for the spline coefficients. Once the coefficients have been calculated, the potential and current on any boundary or interior point may be evaluated. Singularities at corners can be incorporated by constructing singular expansions for regions near the corner. Also, multiply connected regions and curved boundaries may be incorporated

The nonlinear boundary condition associated with surface overpotentials and mass transfer effects necessitates the use of an iterative procedure to solve the potential distributions. Newton's method with Broyden's algorithm to update the Jacobian can be used.

4. Orthogonal Collocation

The orthogonal collocation method uses a series representation for the potential and current and solves for the coefficients of the series by satisfying the governing equations with boundary conditions at each fitting point. Again, treatment of nonconnected regions is difficult, and some judgment and experience is usually required of the user.

Table 2 provides a (somewhat subjective) comparison between the numerical methods.

Table 2.

Comparison between common techniques for solving the Laplace equation.

Feature	FDM	FEM	BEM
Robustness	OK	OK	Best
Speed	Best	Poor	Poor
Applicability to arbitrary boundaries	Poor	Good	Best
Computations primarily oriented to boundaries	No	No	Yes
Computations oriented to internal values	Yes	Yes	No
Applicability to nonlinear boundary conditions	OK	OK	Best
Suitability to moving boundaries	Poor	OK	Best
Order of number of boundary points as a function of the total number of field points (with the field precisely computed)	$N^{1/2}$	$N^{1/2}$	N

FDM finite-difference method, *FEM* finite-element method, *BEM* boundary-element method

XI. NUMERICALLY IMPLEMENTED SOLUTIONS FOR THE CURRENT DISTRIBUTION

Numerical solutions for the current distribution that have been presented in the literature may be divided into three classes: (1) "dedicated" code that has been developed to simulate a specific class of problems, (ii) commercially available software, dedicated to simulating electrochemical systems, and (iii) general scientific CAD software that can be customized for solving electrochemical systems. Each class is briefly discussed below:

1. *"Dedicated" code developed to solve specific problems*: Such programs have been typically developed at academic institutions (e.g., Refs. [31, 44]) or national laboratories.[45] A very broad range of such models have been described in the literature and comprehensively reviewed by Dukovic.[3] The major limitation of this class of numerical solutions is that they are typically not publicly available, often cannot be generalized, lack robustness, are not user-friendly, and typically lack instructions. Therefore, while providing insight into the specific analyzed problem, they usually do not lend themselves to more generalized or broader applications.

2. *Commercial software dedicated to simulating electrochemical systems*: A few software packages dedicated to simulating electrochemical systems became commercially available in recent years. The common denominator of these software packages is that they have been specifically written for electrochemical applications; hence, they incorporate electrochemical boundary conditions and have a user interface dedicated to electrochemical systems. Since these software packages are backed by commercial companies that provide support, documentation, and updates, they can be used by nonexperts. The two most popular packages are briefly reviewed:

   (a) *Cell-Design by L-Chem*: Cell-Design was the first commercially released comprehensive electrochemical modeling software (1985). Developed under a DOE SBIR contract, the software was initially based on a finite-difference algorithm,[24] which was later modified to include higher-order elements[27] and the boundary-elements algorithm. Cell-Design can be used for modeling two-dimensional cross sections of arbitrary cell configurations and axisymmetric bodies of revolution. The software package incorporates a properties database. For chemistries not included in the database, L-Chem offers a special device, the L-Cell,[46] which readily provides the kinetics parameters required for modeling. Cell-Design is offered in a modular form, allowing the user to customize the software. The software focuses on ease of use by nonexperts, and a fast learning curve. Cell-Design has a broad international user base, with most applications in the plating and electrolytic industries.

(b) *Software packages by Elsyca*: Elsyca's software, released commercially in 1998, is based on work at the University of Brussels VUB (Vrije Universiteit Brussel), by Johan Deconinck.[47] The software was originally based on the boundary-element method. Various packages include Ecmmaster, Elsy2D, and corrosion modeling software (CPmaster and Catpro). The major strength of Elsyca's software is the capability of some modules to handle complex three-dimensional configurations. The downside is that the software is more complicated to use. The software is distributed internationally, with primary applications in metal finishing and corrosion.

3. *General scientific CAD software that can be customized for modeling electrochemical systems*: There is a broad range of commercially available scientific and/or engineering software packages that can be customized to model electrochemical systems. The packages range from general-science-based software to software that is directed primarily at modeling fluid flow, stress–strain, heat transport, casting, microelectronics, magnetics, or a combination thereof, among many others. These packages are generally very comprehensive, and, consequently, their use is complex and often requires special training. Since the software in this class is not designed for modeling electrochemical systems (with the exception of a few fuel-cells modules), customizing it (mostly by incorporating the appropriate boundary conditions) for simulating electrochemical systems, is complicated. On the other hand, most of the software in this class is capable of handling

complex geometrical configurations. A very large number of software packages (hundreds) are commercially available and only very few representatives are briefly reviewed.

(a) *Comsol Multiphysics (formerly Femlab) by Comsol*: This general science and engineering oriented software is widely used, particularly in academia. Comsol Multiphysics can handle a large variety of systems with quite complex boundary conditions, including transients. Incorporating the electrochemical boundary conditions is, however, tedious since these must be introduced through generalized functions. Using the software and interpretation of the results requires expertise. A moving-boundaries option has been added recently; however, at this time, its use for electrochemical applications is limited. Introduction of complex geometry is also tedious.

(b) *Fluent, Ansys, and Polyflow by Fluent*: Fluent, among numerous similar companies, offers a powerful and very comprehensive set of software programs. They are typically directed at a class of (nonelectrochemical) applications, and thus incorporating the electrochemical boundary conditions is complex. A special module has recently been developed for modeling fuel cells. Use of the software typically requires training.

XII. DETERMINATION OF THE CURRENT DISTRIBUTION IN SPECIAL APPLICATIONS

A number of applications with specific modeling features are briefly discussed below.

1. Multiple Simultaneous Electrode Reactions, Including Alloy Codeposition and Gas Coevolution

Numerous electrochemical, systems involve multiple simultaneous reactions. Some, such as alloy plating, are deliberate, others, e.g., gas coevolution, may occur owing to the presence of multiple species in solution. The modeling of such systems is complicated by the fact that local fluxes at the electrode and partial currents are associated with each of the reacting species, yet only the total current (and/or potential) can be measured. Occasionally, significant partial reactions may proceed in opposite directions, leading to a very small overall current. A solution procedure has been outlined by Menon and Landau[48] and is based on assuming that each of the possible electrode reactions proceeds in parallel according to a modified Butler–Volmer kinetics, written here in terms of the partial current associated with reacting species j:

$$i_j = i_{0,c_\mathrm{b},j} \left[\left(\frac{a_{\mathrm{e},j}^\mathrm{R}}{a_{\mathrm{b},j}^\mathrm{R}} \right) e^{\frac{\alpha_{\mathrm{A},j} F}{RT} \eta_j} - \left(\frac{a_{\mathrm{e},j}^\mathrm{o}}{a_{\mathrm{b},j}^\mathrm{o}} \right) e^{-\frac{\alpha_{\mathrm{C},j} F}{RT} \eta_j} \right]. \qquad (82)$$

Here, $i_{0,j}$, $\alpha_{\mathrm{A},j}$, $\alpha_{\mathrm{C},j}$, and a_j represent the exchange current density, the anodic and the cathodic transfer coefficients, and the activity of species j, respectively. The superscripts, R and o designate the reduced and oxidized form of the reacting species, j. The overpotential, η_j, is given by

$$\eta_j = V - E_j - \Phi \qquad (83)$$

It should be noted that while the electrode potential, V, and the electrostatic potential in solution, Φ, are identical for all species, the standard potential, E_j, is species-dependent, and hence the driving force for the electrode reaction, η_j, will be different for each species. When the standard potential for the different species are quite different, or when the overpotential $V - \Phi$ is small, η_j may even assume a different direction for different species. The current within the electrolyte is given by the algebraic sum of the partial currents of all species,

$$i = \sum_j i_j \qquad (84)$$

The solution procedure is analogous to the general approach outlined above. For the thin boundary layer approximation, one solves the Laplace equation (25) subject to the boundary conditions (26) and (27). In boundary condition (27), the current on the electrode is given by the sum of the partial currents, as indicated by (82). After a solution has been obtained (through an iterative procedure), and the partial currents have been computed, the deposit composition can be determined from Faraday's law (1). Parasitic hydrogen (or for anodic processes, oxygen) evolution can be evaluated using an identical procedure, providing thereby a quantitative measure of the current efficiency.

Menon and Landau[48] point out a number of complications associated with modeling alloy deposition. First, the values for the kinetics parameters of the alloy species are likely to be different from those measured for the pure components. Furthermore, these parameters may vary with the composition of the alloy. Determination of such parameters can be done experimentally from measured alloy composition in well-controlled deposition experiments. A second issue has to do with the difficulty in determining the activities of the alloy components in the solid phase. The activity, which affects the electrode kinetics equation (82), is unity for single-component deposition; however, in multicomponent alloy deposition it varies with the nature of the deposited alloy.[48]

2. Moving Boundaries in Deposition and Dissolution Applications

Modeling of deposition and dissolution processes, where the thickness of the material added or removed is comparable to the characteristic dimensions of the modeled feature, requires accounting for the changing boundaries. Typical applications include the metallization of small features, e.g., through-holes,[22,49] blind vias, and trenches in microelectronics, pattern plating of printed circuit boards,[43] modeling the propagation of corrosion damage,[50] and, on a larger scale, electrochemical machining. Typically, there is no need to model the truly transient process, and a sequence of pseudo-steady-state steps provides adequate accuracy. The latter is ensured by selecting sufficiently small steps such that each corresponds only to a small fractional increment of the final thickness. To rigorously check if the number of steps taken is sufficient, one may incrementally increase the number of steps, until no further significant effect is noticed. Time steps need not be uniform. Often small steps are advisable in the early stages, when, typically, sharp corners become rounded. Once a radius of curvature has been established, subsequent steps may often be longer. The solution algorithm is typically based on first calculating the current distribution in the initial geometry and then using Faraday's law (1) to calculate the displacement of the boundaries at a number of selected points. A new boundary is then associated with the relocated points and the solution procedure is repeated. Clearly the nature of the solution is such that it lends itself only to numerical procedures;[32] however, on occasion, semianalytical solutions coupled with geometry updating are adequate.[51]

The major challenge for the numerical procedure is the updating of the boundary. Boundary points near a corner or a small curvature region may cross one another if the step size is large in comparison with the distance of the points from the corner. For geometrical fidelity, it is desired to have large point density at high-curvature regions, while still maintaining a step size which may be large compared with the distance of the points, a situation that may lead to "crossover" of boundary points. An algorithm must be provided to properly account for these intersecting regions. Another challenge has to do with surface perturbations. Small computational inaccuracies may give rise to surface perturbations, which, once formed, tend to grow. The numerical algorithm must be capable of treating such features without introducing excessive "smoothing," which may lead to overlooking valid textural changes.

3. Electropolishing, Leveling, and Anodizing

The common denominator to these anodic processes is the presence of an anodic film that is typically highly resistive, thus leading to a more uniform distribution. Electropolishing applications are often carried out in a concentrated electrolyte, e.g., phosphoric acid, which tends to form a sparingly immiscible salt on the surface.[52,53] The high film resistance tends to smooth the current distribution on the nano- and microscales, leading to polishing. Somewhat larger features can be similarly leveled through mass transport controlled dissolution across a concentration boundary layer.[54–56] Anodization processes are similarly controlled by a highly resistive oxide layer, incorporating ionic transport that can be characterized by a Mott–Cabrera-type hopping mechanism. Akolkar et al. have discussed the current distribution in the anodization of aluminum, critically comparing experimental results with numerical modeling.[33]

4. Current Distribution on Resistive Electrodes

In most electrochemical modeling applications, a commonly invoked and valid assumption is that the electrode constitutes an equipotential surface. This is due to the characteristically low electrolyte conductivity ($\kappa \sim 0.1$ S/cm) as compared with metallic conductivities, which are larger by about a factor of one million. However, in some applications, when very long and thin electrodes are present, e.g., in reel-to-reel ("strip") plating, or when only a thin metallic seed is present on an otherwise insulating substrate (such as in the plating of semiconductor wafers or printed circuit boards), this assumption may not hold. In these cases, the voltage within the resistive electrode varies with the distance from the terminal contact point due to ohmic (IR) drop within the electrode itself. Tobias and Wijsman[57] first analyzed this problem by conducting a voltage balance across a parallel electrode cell, accounting also for the voltage drop within the electrodes. They have shown that the current nonuniformity depends on a dimensionless parameter incorporating the ratio of the resistance within the electrode (leading to the terminal effect) into the resistance within the electrolyte. This analysis, which was limited to systems under linear polarization, was later extended by Lanzi and Landau[58] to more complex configurations under Tafel polarization. Recent interest in metallization of interconnects on semiconductor wafers led to further extension of these analyses to the cylindrical wafer configuration[59,60] and to exploitation of this concept to obtain more uniform deposition by introducing a low-acidity, more resistive electrolyte.[61] Clearly more complex, general-geometry configurations necessitate numerical solutions. An example of such an approach, comparing both numerical and analytical solutions, has been presented.[60]

5. Current Distribution in the Metallization of Through-Holes, Blind Vias, and Trenches

The fabrication of printed circuit boards by plating often involves the metallization of through-holes. Because of their relatively small radii (mills) and high aspect ratio (e.g., Refs. [10–20]) it is a challenge to metalize such through-holes without first plating excessively at the hole entrance, blocking further deposition. Kessler and Alkire[62] were the first to present an analysis for the current distribution for a plated through-hole. This was later extended by Lanzi and Landau,[22] who quantified the minimal flow (agitation) required to eliminate mass transport limitations, and furthermore highlighted the presence and analytically characterized an ohmic resistance which limits the degree of achievable uniformity in the plating of such hole. In a subsequent publication,[63] Lanzi et al. analyzed and experimentally demonstrated the use of nonuniformly distributed plating additives, which are aggregated near the via entrance, to control the current distribution.

The recently introduced process of semiconductor interconnect metallization by copper plating[64] critically depends on the ability to achieve a bottom-up fill by plating of submicron (currently down to approximately 40 nm) blind vias and trenches. The bottom-up plating process hinges on the use of a special additives mixture, consisting of an inhibitor (typically a polyether, e.g., PEG), and an antisuppressor [e.g., bis(3-sulfopropyl) disulfide]. The rationale for

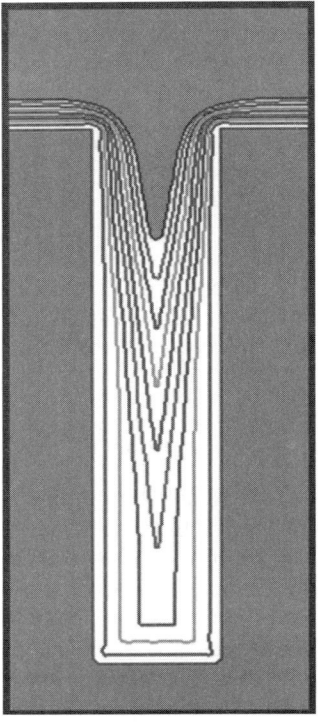

Figure 6. Simulation of bottom-up plating of a 0.5 μm radius × 5 μm deep trench in the presence of PEG and bis(3-sulfopropyl) disulfide (SPS). Nine 1-s time steps, corresponding to the additive concentrations shown in Fig. 7, are shown (Cell-Design software simulations[69]).

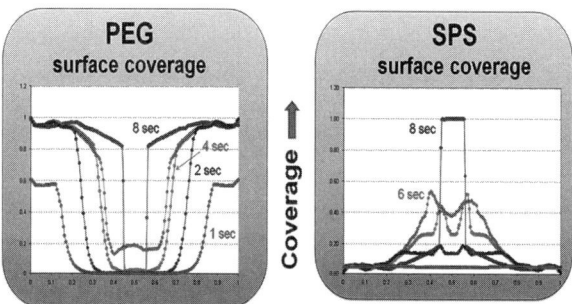

Figure 7. Modeled PEG and SPS coverage on the sidewalls and bottom of a 0.1 μm radius × 0.5 μm deep trench at 1–8 s after immersion/start of plating. Owing to surface shrinkage at the bottom[66,67] and competitive adsorption,[51] the SPS saturates the trench bottom after 8 s, fully displacing the PEG (Cell-Design software simulations[69]).

this combination, which has been empirically formulated,[64] has been provided recently by Akolkar and Landau.[51] They modeled the transient transport and competitive adsorption of the additives mixture, relating the current density and the deposition rate to the position- and time-dependent surface coverage by the additives. The transient transport and kinetics model incorporates also the effect of enhanced antisuppressor concentration at the shrinking via bottom, as introduced by Moffat et al.[65] and West et al.[66] A number of customized numerical models for this process have been reported recently.[67,68] An example of the simulation of the bottom-up fill using a module of Cell-Design software[69] which is capable of simulating transient additive transport and competitive adsorption in conjunction with moving boundaries due to deposition is shown in Figs. 6 and 7.

SYMBOLS

a constant (37), V
a activity, dimensionless
A local electrode area, cm^2
b Tafel polarization slope, $(= RT/\alpha F)$, V
b' linear polarization slope, V cm^2/A
C concentration, g mol/cm^3

d	deposit thickness, cm
D	diffusion coefficient, cm^2/s
E	standard electrode potential, V
F	Faraday constant, 96,487 C/equiv
h	grid size, cm
i	current density, A/cm^2
i_0	exchange current density, A/cm^2
i_L	limiting current density, A/cm^2
i_{ave}	average current density, A/cm^2
i_{crit}	critical current density for smooth deposition, A/cm^2
l, L	length, cm
L	leveling parameter, dimensionless
M	molecular weight of deposited metal, g/mol
N	number of electrons transferred in the electrode reaction, equiv/mol
N	ionic flux, mol/s cm^2
R	universal gas constant, 8.314 J/mol K
R_j	rate of generation of species j, M/s cm^3
R_S^*	specific surface (activation) resistance, Ω cm^2
R_C^*	specific concentration resistance, Ω cm^2
R_Ω^*	specific ohmic resistance, Ω cm^2
R_Ω	ohmic resistance, Ω
t	time, s
T	absolute temperature, K
To	Tobias number, dimensionless
u	ionic mobility, cm^2 mol/J s
v	velocity, cm/s
V	voltage of an electrode, V
Wa	Wagner number, dimensionless
z_j	charge number of the ionic species j, equiv/mol

Greek

α	transfer coefficient
γ	parameter adjusting the exchange current density, (18), dimensionless
δ	parameter adjusting the exchange current density, (18), dimensionless
δ	mass transfer boundary layer thickness, cm
Δx	spacing between grid points in the x direction, cm

Δy	spacing between grid points in the y direction, cm
Δz	spacing between grid points in the axial direction, cm
ε_F	faradaic efficiency
η	total overpotential, V
η_s	surface (activation) overpotential, V
η_c	concentration overpotential, V
η_Ω	ohmic overpotential, V
κ	solution conductivity, S/cm
λ	constant, (79), dimensionless
μ	viscosity, poise = g/cm s
ν	kinematic viscosity, cm^2/s
ν_T	eddy (turbulent) viscosity, cm^2/s
ρ	ADI iteration parameter, dimensionless
ρ	density, g/cm^3
Φ	potential in the electrolyte, V

SUPERSCRIPTS

*	specific, per unit area
o	oxidized species
R	reduced species

SUBSCRIPTS

A	anode
Avg	average value
B	bulk
B	backward
C	cathode
C	mass transport
crit.	critical value
E	at the electrode
Elec	electrode
F	forward
Ins	insulator
J	ionic species j

L linear polarization
N normal to surface
N Nernst
R reactant
S surface overpotential or resistance
T Tafel region (high field)
Ω ohmic

REFERENCES

[1] J. Newman, *Electrochemical Systems*, Prentice-Hall, Englewood Cliffs, NJ, 1973 (Second Edition, 1991).

[2] G. A. Prentice and C. W. Tobias, "A Survey of Numerical Methods and Solutions for Current Distribution Problems," J. Electrochem. Soc., 129(1), 72–78, 1982.

[3] J. O. Dukovic, "Computation of Current Distribution in Electrodeposition, a Review," IBM J. Res. Dev., 34(5), 693–705, 1990.

[4] M. Schlesinger, "Mathematical Modeling in Electrochemistry," in *Modern Aspects of Electrochemistry*, Vol. 43, M. Schlesinger, Ed., Springer, New York, 2008.

[5] R. B. Bird, W. E. Stewart and E. N. Lightfoot, *Transport Phenomena*, Wiley, New York, 1960.

[6] A. J. Bard and L. Faulkner, *Electrochemical Methods*, Wiley, New York, 1980.

[7] E. Gileadi, *Electrode Kinetics*, Wiley-VCH, New York, 1993.

[8] V. G. Levich, *Physicochemical Hydrodynamics*, Prentice-Hall, NJ, 1962.

[9] J. C. Puippe and F. H. Leaman, *Theory and practice of pulse plating*, American Electroplaters and Surface Finishers Society, Orlando, FL, 1986.

[10] K. I. Popov and M. D. Maksimovic, in *Modern Aspects of Electrochemistry*, Vol. 19, B. E. Conway, J. O. M. Bockris and R. E. White, Eds., Plenum Press, New York, p. 193, 1989.

[11] B. K. Purushothaman, P. W. Morrison and U. Landau, "Reducing Mass Transport Limitations by the Application of Special Pulsed Current Modes," J. Electrochem. Soc., **152**(4), J33–J39, 2005.

[12] B. K. Purushothaman and U. Landau, "Rapid Charging of Lithium Ion Batteries Using Pulsed Currents – A Theoretical Analysis," J. Electrochem. Soc., 153(3), A533–A542, 2006.

[13] J. R. Selman and C. W. Tobias, "Limiting Current Mass Transfer Measurements," in *Advances in Chemical Engineering*, Vol. 10, Drew et al. Ed., Academic Press, New York, 1978.

[14] H. Schlichting, *Boundary Layer Theory*, McGraw-Hill, New York, 1979.

[15] U. Landau, *AIChE Symposium Series 204*, Vol. 77, 75–87, 1981.

[16] H. S. Carslaw and J. C. Jaeger, *Conduction of Heat in Solids*, Clarendon Press, Oxford, 1959.

[17] J. Crank, *The Mathematics of Diffusion*, Oxford University Press, London, 1956.

[18] P. Moon and D. Spencer, *Field Theory for Engineers*, Van Norstrand, Princeton, NJ, 1961.

[19] L. M. Milne-Thomson, *Theoretical Hydrodynamics*, The Macmillan Company, New York, 1960.

[20] J. Newman, J. Electrochem. Soc., 113, 1235–1241, 1966.

[21] C. Wagner, J. Electrochem. Soc., 98(3), 116, 1951.

[22] O. Lanzi III and U. Landau, "Analysis of Mass Transport and Ohmic Limitations in Through-Hole Plating," J. Electrochem. Soc., 125(8), 1922–1930, 1988.

[23] A. S. Agarwal, U. Landau and J. H. Payer, "Current Distribution on a Cathode Covered by a Thin Film Electrolyte in Crevice Corrosion – An Analytical Model," ECS Trans., 11, Feb., 2008.

[24] W. J. Cook and U. Landau, in *Engineering of Industrial Electrolytic Processes*, U. Landau, R.E. White and R.J. Varjian, Eds., The Electrochemical Society Softbound Proceeding Series, Pennington, NJ, 1986

[25] C. W. McFarland, "The Finite Difference Method for Modeling Potential and Current Distributions in Electroplating Cells," in *Electroplating Engineering and Waste Recycle – New Developments*, Dexter Snyder, Uziel Landau, and R. Sard, 1982 Symposium Proceedings, The Electrochemical Society, Inc., Pennington, NJ, 1983.

[26] M. Matlosz, C. Creton, C. Clerc, and D. Landolt, J. Electrochem. Soc., 134(12), 3015, 1987.

[27] W. M. Lynes and U. Landau, "A Novel Adaptation of the Finite Difference Method for Accurate Description of Non-Orthogonal Boundaries," Extended Abstract No. 332, Vol. 88–2, The Electrochemical Society, Fall Meeting, Chicago, IL, Oct. 1988.

[28] I. Kadija, J. A. Abys, V. Chinchankar and K. Straschil, "Hydrodynamically Controlled "Hull Cell," Plating and Surface Finishing, July, 1991.

[29] U. Landau, N. L. Weinberg and E. Gileadi, J. Electrochem. Soc., 135(1), 396–403, 1988.

[30] S. Mehdizadeh, J. O. Dukovic, P. C. Andricacos and L. T. Romankiw, J. Electrochem. Soc., 139(1), 78, 1992.

[31] R. Alkire, T. Bergh and R. L. Sani, "Predicting Electrode Shape Change with Use of Finite Elements Methods," J. Electrochem. Soc., 125(12), 1981–1988, 1978.

[32] U. Landau, E. Malyshev, R. Akolkar and S. Chivilikhin, "Simulations of 'Bottom-Up' Fill In Via Plating of Semiconductor Interconnects," Paper 189 d, session TK; Proceedings of the AIChE Annual Meeting, San-Francisco, CA, Nov. 16–21, 2003.

[33] R. Akolkar, U. Landau, H. Kuo and Yar-Ming Wang, "Modeling of the Current Distribution in Aluminum Anodization," J. Appl. Electrochem., 34, 807–813, 2004.

[34] A. C. West, C. Cheng and B. Baker, J. Electrochem. Soc., 145(9), 3070, 1998.

[35] R. Akolkar and U. Landau, manuscript in preparation.

[36] U. Landau, "Novel Dimensionless Parameters for the Characterization of Electrochemical Cells," Proceedings of the D. N. Bennion Mem. Symp., R. E. White and J. Newman, Eds., The Electrochemical Society Proceedings, Vol. 94–9, 1994.

[37] U. Landau, in *Tutorial Lectures in Electrochemical Engineering and Technology – II*, Vol. 79, R.C. Alkire and D.T. Chin, Eds., AIChE Symp. Series 229, 218–225, 1983.

[38] D. Roha and U. Landau, J. Electrochem. Soc., 137(3), 824–834, 1990.

[39] J. H. Shyu and U. Landau, Abstract 163, The Electrochemical Society Extended Abstracts Vol. 79-2, Los Angeles, CA, Oct. 14–19, 1979.

[40] J. H. Shyu and U. Landau, Abstract 401, The Electrochemical Society Extended Abstracts Vol. 80–2, Hollywood, FL, Oct. 5–10, 1980.

[41] U. Landau and J. H. Shyu, in *Electroplating Engineering and Waste Recycle – New Developments*, D. Snyder, U. Landau and R. Sard, Eds., The Electrochemical Society, Inc., Pennington, NJ, 1983.

[42] Y. Oren and U. Landau, Electrochem. Acta, 27, 739–748, 1982.

[43] R. T. Galasco, J. Tang and U. Landau, "Current Distribution in Pattern Plating of Nonuniformly Space and Isolated Lines," Extended Abstract No. 330, Vol. 88–2, The Electrochem. Soc., Fall Meeting, Chicago, IL, Oct., 1988.

[44] M. M. Menon and U. Landau, "Modeling of Electrochemical Cells Including Diffusion, Migration and Unsteady State Effects," J. Electrochem. Soc., 134(8), 2248–2253, 1987.

[45] J. S. Bullock, G. Giles and L. J. Gray, "Simulation of an Electrochemical Plating Process," in *Topics in Boundary Element Research*, Vol. 7, Chap. 7, C. A. Brebbia, Ed., Springer Verlag, Berlin and New York, pp. 121–141, 1990.

[46] U. Landau and E. Malyshev, "The L-Cell- A novel Device for Plating Process Characterization," AESF/SURFIN 2004, Chicago, IL, June 2004.

[47] J. Deconinck, "Electrochemical Cell Design," in *Topics in Boundary Element Research*, Vol. 7, Chap. 8, C. A. Brebbia, Ed., Springer Verlag, Berlin and New York, pp. 142–170, 1990.

[48] M. M. Menon and U. Landau, "Modeling of Cells, with Multiple Electrode Reactions-Thickness and Comp. Variations in Alloy Plating," J. Electrochem. Soc., 137, 445–452, 1990.

[49] E. K. Yung, L. T. Romankiw and R. C. Alkire, "Plating into through-holes and Blind Holes," J. Electrochem. Soc., 136(1), 206–215, 1989.

[50] A. S. Agarwal, U. Landau, X. Shan and J. H. Payer, "Modeling Effects of Crevice Former, Particulates, and the Evolving Surface Profile in Crevice Corrosion," ECS Trans., 3(31), 459, 2007.

[51] R. Akolkar and U. Landau, "A Time-Dependent Transport-Kinetics Model for Additives Interactions in Copper Interconnect Metallization," J. Electrochem. Soc., 151(11), C702, 2004.

[52] J. Mendez, R. Akolkar, T. Andryushchenko and U. Landau, "A Mechanistic Model for Copper Electropolishing in Phosphoric Acid," J. Electrochem. Soc., 155, D27, 2008.

[53] B. Chou, R. Jain, D. McGervey, U. Landau and G. Welsch, "Electropolishing of Titanium," Proc. Electrochem. Soc.: "Chemical Mechanical Polishing," Vol. 2002–1, S. Seal et al., Ed, pp. 126–138, 2002.

[54] S. I. Krichmar, Electrokhimiya, 17, 1444, 1981.

[55] U. Landau and E. Malyshev, "Modeling Copper Planarization under Mass Transport Controlled Dissolution," Extended Abstract # 1656, 210th Electrochemical Society Meeting, Cancun, Mexico, Oct. 29–Nov. 3, 2006.

[56] A. C. West, I. Shao and H. Deligianni, J. Electrochem. Soc., 152(10), C652–C656, 2005.

[57] C. W. Tobias and R. Wijsman, J. Electrochem. Soc., 100(10), 459, 1953.

[58] O. Lanzi III and U. Landau, "Current Distribution at a Resistive Electrode under Tafel Kinetics," J. Electrochem. Soc., 137(4), 1139–1143, 1990.

[59] K. M. Takahasi, J. Electrochem. Soc., 147(4), 1414–1417, 2000.

[60] S. Chivilikhin, U. Landau and E. Malyshev, "Current Distribution On A Resistive Wafer Under Copper Deposition Kinetics," Paper 190 b, session TK; Proc. AIChE Annual Meeting, San-Francisco, CA, Nov. 16–21, 2003.

[61] Uziel Landau, John J. D'Urso and David B. Rear, "Electro deposition Chemistry." U.S. Patent Number 20020063064; May 30, 2002; Related: 6,379,522; April 30, 2002; 6,350,366; Feb. 26, 2002; 6,113,771; September 5, 2000.

[62] T. Kessler and R. C. Alkire, J. Electrochem. Soc., 123, 990, 1976.

[63] Oscar Lanzi III, Uziel Landau, Jonathan D. Reid and Raymond T. Galasco, "Effect of Local Kinetic Variations on Through-Hole Plating," J. Electrochem. Soc., 136, 368–374, 1989.

[64] P. C. Andricacos, C. Uzoh, J. O. Dukovic, J. Horkans and H. Deligianni, IBM J. of Res. and Dev., 42(5), 567, September 1998.

[65] T. P. Moffat, D. Wheeler, W. H. Huber and D. Josell, Electrochemical and Solid-State Letters, **4**(4), C26–C29, 2001.
[66] A. C. West, S. Mayer and J. Reid, Electrochem. Solid-State Lett., 4(7), C50–C53, 2001.
[67] T. P. Moffat, D. Wheeler, S.-K. Kim and D. Josell, "Curvature Enhanced Adsorbate Coverage Model for Electrodeposition," J. Electrochem. Soc., 153(2), C127–C132, 2006.
[68] R. Alkire, X. Li, T. O. Drews, E. Rusli, F. Xue, Y. He and R. Braatz, J. Electrochem. Soc., 154(4), D230–D240, 2007.
[69] Cell-Design, Computer Aided Design Software for Simulating Electrochemical Cells, L-Chem, Inc., Shaker Heights, OH 44120.

Index

A
Abe, C., 411
Abeles, B., 410
Abrantes, L.M., 174
Abruna, H.D., 151
Abys, J.A., 468
Acharyya, M., 274, 275
Achenbach, J.D., 411
Adamo, C., 382
Addison, R.C., 411
Agarwal, A.S., 467, 491
Agrawal, A.K., 291
Agrisuelas, J., 181
Akolkarm, R., 491, 495
Akolkar, R., 468, 470, 491, 492
Ala-Nissila, T., 144
Albano, E.V., 239–285
Albery, W.J., 151, 160, 161
Alder, B.J., 378
Alivisatos, A.P., 115
Alkire, R., 14, 297, 303, 467, 468, 487, 495
Alkire, R.C., 294, 298, 301, 491, 493
Allen, M.P., 133, 143, 253
Almbladh, C.O., 371
Al-Sana, S., 206
Alwitt, R.S., 301
Amemiya, T., 173
Amerio, S., 411
Anderko, A., 291
Anderson, M.R., 152
Andersson, Y., 389
Andrade, E.M., 155, 156
Andricacos, P.C., 468, 494, 495
Andrieux, C.P., 160, 174
Andrieux, C., 152
Andryushchenko, T., 492
Angelopoulos, M., 152
Anisimov, V.I., 343, 376, 386, 402
Anson, F.C., 173, 323, 324
Aoki, K., 163

Arakawa, M., 428
Aramata, A., 242
Arca, M., 173
Argaman, N., 343, 358
Arnau, A., 207
Arnold, W., 411
Arrigan, D.W.M., 242
Aryasetiawan, F., 374
Askeland, D.R., 2
Atalar, A., 421, 422, 438, 440
Audebert, P., 174
Augereau, F., 411
Aulbur, W.G., 374
Auth, T., 393

B
Backman, V., 107, 108
Badiali, J.P., 161, 163
Baerends, E.J., 358, 374
Bagotzky, V.S., 156
Bailey, L., 173
Baker, B., 470
Balduz, J.L., 366, 367, 371
Balkova, A., 387
Bamber, M.J., 411
Bandey, H.L., 172, 175
Banter, J.C., 294
Barbero, C., 173
Bard, A.J., 173, 457–459, 468
Barker, G.C., 161
Barranco, M., 395
Barshad, Y., 282
Bartlett, P.N., 156
Bartlett, R.J., 387
Baskes, M.I., 255, 256
Bathe, K.J., 19
Batra, R., 156
Battocletti, M., 395, 401
Bauermann, L.P., 156
Becke, A.D., 366, 371, 380–382, 384, 388, 389
Beck, T.R., 301

Bedeaux, D., 117, 118
Bendey, H.L., 156
Benito, D., 173
Benyaich, A., 172
Berger, J.A., 400
Bergh, T., 468, 487
Berlin, A., 157
Bertin, J.J., 23
Bertsch, G.F., 399, 400
Bethe, H., 59
Bethe, H.A., 120
Beversluis, M.R., 56
Bhusan, B., 411
Binder, K., 142, 143, 280
Binnig, G., 411
Bird, R.B., 456, 471
Birss, V.I., 172
Bizet, K., 205
Blaha, P., 366, 382, 384
Blanc, C., 312
Blum, L., 136, 137, 252
Bobacka, J., 173, 182, 226
Boccara, A.C., 173
Bockris, J.O.M., 305
Boehmer, V., 118
Bogicevic, A., 143
Bohnke, O., 153
Bohren, C.F., 115
Bolscher, M., 291
Bonello, B., 410
Boren, C.F., 64, 97
Borges, G.L., 252
Bornand, V., 411
Born, M., 65, 96
Bortz, A.B., 144
Bouhelier, A., 56
Bourkane, S., 153
Boyd, G.T., 70
Boyd, J.P., 124, 125
Boyle, A., 152
Braatz, R.D., 298
Braatz, R., 467, 495
Braun, R., 134
Bray, R.C., 411
Brebbia, C.A., 19, 33
Briggs, G.A.D., 411, 436

Brooks, B.R., 134
Brown, G., 137, 138, 142–144, 146
Brown, R., 14
Bruckenstein, S., 152–156, 174–177, 179, 180, 218, 219
Brumleve, T.R., 161, 167
Brune, H., 143
Brusamarello, V., 291
Brzezinska, K., 176, 219
Buck, R.P., 160–163, 166, 167
Bucksbaum, P.H., 125
Budevski, E., 242, 243, 250, 257
Budiansky, N., 291
Buendía, G., 274, 282
Buendía, M., 143, 144
Bueno, P.R., 211, 213, 221, 223, 224
Bullock, J.S., 487
Bund, A., 182, 183
Burgmayer, P., 156
Burke, K., 366, 380, 389, 390, 401
Burstein, G.T., 298
Bussac, M.N., 161
Buttry, D.A., 181
Buttry, D., 174
Buvet, R., 151

C

Cadogan, A., 226
Cafiero, M., 384
Cai, W., 394
Caldwell, J., 133
Calhoun, J., 411
Callaway, J., 344, 393
Calvo, E.J., 173
Cancio, A., 386
Cantaro, F.I., 156
Cao, Y., 143
Capelle, K., 341–402
Carbajal, J., 161
Carrier, G.F., 5, 6
Carslaw, H.S., 466, 493
Carter, E.A., 361
Case, D., 133
Casida, M. E., 373
Ceperley, D.M., 378
Ceresoli, D., 400

Chabalowski, C., 381
Chakrabarti, B., 274, 275
Challener, W.A., 53–110
Chan, G.K.-L., 365
Chaung, J.M., 35
Chayes, J.T., 358
Chayes, L., 358
Cheng, C., 470
Cheng, X., 411
Chen, J., 371, 383, 384, 388
Chen, Z., 107–109
Chermette, H., 402
Chevary, J.A., 139, 366, 381
Chilton, C.H., 309
Chinchankar, V., 468
Chivilikhin, S., 468, 491, 493
Chong, D.P., 343, 376, 386, 402
Chou, B., 492
Chou, M.Y., 386
Christensen, C.R., 323
Chuang, J.M., 14, 31
Chubachi, N., 411, 421, 428
Clementi, E., 386
Clerc, C., 468
Collins, J.A., 289
Collins, J.B., 137
Colwell, S.M., 395, 401
Comtat, M., 173
Cooke, K.E., 411
Cook, W.J., 468, 479
Cooper, J., 173
Cooper, J.M., 173
Cordoba de Torresi, S.I., 173
Cordoba-Torresi, S., 153
Cragnolino, G.A., 291
Crank, J., 466, 493
Cremer, D., 383
Creton, C., 468
Csonka, G.I., 383
Czirok, E., 183

D

Daifuku, H., 152
Daikhin, L.I., 163, 169
Daisley, S.J., 180

Dalcanale, E., 157
Danz, R.W., 397
Dasari, R.R., 62
da Silva, A.B.F., 374, 395, 401
da Silva, A.J.R., 384
Das, M.P., 343, 376, 383, 386, 402
Davids, D.A., 421
Davidson, E.R., 384
Daw, M.S., 255, 256
Dean, W., 14
de Boeij, P.L., 399, 400
Deconinck, J., 488
Degertekin, L., 421
de Levie, R., 328
Deligiann, H., 492
Deligianni, H., 494, 495
Delorme, P., 173
Del Pópolo, M.G., 246, 252, 257
Denuault, G., 173
de Oliveira, M.J., 243, 274, 275
Deore, B., 152
De Paoli, M.A., 173, 218
Derby, B., 411
Desamais, N., 324, 326
Deslouis, C., 163, 172, 173
de Surville, R., 151
de Tacconi, N.R., 173
Devlin, F., 381
Dewhurst, J.K., 395, 401
Diaz, A.F., 218
Dickson, R.M., 394, 401
Dimitrov, N., 319, 321
Ding, H., 172
Dinh, H.N., 172
Dinte, B.P., 389
Diver, A.J., 293
Doblhofer, K., 152
Dobson, J.F., 343, 373, 376, 383, 386, 389, 390, 402
Doghmane, A., 411
Draine, B.T., 116
Dreizler, R.M., 343, 347, 356, 358, 367, 376, 377, 380, 386, 392, 393, 395, 397, 402
Drews, T.O., 467, 495
Duc, L., 183

Duh, J.-G., 410
Dukovic, J.O., 451, 468, 487, 494, 495
Dunn, D.S., 291
Dunsch, L., 161, 165
Durham, P.J., 372
Durliat, H., 173
Durocher, S., 119
D'Urso, J.J., 493

E

Ebert, H., 394, 395, 401
Efimov, I., 156
Egues, J.C., 373
Ehahoun, H., 205
Ehrenbeck, C., 156, 161, 165
Einstein, T.L., 137
Eisenbach, M., 394
Elliott, C.M., 161
El Moustafid, T., 172
Emory, S.R., 62
Emperador, A., 395
Endo, T., 411, 428, 430
Engel, E., 383, 384, 393
Engelhardt, G., 291, 296, 297, 306, 311–312
Enomoto, Y., 436
Eriksson, O., 383, 394
Ernst, A., 372
Ernzerhof, M., 366, 380, 382, 389
Eschrig, H., 343, 344, 357, 358, 376, 386, 393, 402
Evans, G.P., 151

F

Faccio, R.J., 144
Fan, F.R.F., 173
Fano, U., 121
Faulkner, L., 457–459, 468
Fauvarque, J.-F., 152
Favaro, A.P., 361
Fazzio, A., 384
Fei, D., 411
Feldman, B.J., 156, 211
Feld, M.S., 62
Feller, S.E., 134
Feng, B., 411
Ferconi, M., 395

Ferrando, R., 144
Ferreira, J.V.B., 361
Fiaud, C., 323–325, 328
Fichthorn, K.A., 143, 285
Filippi, C., 381
Fiolhais, C., 139, 343, 366, 376, 381, 386, 392, 402
Flannery, B.P., 257
Flatau, P.J., 116
Fleischmann, M., 62
Fletcher, S., 84, 161
Florit, M.I., 155, 156, 172
Flotit, M.I., 155
Fogan, B., 13
Foiles, S.M., 255, 256
Foley, M., 361
Fontana, M.G., 8, 38
Forsen, O., 183
Forsythe, G.E., 19
Foster, 421
Frank, S., 144
Fraoua, K., 157
Freire, H.J.P., 341–402
Freund, L.B., 410
Friedrich, H., 121
Frisch, M.J., 374, 387
Frisch, M., 381
Fuentealba, P., 389
Fujisaka, H., 274
Fujishima, A., 173
Furthmüller, J., 138

G

Gabrielli, C., 151–232, 336–338
Gadegaard, N., 156, 173
Gage, E., 57, 77, 81, 84, 86, 88
Galasco, R., 477, 478, 491
Gallagher, T.F., 123
Galli, G., 394
Gál, T., 358
Galus, Z., 216
Galvele, J.R., 303
Gal, Y.S., 152
Gamboa-Aldeco, M., 136
Gao, Z., 173, 226
García-González, P., 386

Index

Garcia-Jareño, J.J., 173, 181, 206, 211, 213, 221, 223, 224
Garcia-Jareno, J., 169, 207
García, S.G., 252, 274
Gardner, M.K., 293
Gates, D., 84
Gaudet, G.T., 298
Gauitier, B., 411
Geerlings, P., 402
Geldart, D.J.W., 397
Gelten, R.J., 144
Genies, E.M., 152
Gerber, Ch., 411
Gerischer, H., 242, 252
Gersten, J., 70
Ghosez, P., 399
Giacomo, F.D., 387
Gianturco, F.A., 387
Gidopolous, N., 358
Gidopoulos, N.I., 358, 373
Gilbert, T.L., 358
Gileadi, E., 456, 458, 459, 468
Giles, G., 487
Gill, P.M.W., 383
Gilmer, G.H., 144
Giménez, M.C., 239–285
Giménez-Romero, D., 181, 211, 213, 221, 223, 224
Glarum, S.H., 172
Glasser, M.L., 397
Glidle, A., 156, 173
Glosli, J.N., 137
Goad, D., 301, 302, 304
Godby, R.W., 399
Goedecker, S., 383
Goldade, A.V., 411
Gomes, J.A.N.F., 137, 143
Gonsalves, M., 156
Gonzales, C.A., 383
Gonze, X., 372, 399
Goodman, J.J., 116
Gordon, J.G., 174
Görling, A., 373, 384
Gorman, B.M., 274, 275
Gossmann, U.J., 373
Goto, H., 397
Grabo, T., 364, 384
Gräfenstein, J., 383
Graham, M.E., 411
Grahn, H.T., 410
Grande, H.-J., 155
Grande, H., 155, 156
Grayce, C.J., 396, 398
Gray, J.R., 298
Gray, L.J., 487
Green, A.G., 151
Greene, N.D., 8, 38
Greer, J.R., 410
Gregory, J., 181
Grimson, J., 274
Gritsenko, O., 358
Grober, R.D., 88
Groenendaal, L., 157
Gross, E.K.U., 343, 344, 347, 351, 356, 358, 363, 364, 367, 373, 376, 377, 380, 384, 386, 389, 390, 392, 394, 395, 397, 401, 402
Grundmeier, G., 173
Grzeszczuk, M., 172
Gulari, E., 282
Gumbart, J., 134
Gunnarsson, O., 343, 373, 374, 376, 378, 380, 383, 385, 386, 390–392, 401, 402
Gunton, J.D., 137
Guo, Z., 411
Györffy, B.L., 394
Gyurcsanyi, R.E., 182

H

Haas, O., 163, 173
Hadimioglu, B., 421
Hadjoub, F., 411
Hadjoub, Z., 411
Hadyoon, C.S., 173
Hafner, J., 138
Hagness, S., 59, 67
Haile, J.M., 133, 143
Hakkarainen, T., 298, 301, 305
Hall, C.K., 274
Halliday, C.S., 84
Hamad, I.A., 131–147

Hamm, C., 119
Handy, N.C., 365, 381, 395, 401
Hansen, G.D., 137
Hapiot, P., 174
Harb, J.N., 294
Harbola, M.K., 373
Harriman, J.E., 358
Harrison, N.M., 372
Harris, R.A., 396, 398
Hashemi Rafsanjani, S.M., 118
Hashimoto, K., 173
Hassanzadeh, A., 118
Hatton, T.A., 298
Hawrylak, P., 122
Hebert, K., 303
Hebert, K.R., 289–339
Hedin, L., 378, 390–392
Heermann, D.W., 280
Heineman, W.R., 152
Heinonen, O., 363
Heinze, J., 159, 173
Hémery, P., 226
Henderson, E.R., 301, 302
Henderson, M.J., 182
Hendra, P.J., 62
Herzog, W., 242
He, Y., 467, 495
Hilbers, P.A.J., 144
Hillman, A.R., 151–156, 172–177, 179, 180, 182, 218, 219
Hill, T., 251
Hindin, B., 291
Hohenberg, P., 138, 347, 351, 353, 355, 390
Holthausen, M.C., 343, 376, 386, 402
Homescu, C., 298
Hood, R.O., 386
Hood, R.Q., 382, 386
Hoppe, M., 436
Horanyi, G., 173, 183
Hori, H., 56, 76
Horii, K., 428
Horkans, J., 494, 495
Hovorka, F., 294
Hsuing, C.C., 14, 31
Huang, J.-C., 410

Hubbard, A.T., 323, 324
Huckaby, D.A., 136
Huckaby, D., 136
Hudson, A., 172
Hudson, J.L., 291
Huet, F., 153
Huffman, D.R., 64, 97, 115
Hughes, N.A., 175
Hume, E., 14
Humphrey, A.D.W., 134
Hunter, W.R., 65, 93
Hyldgaard, P., 137, 143

I

Iafrate, G.J., 371, 383, 384, 386, 388
Ianno, R.J., 411
Iett, C., 436
Ignaczak, A., 137, 143
Ikarijima, Y., 152
Inganas, O., 157
Innes, R.A., 65
Inokuti, M., 122
Inzelt, G., 152, 161, 163, 167, 173, 183
Isaacs, H.S., 298, 301, 305
Ishiji, H., 428
Ishikawa, T., 411
Isono, Y., 410
Itagi, A.V., 53–110
Itaya, K., 211
Ito, Y., 172
Itzkan, I., 62
Ivaska, A., 173, 182, 226
Iwata, J.-I., 399, 400

J

Jackson, A., 173, 176, 177, 179, 189
Jackson, C.P., 293
Jackson, K.A., 139, 366, 381
Jacobsen, T., 327
Jacoby, M., 342, 398
Jaeger, J.C., 466, 493
Jain, R., 492
Jang, H., 274
Jansen, A.P.J., 144
Jenekhe, S.A., 173
Jenkins, P.J., 436

Jennison, D.R., 143
Jiang, M., 152
Jin, E., 82
Jin, E.X., 56
Jin, J.I., 152
Johansson, T., 157
Johnson, B.G., 383
Johnson, E.R., 366, 381, 388, 389
Jones, R.O., 343, 376, 386, 402
Jönsson, L., 374
Joo, J., 152
Jorcin, J.B., 312
Jørgensen, P., 387
Josell, D., 495
Joulbert, D., 343, 376, 381, 386, 392, 402
Jozefowicz, M., 151
Jupille, J., 118
Jureviciute, I., 176, 177, 180, 218
Jüttner, K., 156, 161, 165, 172, 243
Juwono, T., 131–147

K

Kadija, I., 468
Kadri, A., 323
Kahlert, H., 173
Kalaji, M., 174
Kalia, R.K., 132
Kalos, M.H., 144
Kamrunnahar, M., 298
Kanazawa, K.K., 174
Kaneko, K., 155
Kaneko, M., 155
Kaner, R.B., 152
Kang, H.C., 143, 144
Kaplan, R.D., 152
Karaki, K., 421
Kastner, M.A., 120
Kawagoe, T., 152
Kawashima, K., 410
Keddam, M., 153, 172, 205, 209, 323–325, 327, 328
Kelly, D.M., 156
Kelly, J.J., 296
Kennard, E., 13
Kepas, A., 172

Kertesz, V., 183
Kessler, T., 493
Khuri-Yakub, B.T., 421
Kibler, L.A., 264, 272
Kim, H.M., 152
Kim, J.D., 411
Kim, K., 177
Kim, S.-K., 495
Kim, Y.-H., 382, 386
Kim, Y.T., 187
Kino, G.S., 55, 59
Kishimoto, M., 152
Kleinert, M., 264, 272
Klein, M.L., 134
Klima, J., 324
Klimov, V.V., 116
Kneipp, H., 62
Kneipp, K., 62
Kobayashi, T., 428
Koch, H., 387
Koch, W., 343, 376, 386, 402
Kock, R., 411
Koda, T., 421
Kohanoff, J., 252
Kohn, W., 138, 140, 341, 347, 351, 353, 355, 358, 368, 371, 389, 390
Kolb, D.M., 132, 242, 248, 252, 264, 272
Kolesik, M., 143, 144
Kollman, P., 133
Kolosov, O., 411
Komura, T., 172
Kondratiev, V., 173
Koonin, S.E., 126
Kootstra, F., 399, 400
Koper, M.T.M., 137, 140, 144, 342, 343, 398
Korniss, G., 274
Kotani, T., 384
Kotz, R., 173
Koymen, H., 421
Kraka, E., 383
Kratochvilova, K., 324
Krause, J.L., 126
Kreibich, T., 344, 384
Kreibig, U., 115

Kresse, G., 138
Krichmar, S.I., 492
Krieger, J.B., 371, 383, 384, 386, 388
Krishnaswarmy, S., 411
Krist, K., 291
Kronenberg, M.L., 294
Krtil, P., 175
Kuiken, H.K., 296
Kulesza, P.J., 183, 211
Kundu, T., 411
Kunz, K.S., 58, 67
Kurdakova, V., 173
Kurth, S., 366, 381, 382, 384, 395, 401
Kushibiki, J., 411, 428
Kvarnstrom, C., 182
Kwak, J., 173, 182, 183, 187, 188, 207

L

Lacaze, P.C., 157
Lacroix, J.C., 157
Lagier, C.M., 156
Laird, B.B., 343, 376, 386, 402
Lammert, P.E., 358
Landau, D.P., 142, 143
Landau, U., 451–498
Landolt, D., 468
Lang, G., 161, 167
Langreth, D.C., 389
Lanzi III, O., 467, 470, 491, 493
Lapkowski, M., 152
Lau, K.-H., 140
Laviron, E., 177
Laycock, N.J., 298, 300, 304, 311
Lazzari, R., 118
Leaman, F.H., 461, 472
Lebowitz, J.L., 144
Leburton, J.-P., 382, 386
Lee, A.M., 395, 401
Lee, C.C., 411
Lee, C., 173, 366, 381
Lee, C.Y., 152
Lee, H., 182, 187, 188
Lee, I.-H., 382, 386
Lee, J.-W., 410
Lee, Y.C., 411
Legault, M., 136
Legrand, C., 383
Lein, M., 389
Leite, J.R.R., 70
Leiva, E., 242
Leiva, E.P.M., 239–285
Letheby, H., 151, 172
Levich, V.G., 460, 471
Levi, M.D., 163, 169
Levin, O.V., 159
Levy, M., 355, 365–367, 371, 373
Lewenstam, A., 182, 226
Lezna, R.D., 173
Liang, K., 428
Liao, P.F., 70
Liatsi, P., 226
Libero, V.L., 392, 401
Lieb, E.H., 355, 365
Lien, M.M., 174
Lien, M., 151
Lightfoot, E.N., 456, 471
Ligneres, V.L., 361
Lima, M.P., 384
Lima, N.A., 401
Lindorfs, T., 226
Lin, H.-N., 410
Lipparini, E., 395
Liua, Z.-S., 342, 398
Liu, C., 291
Liu, Y.-C., 173
Li, X., 467, 495
Li, Y., 371, 383, 384, 388
Lizarraga, L., 155, 156
Li, Z.L., 428
Longaygue, X., 324, 326
Lopez, A.C., 274
Lorenz, W.J., 243, 252, 274
Loscar, E., 282
Loveday, D.C., 153, 174
Lüders, M., 372
Ludwig, S., 161, 165
Luebbers, R.J., 58, 67
Lukkien, J.J., 144
Lundin, U., 383
Lundquist, B.I., 143
Lundqvist, B.I., 389
Lundqvist, B., 373, 378, 380, 390–392

Index

Lunt, T.T., 291
Luswig, S., 161, 165
Lwin, T., 84
Lynch, D.W., 65, 93
Lynes, W.M, 468, 487
Lyons, M.E.G., 151, 152, 157

M

MacDonald, A.H., 344, 393
Macdonald, D.D., 291, 297
Machado, E., 274, 282
Madan, A., 411
Madden, P.A., 361
Magnussen, O.M., 132
Magyarfalvi, G., 394, 401
Magyar, R.J., 401
Maia, G., 180
Majewski, J.A., 384
Makarov, D.E., 389
Makov, G., 358
Maksimovic, M.D., 461, 472, 493
Mal, A.K., 411
Malev, V., 173
Malev, V.V., 159
Malhotra, B.D., 156
Malik, M.A., 183
Malinauskas, A., 152
Malinauskiene, J., 152
Malkina, O.L., 394
Malkin, V.G., 394
Malyshev, E., 468, 488, 491–493
Manifar, T., 118
Mann, A.B., 411
Mann, J.A., 319
Mansfield, S.M., 55, 59
Marcasso, T., 395
March, N.H., 343, 361, 376, 386, 402
Maris, H.J., 410
Mark, H.B., 218
Maroulis, G., 342, 343, 398
Marques, M.A.L., 390
Marques, M., 343, 376, 386, 392, 402
Martin, J.M.L., 388
Martin, K., 411
Martin, R.M., 382, 386, 399
Marton, J.P., 117

Martyna, G.J., 134
Masure, M., 226
Matencio, T., 173
Mathias, M.F., 163
Matlosz, M., 468
Matsumoto, T., 87
Mattes, B.R., 152
Maximoff, S.N., 394, 401
Maxwell-Garnett, J.C., 115
Mayer, C., 252, 274
Mayergoyz, I.D., 144
Mayland, J.C., 8, 38
McCartney, L.N., 291
McDaniel, T.W., 54, 55
McFarland, C.W., 468, 487
McGervey, D., 492
McQuillan, A.J., 62
Meerholz, K., 173
Mehdizadeh, S., 468
Meir, Y., 389
Melroy, O., 211
Melroy, O.R., 252
Mendez, J., 492
Menon, M.M., 487, 490, 491
Mesquita, J.C., 174
Meter, L.M., 174
Micka, K., 324
Mie, G., 58, 65, 76
Mikhailov, A.S., 291
Milne-Thomson, L.M., 466, 493
Minouflet, F., 153
Miomandre, F., 161
Miras, M.C., 173
Mirkin, M.V., 173
Mitchell, S.J., 136–138, 140, 142, 143
Mittler, S., 118, 119
Miyasaka, C., 409–449
Miyashita, M., 428
Miyashita, S., 274, 275
Mobed, M., 8, 23, 30, 38
Moffat, T.P., 495
Mohamoud, M.A., 156, 179
Molina, F.V., 155, 156
Monack, M.L., 289
Monkhorst, H.J., 139
Moon, P., 466, 493

Morgan, J.H., 2
Morrison, P.W., 461, 472, 493
Moskovits, M., 62
Mount, A.R., 161
Mo, W.T., 298
Mrozek, P., 136
Mufti, A.A., 6, 11
Muha, M.S., 421
Mundt, C., 161, 162, 167
Murray, R.C., 151
Murray, R.W., 152, 156, 211
Musiani, M.M., 163, 172, 173
Muskat, J., 372
Mutus, B., 119

N
Nadi, N., 172
Nagaoka, T., 152
Nagaraja, S., 382, 386
Nagata, Y., 421
Nagy, Á., 373
Nakajima, T., 152
Nakano, A., 132
Nakano, S., 411
Nalewajski, R.F., 343, 376, 386, 402
Namazu, T., 410
Naoi, K., 174
Nart, F.C., 180
Navarro, J., 211
Navarro-Laboulais, J.J., 173
Nechtschein, M., 174
Nelson, G., 84
Nenciu, G., 399
Neubeck, S., 182
Neudeck, S., 183
Newman, J., 291, 292, 305, 307, 312, 451, 454, 458–460, 463, 467, 470–472, 493
Newman, R.C., 298
Nguyen, P.H., 161, 165
Nie, S., 62
Nieto, F., 144
Nilson, R.H., 14
Nitzan, A., 70
Niu, L., 182
Niu, Q., 400

Nix, W.D., 410
Nogueira, F., 343, 376, 386, 390, 392, 402
Noh, J.S., 298
Nomura, T., 411
Nordström, L., 392
Notten, P.H.L., 296
Novotny, L., 56
Novotny, M.A., 131–147, 274, 275
Nunes, R.W., 399
Nusair, M., 366, 378, 380, 391, 392

O
Odashima, M.M., 365
Odin, C., 174
Ogiso, H., 411
Oh, I., 188
Ohno, M., 421
Ohono, M., 411, 438
Ohsaka, T., 152
Ohtsu, M., 56, 76
Okada, A., 411
Oldham, K.B., 8, 38
Oliveira, L.N., 373, 392, 401
Oliver, G.L., 391
Oliver, W.C., 410
Olsen, J., 387
Onida, G., 374
Opik, A., 182, 183
Orata, D., 174
Orazem, M., 205
Oren, Y., 465, 476
Orestes, E., 374, 384, 395, 401
Organ, L., 291
Ortiz, G., 399
Ostlund, N.S., 363, 375
Otero, T., 156
Otero, T.F., 155
Ovesson, S., 143
Oviedo, O.A., 249
Oxford, S., 365
Oyama, N., 152

P
Paasch, G., 161, 165
Pack, J.D., 139
Pandey, M.K., 156

Pan, Z., 172
Papaconstantinou, P.G., 373
Park, J.W., 152
Park, J.Y., 114
Park, K., 143, 144
Parmon, W., 422
Parr, R.G., 343, 350, 353, 363, 366, 367, 371, 376, 380, 381, 385, 386, 392, 395, 401, 402
Parthasarathi, S., 411
Pasquarello, A., 399
Pastor, R.W., 134
Patchkovskii, S., 383
Pater, E., 155
Patton, D.C., 389
Paulse, C.D., 161
Payer, J.H., 467, 491
Pearson, C.E., 5, 6
Pebere, N., 312
Pederson, M.R., 139, 366, 381, 389
Pedroza, L.S., 384
Peng, C., 57, 67, 77, 81, 82, 84, 86, 88
Perdew, J.P., 139, 365–367, 371, 373, 378, 380–386, 388, 389, 391, 395
Perelman, L.T., 62
Peres, R.C.D., 173, 218
Pereyra, V., 144
Perichon, J., 151
Pernault, J.M., 218
Perreault, S., 152
Perrin, B., 410
Perrot, F., 394, 395
Perrot, H., 151–232
Perry, R.H., 309
Persson, N.K., 157
Peter, L.M., 173
Petersilka, M., 364, 373, 390
Petzold, L.R., 298
Pham, M.C., 218
Phillips, J.C., 134
Philpott, M.R., 137
Pickett, W.E., 357, 358
Pickup, P.G., 161, 173
Pigani, L., 172
Pike, C.R., 144, 146
Pi, M., 395

Piro, B., 218
Pisarevskaya, E.Y., 156
Pistorius, P.C., 298
Pittalis, S., 395, 401
Pivovar, B.S., 154
Plieth, W., 183
Podoprygorina, G., 118
Pople, J.A., 347, 383
Popov, K.I., 461, 472, 493
Porter, J.F., 6, 11
Posadas, D., 155, 156, 172
Pouchana, C., 389
Prentice, G.A., 451
Press, W.H., 257
Proft, F.D., 402
Przasnyski, M., 242, 252
Puippe, J.C., 461, 472
Pulay, P., 394, 401
Punckt, C., 291
Purushothaman, B.K., 461, 472, 493
Pyykkö, P., 394

Q

Quate, C.F., 411
Quinto, M., 173
Qu, Q., 384

R

Rabe, U., 411
Raganelli, F., 387
Rajagopal, A.K., 344, 393
Rajeshwar, R., 173
Ramanavicius, A., 152
Rammelt, U., 183
Rangan, C., 113–129
Rapaport, D.C., 133, 143
Rasing, Th., 70
Rasolt, M., 394–397
Rats, D., 411
Rau, A.R.P., 121, 122, 126
Rear, D.B., 493
Rebinsky, D.A., 411
Reddy, A.K.N., 305
Reinhard, P.-G., 383
Reining, L., 374
Reiss, H., 152
Remita, E., 324, 326

Ren, X., 173
Resta, R., 399, 400
Retter, U., 173
Rezaee, A., 118, 119
Rhee, C.K., 136
Richards, B., 75
Rikvold, P., 274, 282
Rikvold, P.A., 131–147, 274, 275
Rishpon, J., 161
Robb, D.T., 131–147
Roberts, A.P., 144, 146
Robinson, R.A., 308
Rodríguez, J., 155
Rodriguez Nieto, F.J., 172
Rodriguez Presa, M.J., 172
Roha, D., 476
Roig, A.F., 173, 211
Rojas, M., 245
Rojas, M.I., 246, 249, 252
Romankiw, L.T., 468, 491
Romanowski, S., 137, 143
Rooney, P., 118
Ropital, F., 324, 326
Rosberg, K., 161, 165
Rösch, N., 389
Rossignol, C., 410
Rössler, U., 395
Ross, R.B., 343, 376, 386, 402
Rotermund, H.H., 291
Roullier, L., 177
Rousseau, P., 205
Rowe, J.M., 411
Rubashkin, A.A., 161
Rubin, A., 218
Rubinson, J.F., 218
Rubinstein, I., 161
Rubio, A., 374, 390, 399, 400
Runge, E., 351, 363, 373
Ruskai, M.B., 358
Rusli, E., 467, 495
Ryder, K.S., 173

S

Sabatini, E., 161
Sacramento, P., 137
Saga, H., 87
Sahni, V., 371
Saini, K.K., 156
Sakai, M., 411, 421
Salahub, D.R., 394
Salinas, D., 252, 274
Salpeter, E.E., 120
Salsbury, F.R., Jr., 396
Samal, P., 373
Samant, M.G., 252
Sambles, J.R., 65
Sánchez, C., 240
Sánchez, C.G., 252
Sandratskii, L.M., 392
Sani, R.L., 468, 487
Sanmatias, A., 173
Sasaki, Y., 411
Sauerbrey, G., 205
Savéant, J.M., 152, 160
Saville, P.M., 173
Savin, A., 358, 372
Schafer, K.J., 126
Schelkunoff, S.A., 93
Scherer, V., 411
Schiavon, G., 157
Schiff, L.I., 121
Schlesinger, M., 37, 117, 451, 481
Schlichting, H., 462, 493
Schlüter, M., 366, 367
Schmickler, W., 137, 143, 242, 342, 398
Schmidt, E., 243, 252, 274
Schmidt, R., 183
Schmitz, R.H.J., 172
Schneider, F., 387
Scholl, H., 211
Scholll, H., 173
Schönhammer, K., 401
Schreckenbach, G., 394, 401
Schrödinger, E., 355
Schulten, K., 134
Schultze, J.W., 137
Schwerdtfeger, P., 389
Scrosati, B., 152
Scully, J.R., 291
Scuseria, G.E., 366, 374, 381–385, 388, 394, 401
Seeber, R., 172

Segers, J.P.L., 144
Selman, J.R., 462, 493
Seminario, J.M., 343, 376, 386, 402
Sendur, I.K., 67
Sendur, K., 54, 82
Serra, L., 395
Shallcross, S., 395, 401
Sham, L.J., 138, 366–368
Shan, X., 491
Shao, I., 492
Sharland, S.M., 291, 293
Sharma, A.L., 156
Sharma, S., 395, 401
Shaw, J.M., 152
Shay, M., 174
Shen, Y.R., 70
Shia, Z., 342, 398
Shigi, H., 152
Shi, J., 400
Shimano, T., 87
Shi, X., 82
Shyu, J.H., 465, 476
Sides, S.W., 274, 275
Siegenthaler, H., 243
Siegler, K., 173
Sieradzki, K., 319, 321
Silva, M.F., 401
Simonsen, I., 118
Singh, D., 392
Singh, D.J., 139, 366, 381
Sinha, N., 301, 303
Skudlarski, P., 397
Smela, E., 156
Smith, M.L., 23
Smyrl, W.H., 151, 174
Snijders, J.G., 374, 399, 400
Somekh, M.C., 411
Somekh, M.G., 436
Sommerfeld, A., 72
Somorjai, G.A., 114
Spencer, D., 466, 493
Sridhar, N., 291
Stadele, M., 384
Staikov, G., 252, 274
Staroverov, V.N., 366, 382, 384, 385, 388

Stasevich, T.J., 137
Steffens, O., 395
Steinsmo, U., 301, 305
Stephens, P.J., 381
Stewart, W.E., 456, 471
Stickford, G.H., 291
Stocks, G.M., 394
Stokes, R.H., 308
Stolbov, S., 137
Strange, P., 344, 374, 393, 394
Straschil, K., 468
Stratmann, M., 173
Stratmann, R.E., 374
Strehblow, H.H., 296
Suhrke, M., 395
Sukeda, H., 87
Sung, Y.-S., 136
Suraud, E., 383
Sutter, E., 324, 326
Svane, A., 374, 383
Swann, M.J., 173, 175
Syritski, V., 182, 183
Szabo, A., 363, 375
Szotek, Z., 372, 374, 383

T

Taflove, A., 59, 67, 107, 108
Tajkhorshid, E., 134
Takada, Y., 397
Takahasi, K., 172
Takahasi K.M., 493
Takenouti, H., 153, 163, 323
Tak, Y., 301
Tak, Y.S., 301, 302, 304
Tanaka, S., 439
Tansel, T., 132
Tao, J., 366, 382, 385, 388, 395
Taravel-Condat, C., 324, 326
Tasker, P.W., 293
Tatsunada, T., 152
Tauc, J., 410
Taut, M., 381
Telles, J.C.F., 19, 33
Temmermann, W.M., 374, 383
Temmerman, W.M., 372
Terayama, N., 410

Tester, J.W., 298
Teter, M.P., 361
Teukolsky, S.A., 257
Theophilou, A.K., 373
Thomas-Alyea, K.E., 291, 292, 305, 307
Thomsen, C., 410
Thonhauser, T., 400
Ticianelli, E.A., 180
Tildesley, D.J., 133, 143, 253
Tilly, J., 298
Tittmann, B.R., 411, 421, 438, 439
Tobias, C.W., 451, 462, 493
Tobias, D.J., 134
Tomé, T., 243, 274, 275
Tomita, H., 274, 275
Toney, M.F., 252
Torresi, R., 153, 209
Torresi, R.M., 173, 180, 181
Torres, R., 207
Toth, K., 182
Tozer, D.J., 381
Trasatti, S., 242
Trellakis, A., 394
Tribollet, B., 163, 172, 173, 205, 289–339
Trickey, S.B., 365, 389
Troise Frank, M.H., 173
Tsai, C.S., 411
Tsuei, Y.G., 14
Tsukada, A., 163
Tucceri, R.I., 172
Turnbull, A., 291, 293
Tutu, H., 274
Tu, X., 182

U
Uchi, H., 301
Uebing, C., 144
Újfalussy, B., 394
Ujvari, M., 161
Ullrich, C.A., 356–358, 390
Umari, P., 399
Umrigar, C.J., 372, 381, 383
Uppuluri, S.M., 56
Urquidi-Macdonald, M., 291
Uzoh, C., 494, 495

V
Vanderbilt, D., 139, 399, 400
van Faassen, M., 400
van Gisbergen, S.J.A., 374
van Leeuwen, R., 373, 400
van Santen, R.A., 144
van Voorhis, T., 381
Vanysek, P., 172
Varela, H., 181
Varga, R., 127
Vashishta, P., 132
Veillard, A., 386
Verbrugge, M.W., 294
Vercelli, B., 157
Verhoff, M., 297
Verosub, K.L., 144, 146
Vetter, K.J., 137
Vetterling, W.T., 257
Vicente, F., 173, 181, 211, 213, 221, 223, 224
Vieil, E., 161, 163, 173, 182
Vieira, D., 384
Vignale, G., 343, 356–358, 376, 383, 386, 394–398, 400, 402
Villa, E., 134
Vivier, V., 312
Vlieger, J., 117, 118
Vogl, P., 384
Volfkovich, Y.M., 156
Vollmer, M., 115
von Barth, U., 371, 373, 378, 390–392
von Stebut, J., 411
Vorotyntsev, M.A., 159, 161, 163, 169
Vosko, S.A., 139
Vosko, S.H., 344, 366, 378, 380, 381, 383, 384, 391–393
Voss, J.G., 156
Voter, A.F., 133, 143
Vuillemin, B., 153
Vukmirovic, M.B., 319, 321
Vydrow, O.A., 383

W
Waber, J.T., 13
Wagner, C., 467
Waibel, H.F., 272, 273
Wander, A., 372

Index

Wandlowski, Th., 137, 138, 142, 146
Wanga, H., 342, 398
Wang, C.-C., 173
Wang, J., 133, 152
Wang, L., 56
Wang, L.W., 361
Wang, Q., 136
Wang, S., 136, 143
Wang, W., 134
Wang, Y., 62, 139, 366, 378, 380
Wapner, K., 173
Wasow, W.R., 19
Webb, E.G., 298
Webster, J.P.R., 173
Webster, J.R.P., 173
Weglein, D., 422
Weglein, R.D., 411
Weinberg, N.L., 468
Weinberg, W.H., 143, 144, 285
Welsch, G., 492
West, A.C., 470, 492, 495
Westcott, M., 84
West, K., 327
Wheeler, D., 495
White, C.J., 274
White, R.E., 161
White, S.P., 298, 300, 304, 311
Widmann, A., 173
Wieckowski, A., 136
Wijsman, R., 493
Wilde, C.P., 153, 174, 175
Wilkins, J.W., 374
Wilkinson, D.P., 342, 398
Wilk, L., 366, 378, 380, 391, 392
Wills, J., 394
Wilson, P.T., 298
Wilson, R.W., 173
Winter, H., 374, 383
Woakun, A., 70
Woggon, U., 122
Wolf, E., 65, 75, 96
Wolf, R., 133
Woodhead, A.E., 151
Wright, B., 410
Wrobel, L.C., 19, 33
Wrona, P.K., 216

X

Xiang, C., 182
Xiao, D., 400
Xie, Q., 182
Xue, F., 467, 495
Xu, S., 118
Xu, X., 56, 82

Y

Yabana, K., 399, 400
Yakabe, H., 152
Yamagishi, T., 411
Yamaguti, T., 172
Yamamoto, M., 274
Yamanaka, K., 411, 421, 436
Yang, H., 182–184, 187, 188, 207
Yang, K.-H., 173
Yang, N., 173
Yang, W., 343, 350, 353, 363, 366, 376, 380, 381, 384–386, 392, 395, 401, 402
Yao, S., 182
Yasui, T., 274
Yeager, E., 294
Yeu, T., 161
Ying, S.C., 144
Yin, K.-M., 161
Yu, L.T., 151
Yung, E.K., 491

Z

Zabinska-Olszak, G., 172
Zadronecki, M., 216
Zamani, N.G., 1–50
Zanardi, C., 172
Zecchin, S., 157
Zhanga, J., 342, 398
Zhang, F., 411
Zhang, J., 136
Zhang, Y., 134, 182, 366
Zhou, B.J., 361
Zhou, S., 291
Zhou, Y., 303
Ziegler, T., 343, 376, 383, 386, 394, 401, 402
Ziesche, P., 381
Ziff, R., 274, 282

Zigel, V., 173
Zolotova, T.K., 156
Zoski, C.G., 173
Zotti, G., 157

Zunger, A., 378, 380, 382, 383, 386, 388
Zupan, A., 366, 382, 384
Zuppiroli, L., 161

Printed in the United States of America